Apollo 12 – On the Ocean of Storms

David M. Harland

Apollo 12 – On the Ocean of Storms

David M. Harland
Space Historian
Kelvinbridge
Glasgow
UK

SPRINGER–PRAXIS BOOKS IN SPACE EXPLORATION
SUBJECT *ADVISORY EDITOR*: John Mason, M.B.E., B.Sc., M.Sc., Ph.D.

ISBN 978-1-4419-7606-2 e-ISBN 978-1-4419-7607-9
DOI 10.1007/978-1-4419-7607-9
Springer New York Dordrecht Heidelberg London

Library of Congress Control Number: 2011922753

Cover design: Jim Wilkie
Typesetting: BookEns, Royston, Herts., UK

Printed on acid-free paper

Springer is a part of Springer Science+Business Media (www.springer.com)

Contents

Illustrations

Chapter 1

Chapter 2

Chapter 3

Chapter 4

Author's preface

In July 1969 Apollo 11 successfully achieved the first manned lunar landing. A few months later Pete Conrad, Dick Gordon and Al Bean set off to attempt an even more challenging mission. Free of the 'burden of history', these three close friends were determined to have fun. This is the story of their mission. In addition to the official documents issued prior to and after the mission, I have drawn on the flight transcript and debriefing to recreate the drama. I have edited quotations for clarity, for brevity, and to eliminate the intermingling of spontaneous conversation, but have endeavoured to preserve the sense of the moment.

Apollo 12 dramatically showed how conservative Apollo 11 had been. Of course, the rationale for selecting a bland target for the first landing was fully justified, but just imagine if Neil Armstrong had set Eagle down amongst a cluster of craters and he and Buzz Aldrin had walked over to an earlier unmanned lander and snipped off its camera! With its 31.5 hours on the lunar surface, double moonwalk, deployment of geophysical instruments and geological traverse, Apollo 12 undertook what many people had hoped would be attempted by the first landing crew. It set the scene for a program of scientific lunar exploration.

Apollo 12 – On the Ocean of Storms supplements my earlier *The First Men on the Moon – The Story of Apollo 11* in a series of books devoted to the lunar missions.

David M. Harland
Kelvinbridge, Glasgow
August 2010

To

Eric M. Jones

without whose generous assistance this book could not have been written.

Acknowledgements

I must thank, in no particular order, Mike Gentry, Glen Swanson, Marc Rayman, W.D. Woods, Philip J. Stooke, Roland Suhr, Ken MacTaggart, Frank O'Brien, Dave Scott, Eric Jones, Alan Bean, Hamish Lindsay, Lennie Waugh, Richard Orloff, Ken Glover, Ulrich Lotzmann, Ed Hengeveld, Kipp Teague, J.L. Pickering, Frederic Artner, Sy Liebergot, Andrew Chaikin, Mark Gray, Chris Gamble and contributors to the *Apollo Lunar Surface Journal* not already mentioned. And, of course, Clive Horwood of Praxis.

Acronyms

8-ball	see FDAI
AC	alternating current
AGC	Apollo Guidance Computer
AGS	Abort Guidance System
ALS	Apollo Landing Site
ALSCC	Apollo Lunar Surface Closeup Camera
ALSEP	Apollo Lunar Surface Experiments Package
AOT	Alignment Optical Telescope
APS	Ascent Propulsion System
ASA	Abort Sensor Assembly
Capcom	Capsule communicator in the MOCR
CCGE	Cold-Cathode Gauge Experiment (a.k.a CCIG)
CCIG	Cold-Cathode Ion Gauge
CDR	Commander
CM	Command Module
CMC	CM Computer
CMP	CM Pilot
COAS	Crew Optical Alignment Sight
CSM	Command and Service Modules
DAP	Digital Autopilot
db	decibel
DC	direct current
DOI	Descent Orbit Insertion
DPS	Descent Propulsion System
DSN	Deep Space Network
EASEP	Early Apollo Scientific Experiments Package
ECS	Environmental Control System
EECOM	Electrical, Environmental and Communications
ELS	Earth Landing System
EMS	Entry Monitor System
EMU	EVA Mobility Unit

ETB	Equipment Transfer Bag
EVA	Extravehicular Activity
F-1	kerosene-burning engine of the S-IC
FDAI	Flight Director Attitude Indicator (a.k.a the 8-ball)
GET	Ground Elapsed Time
GDC	Gyro Display Coupler
GNC	Guidance, Navigation and Control
Hasselblad	a 70-mm still camera
HF	High Frequency radio
HTC	Hand Tool Carrier
IFR	Instrument Flight Rules
IMU	Inertial Measurement Unit
IU	Instrument Unit of the Saturn launch vehicle family
J-2	hydrogen-burning engine of the S-II and S-IVB
KSC	Kennedy Space Center
LCC	Launch Control Center
LCG	Liquid-Cooled Garment
LEC	Lunar Equipment Conveyor
LGC	LM Guidance Computer
LLTV	Lunar Landing Training Vehicle
LM	Lunar Module
LMP	LM Pilot
LOI	Lunar Orbit Insertion
LPD	Landing Point Designator
LRL	Lunar Receiving Laboratory
LSM	Lunar Surface Magnetometer
LUT	Launch Umbilical Tower
Maurer	a 16-mm movie camera
MCC	Midcourse Correction
MESA	Modular Equipment Stowage Assembly
MLP	Mobile Launch Platform
MOCR	Mission Operations Control Room
MQF	Mobile Quarantine Facility
MSC	Manned Spacecraft Center
MSFN	Manned Space Flight Network
MSS	Mobile Service Structure
NASA	National Aeronautics and Space Administration
Omega	a chronometer
OPS	Oxygen Purge System
Orb Rate	Orbital Rate
ORDEAL	Orbital Rate Drive Earth and Lunar
PAD	Pre-advisory Data
PDI	Powered Descent Initiation
PGA	Pressure Garment Assembly
PGNS	Primary Guidance and Navigation System

PLSS	Portable Life Support System
PSE	Passive Seismic Experiment
psi	pounds per square inch
PTC	Passive Thermal Control
PTT	Push To Talk
RCS	Reaction Control System
RCU	Remote Control Unit
REFSMMAT	Reference to Stable Member Matrix
rock box	see SRC
RTG	Radioisotope Thermoelectric Generator
S-band	A high capacity radio link
S-IB	First stage of the Saturn IB launch vehicle
S-IC	First stage of the Saturn V launch vehicle
S-II	Second stage of the Saturn V launch vehicle
S-IVB	Second stage of the Saturn IB or third stage of the Saturn V
SCE	Signal Conditioning Equipment
SCS	Stabilisation and Control System
SEQ	Scientific Equipment
SIDE	Suprathermal Ion Detector Experiment
SLA	Spacecraft/LM Adapter
SM	Service Module
SPS	Service Propulsion System
SRC	Sample Return Container
SWC	Solar Wind Collector
SWS	Solar Wind Spectrometer
TEI	Transearth Injection
TLI	Translunar Injection
VAB	Vehicle Assembly Building
V_{AC}	Volts AC
V_{DC}	Volts DC
VHF	Very High Frequency radio
Vox	Voice activated transmission

1

Planning and preparations

THE CHALLENGE

On 25 May 1961 President John F. Kennedy delivered a speech to a joint session of Congress on the theme of *Urgent National Needs*. In view of space achievements by the Soviets, he proclaimed, "Now it is time to take longer strides, time for a great new American enterprise, time for this nation to take a clearly leading role in space achievement, which in many ways may hold the key to our future on Earth." Having outlined the political background, he said, "I believe that this nation should commit itself to achieving the goal, before this decade is out, of landing a man on the Moon, and returning him, safely, to the Earth." He had chosen this goal because it would be technically difficult. By literally 'shooting for the Moon', Kennedy was betting that America would not only catch up with the Soviet Union in space, but forge ahead. If space was the arena of superpower politics, then he was challenging his rival, Nikita Khrushchev, for world leadership. He imposed the deadline to ensure that reaching the Moon was perceived as a race. And he also recognised the awesome scale of the task. "No single space project in this period will be more impressive to mankind, or more important for the long-range exploration of space; and none will be so difficult or expensive to accomplish." In order to indicate that it was a matter of national honour, Kennedy said, "In a very real sense, it will not be one man going to the Moon; if we make this judgment affirmatively it will be an entire nation, for all of us must work to put him there." And to emphasise what was at stake, he warned, "If we are to go only halfway, or reduce our sights in the face of difficulty, in my judgment it would be better not to go at all." For Kennedy the Moon was a symbol and, given what he wished to do, it was an excellent symbol. He had the impression that the applause in Congress was "something less than enthusiastic", but there was little opposition in the House and the debate in the Senate ran for less than an hour and NASA's budget was doubled without a formal vote being taken.

On 28 November NASA announced that North American Aviation of Downey,

D.M. Harland, *Apollo 12 – On the Ocean of Storms*, Springer Praxis Books,
DOI 10.1007/978-1-4419-7607-9_1, © Springer Science+Business Media, LLC 2011

California, would develop the Apollo spacecraft.[1] On 7 December NASA decided to develop a two-man precursor spacecraft, the primary objective of which would be to demonstrate that orbital rendezvous was feasible, and a fortnight later it announced that McDonnell Aircraft in St Louis, Missouri, would supply this Gemini spacecraft. On 21 December it decided upon a vehicle configuration that would become known as the Saturn V to launch Apollo lunar missions. On 22 June 1962 it selected lunar orbit rendezvous. As this would require a specialised spacecraft, Grumman Aircraft Engineering Corporation of Bethpage, New York, was hired on 7 November to develop the lunar module. With the development of the main spacecraft already underway, North American was directed to develop the Block I for solo missions in Earth orbit and the Block II fitted with apparatus capable of 'docking' with the lunar module.

CONRAD, GORDON AND BEAN

Pete Conrad was born on 2 June 1930 in Philadelphia, Pennsylvania, of a wealthy family that made its fortune in real estate and investment banking. His father insisted that the boy be named 'Charles Jr', without a middle name, but his mother Frances took to calling her son Peter.

He developed an early fascination for things mechanical. In his teens he carried out various odd jobs at the nearby Paoli Airfield in return for free flights. His formal education was designed to enable him to follow in his father's footsteps, but he was expelled from the private Haverford School. Although his father intended for him to attend Yale, Conrad enrolled at Princeton in 1949 to study aeronautical engineering on a Reserve Officers Training Corps scholarship from the Navy and graduated with a bachelor's degree in 1953.

Upon entering naval service as an ensign he received flight training at the Naval Air Station at Jacksonville, Florida. After transitioning to the F-9 Cougar fighter, he reported to the Navy Test Pilot School at Patuxent River, Maryland, in 1958. Later in that year, he, along with many others, received classified instructions to attend a briefing in Washington DC and checked into the Rice Hotel under the cover name of 'Max Peck'. Only when he got there did he find that he was one of thirty-five by that name – one of whom was an old naval buddy, Jim Lovell. Neither of them made the selection for the Mercury astronauts. On graduating, he remained at Patuxent River as a project test pilot, flight instructor and performance engineer, and then became an instructor for the F4H Phantom II at the Naval Air Station at Miramar, California.

Richard Francis Gordon Jr was born on 5 October 1929 in Seattle, Washington, into a family hit hard by the Great Depression. He didn't have a boyhood dream of

[1] In 1967 North American Aviation merged with the Rockwell Standard Corporation and so became North American Rockwell; in 1973 this became Rockwell International.

becoming a pilot; his aspiration was to join the priesthood. He studied chemistry at the University of Washington, and upon graduating with a bachelor's degree in 1951 was inclining towards dentistry. But he joined the Navy. After gaining his wings in 1953 he attended All-Weather Flight School and was posted to the Naval Air Station at Jacksonville, Florida. In 1957 he attended the Navy Test Pilot School at Patuxent River. When he and Conrad met there, they became great friends. After graduation Gordon did test flight work on the F-8U Crusader, F-11F Tiger, FJ-4 Fury and A-4D Skyhawk, and also served as the first project test pilot for the F-4H Phantom II. In 1960 he became a flight instructor at the Naval Air Station at Miramar, California. His reputation as one of the hottest F-4H drivers enabled him to win the Bendix transcontinental race from Los Angeles to New York in May 1961, in the process of which he established a new speed record.

While Conrad and Gordon were at Miramar, they both responded to NASA's call for a second group of astronauts. When the successful applicants were announced in September 1962 Conrad was on the list,[2] but Gordon was not. When Conrad left, he gave Gordon a picture of himself in a flight suit standing beside a Phantom with the inscription 'To Dick, until we serve together again.' The intensely frustrated Gordon was inclined to leave the Navy and seek a job in commercial aviation, but in a bar with Conrad one night which has been recreated in '*Rocketman*' by Conrad's second wife Nancy, the issue was, "Still crying in your beer, Dickie-Dickie?"

"I'm just crying for you, Pete, ya poor dumb sumbitch. Stuck in a garbage can in space with some Air Force puke while I'm out smoking the field in my Phantom."

"So, Dick," Conrad replied. "They're gonna fill out this Gemini program now that Apollo's approved. At least ten more slots. I think you oughtta apply again."

"And why would I do that?"

"Because you miss me."

Gordon stuck with it, and when NASA announced its third group of astronauts in October 1963 he was on the list.

Alan LaVern Bean was born on 15 March 1932 in Wheeler, Texas. His father was in the Soil Conservation Service and his mother owned her own ice cream parlour. After schooling in Fort Worth he studied aeronautical engineering at the University of Texas at Austin, graduating with a bachelor's degree in 1955. He then joined the Navy on a Reserve Officers Training Corps scholarship, attended flight training, and was assigned to an attack squadron at the Naval Air Station at Jacksonville, Florida. He picked up the obvious moniker of 'Beano'. He was tee-total, which was unusual for a naval aviator, and had little of the macho spirit associated with his profession. Having developed a fascination with art at an early age, he took oil painting evening classes in his spare time. At the Navy Test Pilot School at Patuxent River in 1960 he became a student of Pete Conrad, whom he had met several years earlier while they

[2] The second group of astronauts comprised four Air Force pilots (Frank Borman, Jim McDivitt, Tom Stafford and Ed White), three Navy aviators (Pete Conrad, Jim Lovell and John Young), and two civilians (Neil Armstrong and Elliot See).

were serving in different squadrons at Jacksonville. After 2 years at Patuxent River during which he flew as a test pilot on several types of aircraft, Bean was posted to a squadron at Cecil Field, Florida. At the urging of Conrad he responded to NASA's call for a second group of astronauts, but was unsuccessful. Undeterred, he applied again the following year and was accepted in October 1963 as a member of the third group.[3]

On reporting to the Astronaut Office at the Manned Spacecraft Center, each new astronaut was assigned a technical niche within the vast scope of the space program. Conrad worked on the development of the lunar module and played a key role in the design of its crew cabin. When he started, the control panels were sheets of wood on which drawings of instruments had been pasted. When they arrived later, Gordon was appointed head of the Apollo branch of the astronaut office and liaison for the development of the command and service modules, whilst Bean was given duties pertaining to spacecraft recovery systems.

LEARNING TO FLY IN SPACE

After two unmanned tests, on 23 March 1965 Gus Grissom and John Young took Gemini 3 up for a brief 'shakedown' flight. The Gemini 4 mission by Jim McDivitt and Ed White in early June lasted 4 days and White took a spacewalk. Flight crew assignments were made by Deke Slayton, who had been selected as an astronaut in 1959 and been grounded for medical reasons shortly before he was due to fly one of the Mercury missions. After Gemini 4, White was reassigned as backup commander for Gemini 7 with Mike Collins. In August, Gordon Cooper and Pete Conrad flew an 8-day mission in Gemini 5. Slayton split their backups, assigning Neil Armstrong to command Gemini 8 with Dave Scott, and Elliot See to command Gemini 9 with Charles Bassett. This was because White had experienced difficulty in closing the hatch after his spacewalk, and it had been decided that only the more muscular men would be assigned extravehicular activity. Conrad rotated to backup Gemini 8 with Dick Gordon, putting them in line to fly Gemini 11.

After backing up Gemini 3, Wally Schirra and Tom Stafford had expected to fly in October and rendezvous in space but their Agena target vehicle blew up at launch. However, a bold improvisation enabled Gemini 6 to fly a joint mission with Gemini 7 in December, while Frank Borman and Jim Lovell were demonstrating that the human body could endure 14 days of weightlessness – the longest duration expected for any of the Apollo missions. Then Stafford rotated to backup Gemini 9 with Gene Cernan, setting them up for Gemini 12; Young rotated to Gemini 10 with Mike

[3] The third group of astronauts comprised seven Air Force pilots (Buzz Aldrin, Bill Anders, Charles Bassett, Mike Collins, Donn Eisele, Ted Freeman and Dave Scott), four Navy aviators (Al Bean, Gene Cernan, Roger Chaffee and Dick Gordon), a Marine Corps aviator (Clifton 'C.C.' Williams), and two civilians (Walt Cunningham and Rusty Schweickart).

Collins; and Lovell was assigned to backup Gemini 10 with Buzz Aldrin. But See and Bassett were killed on 28 February 1966 when their jet crashed. This meant Stafford and Cernan advanced to prime crew. Faced with the need for a new backup crew for Gemini 9, Slayton advanced Lovell and Aldrin.

In March, Neil Armstrong and Dave Scott in Gemini 8 achieved the long-awaited rendezvous mission and docked with their Agena target. But soon thereafter a faulty thruster obliged them to undock and return to Earth after a mere 10 hours in space. Slayton rotated Armstrong to backup Gemini 11 with C.C. Williams. Having lost their Agena, Tom Stafford and Gene Cernan rendezvoused Gemini 9 with a passive target in June but a fouled shroud precluded a docking. Cernan began an ambitious spacewalk, but had to curtail it upon discovering that working in weightlessness was rather more difficult than expected.

Slayton had planned to rotate Elliot See to backup Gemini 12 with Al Bean, but the deaths of See and Bassett led to a reshuffle of the remaining assignments. Firstly, advancing Lovell and Aldrin to backup Gemini 9 instead of Gemini 10 gave them the opportunity to rotate to Gemini 12. Although Cernan rotated to backup, he retained his pilot status to enable him to focus on assisting Aldrin in training for a spacewalk. As the program would conclude with Gemini 12, Slayton appointed Cooper to serve as commander for this dead-end slot. To fill the backup slot on Gemini 10 he paired Bean (who would no longer be able to partner See) with Williams, drawn from the Gemini 11 backup crew. Armstrong remained as backup commander of Gemini 11, now with newcomer Bill Anders. As events transpired, the deaths of See and Bassett would have a major impact on the Apollo crewing – one example being that Aldrin would not otherwise have been eligible for Apollo 11, and would not have been a member of the first crew to descend to the lunar surface.

John Young and Mike Collins flew Gemini 10 in July 1966 and utilised their own Agena as a 'shunt engine' to rendezvous with the one left by Gemini 8, then Collins made a spacewalk to retrieve an experiment package. Pete Conrad and Dick Gordon flew Gemini 11 in September and after curtailing an ambitious spacewalk they used their Agena to climb to a record altitude of 740 nautical miles. Jim Lovell and Buzz Aldrin drew the program to a conclusion with Gemini 12 in November, and Aldrin methodically demonstrated how to work purposefully outside a spacecraft.

In 1962, when NASA committed Apollo to orbital rendezvous there were serious doubts that this was even feasible. The Gemini program not only showed that it was, it also demonstrated a variety of ways of achieving it. Another legacy of Gemini to Apollo was a cadre of highly experienced astronauts, training techniques and flight controllers. It was this program which taught NASA how to fly in space.

THE APOLLO PROGRAM

On 1 September 1963 George E. Mueller was hired by NASA headquarters as the Associate Administrator for Manned Space Flight. The schedule drawn up in March 1962 envisaged a series of launches of the Saturn V in which the stages were tested in sequence – with only the first stage being 'live' on the first test scheduled for late 1965. The aim was to 'man rate' this vehicle by the summer of 1967, then use it to launch at least six manned missions in Earth and lunar orbit prior to attempting a lunar landing in late 1968 or early 1969. On 29 October 1963 Mueller decided to reduce this research and development phase by 'all up' testing in which each launch would fly only 'live' stages, modules, systems and spacecraft. And on 30 October he cancelled four manned test flights of the Block I spacecraft that had been scheduled for Saturn I launches in 1965 and transferred them to the Saturn IB, and ordered that the work on the Saturn IB be accelerated and use 'all up' testing. In November 1964 Mueller issued an outline schedule for testing hardware in advance of the introduction of the Saturn V. It had originally been expected that the Saturn IB would be able to lift both spacecraft being developed for Apollo, but they had gained so much weight as to make this doubtful without an intensive weight-saving effort. Nevertheless, he presumed that this weight-loss would have been achieved in time for the first Block II mission.

Outline schedule for Apollo drawn up in November 1964

Mission[4,5]	Payload	Launch Date
AS-201	CSM-009 (unmanned)	1965
AS-202	CSM-011 (unmanned)	1966
AS-203	No spacecraft (S-IVB development flight)	Jul 1966
AS-204	CSM-012 (manned)	Oct 1966
AS-205	CSM-014 (manned)	Jan 1967
AS-206	LM-1 (unmanned)	Apr 1967
AS-207	CSM-101 (manned) and LM-2	Jul 1967

An operational step towards Apollo was realised on 15 December 1965 when Gemini 6 rendezvoused with Gemini 7. The straightforward manner in which this was achieved raised the possibility of side-stepping the issue of the Saturn IB being unable to lift both Apollo craft. On 28 January 1966 Samuel C. Phillips, the Apollo Program Manager in Washington, requested Joseph F. Shea, the Apollo Spacecraft Manager in Houston, to assess the impact, including the effects on ground support equipment and mission control, of a *dual* AS-207/208 mission on the date scheduled for AS-207, which was the Saturn IB that was to have lifted them together. The raw concept was for near-simultaneous launches of AS-207 with CSM-101 and the crew, and AS-208 with LM-2. After rendezvous and docking, the mission would unfold as previously. On 24 February Howard W. Tindall of the mission planning and analysis division in Houston suggested that the CSM be launched first and the LM follow it either 24 hours later or at a recurring daily window. Shea endorsed this on 1 March,

[4] The Saturn IB was the AS-2xx series and the Saturn V was the AS-5xx series.
[5] The Block I spacecraft was the CSM-0xx series and the Block II was the CSM-1xx series.

and one week later Phillips directed the Manned Spacecraft Center, Marshall Space Flight Center and Kennedy Space Center to work to launch the dual mission a month later than that intended for AS-207 on the earlier schedule. When Gemini 8 docked with an Agena target vehicle on 16 March, this reinforced the decision to attempt the AS-207/208 dual mission.

On 21 March 1966 NASA announced that Gus Grissom was to command the first Apollo mission. He would fly CSM-012 with Ed White and Roger Chaffee. They were to be backed up by Jim McDivitt, Dave Scott and Rusty Schweickart. In both cases the commander and senior pilot were Gemini veterans and the third man was a rookie. Slayton had earmarked Grissom for this role immediately after the Gemini 3 flight in March 1965. After commanding Gemini 4 in June, McDivitt was reassigned to back up Grissom. Having flown with McDivitt, White first backed up Gemini 7 in December 1965 and then joined Grissom's crew. Although Slayton was developing a rotation for Gemini in which a pilot could progress through backup to command a mission, after his Gemini 8 flight Scott was assigned to McDivitt's crew to enable them to gain early experience of Apollo training prior to attempting the AS-207/208 mission. In early 1966 it was hoped to launch CSM-012 as Apollo 1 in the final quarter of the year on a test lasting up to 14 days. The crew for the second mission was announced on 29 September 1966. Wally Schirra would command CSM-014 along with Donn Eisele and Walt Cunningham. They would be backed up by Frank Borman, Tom Stafford and Mike Collins. Schirra was the only experienced man on the prime crew, but the backups were all veterans. In fact, Slayton had given Schirra and Borman these assignments in March, when they returned from a 'goodwill tour' after the Gemini 6/7 rendezvous. Stafford and Collins had been assigned following Gemini 9 in June and Gemini 10 in July, respectively. Slayton had planned to assign the rookies Eisele and Chaffee to Grissom, but Eisele injured his shoulder training for weightlessness in a KC-135 aircraft and withdrew from training in late 1965, so Slayton instead assigned White, whom he had earmarked for Schirra. This Apollo 2 mission was to be a straightforward rerun of Apollo 1 to further evaluate the spacecraft.

With the preparation of CSM-012 running seriously late, in early December 1966 George Mueller postponed it to February 1967 and also deleted the Block I reflight to prevent the slippage of Apollo 1 from impacting the Block II missions scheduled for later in 1967. Schirra had hoped to put his crew first in line for the dual mission, but Slayton imposed a rule that the man who would operate the CSM alone while his colleagues flew the LM must be experienced in rendezvous, since if the LM were to become crippled that man would have to perform a rescue. Eisele was a rookie but Scott had performed a rendezvous on Gemini 8. As Slayton had always intended to assign McDivitt to test the LM, in whichever way it was launched, Slayton, much to Schirra's disgust, exchanged the two crews. On the revised schedule, AS-206 would launch LM-1 for an unmanned test as soon as possible after Apollo 1, and if this was successful the dual mission (which was now AS-205/208 because deleting CSM-014 had released AS-205) would fly as Apollo 2 in August. This was publicly announced on 22 December, along with the assignment of Tom Stafford, John Young and Gene Cernan to backup McDivitt's crew. Also, if two unmanned tests proved sufficient to 'man rate' the Saturn V, the plan was to launch AS-503 manned with an integrated Apollo spacecraft. The crew would be Frank Borman, Mike Collins and Bill Anders, backed up by Pete Conrad, Dick Gordon and Clifton Williams. These assignments were made after Young flew Gemini 10 in July, Conrad and Gordon flew Gemini 11 in September, and Cernan backed up Gemini 12 in November. But disaster struck on 27 January 1967 when a 'flash fire' in CSM-012 during what was reckoned to be a routine test killed Grissom, White and Chaffee.

On 25 March George Mueller directed that missions be numbered in the order of their launch, regardless of whether they used the Saturn IB or Saturn V and whether they were manned or unmanned – previously only the manned missions were to be counted. On the 1966 plan Apollo 1 was to be CSM-012 (Grissom), Apollo 2 would be CSM-014 (Schirra) and Apollo 3 would be CSM-101 (McDivitt) flying the dual mission. The cancellation of the Block I reflight advanced CSM-101 (McDivitt) to Apollo 2. After the fire, the desire not to reassign the name Apollo 1 led naturally to the expectation of resuming with Apollo 2. But with paperwork in circulation for a variety of mission plans numbered up to Apollo 3, Mueller precluded the possibility of administrative confusion by directing that the sequence resume at Apollo 4, in the process introducing a minor point of interest for historians. The final report of the investigation into the fire was submitted on 5 April. It was so critical of the Block I spacecraft that NASA decided simply to discard it and instead subject the Block II to the necessary modifications. On 7 April Joseph Shea was made Deputy Associate Administrator for Manned Space Flight in Washington and was succeeded as Apollo Spacecraft Manager in Houston by George M. Low. It was announced on 9 May that Schirra, Eisele and Cunningham would fly CSM-101, backed up by Stafford, Young and Cernan.

On 17 April 1967 the Manned Spacecraft Center proposed a minimum of three manned Saturn V missions involving both the CSM and the LM *prior to* attempting the lunar landing. When George Mueller advocated landing *on* the third mission, Chris Kraft, in charge of flight operations, warned George Low that a landing should not be tried on the first flight to leave the Earth's gravitational field because flying to the Moon was such a great step forward in terms of operational capability that this should be demonstrated separately. Accepting this logic, on 20 September Low led a group to headquarters to argue for an alphabetic sequence of missions: (A) Saturn V and unmanned CSM development; (B) Saturn IB and unmanned LM development; (C) Saturn IB and manned CSM evaluation; (D) Saturn V and manned CSM/LM joint development; (E) CSM/LM trials in an Earth orbit having a 'high' apogee; (F) CSM/LM trials in lunar orbit; (G) the first landing. This was not a list of *flights*, as several flights might be required to achieve one *mission*. Two Saturn V development flights were already listed. Sam Phillips asked whether the second test was essential, and Wernher von Braun of the Marshall Space Flight Center said the second would serve to confirm the data from the first. If Saturn V development were to drag out, the D-mission would be tackled by reinstating the plan in which the CSM and LM would be launched individually by Saturn IB. Most of the discussion was devoted to the proposal for a lunar orbital flight "to evaluate the deep space environment and to develop procedures for the entire lunar landing mission short of

George Low (top left), Sam Phillips, Joseph Shea (bottom left) and Chris Kraft.

LM descent, ascent and surface operations". When Mueller argued "Apollo should not go to the Moon to develop procedures", Low said developing crew operations would not be the main reason for the mission, since a lot remained to be learned about navigation, thermal control and communications in deep space. Although the meeting did not settle this matter, the alphabetic labels soon became common shorthand.

Sam Phillips confirmed on 2 October 1967 that LM-2 should be configured for an unmanned test, and directed that CSM-103 be paired with LM-3 for the first manned test of the lunar module.[6] AS-501 flew unmanned on 9 November as Apollo 4 with CSM-017 and a lunar module test article, and was spectacularly successful. Several days earlier, George Mueller had issued the schedule for 1968: AS-204 with LM-1; then AS-502 as the second unmanned test with CSM-020 and a lunar module test article; possibly AS-503 as the third unmanned test with a 'boilerplate' spacecraft and lunar module test article; possibly AS-206 with LM-2; AS-205 with CSM-101 for the C-mission flown by Wally Schirra, Donn Eisele and Walt Cunningham; and then AS-504 with CSM-103 and LM-3 for the D-mission. It was announced on 20 November that the first manned Saturn V would be flown by Jim McDivitt, Dave Scott and Rusty Schweickart, backed up by Pete Conrad, Dick Gordon and Al Bean. Next would be the E-mission flown by Frank Borman, Mike Collins and Bill Anders backed up by Neil Armstrong, Jim Lovell and Buzz Aldrin.

LM-1 was launched as Apollo 5 on 22 January 1968. It was less than perfect, but on 6 March the second unmanned test was deleted. The second unmanned Saturn V flew on 4 April as Apollo 6, and suffered a number of problems that required several months to rectify. Nevertheless, on 23 April George Mueller called for AS-503 to be manned. Although Sam Phillips directed the next day that AS-503 be prepared with CSM-103 and LM-3 for a manned mission, he also ordered contingency planning to reconfigure it for a third unmanned test should this be required. The Kennedy Space Center replied that, given sufficient notice of the configuration, it could launch the vehicle unmanned in mid-October but a manned flight would not be possible until late November at the earliest.

HOW PETE MISSED BEING 'FIRST MAN'

In view of slippage in the delivery of LM-3, on 7 August 1968 George Low told NASA headquarters that this spacecraft was unlikely to be ready for launch until February 1969. He felt that for Apollo to have a chance of achieving a lunar landing in 1969 the first manned Saturn V must occur in 1968. Aware that in April 1967 the Manned Spacecraft Center had outlined a contingency for a lunar mission involving only the CSM, on 8 August Low asked Chris Kraft to consider sending CSM-103 to the Moon solo. After managers from NASA headquarters and the three field centres

[6] CSM-101 was to fly the Saturn IB and manned CSM evaluation, and CSM-102 was to be retained by North American Aviation as a ground test article.

involved in the Apollo program decided that this concept should be more thoroughly explored, Deke Slayton asked Frank Borman, whose E-mission was to have a high apogee, possibly at near-lunar distance, whether he would be interested in switching to CSM-103 and flying it into lunar orbit. Borman readily accepted. At a meeting in Washington on 14 August Low said that if Apollo 7 was a success they could either fly an impromptu mission in 1968 or await LM-3 and launch in February, and given the deadline for the lunar landing it would make sense to send Apollo 8 to the Moon. Thomas O. Paine, Deputy Administrator since 25 March 1968, said planning should continue. On 17 August Sam Phillips told Low that Administrator James E. Webb had authorised a manned Saturn V launch in December, but the decision on whether to fly to the Moon must await the outcome of the Apollo 7 test flight. On 19 August Phillips directed that if AS-503 was manned and lacked a LM, the mission was to be designated C-prime irrespective of where the spacecraft went. Then McDivitt's crew would ride AS-504 and fly the D-mission with CSM-104 and LM-3. The E-mission had been deleted. Later that day, Phillips told the press that if the flight of CSM-101 was a success, AS-503 would be manned and that since the LM would not be ready in time this would be a CSM-only mission; by not mentioning the option of leaving Earth orbit, he readily conveyed the impression of a mission confined to Earth orbit. On 3 September Low directed that if C-prime flew to the Moon, it would make 10 orbits over a period of 20 hours and then return. On 19 September, with the AS-502 issues resolved, George Mueller declared the Saturn V 'man rated'. On 6 October Webb resigned and Paine was promoted to Acting Administrator.

Apollo 7 lifted off on 11 October and the 11-day shakedown cruise of CSM-101 was, in Wally Schirra's words, a "101 per cent success". On 7 November George Mueller declared that AS-503 was fit for a mission to the Moon. On 11 November Sam Phillips recommended that Apollo 8 enter lunar orbit. That same day, Thomas Paine spoke to Frank Borman by telephone, who confirmed his willingness to fly the mission, and Paine gave the formal go ahead. The next day, NASA announced that Apollo 8 would be launched on 21 December and attempt a lunar orbital mission. Earlier in the year Mike Collins had withdrawn from this crew to undergo a surgical procedure and was replaced by his backup, James Lovell, with Fred Haise joining Neil Armstrong and Buzz Aldrin on the backup crew. Tom Stafford, John Young and Gene Cernan were announced on 13 November as the prime crew of Apollo 10, thereby establishing the precedent of a crew backing up one mission, skipping two and becoming the prime crew of the one after that. However, it had yet to be decided what their mission would be. Apollo 8 launched on time and flew a perfect mission. On 6 January 1969 Deke Slayton told Armstrong, Aldrin and Collins they would fly Apollo 11 and to assume that their mission would involve a lunar landing. Apollo 9 performed the D-mission in low Earth orbit in early March, and on 10 April NASA announced that the prime crew for Apollo 12 would be Pete Conrad, Dick Gordon and Al Bean, with Dave Scott commanding the backup crew of Alfred Worden and Jim Irwin.

At this point, the question was whether to fly the F-mission or to attempt the lunar landing. In fact, it would be impossible for LM-4 to attempt the G-mission since the software for the powered descent was still in development and propellant restrictions in the ascent stage of this somewhat overweight craft meant that it would be unable to lift off and rendezvous. Tom Stafford, the commander, argued against waiting for LM-5 to become available. On 24 March NASA announced that Apollo 10 would fly the F-mission. Its success in mid-May cleared the way for Apollo 11 to attempt a lunar landing in July. If it did not succeed then Apollo 12 would try in September, and if *that* failed Apollo 13 would try in November. So NASA would have three chances to achieve Kennedy's deadline. However, as we all know, Apollo 11 succeeded.

The delay in delivering LM-3 poses an intriguing 'what if', because by prompting the exchange of McDivitt's and Borman's crews it meant instead of Conrad rotating from backing up Apollo 8 and flying Apollo 11, Armstrong did so.[7]

CALL SIGNS AND PATCH

In '*Tracking Apollo to the Moon*', Hamish Lindsay quotes Pete Conrad as saying of the names selected for the two Apollo 12 spacecraft, "I wanted to let the people that built them name them." Specifically, employees were invited to submit names, with a 25-word rationale. Then, Conrad continued, "We picked the final names out of the lists." Appropriately for an all-Navy crew, the chosen names had naval associations. George Glacken, a flight test engineer at North American, which built the command and service modules, suggested 'Yankee Clipper' because he felt those sailing ships had "majestically sailed the high seas with pride and prestige for a new America". Robert Lambert, a planner at Grumman, suggested 'Intrepid' for the lunar module because it denoted "this nation's resolute determination for continued exploration of space, stressing our astronauts' fortitude and endurance of hardship".

The names then inspired the design of the mission patch. In '*All We Did Was Fly To The Moon*', Dick Lattimer quotes Al Bean, "The patch was designed by Pete Conrad, Dick Gordon and myself, with the help of about ten other people around the contractor area at the Cape. The real breakthrough came in our effort to attempt to duplicate our landing site on the Moon, when a couple of engineers came to me and said they thought they could duplicate it exactly. They came back a few days later, and it did look just perfect – all the craters the proper size, shape and the lighting just perfect. It turned out the technique they had used was to get a relief globe of the Moon that was in the library and light it properly, and then take some photographs with a Polaroid at different distances until they got one with just the right curvature that we wanted on the patch. We selected the blue and gold because they are Navy colours. The ship was patterned in the way of a Navy ship. I went to the library and got some clipper ship pictures, gave them to an artist and he drew it.

[7] And if a problem with Apollo 11 had presented the commander of Apollo 12 with the prospect of being the first man to walk on the Moon, Conrad's backup was determined to find a way to take his place!

Pete Conrad (left) and Al Bean in the Mission Operations Control Room on 20 July 1969, shortly after Neil Armstrong and Buzz Aldrin landed on the Moon.

The mission patch for the Apollo 12 mission.

Clifton C. Williams.

After looking it over, we realised there were too many sets of sails outboard of the hull, so we asked him to redesign it. The original was like that which Jason and the Argonauts used to search for the Golden Fleece and we didn't feel that was American enough, whereas we felt that the clipper ship was definitely an American symbol."

The patch depicted the ship arriving in lunar orbit. Three stars were placed in the lunar sky to symbolise the members of the crew and, at Bean's suggestion, a fourth star for the man he had replaced.

When Deke Slayton asked Conrad in late 1966 to choose a lunar module pilot for his Apollo crew, Conrad asked for Bean but was told he could not have him because Bean was earmarked for the Apollo Applications Program, a mishmash of plans that George Mueller was developing and which Conrad regarded as "Tomorrow Land". Instead, Conrad chose Clifton Williams, an aviator in the Marine Corps and a former student at the Navy Test Pilot School at Patuxent River who had been selected as an astronaut in October 1963 – but only after compressing his spine the night before the height measurement in order not to exceed the 6-foot height limit. Unfortunately, Williams lost his life on 5 October 1967 flying from the Cape to Mobile, Alabama, to visit his dying father. His NASA T-38 jet went into an uncontrollable aileron roll, lost lift, and crashed near Tallahassee, Florida, at too low an altitude for his parachute to function. Shortly afterwards, Conrad sought out Bean to ask him if he would like to join his crew. Bean said he would. This time Slayton relented. Bean felt it only right to honour his predecessor.

LUNAR SURFACE EXPERIMENTS

In early 1964 the Space Science Board of the National Academy of Sciences drew together experts in a range of scientific disciplines to form science planning teams to assist NASA by compiling preliminary lists of instruments to be placed on the Moon by Apollo missions. On 19 November, after tests conducted on an aircraft providing one-sixth gravity had shown that astronauts would be able to offload packages from the descent stage of their craft onto the lunar surface, the Manned Spacecraft Center ordered a study of how scientific instruments might be powered. This recommended a radioisotope thermoelectric generator (RTG). In January 1965 NASA undertook a time-and-motion investigation to determine how to make the best use of the limited time that would be available to the first Apollo crew to land on the Moon. In May, a preliminary list of surface experiments was drawn up, and George Mueller initiated a two-phase procurement process: the definition phase was to be done in parallel by a number of companies, one of which would be selected to develop the hardware for flight. In June the Manned Spacecraft Center set up the Experiments Program Office with Robert O. Piland in charge, to manage the development of all experiments for manned flights. On 7 June, Mueller approved the procurement of the Lunar Surface Experiments Package (LSEP). This suite of instruments had to be capable of being deployed on the Moon by two astronauts in less than an hour, and to transmit data to Earth for at least a year.

On 3 August NASA issued Bendix Systems, TRW Systems and the Space–General Corporation each a 6-month $500,000 contract to propose designs. On 14 October the space agency hired the General Electric Company to supply the RTG under the supervision of the Atomic Energy Commission. By early 1966 the instrument suite had been renamed the Apollo Lunar Surface Experiments Package (ALSEP). On 16 March Bendix of Ann Arbor, Michigan, was awarded the contract to design, manufacture, test and supply the ALSEP. By the end of 1966 the data-processor of the central station that would distribute power to the instruments, relay commands to them and transmit their data to Earth was proving an extreme challenge. With the magnetometer instrument facing serious development issues, the decision was taken to devise a simpler magnetometer as a possible replacement. In December the Manned Spacecraft Center established its Science and Applications Directorate. This took over the tasks of the Experiments Program Office. Wilmot N. Hess, formerly of the Goddard Space Flight Center, was appointed as Director, with Piland providing continuity as his deputy.

On 4 January 1967 Chris Kraft argued that if a lunar landing was to involve two periods of extravehicular activity on the surface, then the first one should enable the astronauts to familiarise themselves with the lunar environment, make an inspection of their lander, take photographic documentation and collect a contingency sample. On the second excursion they should deploy the ALSEP and perform a systematic survey of the site. However, if only one excursion was planned, that mission should *not* be provided with an ALSEP, as its deployment would take a disproportionate amount of the time. In June 1967, Sam Phillips created an *ad hoc* team to review the status of the magnetometer. It was concluded that while the technical problems were sure to be resolved eventually, the instrument was unlikely to be ready for the first landing, which at that time seemed possible in the latter part of 1968. Unfortunately, the simpler magnetometer would not be ready by then either, so work on this was terminated. Leonard Reiffel, a member of Phillips' science staff, recommended on 20 June that considering the uncertainty about an astronaut's ability to work in one-sixth gravity, "an uncrowded time line" would be "more contributory to the advance of science than attempting to do so much that we do none of it well". By September 1967, presuming that the LM would spend 22.5 hours on the lunar surface, mission planners recommended that two excursions be defined, but the second, which would follow a sleep period, should not be listed as a primary objective. The decision on whether to make a second excursion and deploy the ALSEP should be based on the performance of the astronauts during their first moonwalk. However, a year later, on 6 September 1968, with the RTG behind schedule, Robert R. Gilruth, Director of the Manned Spacecraft Center, recommended that the first landing should make a single excursion of 2.5 hours, the geological activities be restricted to the 'minimum lunar sample' and the ALSEP not be carried. Wilmot Hess then argued for a compromise in which the lander carried a smaller package of instruments that would not require the RTG and would be easier to deploy. The Manned Space Flight Management Council at headquarters, which Mueller chaired, agreed on 9 October to the development of three lightweight experiments for the first lunar landing, namely a solar powered passive seismometer, a laser reflector that did not require power, and a solar wind composition experiment that would be retrieved for return to Earth. As a result, the ALSEP was deferred to the second lunar landing.

LANDING SITE SELECTION

After so many years in development, the Apollo program had sprung forth from the starting block with a lunar orbital mission at the end of 1968 and a schedule for 1969 which called for missions at bi-monthly intervals. It was believed that if Apollo 11 achieved the first manned lunar landing there would be enough Saturn V rockets for a total of ten such missions, and the schedule published a week prior to the launch of Apollo 11 envisaged a mission every 11 weeks. Scientists opposed this rapid pace, which would place the final mission in June 1971, arguing instead for 6–12 months between missions in order to provide time to feed the results of one mission into the next. But it would have been impractical to reduce the rate to one mission per year, as it would have meant maintaining the entire infrastructure in an underutilised state. Immediately after the return of Apollo 11, George Mueller ordered that the program be drawn out through 1972. As a result, the launch of Apollo 12 was postponed from September to November 1969.

The scheme introduced to describe missions leading up to the first landing ended with the G-mission. In March 1969, the planners introduced four H-missions. Unlike the first landing, which would spend less than 24 hours on the lunar surface, the new missions were to involve two periods of extravehicular activity, with the astronauts sleeping in between. The first outing would deploy an ALSEP and the second would sample the geological features which had attracted the attention of the site selectors. The Apollo 11 mission went well, but uncertainty in the LM's position prior to the powered descent resulted in the entire profile being displaced 4.3 nautical miles westward. As people in Mission Control tried to figure out in real-time where it had touched down, Sam Phillips remarked to Bill Tindall of the mission planning and analysis division, "Next time, I want a pinpoint landing." As the operating radius for astronauts travelling on foot was severely limited, an accurate landing would be essential if they were to investigate a specific geological feature. Afterwards, Tindall convened a group to improve landing accuracy. It was soon determined that several procedural factors had contributed to Apollo 11's navigational error. These could be readily solved. Although the gravitational irregularities caused by the lunar mascons had not been mapped sufficiently to permit the required navigational accuracy, Emil Schiesser, a deep-space navigation expert, drew attention to the fact that when the vehicle made its appearance around the eastern limb it was heading directly towards Earth, and when it was midway across the lunar disk its travel was perpendicular to the line to Earth. He rejected trying to model the mascons in ever-greater detail, and argued instead for monitoring in real-time how the Doppler on the spacecraft's radio signal in the descent orbit differed from the predicted profile. As the divergence was a measure of the trajectory error, if this could be calculated by Mission Control prior to the powered descent, then it ought to be possible to undertake a correction early in that operation, where such manoeuvring would be most efficient in terms of propellant. The real beauty of this 'correction' was that it overcame *all* sources of navigational error. Tindall described the simplicity of this solution as "astounding". One option for making the correction was to update the state vector, informing the computer of its true position and velocity, but to directly uplink the state vector would require the computer to be idle, which was incompatible with performing the powered descent, and having the crew make the update manually by rapidly keying in seven multiple-digit 'numbers' raised the prospect of their making an irrecoverable error. It was decided not to tell the computer where it actually was, and instead to make it believe that the target had been changed, so that it would revise its trajectory to aim for that point. As Tindall put it, "If they found that the LM was coming down 800 feet short of the landing site, they would tell it to land 800 feet further downrange than it was planning." To implement this procedure required the crew to key only one 'number'. The planners were so confident that this technique would work that they reduced the size of the target ellipse. And given the relaxation in the pace of the program, it was decided to reduce the requirement for backup sites from two to one.

But where to go? In choosing the target for Apollo 11, the site selectors had opted for an open plain and (as far as they could) avoided craters. But for Apollo 12 the scientists wished to examine a substantial crater. On 15 December 1967 the Apollo Site Selection Board had convened to decide the sites for the first two landings. The scientists argued that if the first landing were to occur on an eastern mare plain, then the second should be made on a western mare. The shortlist for the first landing had contained five sites. The easterly ALS-1 and ALS-2 sites in the Sea of Tranquillity were backed up by ALS-3 in the Central Bay in the event of a 2-day launch delay, and by ALS-4 and ALS-5 further west in the Ocean of Storms in case of a longer delay. After Apollo 11 visited ALS-2 it would have been natural to send Apollo 12 to one of the westerly sites, but the geologists were eager to sample something more interesting. Inevitably, this meant landing as *close as possible* to a sizeable crater. In fact, even before Apollo 11 flew, the selectors had drawn up a list of craters for this eventuality, starting by re-examining the sites they had rejected for the first landing due to the *inconvenient* proximity of a crater. A relaxation of operational constraints enabled ALS-6 to be reinstated. This was in the Flamsteed Ring, an ancient crater that had been almost submerged by the Ocean of Storms. Its geological context was well understood, but the fact that it was not far from where Surveyor 1 had landed prompted the flight dynamics team to propose making *this* the target. Of course, it would be embarrassing if Apollo 12 missed, but to land within walking distance of another craft would showcase landing precision. Unfortunately, Surveyor 1 had been deliberately sent to an area devoid of craters, and it was so far west that it could not accommodate a backup. Jack Sevier, the planning coordinator, proposed Surveyor 3, which had landed *inside* a 600-foot-diameter crater in a cluster of somewhat smaller craters. It was not so far west as to preclude a backup. By the time of the Apollo Site Selection Board meeting on 10 July 1969 George Mueller had agreed to Surveyor 3 being considered, and it was added to the list as ALS-7. Geologists were in favour because the site was crossed by a bright ray from the prominent crater Copernicus, situated some 200 nautical miles to the north. Lou Wade of the mapping sciences branch of the Manned Spacecraft Center was in favour because it would

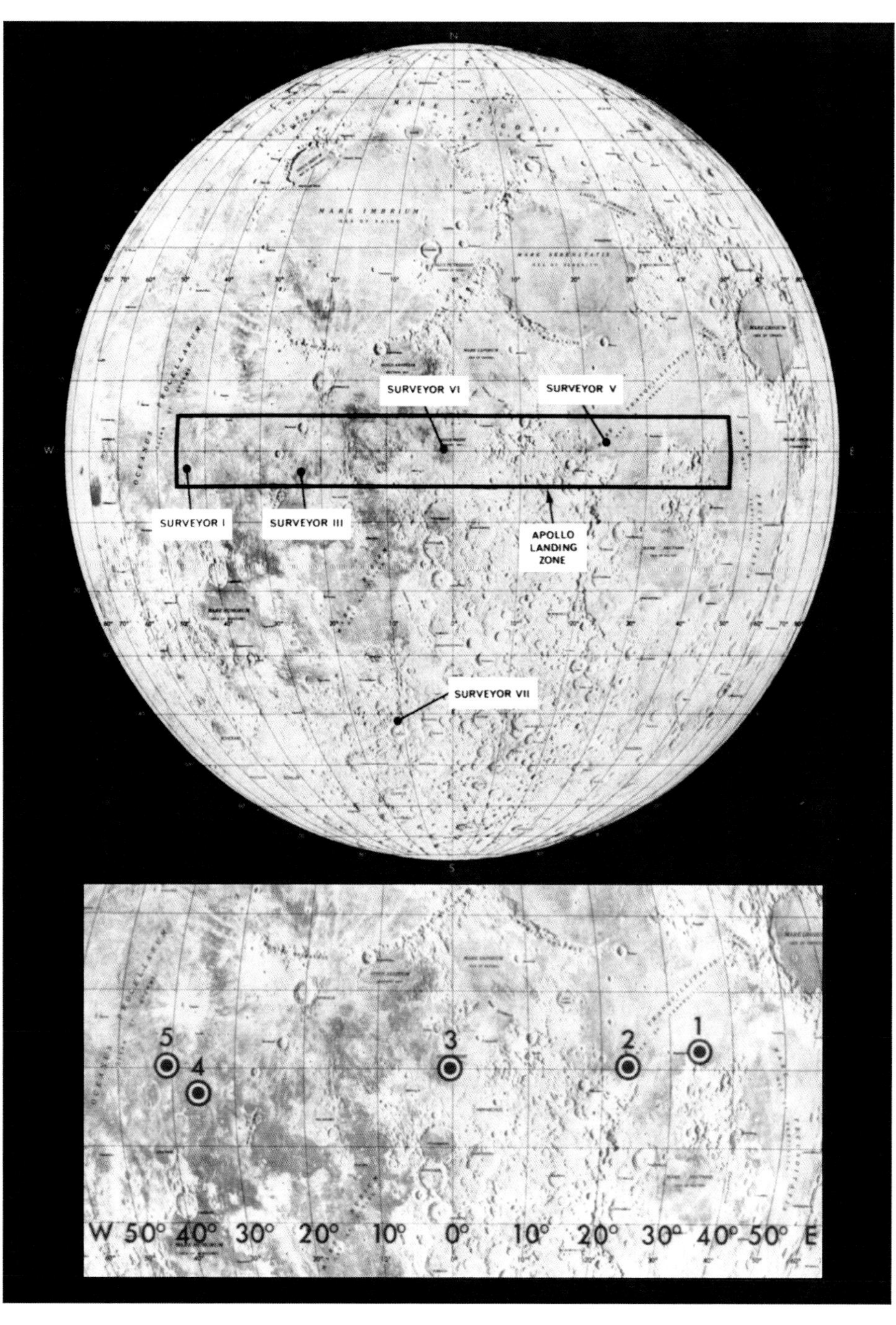

The sites of the Surveyor landers, and the equatorial 'Apollo zone' featuring five candidate sites for the first Apollo landing, which was actually made at ALS-2.

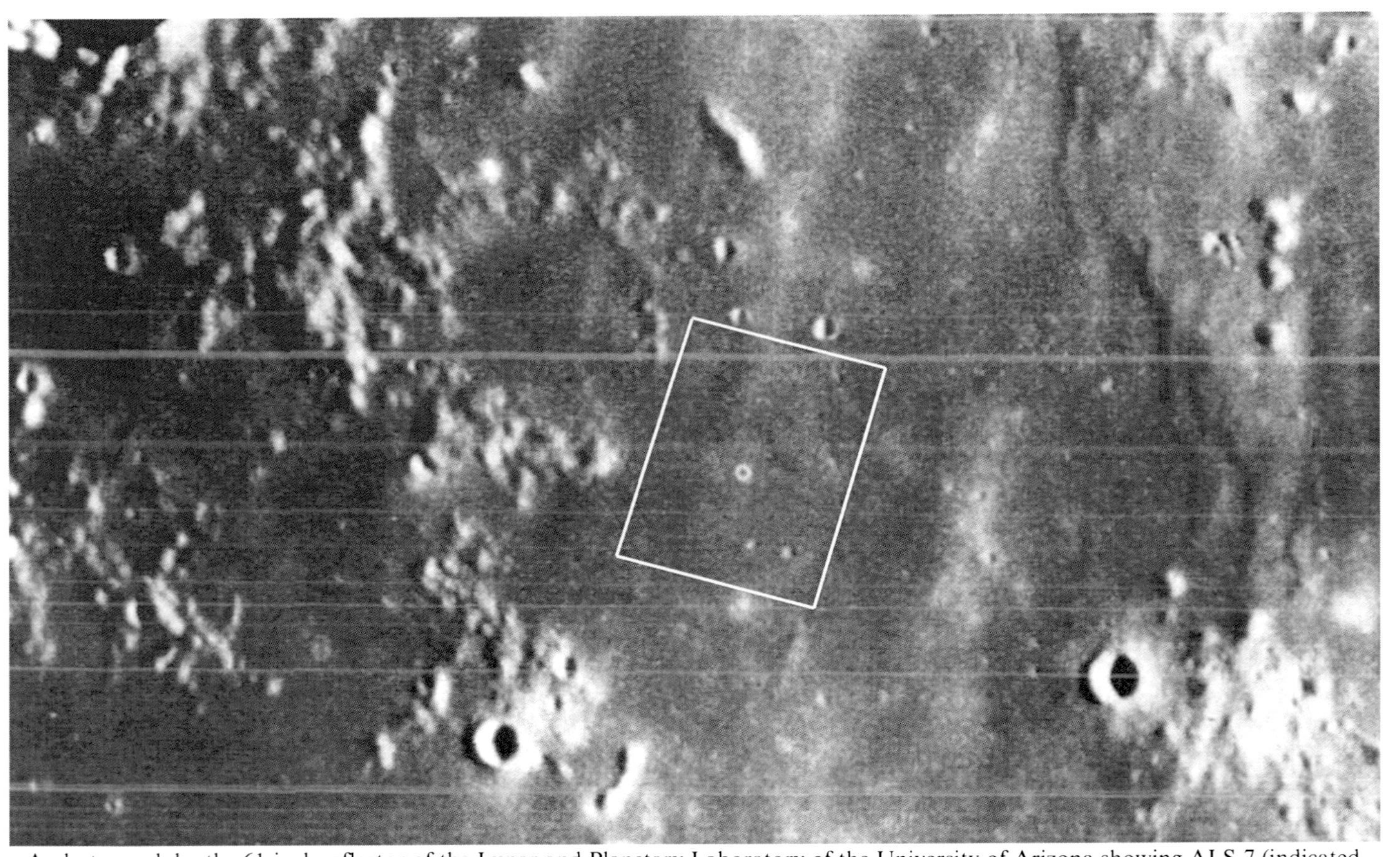

A photograph by the 61-inch reflector of the Lunar and Planetary Laboratory of the University of Arizona showing ALS-7 (indicated by a small circle) in the Ocean of Storms. The outline shows the area covered by the next illustration.

Frame M-154 taken by Lunar Orbiter 3 on 20 February 1967 showing the general target for Surveyor 3. The outline shows the area covered by the next illustration.

A portion of Lunar Orbiter 3 frame H-154 showing the specific area in which Surveyor 3 landed. The outline shows the area covered by the next illustration.

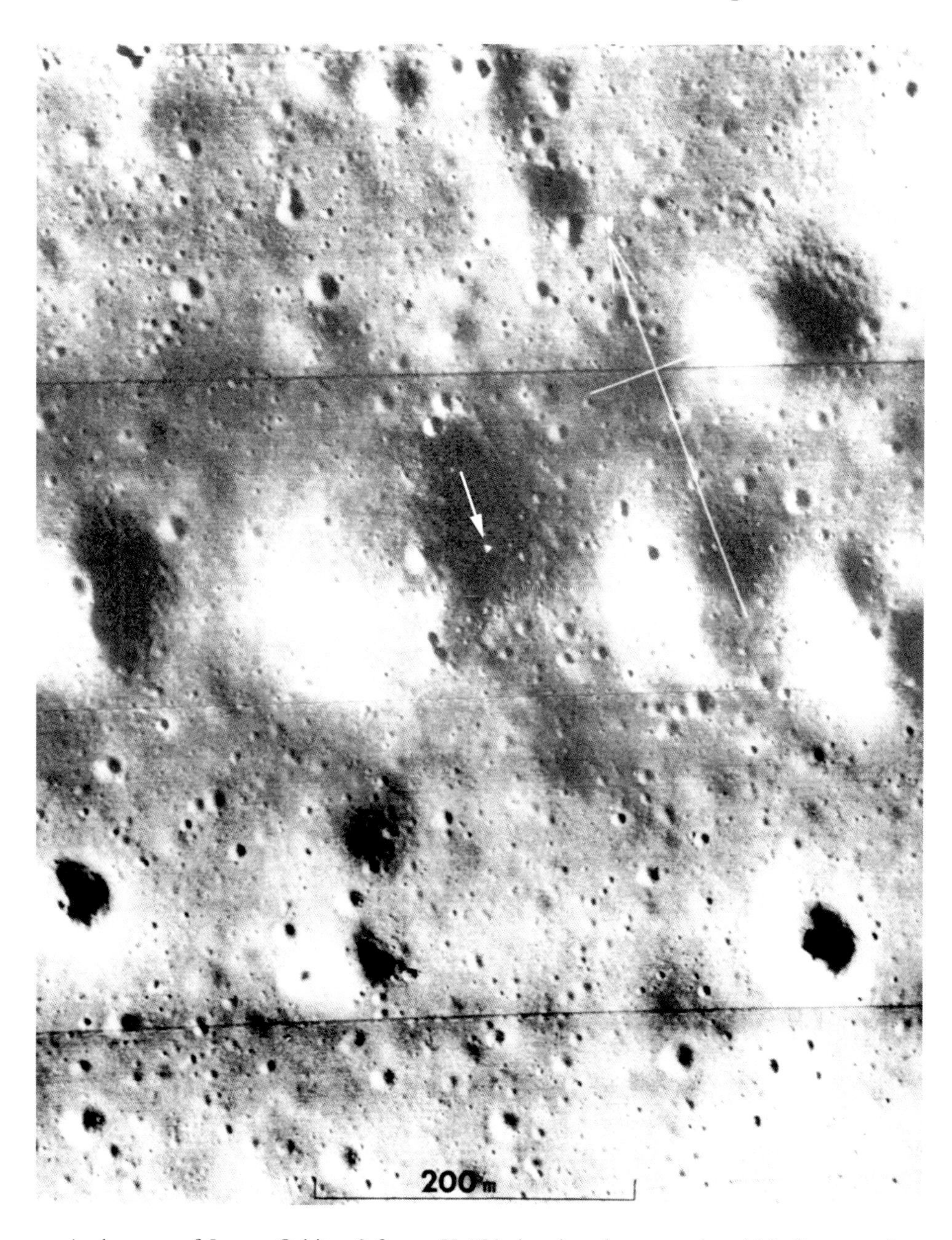

A close up of Lunar Orbiter 3 frame H-154 showing the crater in which Surveyor 3 landed, with the inferred position of the lander indicated by the arrow. (The lander was not present at the time, however.)

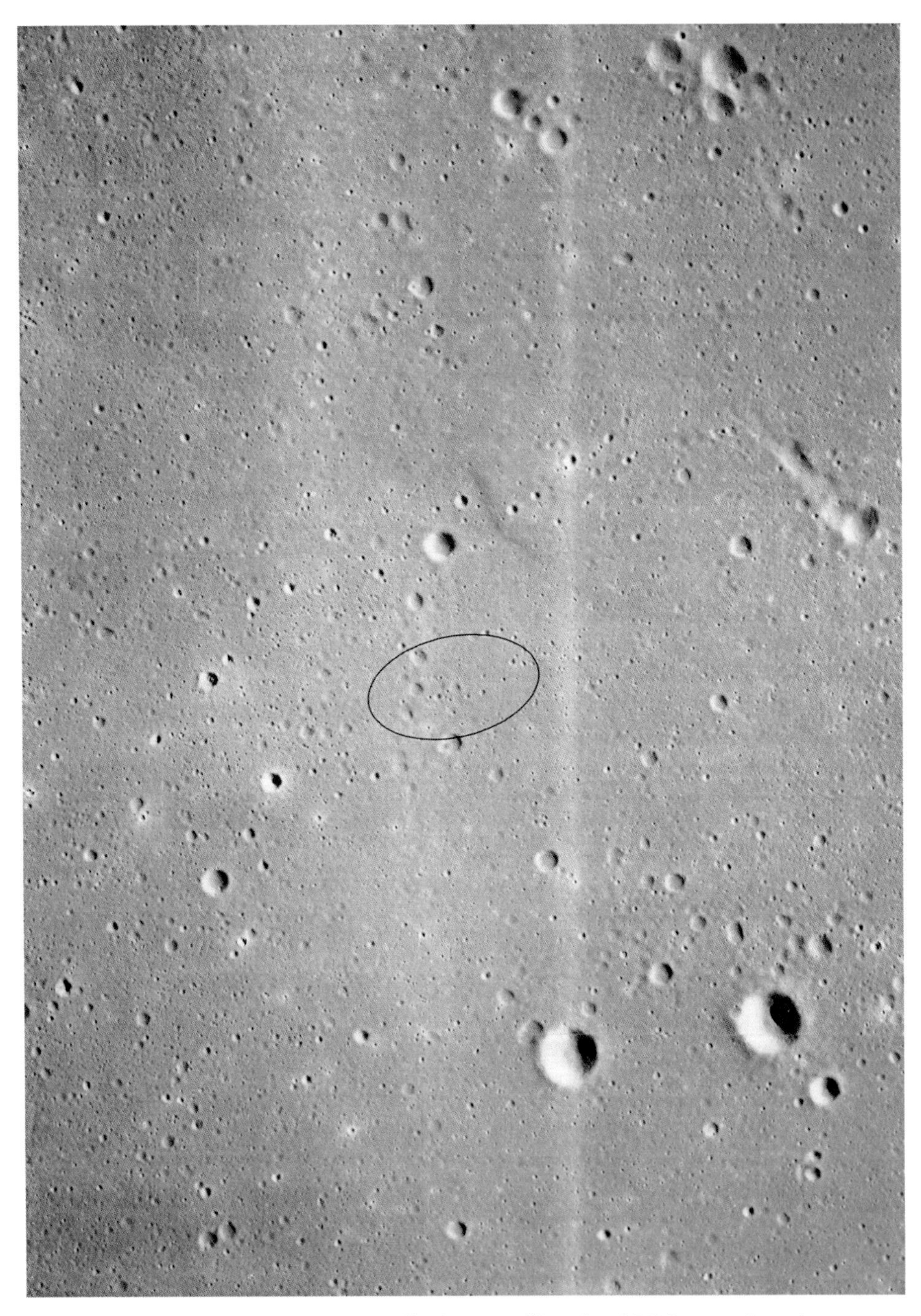

A Lunar Orbiter image showing the final target ellipse for ALS-7 centred on the crater cluster named the Snowman.

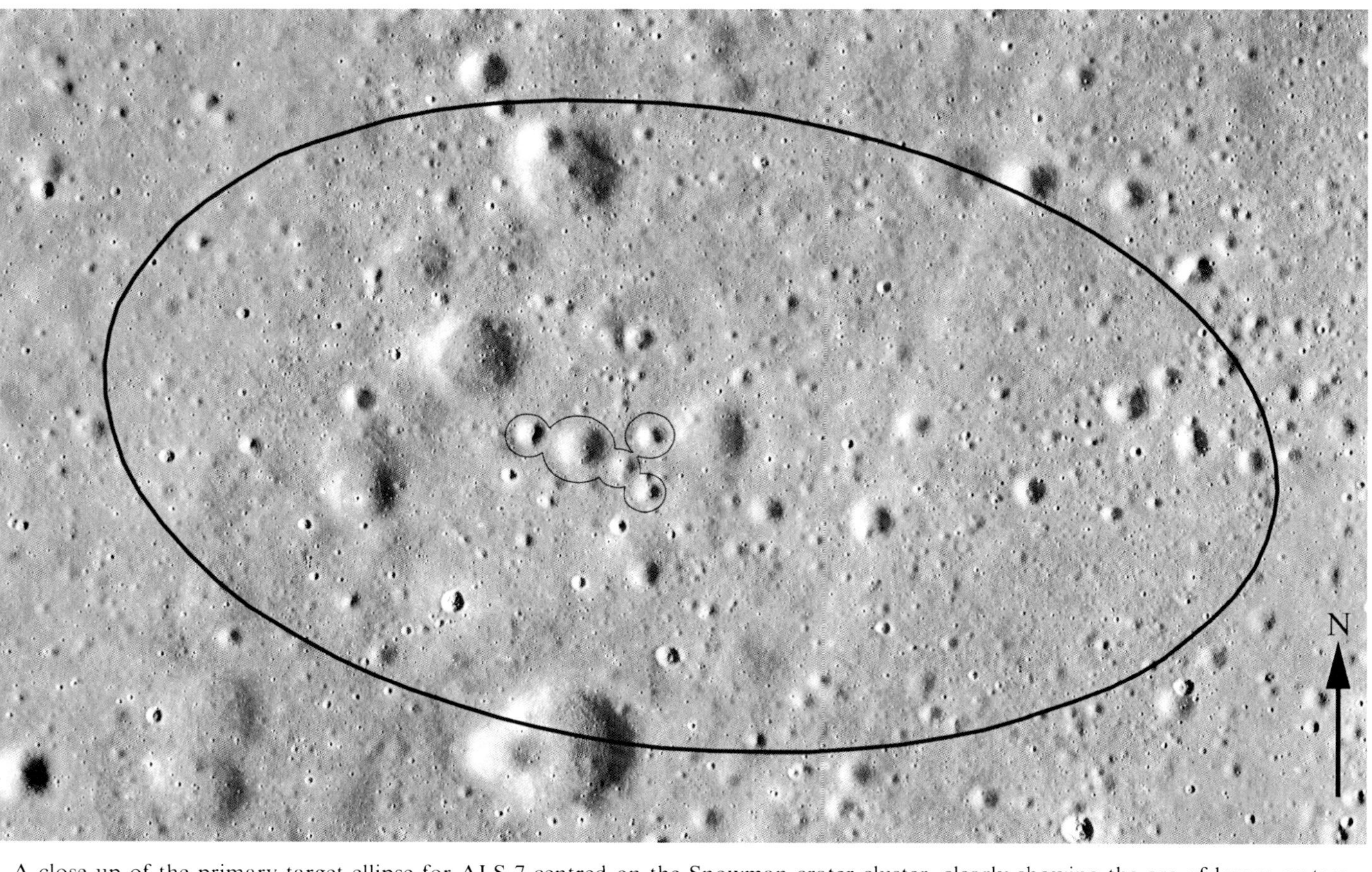

A close up of the primary target ellipse for ALS-7 centred on the Snowman crater cluster, clearly showing the arc of larger craters dubbed the Crescent.

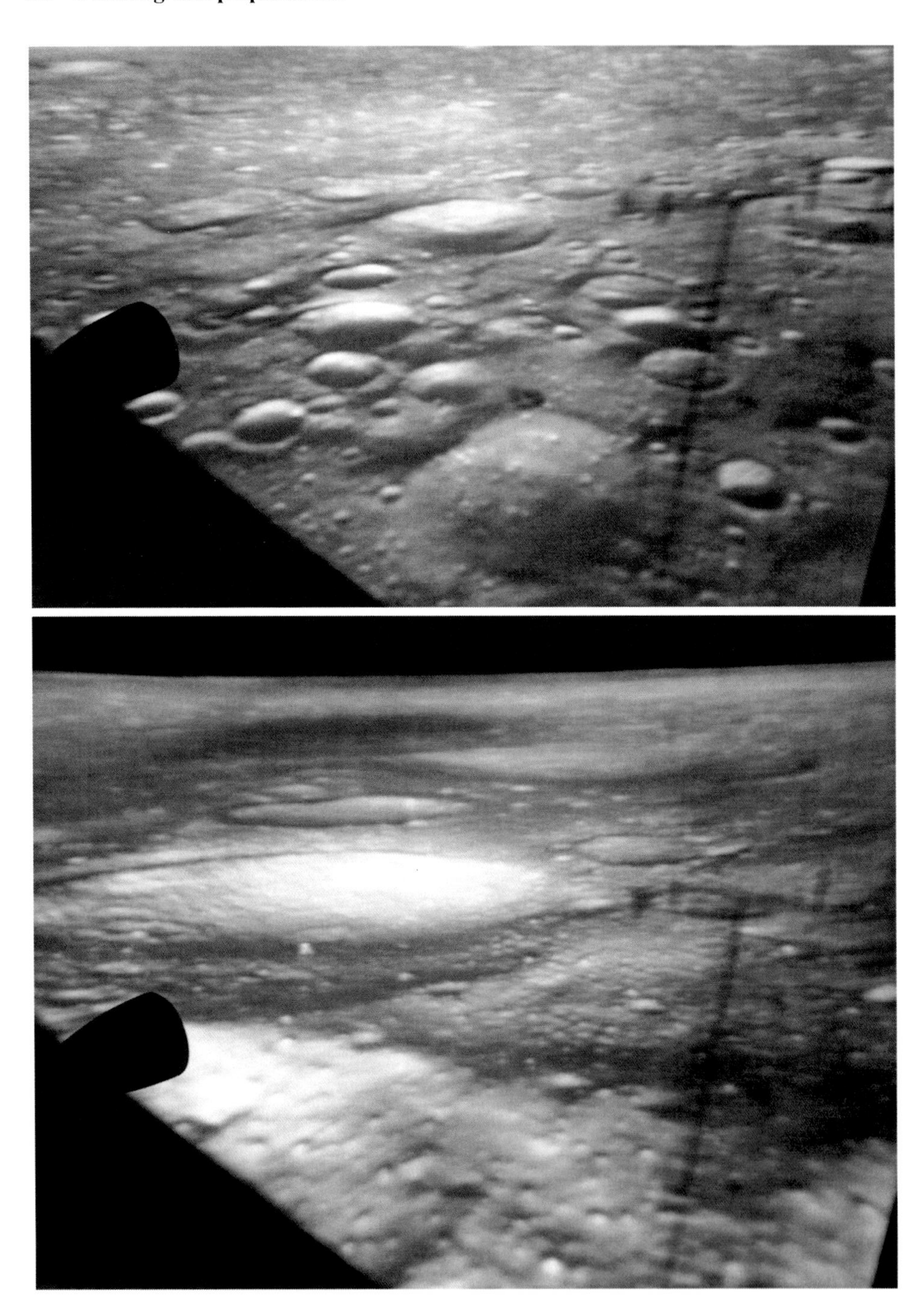

A lunar module simulator view of the Snowman crater cluster as Pete Conrad would see it in descending through altitudes of 1,800 feet (top) and 250 feet.

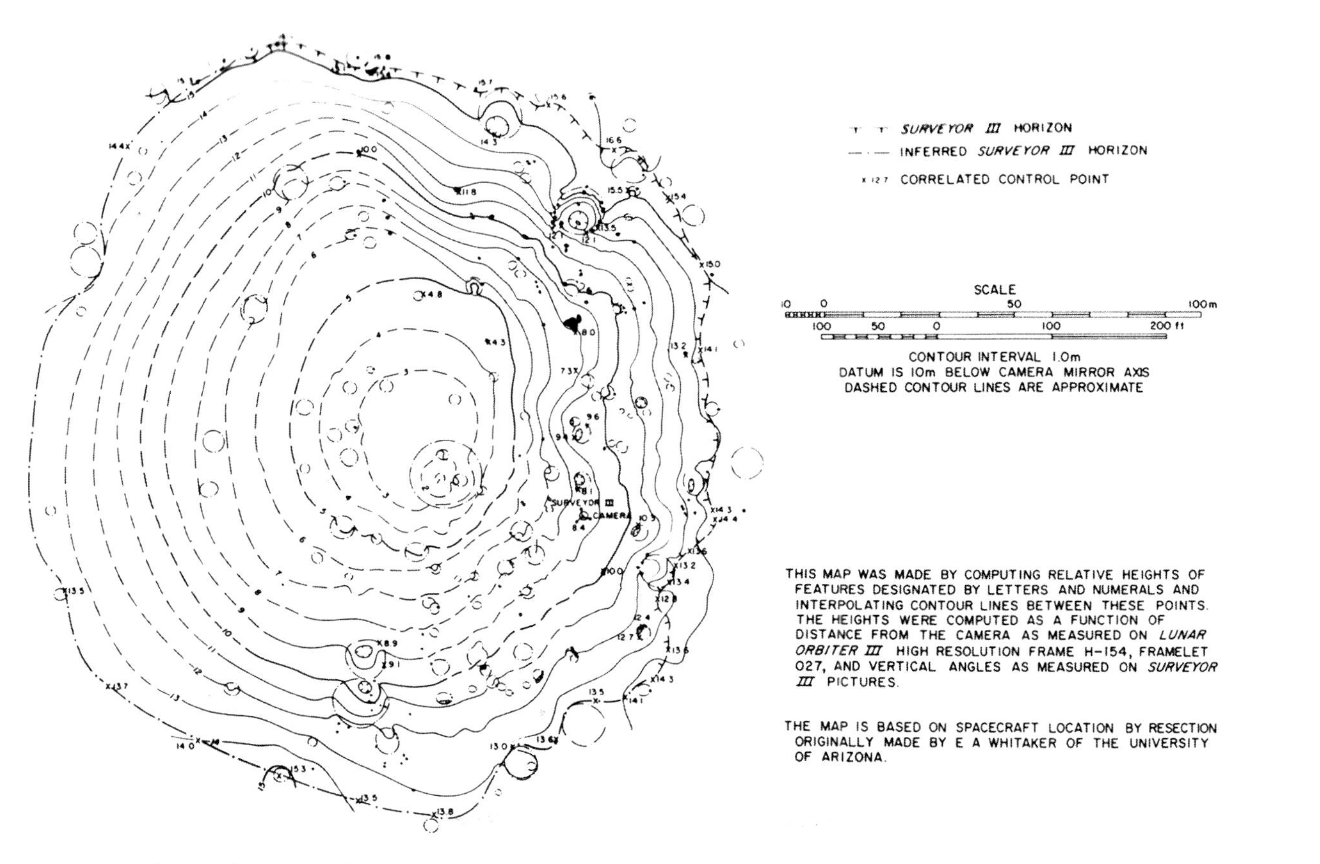

A contour map of the 600-foot crater in which Surveyor 3 landed. The contours were drawn by interpolating between control points derived by the photographic trigonometry method. The probable vertical accuracy is ± 0.5 metre.

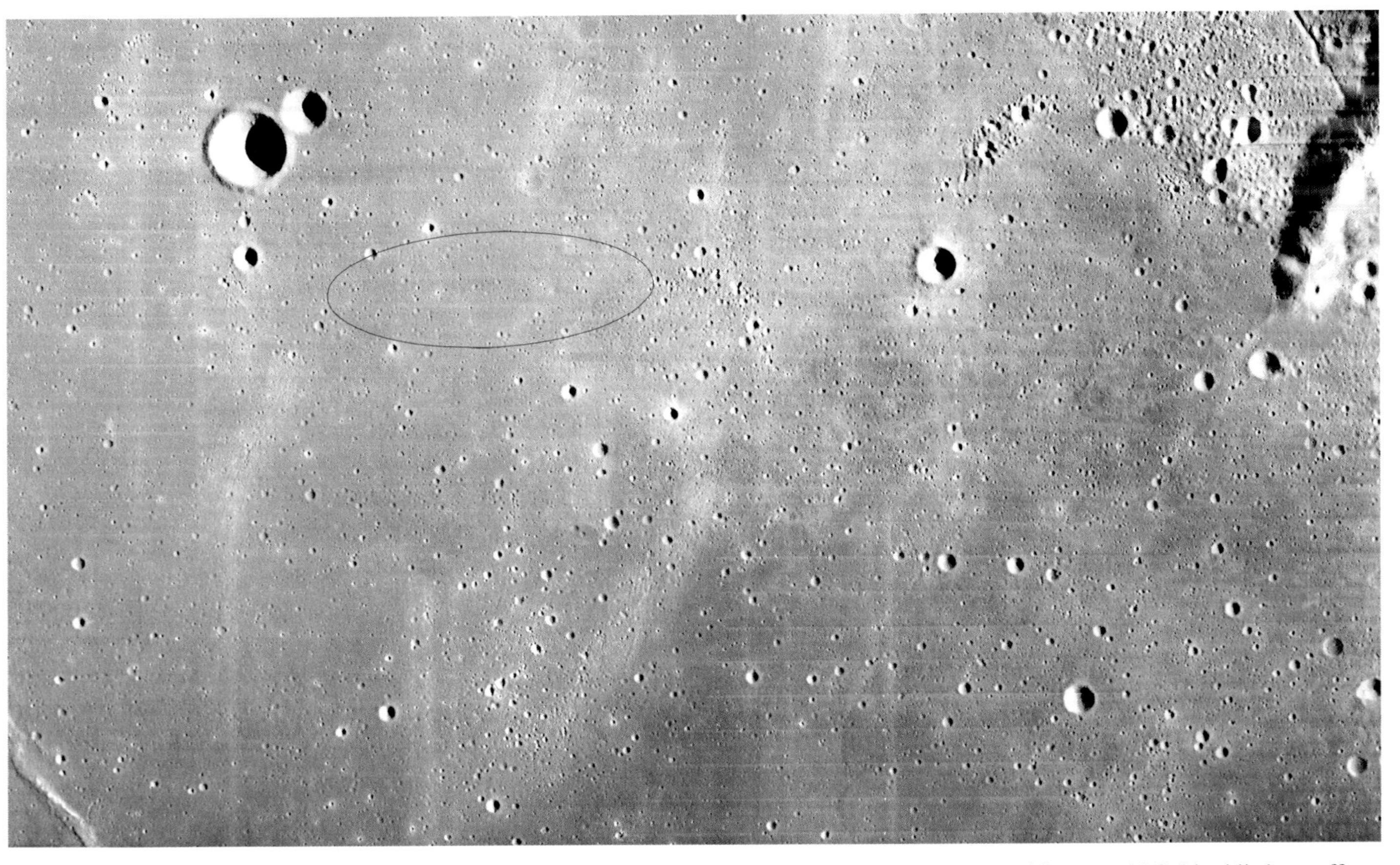

Having been selected as a backup target for the first Apollo landing as a bland patch of the Ocean of Storms, ALS-5 had little to offer Apollo 12.

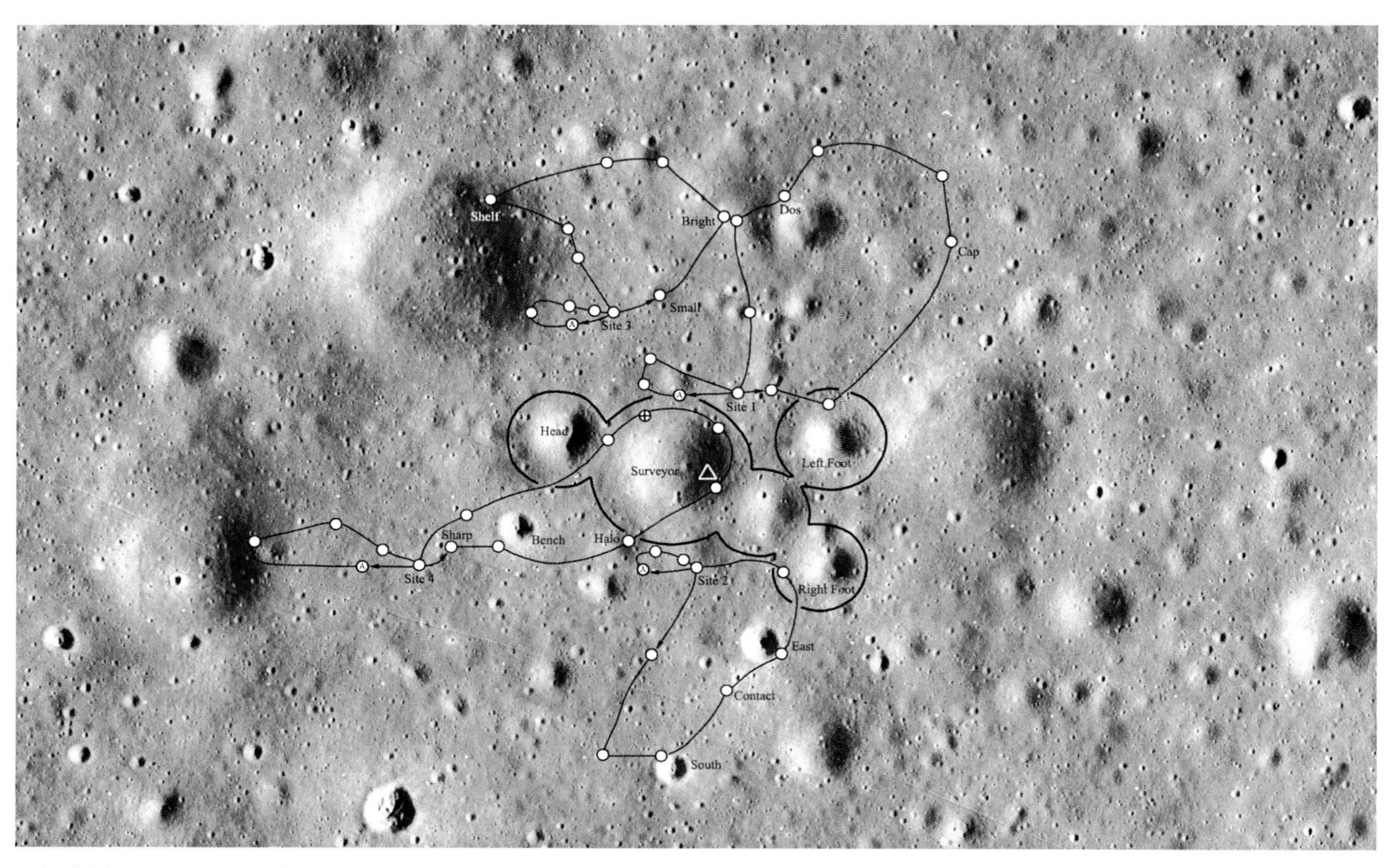

Activities were planned for four landing sites near the Snowman. In each case the ALSEP would be deployed at the position marked by an A in a circle. The triangle marks Surveyor 3.

enable the mothership to photograph the Fra Mauro site that was being considered for Apollo 13. On 29 July, after the return of Apollo 11, Mueller endorsed Surveyor 3 as the primary target for Apollo 12. The backup site for a 2-day launch delay would be ALS-5, in whose ellipse several small fresh-looking craters were selected as potential pinpoint targets.

Mission reviews on 15 August and 2 October discussed progress in the effort to permit a precision landing, simulations of deploying the ALSEP, the planning of the geological traverse, and the visit to Surveyor 3 to retrieve items for return to Earth. The Hughes Aircraft Company, which built Surveyor, was providing advice on what the astronauts should try to retrieve, how each item should be detached, and the tools required. Despite the role of the unmanned lander in demonstrating precision in the descent, the priorities for the surface activity were the deployment of the ALSEP on the first moonwalk and a geological traverse amongst the craters on the second. The visit to Surveyor 3 was something that would be done if time allowed.

In the case of Apollo 11, the 3-sigma ellipse within which there was a 99 per cent chance of achieving a landing was 10 × 2.6 nautical miles. For ALS-7 this was trimmed to 2.3 × 1.5 nautical miles. In planning, the crater in which Surveyor 3 had landed was referred to as 'Surveyor'. The cluster of craters became the 'Snowman', with Surveyor as the belly of the figure and other craters being named 'Head', 'Left Foot' and 'Right Foot'. Because the ellipse encompassed the entire Snowman, it was not possible to say precisely where Intrepid would land. The furthest the astronauts could reasonably be expected to walk to visit the unmanned lander was 2,000 feet, so activity plans were drawn up for four potential landing points.

Pete Conrad was initially sceptical of Emil Schiesser's plan to refine the descent trajectory. When Flight Dynamics Officer Dave Reed asked him where to place the aim point for the computer, Conrad, doubting that it would make much difference in practice, suggested landing 'long', since this would enable him to use the distinctive Snowman pattern to verify his line of approach. However, after simulations in which a television camera flew over a plaster model of the target, Conrad asked Reed to move the aim point onto what appeared to be a smooth patch northeast of Surveyor crater and was astonished when Reed said he would calculate the required changes. On reflection, Conrad suggested it would be simpler if the computer were to aim for the centre of Surveyor crater itself.

By 4 November Bill Tindall was sufficiently confident to say, "For whatever it is worth, my feeling is that as long as the systems work as well as they did in the past, we have a pretty good chance of landing near the Surveyor."

PREPARING THE SPACE VEHICLE

Even after the production lines had been established, it took several years to produce the launch vehicle and spacecraft for the Apollo 12 mission.

The Saturn V launch vehicle comprised the S-IC first stage, the S-II second stage, the S-IVB third stage and the instrument unit.

Starting at the base, the S-IC comprised the thrust structure, fuel tank, intertank, oxidiser tank and forward skirt. It stood 138 feet tall and its cylindrical body was 33 feet in diameter.

The thrust structure was a ring that housed the cross-beams on which the five F-1 kerosene-burning engines were mounted. The inboard engine was fixed on the axis, but the outboard engines were hydraulically gimballed to steer the vehicle by thrust vector control. Four half-cone fairings on its exterior smoothed the airflow over the engines. Each had a rigid triangular fin, and two solid-fuel rockets to retard the stage when jettisoned. The thrust structure was 19 feet 6 inches from the base of the fins to the top of the ring. The fuel tank was 43 feet tall including its dome caps, and held 216,000 gallons of kerosene. The intertank ring was 21 feet 10 inches tall. It enclosed the dome caps at the top of the fuel tank and at the base of the oxidiser tank with a narrow gap in between, and had a hatch to allow personnel access to its interior. The oxidiser tank was 64 feet tall including its dome caps, and held 345,000 gallons of liquid oxygen. The forward skirt was a ring 10 feet tall that enclosed the dome cap at the top of the oxidiser tank. The S-IC had an independent pair of 28 V_{DC} batteries. One battery, rated at 640 ampere-minutes, energised the solenoids of the valves and other stage controls. The other one, rated at 1,250 ampere-minutes, ran the flight instrumentation and the redundant systems that backed up the primary systems. (Silver-zinc batteries were used throughout the Saturn V because their rating was about four times greater than lead-acid or nickel-cadmium batteries of the same weight.) For redundancy, the range safety equipment could draw from either battery. If commanded by radio from the ground, the range safety system would shut down the engines and detonate ordnance to open the propellant tanks.

Each F-1 engine stood 19 feet tall and the open end of its nozzle extension was 12 feet 4 inches in diameter. The fuel and oxidiser turbopumps of each engine were on a single shaft which was driven by a turbine. Ten suction lines ran from the base of the fuel tank to the turbopumps (two lines per engine) and five suction lines passed through the fuel tank to feed liquid oxygen. Fuel and oxidiser were drawn off to a secondary combustion chamber to burn in a fuel-rich mixture that produced a 'cool' (815°C) exhaust to drive the turbine. The turbine operated at 5,500 revolutions per minute and was rated at approximately 60,000 brake horsepower. The pressure in the main combustion chamber was 965 psia and the temperature was 3,300°C. The expansion ratio between the area of the throat of the chamber and the aperture of the nozzle extension was 16, giving a sea-level thrust of 1,522,000 pounds. The interior surface of the nozzle extension was protected from the engine exhaust by injecting the 650°C exhaust from the turbines into the nozzle extension. This 'film cooling' formed a 'boundary layer' which prevented the hotter gases from impinging on the metal. Although the propellant flow rates were fixed, the thrust would increase as the vehicle climbed into the rarefied upper atmosphere. In its flight configuration, an F-1 engine weighed 18,500 pounds.

The S-IC had a dry weight of 300,000 pounds, and when loaded with 4,492,000 pounds of propellants had a launch weight of 4,792,000 pounds and a liftoff thrust of 7,610,000 pounds.

Working from the bottom, the S-II consisted of the aft interstage, aft skirt, thrust structure, oxidiser tank, fuel tank and forward skirt. It stood 81 feet 7 inches tall and its cylindrical body matched the S-IC at 33 feet in diameter.

The aft interstage was a ring 18 feet tall that enclosed the engine cluster and thrust structure. It had four 'positive acceleration' solid-fuel rockets at 90 degrees around its exterior and a hatch to allow personnel access to its interior. The aft skirt was a ring 7 feet tall that enclosed the bottom dome cap of the oxidiser tank and contained all equipment required for flight apart from that mounted on the aft interstage. The thrust structure was an inverted cone which narrowed to a diameter of 18 feet for the five J-2 hydrogen-burning engines. As with the S-IC, the inboard engine was fixed on the axis and the outboard engines were hydraulically gimballed for thrust vector control to steer the vehicle. Because the propellant tanks of the S-II had a common bulkhead there were domical caps only at the bottom of the oxidiser tank and top of the fuel tank. The integrated tanks were 78 feet tall. If a conventional arrangement had been employed, the additional end caps would have made the stage some 10 feet longer and 8,000 pounds heavier, taking it beyond its specifications. The common bulkhead was both a structural unit and a thermal shield: if the liquid hydrogen at a temperature of –253°C had been allowed to chill the liquid oxygen at –183°C, this would have solidified the latter! The fuel tank had a capacity of 282,300 gallons and the oxidiser tank 95,340 gallons. The forward skirt was a ring 11 feet 6 inches tall that enclosed the upper dome cap of the fuel tank. The S-II had four 28 V_{DC} batteries rated at 35 ampere-hours with which to energise the stage's systems. There were also two 28 V_{DC} batteries wired in series on the aft interstage for the pumps of the liquid hydrogen recirculation system that sustained the 'chill' in flight prior to stage ignition.

Each J-2 engine stood 11 feet tall and the open end of its nozzle extension was 6 feet 8 inches in diameter. The pressure in the combustion chamber was 763 psia and the temperature was 3,176°C. The expansion ratio between the area of the throat of the chamber and the aperture of the nozzle was 27.5. An oxidiser/fuel mixture ratio of 5.5:1 by weight provided a maximum thrust in vacuum of 230,000 pounds, and a ratio of 4.5:1 gave a thrust of 175,000 pounds. A propellant utilisation system would vary the mixture dynamically to suit flight requirements. In its flight configuration a J-2 engine weighed 3,480 pounds.

The S-II had a dry weight of 95,000 pounds (including the aft interstage). When loaded with 942,000 pounds of propellants, its launch weight was 1,037,000 pounds.

The S-IVB comprised the aft interstage, aft skirt, thrust structure, oxidiser tank, fuel tank and forward skirt. It stood 58 feet 7 inches tall and its cylindrical body was 21 feet 8 inches in diameter.

The aft interstage was a truncated cone weighing 7,700 pounds. It was 18 feet 8 inches tall and narrowed from 33 feet in diameter at its base to match the diameter of the stage's body. It enclosed the stage's thrust structure and the single J-2 hydrogen-burning engine. It was to be shed with the S-II, and had four solid-fuel retrorockets at 90 degrees to retard that stage. The aft skirt was a ring 7 feet 2 inches tall that enclosed the bottom dome of the oxidiser tank and held the stage's electronics. It had two solid-fuel 'positive acceleration' rockets on its exterior, 180 degrees apart, to settle the propellants in their tanks prior to igniting the J-2 engine for insertion into orbit. It also had two auxiliary propulsion system units, offset 90 degrees from the two solid rockets. Each such unit had four liquid rockets with common tanks for monomethyl hydrazine and nitrogen tetroxide, which were hypergolic propellants. Three rockets of 150 pounds thrust (one left, one right and one perpendicular) were for control of pitch, yaw and roll in space, and an aft-pointing rocket of 70 pounds thrust was to settled the propellants in the main tanks prior to reigniting the J-2 for translunar injection. The thrust structure held ten external helium bottles in addition to the J-2 engine, which was hydraulically gimballed for steering by thrust vector control. As in the case of the S-II, the cryogenic propellant tanks of the S-IVB had a common bulkhead, with the liquid oxygen tank below the liquid hydrogen tank. The integrated tanks were 44 feet tall, the fuel tank had a capacity of 78,000 gallons and the oxidiser tank 21,200 gallons. The forward skirt was a ring 10 feet 2 inches tall that enclosed the forward dome of the fuel tank. Its exterior held the non-propulsive oxygen vent values, and a pair of aft-facing hydrogen vents set 180 degrees apart to provide low-thrust continuous propulsive venting to maintain the propellants settled in their tanks while coasting in orbit. It was powered by three 28 V_{DC} batteries rated at 228 ampere-hours for stage operations, and the chill-down pumps operated off a pair of 28 V_{DC} batteries wired in series that were together rated at 67 ampere-hours.

The S-IVB had a dry weight of 34,000 pounds, including the aft interstage, which would be shed with the S-II. When loaded with 228,000 pounds of propellants, its launch weight was 262,000 pounds.

The instrument unit was a ring 3 feet tall that was 21 feet 8 inches in diameter, matching the third stage, and had a hatch to provide access to its interior and also to the forward skirt of the S-IVB. The fabrication, assembly, system testing, integration and final checkout was by IBM. The instrument unit weighed 4,500 pounds, and its interior surface carried guidance and flight control systems plus systems to collect, condition and transmit measurements to Earth. It had four 28 V_{DC} batteries rated at 350 ampere-hours. For extreme reliability, the launch vehicle's digital computer had triple-redundancy and voting logic.

From the perspective of the launch vehicle, everything above the instrument unit was the 'spacecraft'. It comprised the adapter, lunar module, command and service modules, and launch escape system. It stood 80 feet tall from the base of the adapter to the aerodynamic sensor at the tip of the escape tower.

In terms of dry weights, the command module was 12,365 pounds, the service module was 10,510 pounds, the descent stage of the lunar module was 4,875 pounds and its ascent stage was 4,760 pounds. The adapter weighed 3,960 pounds, and the launch escape system on the apex weighed 8,963 pounds. The total dry weight of the 'spacecraft' was therefore 45,433 pounds. The propellant loads were 40,595 pounds for the command and service modules, 17,925 pounds for the descent stage of the lunar module and 5,765 pounds for its ascent stage. Hence, the launch masses were 63,470 pounds for the command and service modules, and 33,325 for the lunar module.

The integrated launch vehicle and spacecraft was called the space vehicle. It stood 363 feet tall, had a liftoff weight of 6,200,000 pounds, and could insert a total mass of

250,000 pounds into a circular orbit at an altitude of 600,000 feet – on an Apollo lunar flight the insertion mass comprised the S-IVB and the spacecraft.

The space vehicle for the Apollo 12 mission would use the AS-507 launch vehicle and a spacecraft comprising CSM-108 and LM-6.

The first stage of the Saturn V, the S-IC, was produced by Boeing at the Michoud Assembly Facility on the Louisiana coast in a plant with 2 million square feet of manufacturing floor space. The Vertical Assembly Building, purpose-built separate from the Manufacturing Building, was a single-storey shell about 180 feet tall in which an overhead crane with a capacity of 180 tons stacked the thrust structure, the fuel tank, the intertank, the liquid oxygen tank and the forward skirt in an upright configuration. The nearby Stage Test Building housed four test cells, each having an area of 83 by 191 feet and a clear height of 51 feet and equipped to test the electrical and mechanical systems of a completed stage.

Component assembly of S-IC-7 started in January 1967. Stage assembly began on 6 March with the transfer of the thrust structure to the Vertical Assembly Building. A week later the fuel tank was mated with the thrust structure. After the intertank and the liquid oxygen tank were in place, the forward skirt was added on 30 March. Vertical assembly was completed on 20 April. The stage was then transferred to a horizontal trailer and taken to the Manufacturing Building for the installation of its components. The five F-1 engines were manufactured by Rocketdyne, which was a subsidiary of North American, and installed between 4 May and 19 June. After the installation of hardware was completed on 11 August, the stage was moved in the horizontal configuration to Test Cell no. 2 in the Stage Test Building for a full post-manufacturing checkout, concluding with a simulated static firing on 10 November. It was placed into storage on 22 November to enable Michoud to undertake higher priority work. It was retrieved on 1 April 1968 and installed in Test Cell no. 1 of the Stage Test Building to undergo some final modifications, then transferred to an assembly position in the Manufacturing Building on 23 August to await shipment.

The S-IC was shipped to the Mississippi Test Facility by the barge *Pearl River* on 12 September 1968 and installed in the B-2 test stand, conveniently adjacent to the barge turning basin, the next day. The system developed to suppress the 'pogo' that afflicted the S-IC stage on the Apollo 6 unmanned test flight earlier in that year was installed and then the pre-static firing tests were performed, essentially to verify the post-manufacturing checkout by Michoud. On 18 October NASA and Boeing held the pre-static firing review, which decided to proceed with the test. The firing was scheduled for 23 October but a malfunction of the liquid oxygen depletion cutoff system during a simulated static test on that date delayed the acceptance firing. The countdown began on 28 October, one day prior to the rescheduled test, but after the kerosene had been loaded a leak developed which required the tank to be drained in order to fix it. The fuel was loaded on 29 October and after the integrity of the tank was verified it was decided to proceed. The cryogenic liquid oxygen was loaded on 30 October and the five F-1 engines ignited in a 1-2-2 sequence. The engine gimbal program was initiated at T + 3 seconds and completed at T + 111 seconds. Cutoff was initiated automatically after a main-stage of 126.46 seconds, which, because main-stage had to precede the commitment to lift off, corresponded to a flight of 125 seconds. The engines and stage operated within their specifications. The post-firing checkout established that the stage was in good condition.

Soon after S-IC-7 was installed on the B-2 stand in September, the motor of the derrick had developed a fault. As a result, the stage could not be removed until this was fixed – which would require a component to be specifically manufactured. The repair was finished on 8 November and the stage was loaded onto a barge later that day. After arriving at the Michoud Assembly Facility on 9 November the stage was refurbished. A simulated flight on 20 January 1969 drew the post-static checkout to a successful conclusion. NASA took formal acceptance of the stage on 17 February. It left aboard the barge *Orion* on 29 April and arrived in the turnaround basin of the Kennedy Space Center on 3 May, where it was immediately offloaded onto a tow-trailer and taken to the Vehicle Assembly Building (VAB).

The S-II was manufactured by North American. The main fabrication and testing facilities were purpose-built at Seal Beach south of downtown Los Angeles, about 15 miles from Downey, where the company built the command and service modules of the Apollo spacecraft.

Work on S-II-7 began on 28 July 1966 in the Bulkhead Fabrication Building. On 27 January 1967 work was paused for a fortnight after seismic activity required the thrust structure assembly fixture in the 125-foot-tall Vertical Assembly Building to be restored to its precise alignment. Vertical assembly finally started on 31 March, but was delayed by a variety of problems. The main structure was completed on 12 January 1968 by the addition of the forward skirt. The five J-2 engines supplied by Rocketdyne were installed on 10 February. Systems integration was completed on 6 May, and systems checkout on 27 June.

The preparation of the stage for shipment began on 8 July. It left aboard the barge *Point Barrow* on 29 October and arrived at the Michoud Assembly Facility on 11 November. The next day it was transferred to *Little Lake* for the journey up-river to the Mississippi Test Facility. It was hoisted onto the A-1 test stand on 13 November. After the post-shipment systems checkout a number of modifications were made. A combined tanking and cryogenic proof test was made on 15 January 1969. The pre-static checks were completed on 20 January. The two propellants were loaded on 22 January, the day of the test. The engines were fired for 364 seconds, and achieved all of the test objectives. After the post-static checkout the stage was transferred from the stand into the S-II Stage Storage Building at the Mississippi Test Facility. It left by barge on 15 April and arrived at the Kennedy Space Center on 21 April, where it was stored in the VAB.

The S-IVB third stage was manufactured by Douglas Aircraft in various company facilities in California.[8] The fabrication of subassemblies was in Santa Monica, west of Los Angeles. The final assembly and factory checkout was in Huntingdon Beach, down the coast, where there were towers for vertical assembly and checkout.

[8] In 1967 Douglas Aircraft merged with McDonnell Aircraft to form McDonnell Douglas.

The S-IC stage of the Apollo 12 space vehicle in the Vehicle Assembly Building on 7 May 1969.

The S-IC stage of the Apollo 12 space vehicle is hoisted upright in the Vehicle Assembly Building on 7 May 1969.

The S-II stage of the Apollo 12 space vehicle arrives at the Vehicle Assembly Building on 22 April 1969.

The S-II is hoisted and stacked on top of the S-IC stage of the Apollo 12 space vehicle in the Vehicle Assembly Building on 21 May 1969.

The S-IVB of the Apollo 12 space vehicle arrives at the Kennedy Space Center aboard a Super Guppy on 10 March 1969.

The stacked S-IC and S-II stages of the Apollo 12 space vehicle in the Vehicle Assembly Building on 21 May 1969.

Work on S-IVB-507 began in early 1967. Assembly and systems integration was finished on 15 November, and systems checkout was completed on 24 January 1968. After the post-manufacturing checkout and final inspection on 28 February, it was placed into storage. On 7 August the stage was transferred by road to the nearby Los Alamitos Naval Air Station for a flight to Mather Air Force Base aboard the Super Guppy of Aero Spacelines. On 9 August it was installed on the Beta I test stand of the manufacturer's Sacramento Test Center, only a short distance by road from the airfield. The integrated systems checkout was completed on 15 October. The static test firing on 16 October lasted 433 seconds and simulated orbital insertion, a hiatus in parking orbit and then translunar injection. The post-static firing checkout ended on 29 October. The stage was transferred to the Vertical Checkout Laboratory on 30 October for modifications. The final checkout was completed on 10 February. On being formally accepted by NASA, it was delivered to the Kennedy Space Center on 10 March 1969 and was stored in the VAB.

The instrument unit was made by IBM in its own facility in Huntsville, Alabama. The systems checkout was automated, with each of the six subsystems being tested individually and again as an integrated unit. It was delivered to the Kennedy Space Center on 8 May 1969.

The S-IC was installed on mobile launch platform no. 2 on 7 May. The S-II was stacked on 21 May and the S-IVB/IU on 22 May.

The mobile launcher comprised a two-level steel base structure 25 feet high, 160 feet long and 135 feet wide. There was a 45-foot-square hole offset to one end for the F-1 engine cluster. In addition to the four hold-down arms and three tail-service masts for the vehicle, the upper deck (referred to as level zero) had the permanently installed umbilical tower, which was an open steel structure 380 feet tall. This had nine 'swing arms' to provide access to the space vehicle. Arm 1, S-IC Intertank, had liquid oxygen fill and drain interfaces. It was withdraw at T-30 seconds, taking 13 seconds to retract and lock against the tower. Arm 2, S-IC Forward, had pneumatic, electrical and air-conditioning interfaces. It would withdraw at T-16 seconds, taking 6 seconds to lock against the tower. Arm 3, S-II Aft, provided personnel access. It would be retracted prior to liftoff as required. Arm 4, S-II Intermediate, primarily provided liquid oxygen and liquid hydrogen. It would not be retracted until the vehicle was in the process of lifting off, taking 6.4 seconds to lock against the tower. Arm 5, S-II Forward, provided gaseous hydrogen venting, electrical and pneumatic interfaces. It would retract at liftoff, taking 7.4 seconds to lock against the tower. Arm 6, S-IVB Aft, primarily provided liquid oxygen and liquid hydrogen. It would retract at liftoff, taking 7.7 seconds to lock against the tower. Arm 7, S-IVB/IU Forward, had gaseous hydrogen venting, electrical and pneumatic interfaces for the S-IVB and air-conditioning, electrical and pneumatic interfaces for the instrument unit. It would retract at liftoff, taking 8.4 seconds to lock against the tower. Arm 8, Service Module, had air-conditioning, vent line, coolant, electrical and pneumatic interfaces. It would retract at liftoff, taking 9 seconds to lock against the tower. Arm 9, Command Module Access, held the White Room that provided access to the crew hatch. It would be retracted in two stages, first 12 degrees to a standby position, ready for rapid reinstallation, and then all the way immediately prior to launch, but

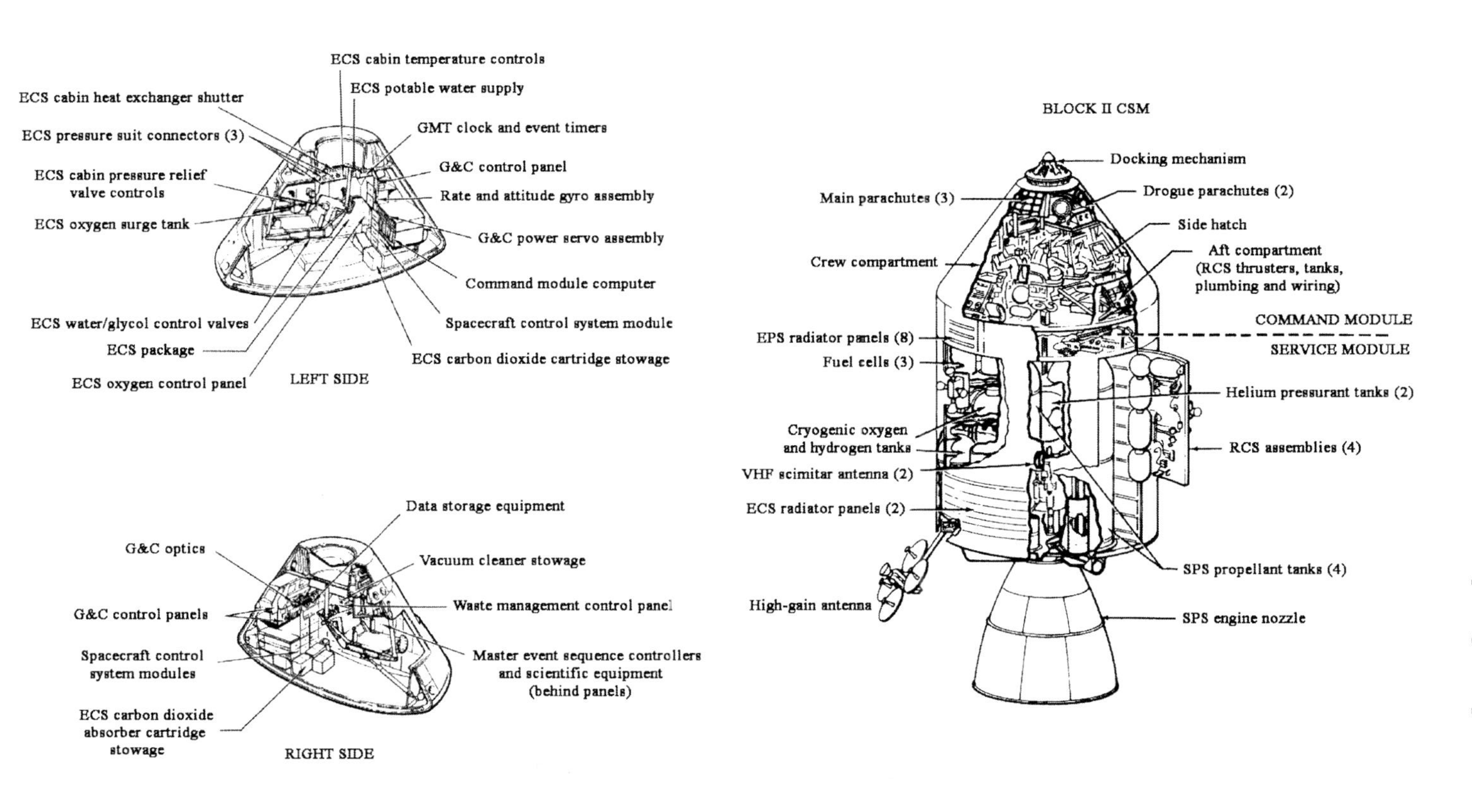

Details of the command and service modules of the Block II Apollo spacecraft.

Pete Conrad (left) and Dick Gordon inspect their command module during its manufacture by North American Rockwell in January 1969.

The Apollo 12 command and service modules on 13 June 1969 awaiting the engine bell of its service propulsion system.

The Apollo 12 command and service modules on 30 June 1969.

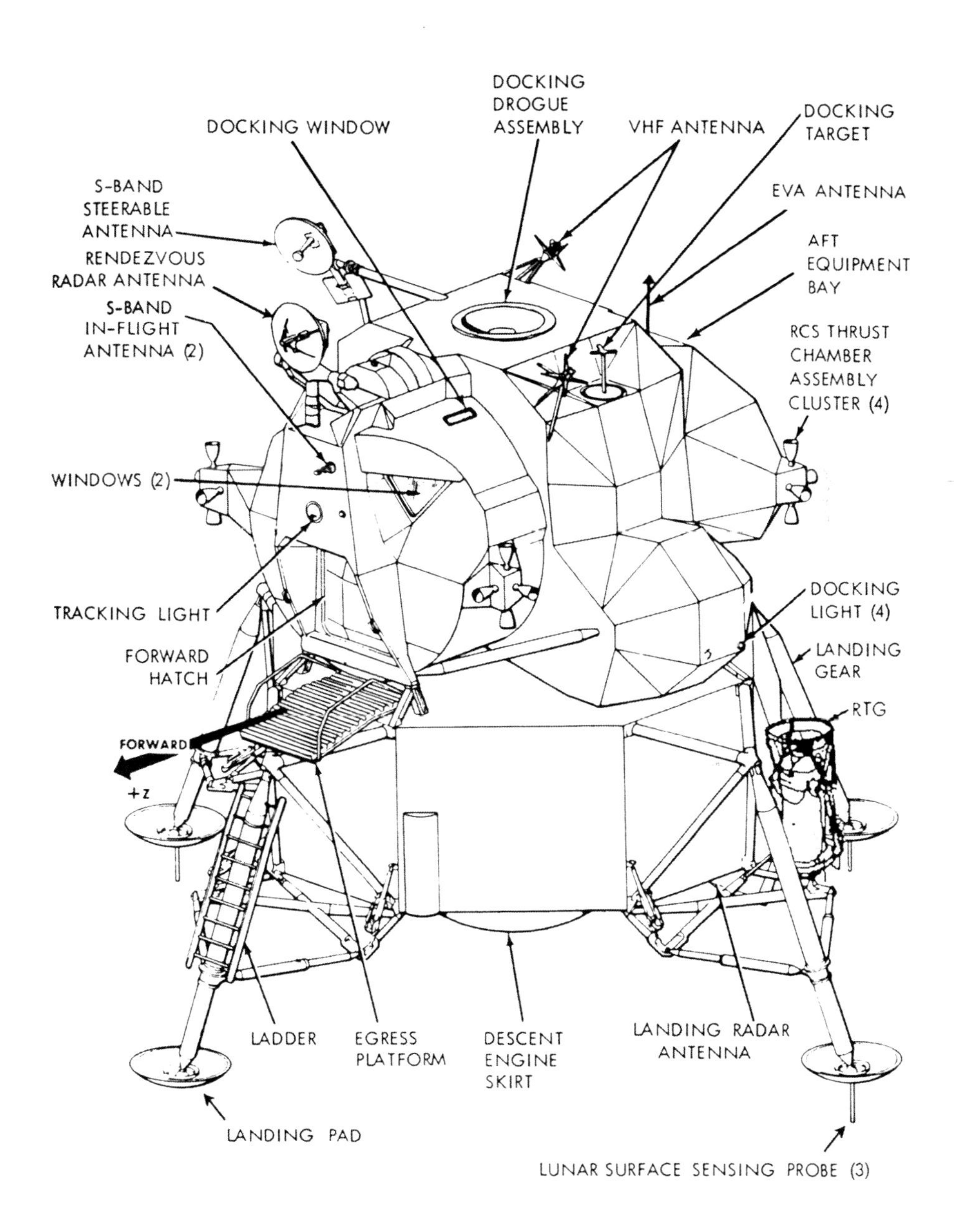

Details of the Apollo lunar module.

The two modules of the Apollo 12 lunar module arrive at the Kennedy Space Center in a Super Guppy on 24 March 1969.

Checking out the ascent stage of the Apollo 12 lunar module on 26 March 1969.

Checking out the descent stage of the Apollo 12 lunar module on 26 March 1969.

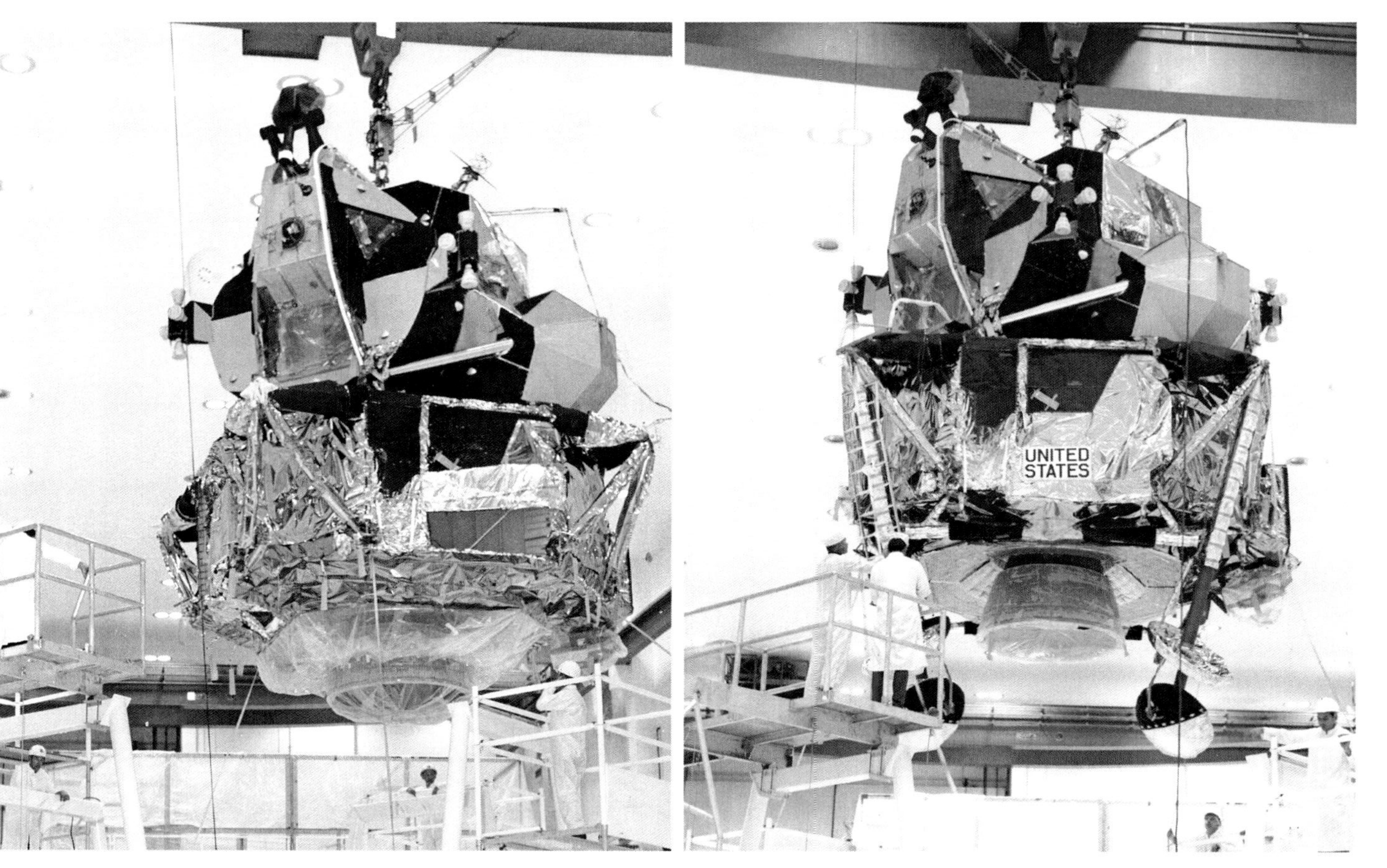

The Apollo 12 lunar module on 16 June 1969 awaiting its legs, and complete on 23 June (right).

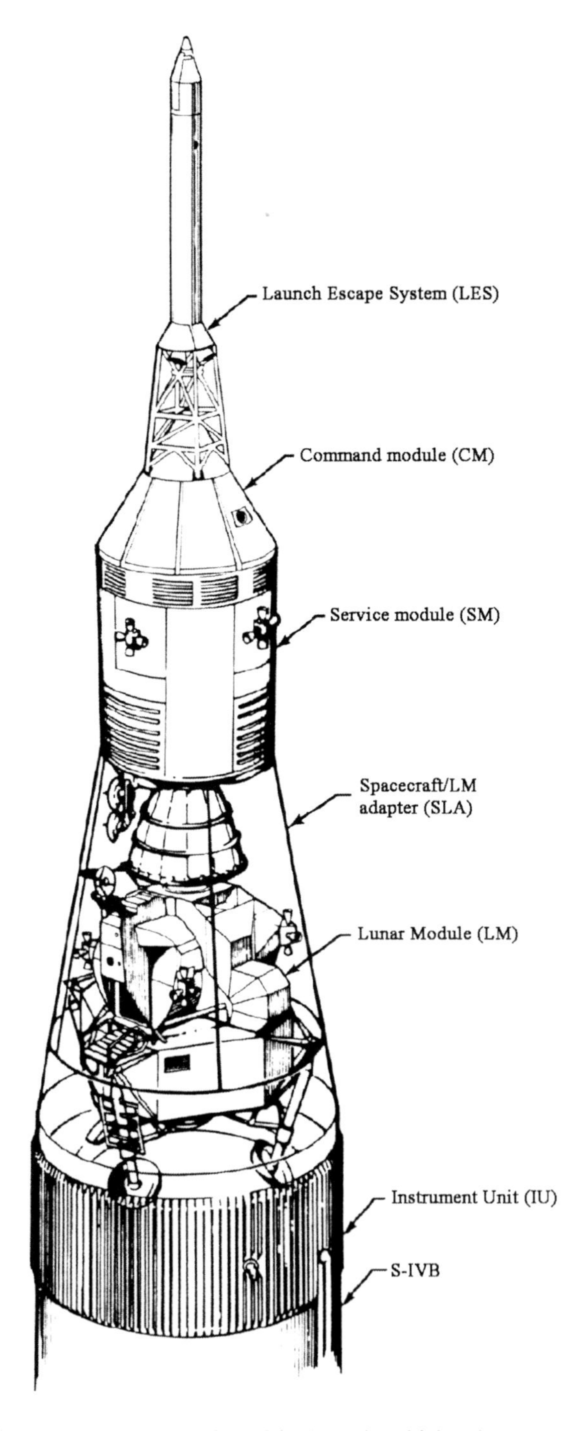

Details of the Apollo spacecraft and its launch vehicle adapter.

Installing the Apollo 12 lunar module in the launch vehicle adapter on 23 June 1969.

The Apollo 12 spacecraft is hoisted on 30 June 1969 ready for stacking with its launch vehicle.

The Apollo 12 launch vehicle on 28 May 1969 (left) and stacking the spacecraft on 1 July.

A view from the roof of the Vehicle Assembly Building of the Apollo 12 space vehicle's rollout on 8 September 1969.

The Apollo 12 space vehicle's rollout from the Vehicle Assembly Building seen from alongside the crawler on 8 September 1969.

Apollo 12 on Pad A of Launch Complex 39 on 28 October 1969, during the 'wet' part of the Countdown Demonstration Test.

The Firing Room of the Launch Control Center on 29 October during the Countdown Demonstration Test.

Unsurprisingly, the command module simulator at the Kennedy Space Center (top) was referred to as the 'train wreck'. Al Bean (bottom left), Dick Gordon and Pete Conrad in the command module simulator on 22 October 1969.

The Apollo 12 crew on the steps of the command module simulator at the Kennedy Space Center on 22 October 1969.

Dick Gordon (top) in the command module simulator on 6 November 1969 checking out the cameras for the photography he will undertake during the solo portion of the mission. Pete Conrad (bottom left) and Al Bean in a lunar module simulator on 22 October 1969.

Al Bean at the Manned Spacecraft Center on 24 October 1969 in the suspension rig with six degrees of freedom designed to simulate walking in lunar gravity.

Al Bean carries the hand-tool carrier while rehearsing a geological traverse at Flagstaff, Arizona, on 10 October 1969.

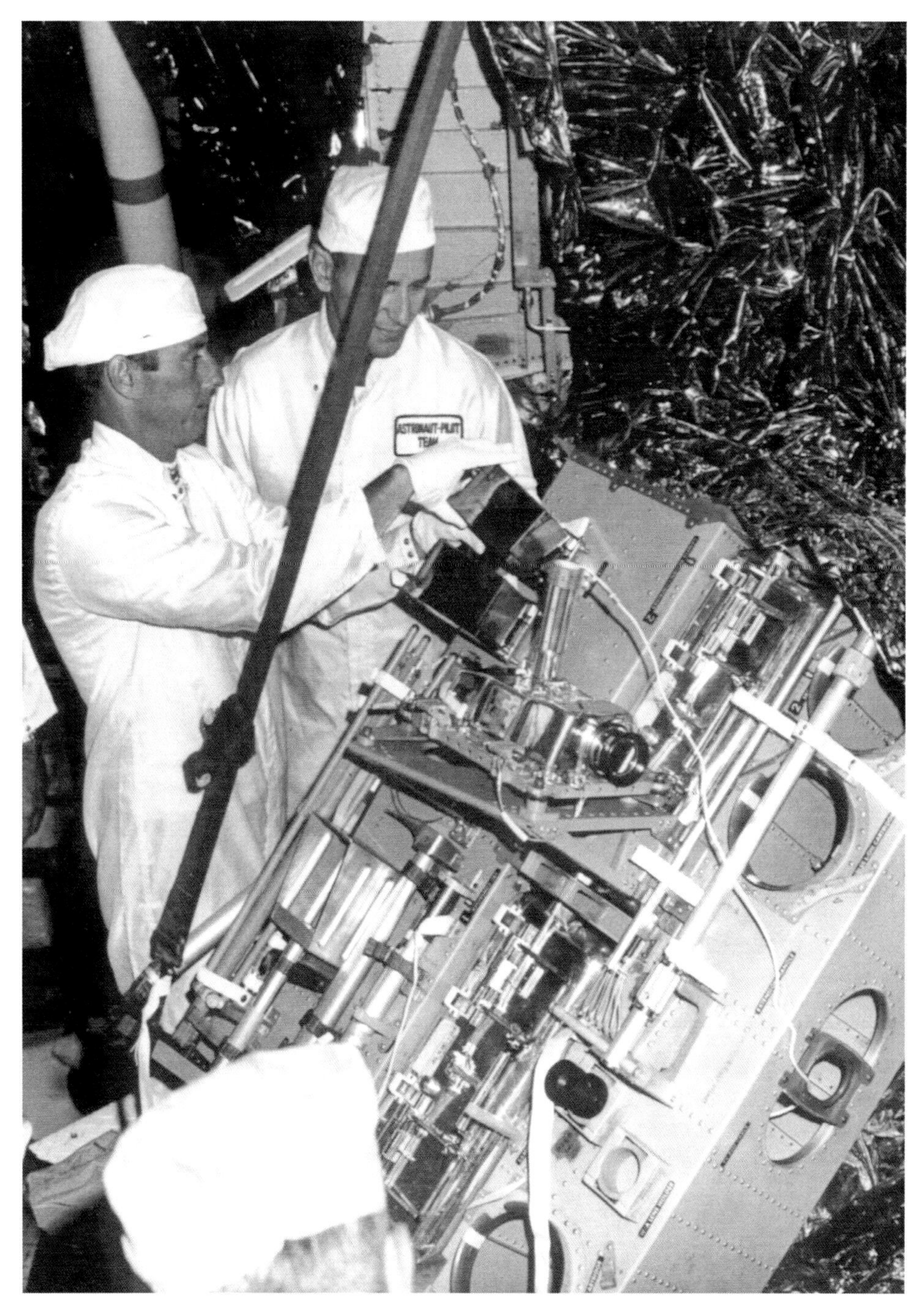

In training on 20 June 1969, Pete Conrad (left) and Al Bean examine the lunar module MESA. Note the television camera affixed to the small platform.

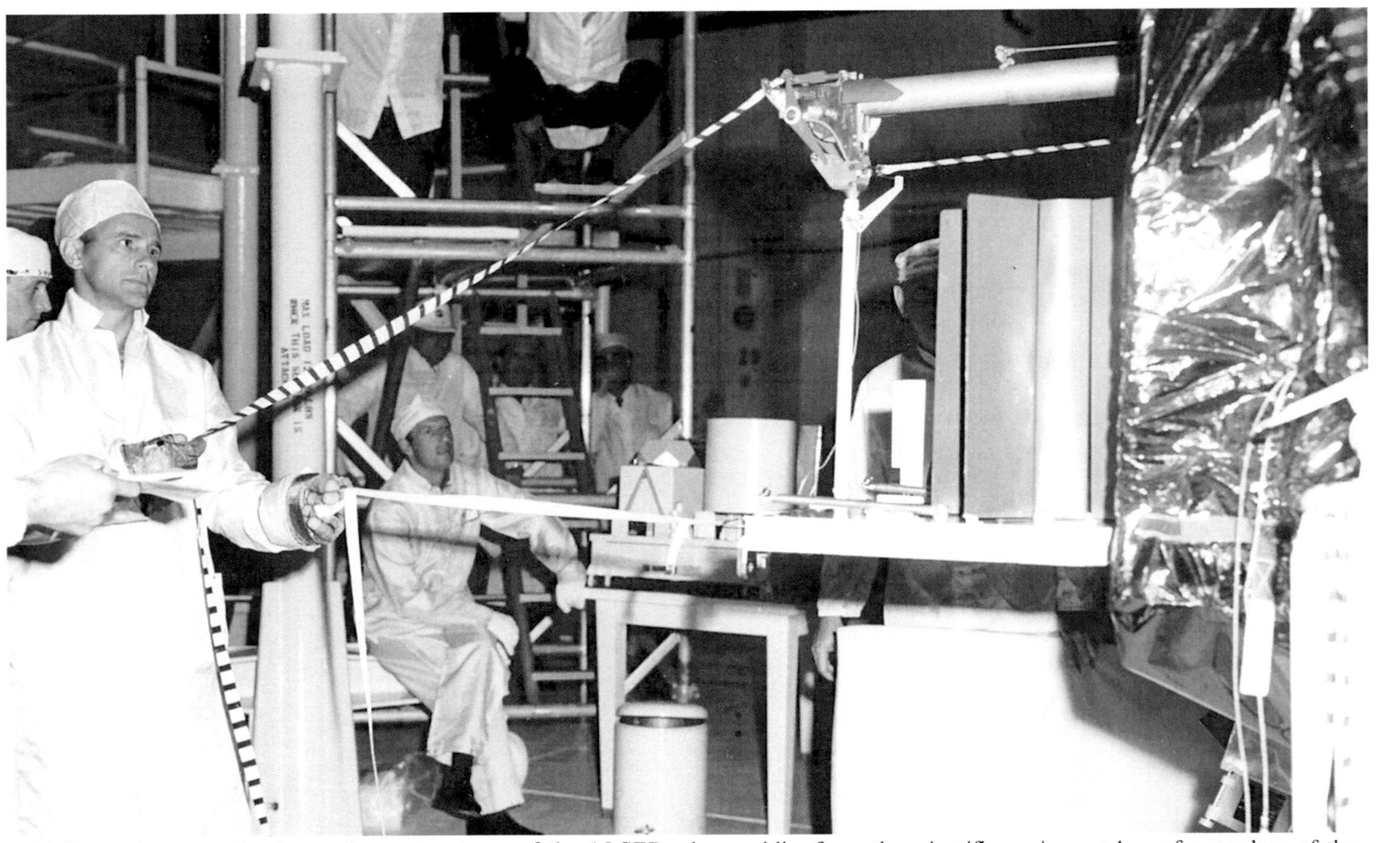

Al Bean rehearses using lanyards to extract one of the ALSEP subassemblies from the scientific equipment bay of a mockup of the lunar module on 23 April 1969. Pete Conrad (seated) looks on.

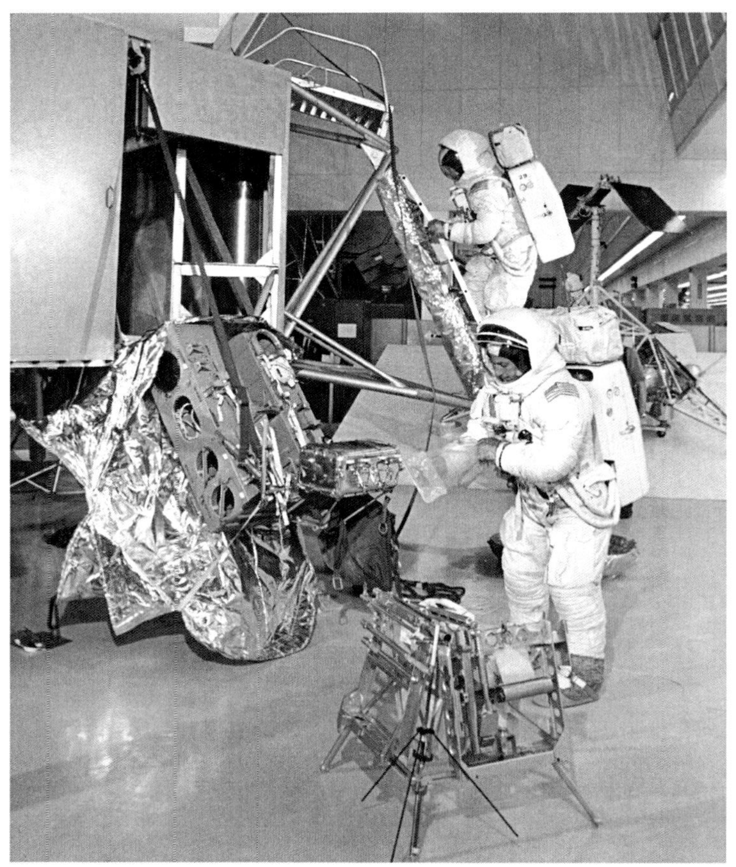

Al Bean on the ladder of the lunar module mockup in training on 6 October 1969. Standing by the hand-tool carrier, Pete Conrad holds a photographic calibration chart (left) and a sample bag (right). Ncte the MESA with a 'rock box' on the 'table'.

In training on 6 October 1969 Pete Conrad rehearses using the lunar equipment conveyor with Ed Gibson watching. (Note the ramp with the Surveyor model in the background.)

In training at the Kennedy Space Center on 6 October 1969, Pete Conrad (left, with the Surveyor parts bag on his backpack) and Al Bean rehearse sampling. Conrad lifts a rock using tongs while Bean holds a sample bag ready (top) and passes the rock to Bean (bottom). Note the chest-mounted Hasselblad cameras and the tripod gnomon.

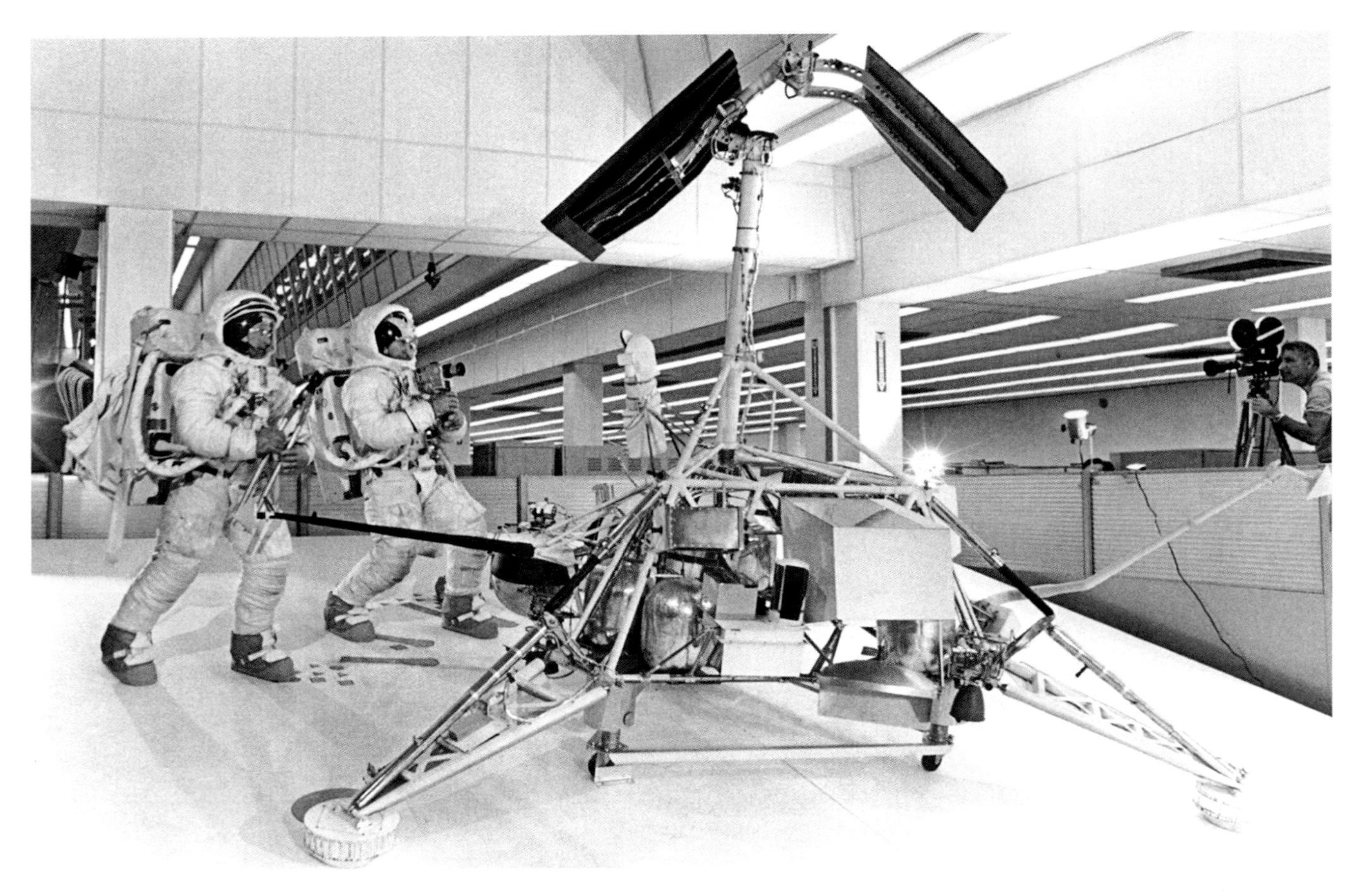

In training on 6 October 1969 Pete Conrad (left) is holding the cutters about to rehearse Surveyor tasks. The model is on a ramp inclined to match the slope of its counterpart on the Moon.

The Apollo 12 crew rehearse post-splashdown egress in the Gulf of Mexico on 20 September 1969 using full-body Biological Isolation Garments.

Pete Conrad and Al Bean review the scientific objectives of their mission with Jack Schmitt (standing) and Ed Gibson (left) on 7 November 1969.

Al Bean (left), Dick Gordon and Pete Conrad at a press conference in early October 1969.

Al Bean shows off models of the central station of the ALSEP and some of its instruments at a press conference in early October 1969.

Pete Conrad piloting the Lunar Landing Training Vehicle on 25 October 1969.

in contrast to the other arms it hinged in the opposite direction to position the White Room as far away as possible from the exhaust of the rising vehicle. There were two high-speed elevators in the interior of the tower structure (accessed from the upper interior level of the platform) and there was a hammerhead crane on top which had a 50,000-pound load capacity.

North American completed the individual command module and service module systems tests on 20 January 1969, and the integrated systems test on 3 February. The modules arrived individually at the Kennedy Space Center on 28 March. They were mated on 2 April, the combined system test was completed on 21 April, and altitude chamber testing was completed on 10 June.

Grumman completed the integrated test of LM-6 on 31 December 1968. After the final engineering evaluation acceptance test on 18 February 1969, the stages were delivered individually to the Kennedy Space Center on 24 March. They were mated on 28 April, the altitude chamber tests were completed on 16 June, and the landing gear was installed on 22 June. The adapter was a 28-foot-long truncated cone which tapered from the diameter of the S-IVB/IU to that of the service module, and housed the lunar module. It arrived at the Kennedy Space Center on 6 May, and was mated with the LM on 23 June.

The complete spacecraft was integrated on 27 June, and mated with the launch vehicle on 1 July. The overall test of the space vehicle was completed on 17 July. At this point Apollo 12 was still officially expected to fly in September, but by the end of the month it had been rescheduled to November.

The 6-million-pound transporter for the mobile launch system was 131 feet long and 114 feet wide, and travelled on four independent double-tracked crawlers, each 'shoe' of which weighed about 1 ton. The access road was comparable in width to an 8-lane highway. It comprised three layers, averaging a total depth of 7 feet. The base was a 2-foot-6-inch-thick layer of hydraulic fill. Next was a 3-foot-thick layer of crushed rock. This was sealed by asphalt. On top was an 8-inch layer of river rock to reduce friction whilst steering. The crawler was operated jointly by drivers in cabs on opposite diagonals, linked by intercom. It had a turning radius of 500 feet. When loaded, it had a maximum speed of 1 mile per hour.

On 8 September the crawler drove the Apollo 12 space vehicle out to Pad A, the southernmost of the two facilities of Launch Complex 39. Because the concrete pad was built above ground level in order to accommodate a 43-foot-tall deflector in the flame trench, the crawler had to climb a 5 per cent gradient, tilting the platform such that the tip of the launch escape system tower did not diverge more than 1 foot from the vertical alignment. Once in position, hydraulic jacks lowered the platform onto six 22-foot-high steel pedestals on the pad with an accuracy of 2 inches. In all, the 'roll out' lasted 6 hours. In its final orientation, the tower was towards the north, with the axis of the trench aligned north and south. After the crawler had withdrawn, the flame deflector was rolled in beneath the hole in the platform. Two days later, the crawler collected the mobile service structure from its nearby parking place and delivered it to the south side of the pad. This 400-foot-tall open steel structure gave access to the portions of the vehicle which were not accessible from the main tower.

After preliminary checks, the space vehicle flight readiness test was completed on 30 September. The 'wet' part of the countdown demonstration test was completed on 28 October and the 'dry' part, with the crew onboard, was conducted the next day with rain and high winds raging at the time of simulated liftoff. As Kurt H. Debus, Director of the Kennedy Space Center, once said only half in jest, "When the weight of the paperwork equals the weight of the stack, it is time to launch!"

CREW TRAINING

In backing up the D-mission, Pete Conrad, Dick Gordon and Al Bean received in excess of 1,500 hours of training, so when assigned to Apollo 12 they were already familiar with the vehicles and flight operations. Underwater training at the Manned Spacecraft Center used mockups to rehearse all aspects of the intravehicular transfer through the docking tunnel, as well as an external transfer wearing pressurised suits. Donning and doffing of the pressure suits was rehearsed in a KC-135 aircraft which provided short periods of weightlessness. Conrad and Bean received more mission-specific training than had Neil Armstrong and Buzz Aldrin. Particularly, they were the first to receive a revised geology training program that stressed site exploration. The US Geological Survey trained them in field sampling procedures and provided rehearsals on an artificial crater field laid out near Flagstaff, Arizona. Their specific lunar surface training included one-gravity suited walk-throughs of the deployment of the S-band antenna and the ALSEP. Their first briefing on the Surveyor lander was a visit to JPL in mid-August, where there was an engineering model available. This was flown to the Cape to assist in training and installed on a slope to match the interior wall of the crater in which Surveyor 3 stood. In September they spent a day in the Lunar Receiving Laboratory with some of the Apollo 11 samples. The training plan was completed on 1 November, as scheduled. After that, the crew were able to relax, training only as required for refresher purposes.

2

Moonbound

COUNTDOWN

The countdown for Apollo 12 started on 8 November, with a view to lifting off at the opening of the launch window on 14 November. It began with the clock at T-98 hours, and a 'pre-count' in which the launch vehicle and spacecraft activities were undertaken independently. The 'terminal count', with coordinated activities, began at 9:00 p.m. EST on 12 November with the clock at T-28 hours. Two holds were planned, the first at T-9 hours for 9 hours 22 minutes and the second at T-3 hours 30 minutes for 1 hour. The work progressed smoothly until technicians tried to load liquid hydrogen fuel cell reactant into the service module and discovered that tank no. 2 was not chilling down.[1] When the hydrogen flow was halted, the level in the tank declined rapidly. Upon peering in through an inspection panel, a technician observed frost on the exterior of the tank. A flaw in the outer shell had ruined the vacuum insulation. This issue was new to the manufacturer's field team at the Cape. After consulting with the Manned Spacecraft Center, John Williams, the Spacecraft Operations Director, decided to replace the faulty tank with one from CSM-109, which was on-hand at the Cape for Apollo 13. An unscheduled hold was initiated at T-17 hours.[2] After an access panel had been removed, the tank was cryogenically and electrically isolated from the hydrogen subsystem shelf, and the new tank fitted. Once this was done, the cryogenics were successfully loaded. To preclude delaying the launch, the scheduled hold at T-9 hours was reduced by 6 hours to compensate. Meanwhile, at T-15 hours the cask of plutonium which was to power the scientific station that the astronauts were to deploy on the lunar surface was affixed to the descent stage of the lunar module, and at T-10 hours the crawler began to withdraw the mobile service structure from the pad.

The loading of kerosene into the fuel tank of the S-IC began at T-12 hours. This was

[1] This problem was actually discovered late in the pre-count and work to resolve it continued into the terminal count.

[2] It was 8:00 a.m. EST on 13 November.

D.M. Harland, *Apollo 12 – On the Ocean of Storms*, Springer Praxis Books,

DOI 10.1007/978-1-4419-7607-9_2,

done through a 6-inch duct in the bottom domical cap, with access via the thrust structure's umbilical. The fill rate was 200 gallons per minute until the 10 per cent level, then 2,000 gallons per minute until the 98 per cent level, at which point it resumed the slower rate until the tank was at 102 per cent of the flight requirement. In the final hour of the countdown the tank would be drained to the required level and then pressurised.

At 1:22 a.m. on 14 November the clock resumed after the abbreviated hold at T-9 hours, and the pad was cleared to perform the final propellant loading of the launch vehicle. First liquid oxygen was loaded into all three stages. In the case of the S-IC, this was by a pair of 6-inch ducts in the bottom domical cap of the tank, with access via the interstage umbilical. The fill rate was 1,500 gallons per minute until the 6.5 per cent level, then 10,000 gallons per minute until 95 per cent, at which time it reverted to the slower rate until a level sensor in the upper dome indicated it was at the desired level. The liquid boiled continuously at –183 °C, and the gaseous oxygen was vented via the forward skirt. Because boil-off was the process of evaporation, it served to cool the remaining liquid. The venting was compensated by topping off. The cryogenic loading of the upper stages had to take care not to stress the common bulkheads, which had to cope with a differential of 70°C. In each stage, it began by feeding liquid hydrogen at a slow rate in order that when the cryogenic fluid boiled upon coming into contact with the warm structure this would chill the metal. In the case of the S-II, this chill-down at 1,000 gallons per minute continued to the 5 per cent level. Next, the liquid oxygen tank was loaded. It began at a rate of 500 gallons per minute to the 5 per cent level, 5,000 gallons per minute to the 90 per cent level, and 1,000 gallons per minute to 100 per cent. The liquid hydrogen loading resumed at 10,000 gallons per minute until the 90 per cent level, and then the slower rate was resumed. In the case of the S-IVB, with its smaller tanks, the liquid hydrogen chill-down rate was 500 gallons per minute to the 5 per cent level. Next the liquid oxygen tank was loaded at this same rate until it attained the equivalent level. It continued at 1,000 gallons per minute to the 98 per cent level, and concluded at 300 gallons per minute. The liquid hydrogen loading resumed at 3,000 gallons per minute to the 90 per cent level, and concluded at 500 gallons per minute. In both of these stages the liquid hydrogen was in the upper tank because, whilst it was voluminous, it was less dense than liquid oxygen. As a result, the S-IC was loaded with 1,359,000 pounds of kerosene and 3,133,000 pounds of liquid oxygen (a total of 4,492,000 pounds); the S-II had 153,000 pounds of liquid hydrogen and 789,000 pounds of liquid oxygen (a total of 942,000 pounds); and the S-IVB had 37,000 pounds of liquid hydrogen and 191,000 pounds of liquid oxygen (a total of 228,000 pounds). Cryogenic propellant loading was finished by the start of the planned 1-hour hold at T-3 hours 30 minutes, at 6:52 a.m. For the remainder of the count, a very slow rate of replenishment would compensate for the venting of boil-off.

LAUNCH DAY

On the eve of launch, NASA Administrator Thomas O. Paine joined Pete Conrad, Dick Gordon and Al Bean for dinner in the Crew Quarters of the Manned

Rolling away the mobile service structure in preparation to launch Apollo 12.

Spacecraft Operations Building. As he had with Neil Armstrong prior to Apollo 11, Paine told Conrad that if the mission had to be aborted, his crew would be assigned another mission so that they could try again, which was Paine's way of urging Conrad not to take any undue risks in attempting to achieve a lunar landing.

After flying Apollo 10, Tom Stafford was made head of the Astronaut Office so that Al Shepard could begin training for an Apollo mission. At 6:05 a.m., Stafford awakened Conrad, Gordon and Bean. They had managed a good 8 hours of sleep. A final physical examination by doctors Alan C. Harder and John T. Teegan declared them fit to fly. Stafford provided a weather forecast. The broad band of cloud and precipitation, punctuated by thunderstorms, which had rolled across central Florida the previous day would have ruled out a launch if one had been scheduled. The thunderstorms had ended by nightfall and the sky had cleared. However, a cold front approaching from the northwest contained showers of rain. The prediction for the planned time of liftoff was scattered clouds at 2,500 feet, a broken ceiling at 10,000 feet, winds from the southwest of 15 knots, gust at 25 knots, and a temperature of about 20°C. This was acceptable. The weather in the abort contingency areas was also acceptable. The launch window would open at 11:22 a.m., and last 3 hours 5 minutes. If it were missed, a 2-day recycle would require reassignment to the backup target on the Moon.[3] However, the fact that Apollo 11's success had greatly relaxed the sense of urgency meant that it would be more sensible to wait and try again for the primary target at the next lunation.

When they arrived for breakfast at 6:52 a.m., Conrad was presented with a mascot in the form of a stuffed adult gorilla named 'Irving', sent by one of his friends. It had been dressed in a flight smock and a crash helmet, and was seated on a chair against the wall. They were joined by Stafford, Jim Irwin of the backup crew, Paul Weitz of the support crew, who would be the Capcom in the firing room, Chuck Tringali, leader of the crew support team for this mission, and Jim McDivitt, currently Apollo Spacecraft Manager in Houston. The low-roughage fayre consisted of orange juice, toast, steak and eggs, and coffee.

Hamilton Standard of Windsor Locks, Connecticut, was prime contractor for the Apollo space suit, or pressure garment assembly. Earth's atmosphere has a sea-level pressure of about 15 psi, and a gas mix of approximately 80 per cent nitrogen and 20 per cent oxygen. The International Latex Corporation of Dover, Delaware, made the airtight bladder to hold pure oxygen at a differential pressure of 3.7 psi. Although contoured to the human shape, the extremely flexible material of the bladder would tend to 'balloon' when pressurised. It was therefore restrained by a complex system of bellows, stiff fabric, inflexible tubes and sliding cables which, although they held the shape of the suit, impaired the mobility of the occupant and, in particular, made the waist inflexible. The bladder incorporated a network of ventilation tubes to cool

[3] For a launch on 14 November, the elevation of the Sun above the local horizon at the time of the scheduled landing at ALS-7 would be 5.1 degrees. A slip to the launch window that would open at 14:09 on 16 November and last 3 hours and 18 minutes would oblige landing at the more westerly ALS-5 with the elevation of the Sun at 10.7 degrees.

Breakfast: Tom Stafford (facing, left), Pete Conrad, Dick Gordon and Chuck Tringali; Jim McDivitt (back to camera, left), Al Bean and Paul Weitz. Also present was Jim Irwin (out of frame to the right). Seated in the background is 'Irving'.

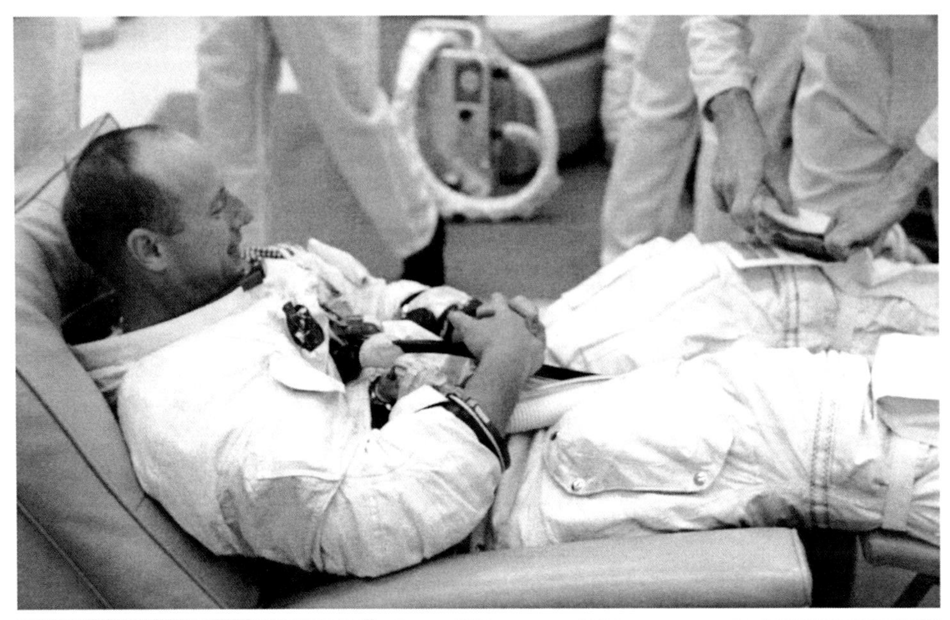

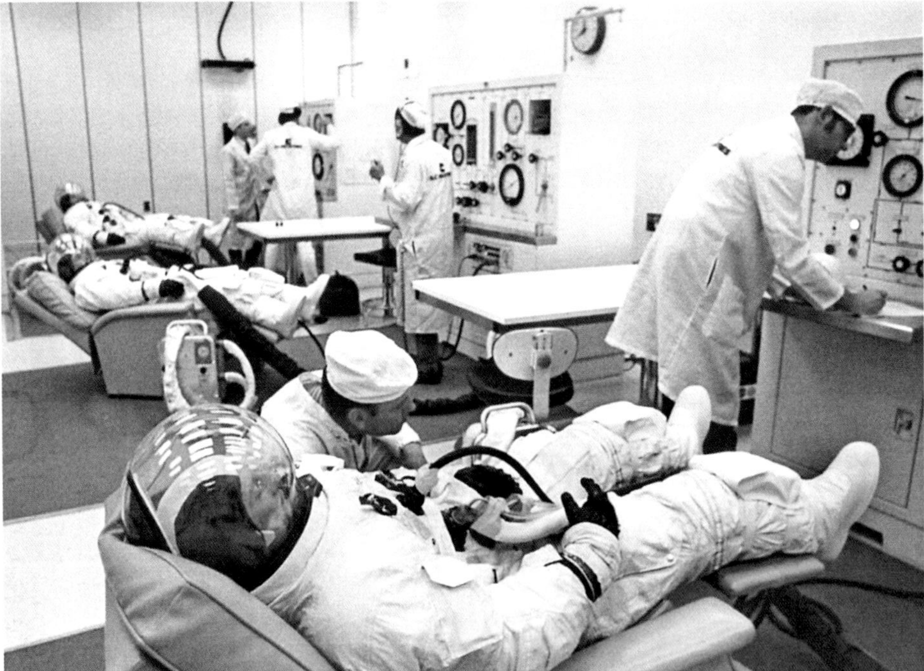

As Pete Conrad suits up in the Manned Spacecraft Operations Building, a technician stuffs a sandwich into his shin pocket. Once suited, the astronauts breathe pure oxygen in order to cleanse the nitrogen from their bloodstream.

the occupant and preclude the build-up of moisture. There were two versions of the suit: one for use inside the spacecraft as protection against loss of cabin pressure, and the other with the thermal and micrometeoroid protection required for operating on the lunar surface.

The space suits varied in certain respects:

- Both suits shared a nomex inner layer, a neoprene-coated nylon pressure bladder, and a nylon restraint layer.
- The outer layers of the intravehicular suit comprised nomex and a double layer of teflon-coated beta cloth.
- The integral thermal and micrometeoroid protection for the extravehicular suit had a double-layer liner of neoprene-coated nylon, a number of layers of beta-kapton laminate and a teflon-coated beta cloth surface.
- The intravehicular suit had one pair of umbilical connectors installed on the chest to circulate oxygen from the cabin system.
- The extravehicular suit had two pairs of such connectors, one pair as on the intravehicular suit, and the other pair for the portable life-support system.
- The extravehicular suit also had a coolant water loop.
- Both suits had a connector for electrical power and communications.

The boots were part of the bladder, but the helmet and gloves used aluminium locking rings to maintain the integrity of the bladder. The helmet was a transparent polycarbonate 'bubble', with adequate air flow to prevent a build-up of carbon dioxide. The gloves were required to support a natural range of bending and rotating motions of the wrist, with a finger-covering material that was sufficiently thin and flexible to allow the manipulation of switches. Each astronaut had three individually tailored suits – a training suit for use in simulations, during which it was likely to suffer wear and tear; and two flight suits (one prime, the other backup) which, after integrity tests, were reserved for countdown demonstrations and the actual mission. Each suit had a US flag on the left shoulder, a NASA 'meatball' on the right breast and the mission patch on the left breast.[4]

Joseph W. Schmitt led a four-man team. He had supervised the suiting-up of each American astronaut since Al Shepard in 1961. This laborious process started with the astronaut rubbing his posterior with salve prior to donning a diaper that would contain both fecal matter and associated odours. This was a precaution against a loss of pressure in the cabin when retrieving the lunar module from the final stage of the launch vehicle after translunar injection, in which event the crew might require to spend several days in their suits. Next was a prophylactic-style urine collector, with a collection bag worn around the waist. A connector on the thigh of the suit enabled the bag to be emptied while the astronaut was suited. Biosensors were attached to the

[4] The Stars and Stripes shoulder patch was introduced by Jim McDivitt and Ed White after being prohibited from naming their Gemini 4 spacecraft 'American Eagle'. In addition to retaining the flag, for their Gemini 5 flight Gordon Cooper and Pete Conrad introduced a mission patch. Both became standard adornments.

chest, and linked to a signal-conditioning package which would supply telemetry via the electrical umbilical. After donning cotton long-johns, which NASA referred to as a constant-wear garment, each man was assisted into his one-piece pressure garment assembly. Conrad and Bean were to wear the 55-pound extravehicular suit and Gordon the 35-pound version. In the suiting-up procedure, the astronaut sat on a reclining couch, inserted his legs into the suit's open rear, inserted his arms, bent forward to ease his head through the rigid metal neck ring, then stood and shuffled until the suit felt comfortable, whereupon a technician would seal the bladder and zipper. The next item was the brown-and-white soft communications carrier, dubbed a 'Snoopy hat', with its integrated earphones and microphones. Once the gloves were fastened to the wrist rings and the helmet was in place, oxygen umbilicals were attached to the sockets on one or other side of the chest and the suit was pumped to above-ambient pressure to test the integrity of the bladder, helmet and gloves. There was a pressure gauge on the right arm of the suit. The Omega watches on the suit arms were set to Houston time, one hour behind the Cape. They would breathe pure oxygen at sea-level pressure to purge nitrogen from their blood stream, and thereby preclude 'the bends' when the pressure was reduced during the ascent to orbit. With the suit sealed, communication was by umbilical intercom.

At 7:52 a.m. the countdown resumed with the clock at T-3 hours 30 minutes. At 8:10 a.m., the suited astronauts left the Manned Spacecraft Operations Building in the transfer van for the 5-mile drive to the pad. After riding the elevator to the 320-foot level of the launch umbilical tower they walked across the swing arm to the White Room. Guenter F. Wendt's job title was Pad Leader, but John Glenn had dubbed him *der pad fuehrer* on account of his Teutonic accent being as thick as the lenses of his spectacles. Although from Germany, he was not one of Wernher von Braun's rocket team; he had flown night-fighters for the Luftwaffe as an engineer. After the war he emigrated to the United States, was granted citizenship, and joined the McDonnell Aircraft Company. When McDonnell won the contract to build the Mercury spacecraft, Wendt was given the task of ensuring that the craft was ready for launch – supervising it from the moment that it arrived at the Cape, to the sealing of its hatch. When the company produced the Gemini spacecraft, he remained at the Cape. When the contract for the Apollo spacecraft was awarded to North American, they formed their own pad team. However, after the loss of the Apollo 1 crew in a fire on Pad 34, Wally Schirra insisted that Wendt be rehired. While the Apollo 12 flight crew had been suiting up, backup command module pilot Alfred Worden, an Air Force pilot selected as an astronaut in April 1966, had been in the spacecraft configuring its switches, and was waiting in the lower equipment bay to assist the men into their couches. Conrad entered first, taking the left couch at about 8:30 a.m. Six minutes later, Bean took the right couch. Seven minutes after that Gordon joined them. After Worden scrambled out, the hatch was closed and checked for leaks, and then the cabin purge was initiated to replace the sea-level atmosphere with a 60/40 mixture of oxygen/nitrogen. The astronauts, of course, were on 100 per cent oxygen, as they had been since suiting up. At T-1 hour 28 minutes, Wendt was preparing the White Room for the retraction of its swing arm.

Walter J. Kapryan was the Launch Director at the Cape. Born in Flint, Michigan,

Pete Conrad leads Dick Gordon and Al Bean from the Manned Spacecraft Operations Building into the transfer van. Chuck Tringali, leader of the flight crew support team, looks on.

At Launch Complex 39 the crew disembark from the transfer van in the rain.

For launch Al Worden (top left) and Guenter Wendt were in the White Room, and Jack King (bottom left) and Walter Kapryan were in the Launch Control Center.

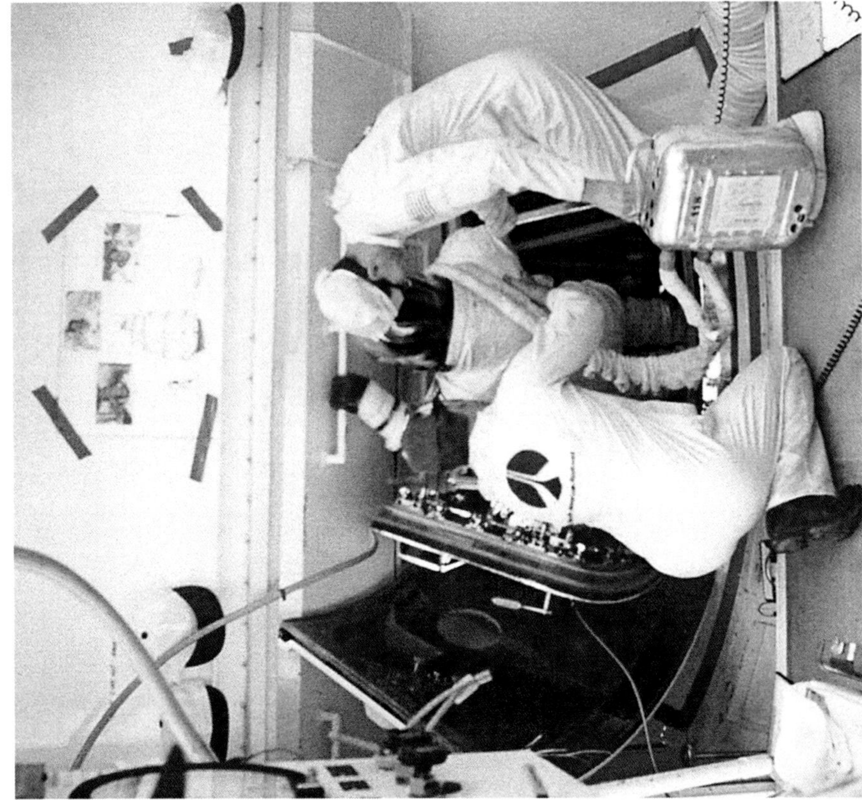

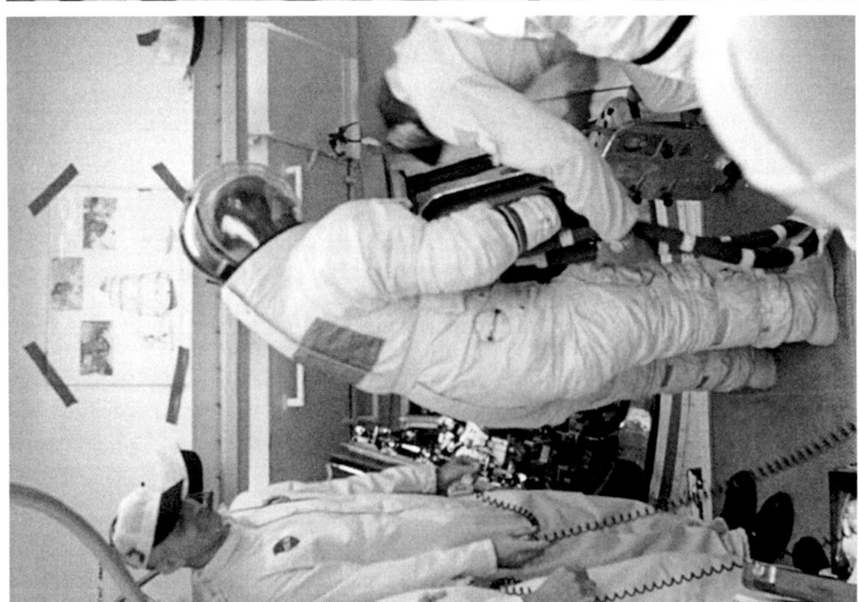

The White Room team assist Pete Conrad into the command module (left). As soon as Al Bean was in, Dick Gordon joined them (right). Note the hatch in the boost protective cover that will close over the command module hatch.

he graduated from Wayne State University, served as a B-29 flight engineer during World War II, joined the Langley Aeronautical Laboratory of the National Advisory Committee for Aeronautics in 1947, and became one of the original members of the Space Task Group of the newly created NASA in 1959. He was assigned to the Cape in 1960 as a project engineer for the Mercury program and remained for Gemini. In 1966 he was made Assistant Apollo Spacecraft Program Manager, and later Deputy Director of Launch Operations. After Apollo 11, his boss was made Apollo Program Director in Washington and Kapryan moved up a notch. Apollo 12 was therefore his first Saturn V launch in this capacity. When a pump in the ground equipment that was topping off liquid oxygen in the launch vehicle failed at T-1 hour 22 minutes, Kapryan decided to proceed on the backup pump. His reasoning was that the space vehicle itself was ready, and if he were to call a 2-day recycle things might not go as well the second time around. His concern was the weather. By T-1 hour, the 'rain line' of the front was 8 miles west of the launch area and the revised forecast was "a very good possibility" of rain at launch time.

Although the weather was deteriorating, the clock was still running. The White Room team was ready to depart the 320-foot level. Vice President Spiro T. Agnew had just arrived in the VIP gallery of the Firing Room of the Launch Control Center. At this same time, Air Force One landed at Patrick Air Force Base, down the coast from the Kennedy Space Center, bearing President Richard Nixon, his wife Patricia and their daughter Tricia, who were to be helicoptered to the VIP area 3 miles from Pad 39 – this distance having been calculated to be as far as an exploding Saturn V could shoot a 100-pound fragment.

By T-49 minutes, Kapryan was receiving weather reports in preparation for taking the Go/No-Go decision. His concern was not the rain, but high winds and lightning. There had been weather balloons at various altitudes above the Cape throughout the morning. An Air Force weather aircraft reported the ceilings to be acceptable, winds to be light at all altitudes, and no evidence of lightning within 20 miles of the pad. Air Force One had flown through the front inbound, without suffering turbulence. But the front was approaching more rapidly than predicted earlier in the morning. As the local Public Affairs Officer, Jack King, put it, conditions were "touch and go". If a hold were to be declared, most probably at T-24 minutes, the crew would remain in the spacecraft to await the outcome, but if it seemed the front contained lightning they would be ordered out.[5] Meanwhile, the

[5] For a hold that became a scrub prior to T-22 minutes, the turnaround procedures would be initiated from the point of hold. A hold called between T-22 minutes (chilling the 'start bottle' of the S-II) and T-16.2 seconds (disconnecting the forward S-IC umbilical) would involve a recycle to T-22 minutes as a preliminary to determining whether to hold or to scrub. Depending on the circumstances, a hold that was initiated between T-16.2 seconds and T-8.9 seconds (ignition sequence start) could result in either a recycle or a scrub. There was a 48-hour turn-around capability from any point prior to T-8.9 seconds that would involve reservicing all the vehicle's cryogenics and resuming at T-9 hours. But a manual or automatic sequencer cutoff after T-8.9 seconds would require a scrub and involve a much longer postponement.

range safety checks of the launch vehicle's command destruct system were completed, and the launch escape system was armed.

With 30 minutes to go, the weather at the Cape was 14 knot winds, light showers, overcast skies at 10,000 feet and broken cloud at 800 feet – which was only twice the height of the derrick on the launch umbilical tower. Nevertheless, the surface winds were within acceptable limits, the clouds were above the 500-foot minimum, and the vehicle design permitted a launch in rain. Kapryan opted to continue to T-10 minutes before making the Go/No-Go decision, although a hold might have to be called prior to that if the conditions predicted for the time of launch fell below minimums. In the meantime, the White Room swing arm was withdrawn 12 degrees to its standby position, 6 feet from the hatch of the spacecraft – in the event of a decision to extract the crew, it could be rapidly swung back into position.

After attending the Navy's electronics and guided missile technical schools, several engineering assignments and a period at the National Hydraulic Laboratory, Paul C. Donnelly joined NASA in 1958 as the Chief Test Conductor at the Cape for manned spacecraft. In 1964 he was promoted to Launch Operations Manager, in which role he was responsible for the checkout of manned space vehicles. By 11:00 a.m. the rain was worsening. At about T-15 minutes, after discussing the situation with Donnelly, Kapryan decided that although conditions were marginal at best, they did not violate Mission Rule 1-404 that a Saturn V must not be launched if its flight path would carry it through a cumulonimbus (thunderstorm) cloud formation.[6] The option to wait until later in the window was constrained by the fact that the space vehicle, which was itself in excellent health, was being replenished by a backup pump in the ground equipment that could turn a prolonged hold into a scrub for the day. As the countdown continued, the spacecraft was transferred to internal power. After a final status check of the lunar module's systems, it was powered down again.

Despite the deteriorating weather, some 3,000 guests were present to observe the launch. NASA Administrator Thomas Paine stood alongside the Nixons, wearing a transparent plastic rain coat and holding an umbrella – this was the first time that a sitting president had come to witness an Apollo launch.[7]

As the astronauts worked through the pre-launch checklist, a hard rain lashed the vehicle and Conrad saw water trickling down across the command module windows, inside the boost protective cover. As he recalled in the post-flight debrief, "I could see water on windows 1 and 2. We experienced varying amounts passing across these

[6] As Kapryan told a post-launch press conference, "We were within our minimums. [...] Number one, we would not launch into a thunder cloud. Number two, we would not launch when we had lightning in the system. There was some concern. We had very unpredictable weather [...]. The weather was deteriorating." There was speculation that President Nixon's presence placed undue pressure on the managers, but Kapryan denied that this was a factor in his decision to proceed with the launch.

[7] Deke Slayton, Jim McDivitt and Rocco Petrone (until recently Director of Launch Operations at the Cape) all played roles in Nixon's visit.

President Richard Nixon arrives at the Kennedy Space Center after a short helicopter ride from Patrick Air Force Base.

Frank Borman (left) and Thomas Paine in the rain-soaked VIP stand of Launch Complex 39.

Skip Chauvin (top) and Paul Donnelly in the Launch Control Center.

Apollo Spacecraft Manager Jim McDivitt in the Launch Control Center.

In the Firing Room of the Launch Control Center at 10:45 a.m. local time on 14 November 1969.

Jim Lovell (left), Bill Anders and Vice President Spiro Agnew watch the launch of Apollo 12 from the Launch Control Center.

windows, depending on how heavily it was raining."[8] In fact, the previous evening he had asked about the effect of the heavy rain of 13 November and been assured that the cover was waterproof! He had also expressed concern about water accumulating in the upward-facing thrusters of the service module, and been told this would not pose a problem. At T-8 minutes Spacecraft Test Conductor Clarence A. 'Skip' Chauvin polled his team and received a 'Go' from each man. After the final space vehicle status checks, the destruct system was armed and the White Room swing arm was fully retracted.

"Have a good trip, Pete," Chauvin called.

"We appreciate everything everyone has done," Conrad replied.

"Hold off the weather for 5 more [minutes], will you," Gordon chipped in.

At T-4 minutes Conrad reported that the lights on the spacecraft's abort panel had illuminated – serving as visual cues to the crew, they were to remain lit until the five F-1 engines had developed sufficient thrust for liftoff, then the lights would go out. At T-3 minutes 6 seconds, the firing command was issued to the electromechanical terminal countdown sequencer in the mobile launch platform. Once initiated, this sequence could be halted only by issuing a cutoff signal – there was no provision for holding the clock. It closed the vents on the propellant tanks and refined their levels. The kerosene tank of the S-IC was drained to the required level and pressurised, and the liquid oxygen tank was replenished and pressurised using helium supplied by ground equipment to provide the ullage required by the suction lines at ignition. Topping off of the cryogenic propellants of the S-II was concluded at T-160 seconds for the liquid oxygen and at T-70 seconds for the liquid hydrogen. In the final phase of the countdown, both tanks were pressurised using helium from ground equipment. By T-90 seconds the two tanks of the S-IVB had been replenished and pressurised using helium from ground equipment. Cryogenic propellants were recirculating through the lines, pumps and valves of the J-2 engines of the S-II and S-IVB to chill these systems – a process that would continue in flight until the prevalves were opened to ignite the engines.

"The launch team wishes you good luck," called Donnelly, in concluding the final crew checks. "May the winds be always behind you."

Conrad replied, "Thank-you very much."

At T-90 seconds, Bean brought the command module's batteries on line. With the clock at T-1 minute, Conrad thrust out his gloved right hand and all three men briefly touched hands as a gesture of solidarity. At T-50 seconds the launch vehicle went on internal power. At T-40 seconds, Conrad's final task was to press the button to align the stabilisation and control system of the spacecraft, and at T-17 seconds the guidance system of the instrument unit of the launch vehicle was 'released' in readiness for use in flight.

[8] In the command module, window no. 1 was a square on the left side (from the crew's perspective), no. 2 was the left forward-facing rendezvous window, no. 3 was a circle on the hatch, no. 4 was the right rendezvous window, and no. 5 was a square on the right side.

Viewed from the top of the tower, the S-IC has lit its five F-1 engines, but the vehicle has yet to be released. Note the droplets of rain on the camera's lens.

The swing arms rotate clear as Apollo 12 initiates its ascent.

Apollo 12 is launched into a dark sky.

AN ELECTRIFYING ASCENT

The ignition sequence started at T-8.9 seconds, with the five F-1 engines firing up in a 1-2-2 sequence at 300-millisecond intervals; first the centre engine of the cluster and then opposed pairs. In each engine the main liquid oxygen valve opened first, to permit oxidiser to surge into the main thrust chamber. Then the gas generator and the turbopump started. A diaphragm which was burst by the initial flow of kerosene enabled a hypergolic fluid to enter the chamber and react with the liquid oxygen to initiate combustion. Finally, the main fuel valve let kerosene enter the chamber at a high rate to maintain combustion. In the spacecraft, Conrad watched the five lights of the abort panel wink out in sequence to indicate that the engines had achieved the desired thrust.

"Commit," announced Jack King. "Liftoff! We have liftoff, 11:22 a.m. Eastern Standard Time."

The four arms which had held the vehicle down as the thrust developed were now released. The acceleration was slowed by drawing tapered pins up through holes, in a 'soft release' that took about half a second.

"Liftoff", announced Conrad onboard.

"The clock's running," Gordon added.

"Liftoff. The clock's running," Conrad reported over the radio.

About 2.5 seconds after being released, the vehicle initiated the yaw manoeuvre to 'side step' away from the launch umbilical tower as a precaution against a gust of wind causing a collision. "I've got a yaw program," Conrad reported.

By T+10 seconds the vehicle had resumed its vertical stance, and 2 seconds later its tail cleared the top of the tower.

The Saturn V had launched with its guidance system aligned to a pad azimuth of 90 degrees east of north, and now it initiated a roll designed to rotate onto the flight azimuth. This was determined by the month, day and time of launch, and optimised the translunar injection parameters for a free-return trajectory. The azimuth range was 72 to 96 degrees, and for liftoff at the start of this particular window the angle was 72 degrees. At the same time, the vehicle initiated a pitch manoeuvre to transition from climbing vertically to travelling along the flight azimuth.

"I've got a pitch and a roll program," Conrad reported. He then ebulliently added, "This baby's really going."

"Man, is it ever," Gordon agreed onboard.

"Roger, Pete," acknowledged Gerry Carr, at Mission Control in Houston.

Gerald Paul Carr was a Marine Corps aviator who was selected as an astronaut in April 1966. As a member of the mission's support crew, he was at a console located below and to the left of the Flight Director, serving as the singular communications link to the spacecraft.[9]

[9] For historical reasons this assignment was known as the Capcom.

Gerry Griffin (left) and Gerry Carr.

"It's a lovely liftoff," Conrad radioed. "It's not bad at all."

"Everything's looking great," Gordon said several seconds later. "Sky's getting lighter."

"Thirty seconds," noted Bean.

"Looks good," said Conrad, monitoring the 8-ball of the Flight Director Attitude Indicator for the termination of the vehicle's slow axial roll. "Roll's complete."

"Roger, Pete," Carr acknowledged.

With Apollo 12 at an altitude of 6,600 feet and climbing through cloud, at T + 36.5 seconds observers on the ground saw lightning strike the pad. Inside the spacecraft, Conrad saw a bright flash. At the same time there was an audible click and a burst of static in his earphones, a number of lights illuminated on the caution and warning panel and the master alarm sounded.

"What the hell was that?" Gordon asked onboard. "I lost a whole bunch of stuff."

The spacecraft's power system provided energy sources and the associated power generation, control, conversion, conditioning and distribution functions. The primary sources of power were the fuel cells in the service module, which reacted cryogenic hydrogen and oxygen to provide electricity and water using the inverse process to electrolysis. The fuel cells were supplemented at periods of peak power demand, and in particular during the launch phase, by up to three rechargeable silver-zinc oxide storage batteries. The voltage transient induced on the battery relay bus by the static discharge of the lightning strike exceeded the current rate-of-change characteristics of the silicon-controlled rectifiers in the fuel cell overload sensors and disconnected the fuel cells from the two main power buses, illuminating warning lights associated with total fuel cell disconnection and also causing the silicon-controlled rectifier in the overload circuit to indicate an AC overload. The disconnection of the fuel cells left the entire electrical load on the two batteries at that time online in reserve. The step-increase in current from approximately 4 amperes to 40 amperes on each of the batteries resulted in the main bus voltages dropping momentarily from 28 volts of direct current to approximately 18 or 19 volts. The voltages recovered within several milliseconds to 23 or 24 volts, but the transient triggered the undervoltage warning lights. Amongst other things, this also caused a lower voltage input to the inverters, which in turn caused a low output voltage that tripped the AC undervoltage sensor and lit the AC bus 1 failure warning light. Even in the worst-case launch simulation, Conrad had never faced such a situation. As he would reflect later, the caution and warning panel was "a sight to behold".

As the crew tried to make sense of the lights, Gerry Carr made the routine call to indicate the next stage in the abort mode sequence, "One Bravo."

"Roger," acknowledged Conrad against a background of heavy static and a slow whistling tone on the downlink. Then he gave the first indication of a problem, "We had a whole bunch of buses drop out."

At T + 52 seconds, with the vehicle climbing through 14,500 feet, it was struck by lightning again, although this time no-one onboard saw the flash. The 8-ball began to tumble, indicating that the platform of the inertial measurement unit had lost its reference. Conrad, his voice showing only mild concern, radioed, "We just lost the platform, gang. I don't know what happened here. We had everything in the world drop out."

"Roger," replied Carr. It was a bland response, but the truth was he had nothing else to say.

Onboard, Gordon advised, "There's nothing I can tell is wrong, Pete." The fuel cells had disconnected, but there was no indication of a short circuit. The platform had tumbled and several gauges were showing anomalous readings, but otherwise everything seemed normal.

Thirty-six seconds after the alarm sounded, Conrad recited the lights to appraise the flight controllers of the situation, "I've got three fuel cell lights, an AC bus light, a fuel cell disconnect, AC bus overload 1 and 2, main bus A and B out."

Bean was mystified by the fuel cell disconnection. As he explained in the post-mission debriefing, "My first thought was that we might have aborted, but I didn't feel any g's, so I didn't think that was what had happened. My second thought was that somehow the electrical connection between the command and service modules had separated, because all three fuel cells had dropped off and everything else had gone. I immediately started working the problem from the low end of the pole."

Having used his meters to examine the various circuits Bean announced onboard, "I got AC."

"We got AC?" Conrad queried.

"Yes," said Bean. Power was present on both AC buses, and both main DC buses read 24 volts. As he reflected afterwards, "Usually when you see an AC overvoltage light, either an inverter goes off or you have one of the AC phases reading zero and you have to take the inverter off. In this case they all looked good, and that was a bit confusing. I switched over and took a look at the main buses. There was power on both, although it was down to about 24 volts, which was a lot lower than normal. I looked at the fuel cells and they weren't putting out a thing. I looked at the battery buses and they were putting out the same 24 volts. They were hooked into the mains and it turned out that they were supplying the load." Also, "One of the rules of space flight is you don't make any switch-a-roos with that electrical system unless you've got a good idea why you're doing it. If you don't have power at all, you might change a couple of switches to see what will happen [but if] you have power and everything is working you don't want to switch too much. I didn't have any idea what had happened. I wasn't aware anything had taken place outside of the spacecraft. I was visualising something [having occurred] in the electrical systems."

The Mission Operations Control Room did not show launch video, the only 'view' that flight controllers had of the space vehicle was the telemetry downlinked to their consoles. As a result, they had no idea that lightning had struck the pad shortly after liftoff.

Gerald D. Griffin was in the Flight Director's chair. After serving as a pilot in the Air Force, he got a degree in aeronautical engineering from Texas A&M University and joined NASA in 1960, where he specialised in guidance and navigation systems. In 1967 he was promoted from the ranks of the flight controllers to Flight Director. This was his first Saturn V launch in that role. Upon hearing that the spacecraft had lost its platform, he called over his intercom loop, "How's it looking EECOM?"

On graduating with an engineering degree from Oklahoma's Southwestern State College in 1964, John W. Aaron had speculatively applied to NASA and promptly been hired to work in Mission Control. A specialist in the systems of the command and service modules of the Apollo spacecraft, he was the Electrical, Environmental and Communications officer on Griffin's team.

Expecting Aaron immediately to recommend an abort, Griffin was surprised when there was no reply. At T + 36 seconds, Aaron had been monitoring the cabin pressure. As a precaution after the pure-oxygen fire which had killed the Apollo 1 crew during what was regarded as a routine test, it had been decided to launch with a mixture of nitrogen and oxygen in the cabin and then reduce the pressure during the ascent to the partial pressure of pure oxygen intended for flight, purging the nitrogen in the process. One effect of the low-voltage transient was to cause the signal conditioning equipment to cease operating. This apparatus ran a repeating cycle of sampling the signals from a variety of sensors, scaling each against a single reference voltage for measurement and real-time transmission as telemetry. Its loss curtailed the flow of telemetry. Aaron had been about to tell Griffin how the purge was progressing when his console display ceased to update, and when it resumed a few seconds later much of the data was nonsense. When Griffin called him, Aaron was conferring with his colleagues in a support room.[10]

"EECOM what do you see?" Griffin persisted when Conrad called down the list of caution and warning lights.

Fortunately, Aaron obtained a crucial insight from the 'nonsense' readings on his console. A year earlier, while monitoring an engineering test being conducted at the Cape, his console had temporarily become incomprehensible. When he later asked what had happened, he was informed that an engineer had inadvertently reduced the voltage of the spacecraft's power supply. After investigating why the telemetry had shown nonsense values rather than simple zeros, Aaron found that the low voltage had 'upset' the signal conditioning equipment. He noted that the unit had two power settings and that whereas in its normal setting it would trip off if the voltage fell, in the auxiliary setting it would attempt to continue at a lower voltage. Given that the spacecraft seemed to be undergoing a power crisis, the logical action was to switch the signal conditioning equipment from normal to auxiliary.

"Flight, EECOM," Aaron replied. "Try S-C-E to Aux."

Griffin had never heard of this switch. "Say again, S-C-E to Off?"

"Aux," Aaron corrected.

"S-C-E to Aux?" Griffin asked, still puzzled.

"Auxiliary, Flight," Aaron clarified.

Griffin had to trust his team. "S-C-E to Aux, Capcom."

At the console across the isle to Aaron's left, Carr turned in his seat and looked up at Griffin, his expression indicating bafflement, but he made the call in the hope that the crew would know what was required: "Apollo 12, Houston. Try S-C-E to Auxiliary." Precisely one minute had elapsed since the onset of the crisis.

"N-C-E to Auxiliary?" replied Conrad having misheard, and then onboard asked, "What the hell is that?"

"S-C-E," Aaron called urgently to Carr. "S-C-E to Auxiliary."

"S-C-E," Carr repeated enunciating each letter distinctly.

Meanwhile, Conrad told Bean, "Try the buses. Get the buses back on the line."

"Everything looks good," Bean assured.

Having heard Carr, Conrad repeated onboard, "S-C-E to Aux." He hoped one of his colleagues knew where the switch was.

Bean recalled there being a switch on his panel labelled SCE. It was rarely used, even in simulations. On previous missions it had been placed at its Normal setting prior to launch by whoever configured the spacecraft and never touched again. He made the change, and several seconds later Aaron's console came to life.

"We got it back, Flight!" Aaron announced. "Looks good." He immediately saw that the fuel cells had disconnected, leaving the spacecraft running on batteries, and the low voltage explained why the SCE in its Normal setting had tripped off, but he

[10] This account of the MOCR draws upon '*Apollo*' by Charles Murray and Catherine Bly Cox, who evidently had access to tapes of the Flight Director's intercom loop; '*Failure Is Not An Option*' by Gene Kranz, who was there; and '*Tracking Apollo to the Moon*' by Hamish Lindsay, who features retrospective interviews.

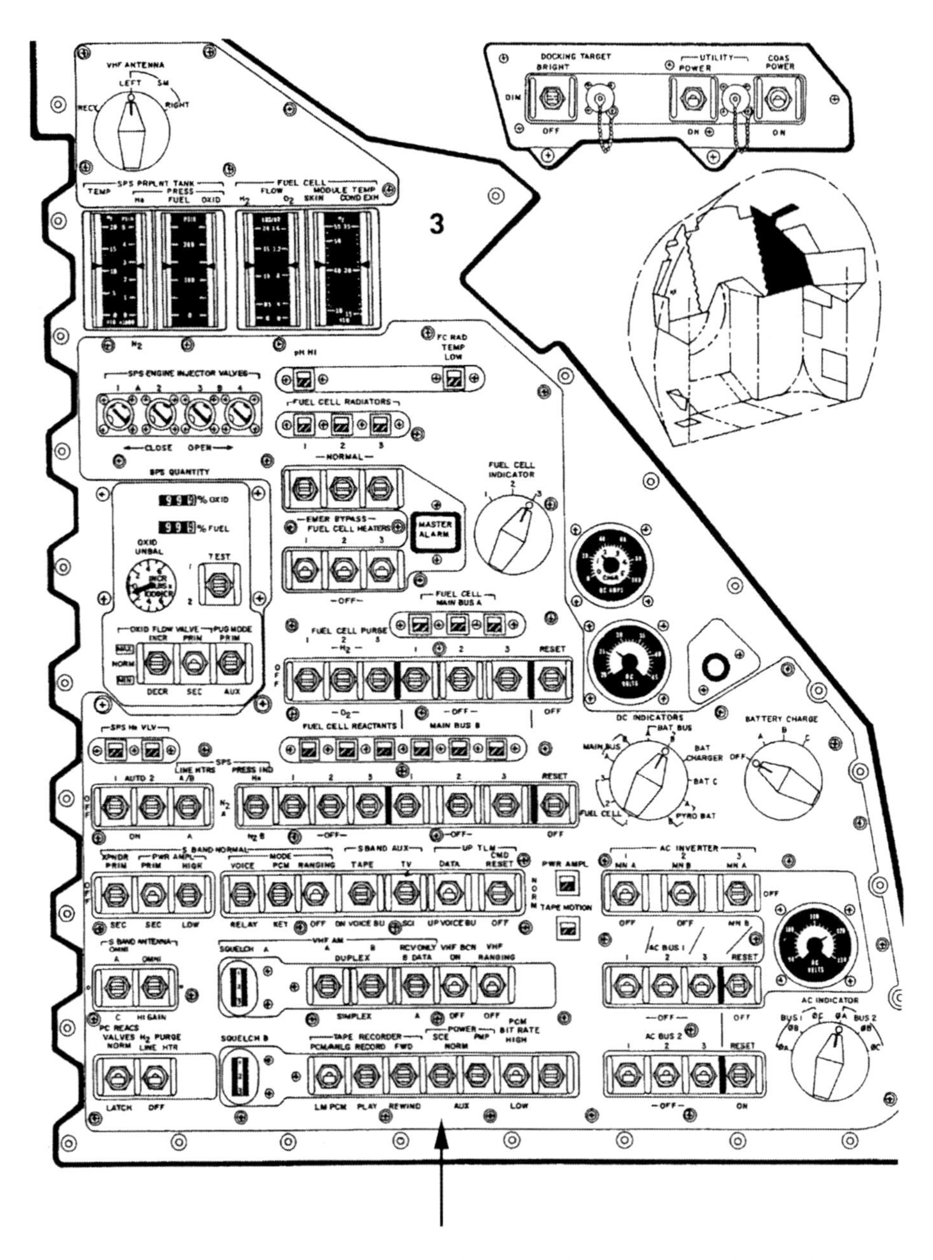

The Signal Conditioning Equipment switch (arrowed at bottom) on Al Bean's portion of the main control panel in the command module.

saw no explanation for *why* the fuel cells had disconnected. His immediate concern, however, was that the spacecraft was running on its batteries, as they would not last long in the power-hungry launch configuration. If the reactant valves to the fuel cells had closed and could not be restored within a few minutes, then the lunar mission would be off. Fortunately, the valves were still open. But if the fuel cells could not be brought back on line soon, the mission would have to abort because the batteries, which could not be recharged without the fuel cells, were the only source of power for re-entry.

Approaching the 2-minute mark, Carr, following procedure, further advanced the abort mode, "One Charlie."

Griffin now polled his flight controllers about S-IC staging. With the Saturn V on course and safely through the region of maximum aerodynamic pressure, there was no imperative to call an abort – so the recommendation was to proceed.

"Apollo 12, Houston. Go for staging," Carr called.

"Roger. Go for staging," Conrad acknowledged. "We had some really big glitch, gang."

With most of the S-IC's propellants consumed and the F-1 engines operating at a fixed thrust, at 2 minutes 15 seconds the instrument unit (as per plan) shut down the centre engine to limit the peak acceleration, as it was an operating requirement that the S-IC not exceed 4.0 g.

"Centre engine," Conrad confirmed.

Aaron, anxious to relieve the drain on the batteries, called Griffin, "Try to put the fuel cells back on the line – and if not [then] tie battery C to main [buses] A and B." Adding the third battery would share the load.

Carr made the call, "Apollo 12, Houston. Try to reset your fuel cells now."

To Aaron's relief, several seconds later two of the fuel cells were brought back on line and the voltages were at their normal levels.

The instrument unit terminated the pitch manoeuvre in readiness for staging. The primary trigger for S-IC cutoff was liquid oxygen depletion, which was indicated by the exposure of sensors near the top of the suction lines of the outboard engines. The backup trigger was fuel depletion, indicated by a sensor on the lower bulkhead of the kerosene tank. Upon receiving a 'dry' signal from any pair of oxygen sensors, the instrument unit would shut down the outboard engines. Some 600 milliseconds later, the instrument unit simultaneously fired the eight retrorockets in the tail fairings and commanded stage separation in order to shed the 300,000-pound dead-weight of the spent first stage. Each solid-rocket was 7 feet 2 inches long, 15 inches in diameter, weighed 504 pounds, and delivered a thrust of 86,600 pounds for 0.6 second – in the process blowing off the tips of the fairings. The system was designed to deliver an impulse sufficient to open a separation of at least 6 feet between the two stages in less than 1 second. While this was occurring, four solid-rockets on the aft interstage of the S-II fired to settle the propellants in their tanks in advance of igniting the five J-2 engines. Each of these rockets was 7 feet 5 inches long, 12.5 inches in diameter, weighed 336 pounds, and delivered a thrust of 22,500 pounds for 4 seconds. During the first phase of the flight, the ongoing boil-off of the cryogenic propellants had maintained the desired pressures in the S-II's tanks. The recirculation that had been chilling the feed lines, pumps and valves of the J-2 engines was terminated when the prevalves opened to let the propellants reach the combustion chambers for ignition. There was a 4.4-second interval between S-IC cutoff and the S-II achieving 90 per cent of its nominal thrust. A valve controlled the ratio of oxidiser to fuel. This was in its null position at stage ignition, providing a 5:1 ratio of oxidiser to fuel by weight, but some 5.5 seconds into the burn it was adjusted to a ratio of 5.5:1 to raise each J-2 engine to its rated thrust of 230,000 pounds and deliver a stage thrust of 1,150,000 pounds. In S-II flight, both liquid propellants were siphoned off on their way to the engines, heated to gas and then fed back into the tanks in order to act as pressurant.

During the S-IC burn, the acceleration had compressed the vehicle longitudinally. At shutdown it had 'snapped back', slamming the crew against their harnesses. The ignition of the S-II slammed them right back into their couches again.

"Got a good S-II, gang," Conrad reported.

"Roger. We copy, Pete," Carr acknowledged. "You're looking good."

Bean brought the third fuel cell back on line.

As they began to accelerate once more, Conrad speculated to Carr, "I don't know what happened; I'm not sure we didn't get hit by lightning."

"Your thrust is looking good, Pete," Carr replied.

"I've a good GDC," Conrad said. While the inertial platform was unavailable, the gyro display coupler was driving the 8-ball using the stabilisation and control system which employed less sophisticated body-mounted gyroscopes. "Al has got the fuel cells back on, and we will soon be working on our AC buses."

"Your fuel cells look good down here," Carr assured.

Some 30 seconds into the S-II burn, the aft interstage was jettisoned to shed its dead weight, including the spent solid-rockets. In addition to serving as a structural component of the integrated launch vehicle, the interstage implemented a dual-plane separation designed to protect the cluster of J-2 engines from damage in the event of a separation system failure which resulted in the S-IC re-establishing contact with the S-II.

"I think we need to do a little more all-weather testing," Conrad radioed, speaking for his all-Navy crew and referring to the fact that naval aviators operating off carriers have to be able to launch and recover in all weathers.

"Amen," replied Carr, prompting howls of laughter onboard.

After several seconds to ensure that discarding the interstage caused no problems, the crew jettisoned the launch escape system from the apex, which took with it the boost protective cover, thereby exposing the rest of the cabin windows. This system incorporated three solid-rockets: an escape motor with a thrust of 147,000 pounds, a pitch control motor with a thrust of 2,400 pounds, and a jettison motor with a thrust of 31,500 pounds. If the escape system had been used in the lower atmosphere, a pair of canard vanes would have deployed to ensure that the command module faced its basal heat shield in the direction of travel when it was released. During an abort at high altitude, the pitch motor would have served this purpose. The system was 33 feet tall and weighed just short of 9,000 pounds. It was the jettison motor that drew the structure clear of the vehicle. If an abort was required during the ascent after this point, the service propulsion system would boost the command and service modules

The Firing Room with the clock showing 2 minutes 54 seconds into the ascent.

clear, then the command module would separate and use its own thrusters to orient itself for re-entry. Unfortunately, a lot of trapped rain ran down the windows as the cover lifted off, caught some particulate efflux from the jettison motor, and froze in place. Although the water sublimed when exposed to sunlight, it left behind a white powdery deposit that impaired the optical quality of the windows for the remainder of the mission.

With the aft interstage and launch escape system jettisoned, the instrument unit initiated its iterative guidance mode. The S-IC had flown a programmed trajectory designed to preclude vehicle motions that might cause the stack to break up while attempting to counter winds, gusts and jet streams. Its job was to climb to an altitude of 200,000 feet, above what was referred to as the 'sensible atmosphere'. The S-II, operating in near-vacuum, was free to manoeuvre, and its task in this guidance mode was to correct any deviations inherited from the first stage, continuing to climb and increasing the horizontal velocity.

"You're in Mode 2," Carr announced, further advancing the abort mode.

"Roger, Mode 2," replied Conrad. "No sweat."

The vehicle was now 122 nautical miles from the launch site, at an altitude of 61 nautical miles and travelling at a velocity of 10,000 feet per second.

After a check of the power system, Conrad reported, "We have all the buses back on the line, and we'll just square up the platform when we get into orbit." Referring to the situations thrown at them in training simulations, he quipped, "Hey, that's one of the better sims, believe me!"

"We've had a couple of cardiac arrests down here, too, Pete," Carr replied.

"There wasn't any time for that up here," retorted Conrad.

At about this time, 4 minutes 35 seconds into the mission, the spent S-IC reached the peak of its ballistic arc at an altitude of 366,000 feet and began to fall back.

As the crew discussed the state of the inertial measurement unit, Carr called with an abort advisory, "If you do a Mode 4, it'll be on the backup."

"Yes, no sweat," Conrad replied. "I've got a good SCS."

The inertial reference provided by the stabilisation and control system would be adequate for an abort.

"Okay. Good show," said Carr.

The thrust of an S-II was normally very smooth. This one, however, developed an intermittent low-frequency (16 hertz) oscillation. "I've got a little vibration of some kind – she's chugging along," Conrad reported. "She's minding her own business though."

At 7 minutes, Carr announced, "You're right smack dab on the trajectory; your IU is doing a beautiful job."

In a planned action, the instrument unit shut down the centre engine to preclude the longitudinal oscillations (pogo) that prior experience had shown could occur late in this burn. This marked the maximum inertial acceleration of 1.83 g on the S-II. Soon thereafter, having monitored the propellant consumption rates, the instrument unit adjusted the mixture ratio to 4.5:1 in order to ensure simultaneous and precise depletion of oxidiser and fuel. With each of the four outboard engines now running at 175,000 pounds, the stage delivered a thrust of only 700,000 pounds.

Just after the vehicle passed over the tracking station on the island of Bermuda, Carr was back, "Apollo 12, Houston. Go for staging."

To preclude the separation system triggering a premature operation, it was not activated until 10 seconds prior to the predicted propellant depletion. The engine cutoff system monitored five sensors in each propellant tank, and when any two in the same tank indicated 'dry' it sent a signal to the instrument unit. Having used its 942,000 pounds of propellants, the 95,000-pound dead-weight of the S-II was shed. The stage did not have any retrorockets of its own, but the S-IVB would leave its aft interstage behind with four retrorockets. Each solid-rocket was 8 feet 8 inches long, 9 inches in diameter, weighed 380 pounds, and delivered a thrust of 35,000 pounds for 1.5 seconds in order to retard the S-II and allow the S-IVB to draw its single J-2 out of the conical interstage. Meanwhile, the S-IVB fired two solid-rockets to settle its propellants in their tanks in advance of igniting the main engine. These rockets were mounted on brackets on the aft skirt. Each provided 3,390 pounds of thrust for 4 seconds and was then jettisoned together with its bracket to shed the dead-weight.

Although the S-II shut down at an altitude of 600,000 feet, which was essentially that desired for the parking orbit, its velocity of 22,890 feet per second was short of that for orbit. The S-IVB was to achieve orbital insertion. During the S-IC and S-II phases, the pressure in the liquid oxygen tank of the S-IVB had been maintained by helium drawn from eight bottles within the liquid hydrogen tank, and in the liquid hydrogen tank the ongoing boil-off was sufficient to maintain the desired pressure. The recirculation that had been chilling the feed lines, pumps and valves of the J-2 engine was halted when the prevalves opened to let the propellants reach the combustion chamber for ignition. There was an interval of 6.5 seconds between S-II cutoff and the S-IVB achieving 90 per cent of its nominal thrust. During the initial burn of the S-IVB, the helium was regulated to sustain the desired liquid oxygen pressure. Some of the liquid hydrogen reaching the engine was siphoned off, heated to gas and fed back into the tank in order to act as pressurant. In powered flight, the S-IVB was steered in pitch and yaw by gimballing the J-2 and in roll by the liquid motors of the auxiliary propulsion system pods that were mounted 180 degrees apart on the aft skirt. The propellant utilisation valve was set to its null position with a ratio of liquid oxygen to hydrogen by weight of 5:1 for ignition, and remained at this setting, delivering a sustained thrust of about 206,000 pounds.

"Got a good S-IVB," Conrad announced. "Nice smooth staging."

"Roger, Pete. Your thrust looks good," Carr replied.

Meanwhile, the spent S-IC splashed into the Atlantic at 30°N, 74°W.

"Mode 4," called Carr, advancing the abort mode to indicate that Apollo 12 could reach orbit using the service propulsion system.

But all was going well, and at 11 minutes 34 seconds the instrument unit, sensing it had achieved the desired altitude and velocity, shut down the S-IVB's engine. The maximum inertial acceleration on this stage of 0.69 g occurred at this time. After allowing 10 seconds for the thrust to tail off, the insertion velocity was measured. The total space-fixed velocity of 25,565.9 feet per second had achieved an orbit of 97.8 × 100.1 nautical miles inclined at 32.5 degrees to the equator and with a period of 88.2 minutes. Insertion occurred 1,430 nautical miles downrange, with a velocity

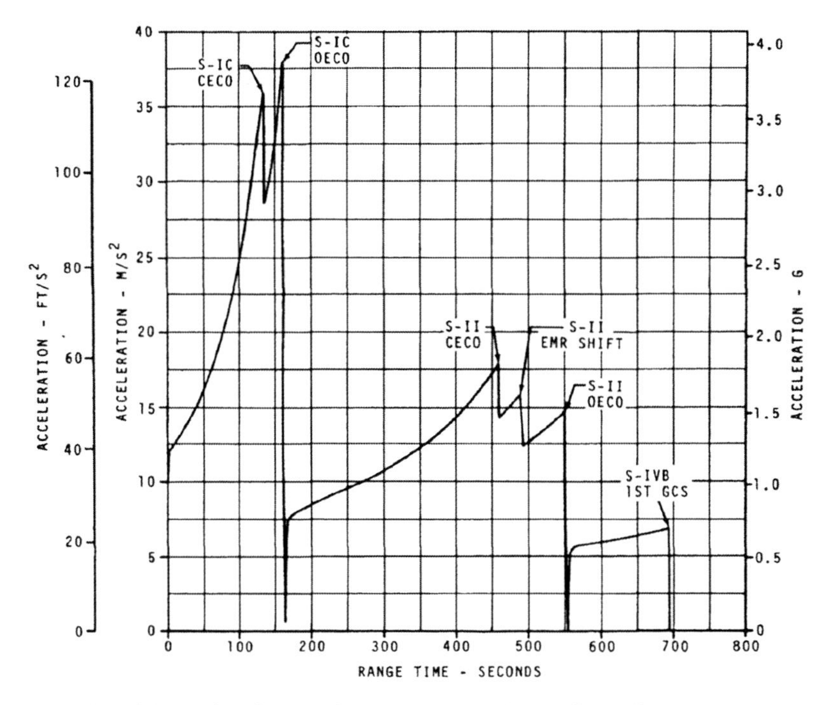

The total inertial acceleration during the ascent phase of Apollo 12. It shows how shutting down the centre engine of the S-IC maintains the load within the operating limit. Also shown are the propellant utilisation shift during the final phase of the S-II burn and achievement of parking orbit by the S-IVB and the spacecraft.

shortfall of 1.7 feet per second, a flight path angle 0.014 degree less than nominal, a perigee 2.2 nautical miles less than nominal and an apogee 0.1 nautical mile higher; all of which were trivial deviations from the desired parking orbit. The 70-pound-thrust aft-pointing motors of the auxiliary propulsion system fired for 90 seconds in order to maintain the propellants settled in their tanks during the tail-off, after which this state was to be maintained by continuous propulsive venting of the excess boil-off from the hydrogen tank through a pair of aft-pointing valves on the forward skirt which were opened at this time. The liquid oxygen pressure was allowed to decay by venting the boil-off in a non-propulsive manner. The auxiliary propulsion system would provide 3-axis control of the vehicle during the orbital coast, maintaining its long axis parallel to the horizon with the spacecraft facing the direction of travel.

"Your S-IVB is looking good," Carr called. "You are configured for orbit."

"Roger, Houston," acknowledged Conrad matter-of-factly.

Although this had been the most serious launch anomaly of any manned mission so far, the VIPs in the gallery at the rear of the Mission Operations Control Room were largely unaware of the drama, simply being delighted that Apollo 12 launched on time. The simulation supervisors were generally ruthless in devising scenarios to test the skills of the astronauts and flight controllers, but the launch rules prohibited

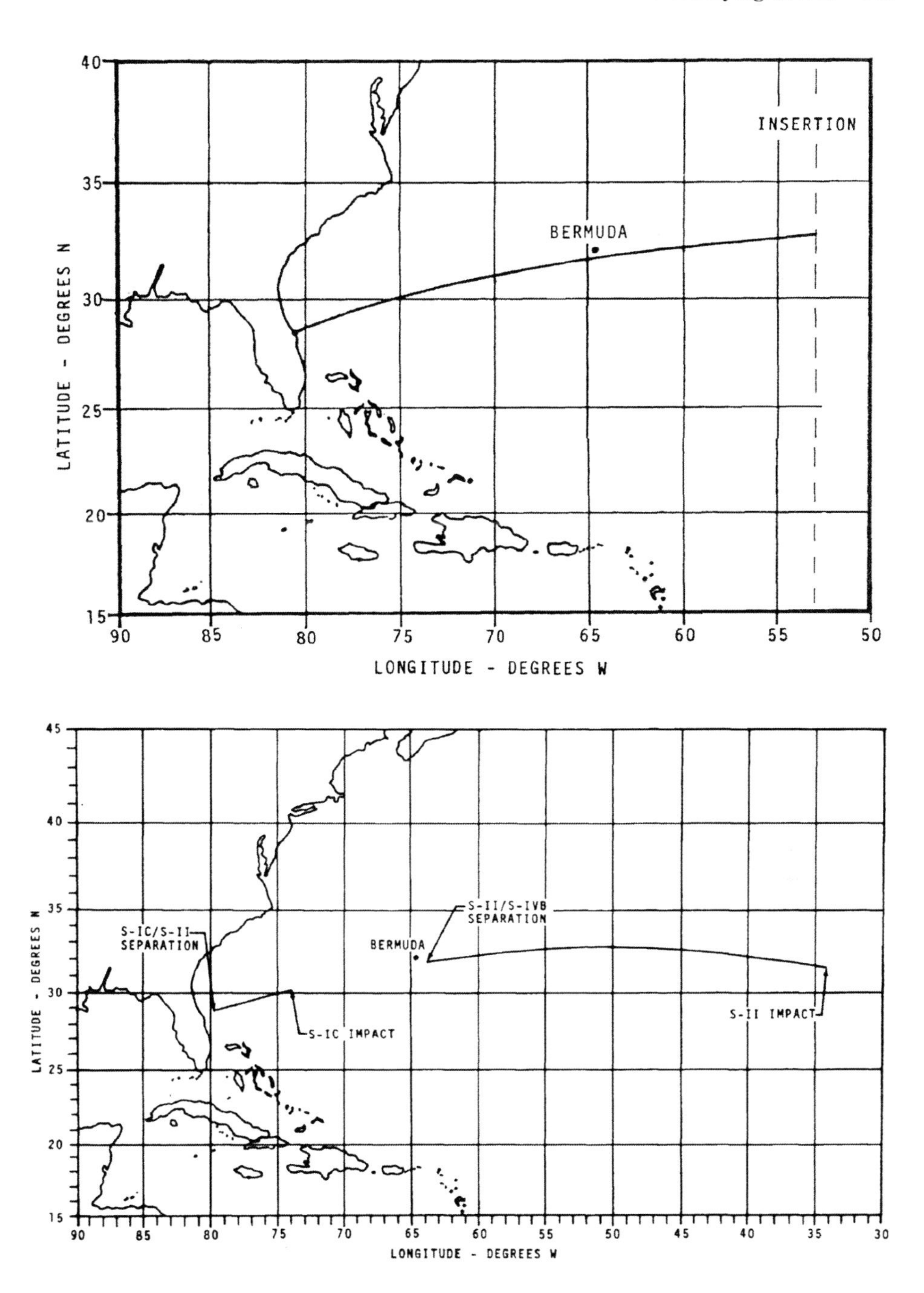

The track across the North Atlantic of Apollo 12's ascent from the Kennedy Space Center to insertion into parking orbit (top). The ballistic flights of the discarded S-IC and S-II stages and the points at which they splashed into the North Atlantic (bottom).

launching into a lightning storm and so no-one had considered what would happen if the spacecraft were to be struck by lightning.

In fact, it was a while before it was confirmed that the space vehicle had been hit by lightning. Journalists at the post-launch press briefing enquired about reports of lightning striking the pad. Tom Stafford, who was in the Firing Room for the launch, said this was mere "speculation". Launch Director Walter Kapryan was assertive, "I think we're pretty certain that it was *not* lightning. If the vehicle had been struck by lightning then the damage would have been quite severe rather than a momentary [telemetry] drop out." However, Jim McDivitt, who was also in the Firing Room for launch but not attending the press conference, had seen lightning. An investigation of automatic cameras in the vicinity of the pad eventually established that two discharges occurred at T + 36.5 seconds, each lasting about 50 milliseconds and separated by about 60 milliseconds. The first grounded 1,500 feet from the launch umbilical tower and showed pronounced downward branching. The second, which was partly obscured by steam rising from the pad, grounded just 100 feet from the tower. An analysis of the downlinked telemetry confirmed that the vehicle had indeed been hit. The incident was discussed by meteorologists at the American Geophysical Union in December. Because there were no indications of thunder or lightning at the Cape for 6 hours prior to or following launch, it was evident that the vehicle had caused the lightning. It had entered an electrified cloud and distorted the field sufficiently for a breakdown to occur. The 363-foot-long metal vehicle was trailing a 1,600-foot-long plume of ionised exhaust gas that had provided an excellent conductor for a field that was otherwise too weak to discharge to the ground. In effect, Apollo 12 had become was the world's largest lightning rod! The discharge at T + 52 seconds was from cloud to cloud.

Fortunately, the Saturn V was not adversely affected by the lightning strikes. In particular, because the instrument unit did not use as much solid-state electronics as the spacecraft it was not as susceptible to induced electrical currents. The electrical system of the F-1 engines was also rather simple, and the wiring harnesses were not only under thermal insulation but also protected by the supporting frames. And the stage separation systems were not armed until shortly before they were needed. Of course, the launch escape system was armed, but fortunately this was not triggered. It was the high-technology spacecraft which suffered, but redundancy in its design enable it to survive. As Conrad reflected in the post-mission debriefing, "I never considered any kind of an abort. The only concern that passed my mind was winding up in orbit with a dead spacecraft. As far as I could see, as long as Al said he had power on the buses and the communication was good, we'd press on." Even when on its stabilisation and control system, the spacecraft was capable of backing up the launch vehicle's guidance in powered flight. On a nominal mission, steering doesn't begin until after S-IC staging. In the event of the inertial platform of the instrument unit failing during the S-IC phase, the crew could switch guidance to the command module computer and complete the programmed trajectory. In the iterative guidance mode, the spacecraft's computer could steer the vehicle by providing inputs to the instrument unit in response to the commander's hand-controller. If an early abort had been called, it would have required the Range Safety Officer at the Cape to issue

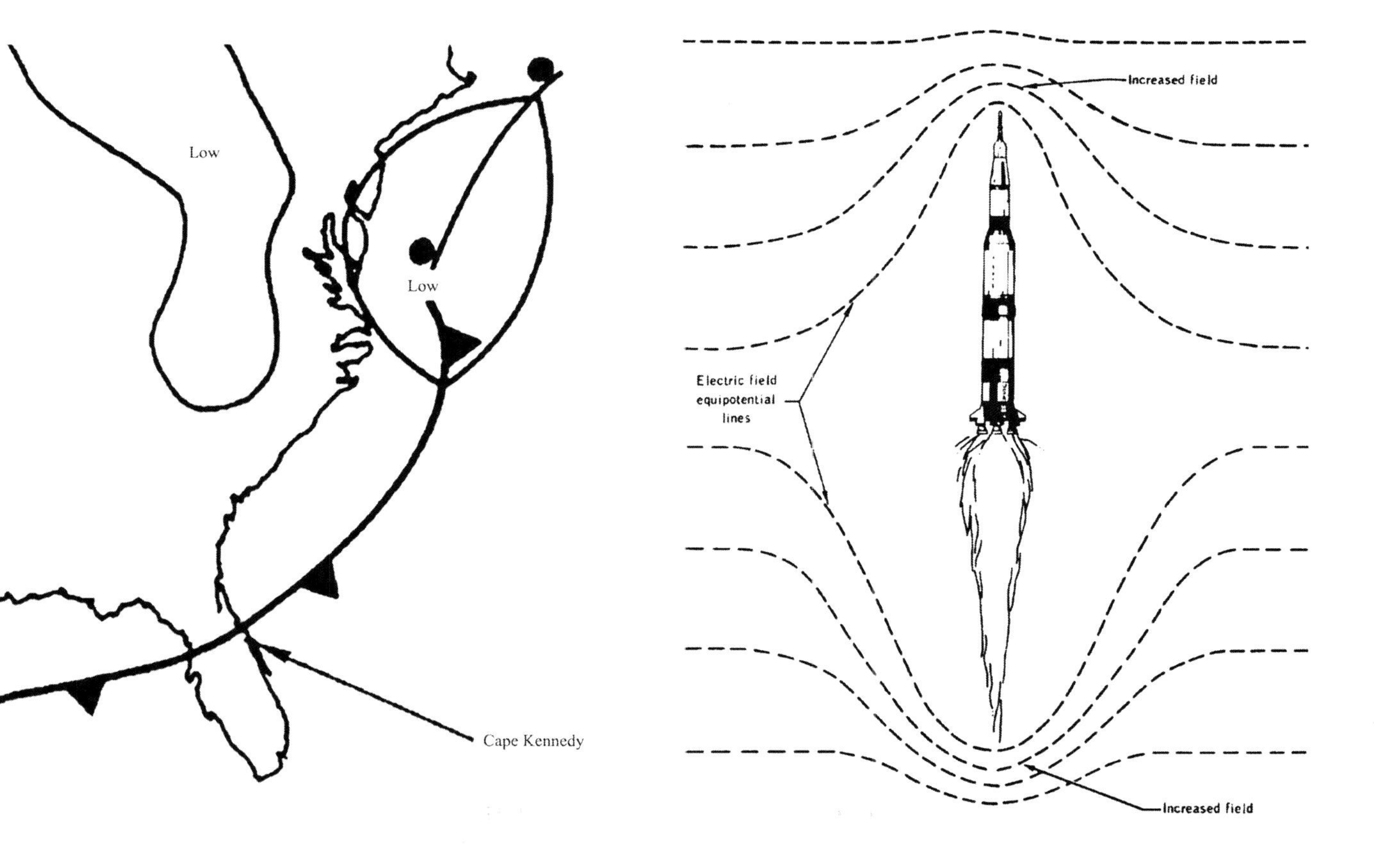

The weather front that threatened the launch (left), and the manner in which the space vehicle and its exhaust plume intensified the electric fields of the clouds through which it ascended sufficiently to cause lightning discharges.

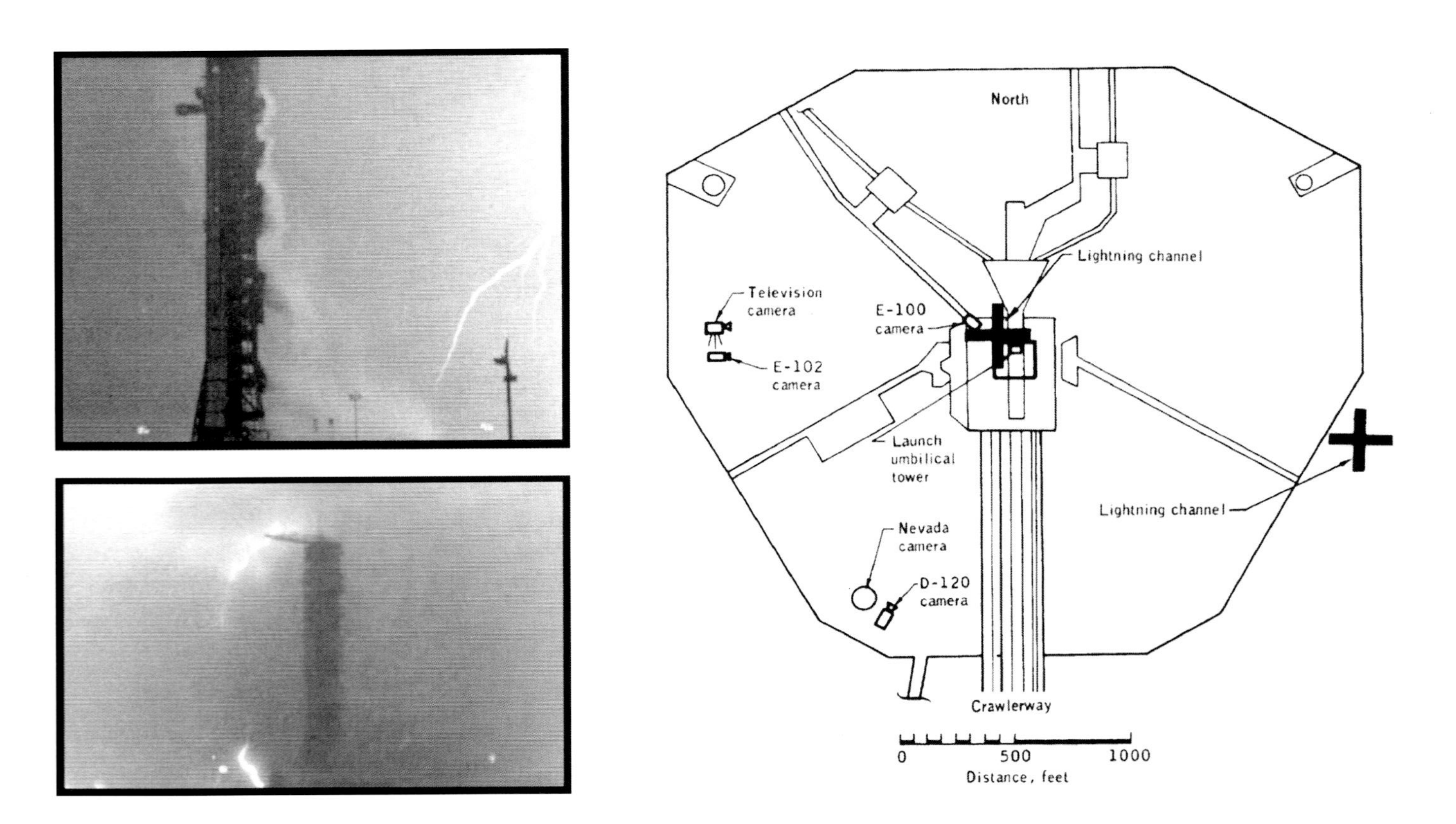

Two lightning bolts striking the Pad 39A area at T + 36.5 seconds, one grounding 100 feet from the launch umbilical tower and the other 1,500 feet away, near the perimeter.

the radio command to destroy the ailing vehicle as soon as the launch escape system was confirmed to have drawn the command module clear.

PARKING ORBIT

The Manned Space Flight Network (MSFN) was operated by the Goddard Space Flight Center at Greenbelt in Maryland. It linked Mission Control in Houston to the thirteen ground stations that were equipped with 30-foot and 85-foot antennas, the instrumentation ship USNS *Vanguard* (at longitude 49 degrees west, due east of the launch site and some 1,000 miles beyond Bermuda to monitor orbital insertion), and four Apollo Range Instrumentation Aircraft (ARIA). In addition, this mission would be supported by two Intelsat geostationary communications satellites. One above the Atlantic was to relay between Goddard, the *Vanguard* and the stations on the Canary Islands, Ascension Island and near Madrid in Spain. Another satellite was above the Pacific to link Goddard to the stations in Hawaii, Carnarvon in western Australia, and Honeysuckle Creek and Canberra in southeastern Australia.

The parking orbit checkout was similar to the previous mission, except for special attention in Mission Control given to determining the effects of the lightning strikes. Onboard, the key task was to regain the inertial reference. The two circuit breakers were withdrawn to remove power from the platform in readiness for a restart. After 3 minutes the breakers were reinserted, ready for realignment by the time-honoured method of taking star sightings. As the vehicle approached the terminator, Gordon unstrapped and weightlessly floated into the lower equipment bay to prepare the optics. The platform of the inertial measurement unit determined the attitude of the spacecraft with respect to the three orthogonal axes of a coordinate system in a given frame of reference. As the spacecraft rotated, motorised gimbals turned the platform in order to maintain its orientation fixed in inertial space. The attitude of the vehicle was measured by 'reference to a stable member matrix' (REFSMMAT), which was a specified vector in inertial space. Different vectors were to be used for the various phases of the mission. For the start of the journey, the REFSMMAT was defined (in part) by the line from the centre of the Earth through the launch site at the time of liftoff and also with respect to the flight azimuth. This frame would be used through to translunar injection.

The digital computer for the Apollo program was a technological marvel. Just a decade earlier, navigating in space would have been a considerably more manual process. Indeed, it could be argued that a lunar landing mission would not have been a practicable venture prior to the 1960s.

Software by the Instrumentation Laboratory of the Massachusetts Institute of Technology used 'programs', 'nouns' and 'verbs'. Verbs were the instructions to do something, while nouns represented the structures that were to be operated upon. For example, if Verb 06 Noun 62 was entered, the verb meant "display in decimal" and the noun indicated the quantity to be displayed (in this case three numbers of the spacecraft's inertial velocity, vertical speed and altitude). Some programs had many verb and noun pairings. The display and keyboard (DSKY, pronounced 'disky') was supplied by the Raytheon Company. It had a power supply, decoder relay matrix, status and caution circuits, a 21-character display unit, and a 16-button key-pad with the digits '0' through '9', 'VERB', 'NOUN', 'CLEAR', 'ENTER', 'PROCEED' and 'KEY RELEASE'. Numerical data took the form of five-digit quantities, scaled to fit and with an implied decimal point. When a key was depressed, the appropriate item illuminated on the display. A verb would be flashed if the computer wished to attract attention when awaiting crew input. There were two units in the command module (one on the main control panel and the other in the lower equipment bay by the navigational instruments) and one in the lunar module.

As a precaution against the REFSMMAT stored in the memory of the computer having been corrupted by the lightning strikes, it was decided to uplink this directly into the computer while the spacecraft was in range of Canary, in readiness for the platform alignment. This involved placing the computer into Program Zero (P00), which was essentially its idle state, and changing a switch on the control panel from Block to Accept to permit the upload. Once the stored data had been verified, the switch was reverted to Block to inhibit further access. This protocol was designed to ensure that Mission Control could not change the computer without the consent of the crew.

"Hey, Houston," Conrad called. "The LMP is no longer a rookie, and you can tell Simsup he gave him a heck of a one to break in on. We're still laughing." This was a reference to the simulation supervisor.

"Roger, Pete. We'll tell him," Carr replied.

With that, they left Canary behind. The next contact would be Carnarvon, after a gap of 30 minutes.

Meanwhile at the Kennedy Space Center, Richard Nixon, his wife and daughter, and NASA administrator Thomas Paine, had moved from the VIP viewing stand to the Firing Room, where Nixon congratulated the launch team. Also, the spent S-II had splashed into the Atlantic some 2,420 nautical miles downrange at 32°N, 34°W.

The attitude of Apollo 12 was such that the optics on the wall of the command module faced away from Earth, at the sky. Once in the Earth's shadow, Gordon took a look through the telescope but could not see anything. He realised he would have to wait for his eyes to adapt to the darkness. In due course, Bean, looking out of his window, informed Gordon that the constellation of Orion should be entering the field of the telescope. To provide a wide field of view, this did not magnify. The viewing axis was south of Orion, but Gordon could see part of it at the upper edge of the field, so he drove the optics in that direction and saw the three stars of Orion's belt. This enabled him to identify first Rigel and then Sirius. He then switched to the boresighted sextant, which had a much narrower field of view and a magnification factor of 28, and adjusted the optical alignment to centre each star in turn. P51 used these optics angles to calculate the vehicle's attitude relative to the REFSMMAT. With its platform reinitialised, the inertial measurement unit could keep track of further spacecraft manoeuvres. As a check, Gordon ran P52, in which the computer aimed the optics at another star, in this case Canopus, and he precisely centred the star so that the computer could refine its calibration.

George Mueller congratulates the Firing Room team after the launch.

Apollo navigation stars

Popular Name	Code (octal)	Official Name
Alpheratz	*01*	*alpha Andromedae*
Diphda	*02*	*beta Ceti*
*Navi**	*03*	*epsilon Cassiopeiae*
Achernar	*04*	*alpha Eridani*
Polaris	*05*	*alpha Ursa Minoris*
Acamar	*06*	*theta Eridani*
Menkar	*07*	*alpha Ceti*
Mirfak	*10*	*alpha Persei*
Aldebaran	*11*	*alpha Tauri*
Rigel	*12*	*beta Orionis*
Capella	*13*	*alpha Aurigae*
Canopus	*14*	*alpha Carinae*
Sirius	*15*	*alpha Canis Majoris*
Procyon	*16*	*alpha Canis Minoris*
*Regor**	*17*	*gamma Velorum*
*Dnoces**	*20*	*iota Ursae Majoris*
Alphard	*21*	*alpha Hydrae*
Regulus	*22*	*alpha Leonis*
Denebola	*23*	*beta Leonis*
Gienah	*24*	*gamma Corvi*
Acrux	*25*	*alpha Crucis*
Spica	*26*	*alpha Virginis*
Alkaid	*27*	*eta Ursae Majoris*
Menkent	*30*	*theta Centauri*
Arcturus	*31*	*alpha Boötis*
Alphecca	*32*	*alpha Coronae Borealis*
Antares	*33*	*alpha Scorpii*
Atria	*34*	*alpha Trianguli Australis*
Rasalhague	*35*	*alpha Ophiuchi*
Vega	*36*	*alpha Lyrae*
Nunki	*37*	*sigma Sagittari*
Altair	*40*	*alpha Aquilae*
Dabih	*41*	*beta Capricornus*
Peacock	*42*	*alpha Pavo*
Deneb	*43*	*alpha Cygni*
Enif	*44*	*epsilon Pegasi*
Fomalhaut	*45*	*alpha Piscis Austrinus*

* Three of these names were coined by Gus Grissom to celebrate his Apollo 1 crew: 'Navi' was his middle name 'Ivan' spelt in reverse; 'Dnoces' was the reverse spelling of 'second', as in Edward H. White II; and 'Regor' was the reverse spelling of 'Roger', as in Roger B. Chaffee. But as far as the International Astronomical Union was concerned, these names had no official standing.

On coming into range of Carnarvon, Conrad reported that they had successfully aligned the platform and were undertaking the checks which had to be done before a decision could be made on whether to attempt translunar injection. As part of this expanded checkout, Carr requested that they run a computer self-check and 'dump' its erasable memory to the ground so that the flight controllers could examine it for any changes arising from the lightning strikes.

One hour into the mission, relaying through Honeysuckle Creek, Mission Control uploaded the state vector into the computer. This was a set of seven numbers which specified where the vehicle was in space at any given moment, and how fast it was travelling. The original state vector been lost when the second lightning strike upset the inertial system. During this initial orbit, various systems were exercised and their telemetry monitored to verify their functionality. Fuel cell purging was tested while over Carnarvon and the extension of the docking probe on the apex of the command module while over Honeysuckle Creek.

Apollo 12 was out of contact for 24 minutes while crossing the Pacific Ocean to the MSFN station at Guaymas in northwestern Mexico. During this time, Mission Control analysed the telemetry available from the launch phase. This revealed that the lightning strike at 36.5 seconds had disabled five thermal transducers, with the most probable cause being an electrical overstress of a diode or resistor. In addition, it disabled four RCS propellant quantity measurements, apparently due to electrical overstress of the semiconductor strain gauges on the pressure-sensing diaphragms. All of the failed sensors were located on the service module just below the edge of the boost protective cover. The skin sensors were not essential for crew safety, and their loss did not pose a problem. Likewise the loss of the RCS propellant sensors was not a show-stopper, as the amounts of propellant remaining would be able to be estimated by the ground using pressure and temperature measurements. It was also decided that while the spacecraft was in continuous communication passing over the continental United States the crew should run through a section of the checklist that would normally be performed in preparing for lunar orbit insertion, to obtain data to enable Mission Control to evaluate the health of the command and service modules. When Guaymas had acquisition of signal, Gerry Carr had been joined at the Capcom console by Neil Armstrong and Buzz Aldrin, in case there should be a need to call upon their experience.

"We've got a few words for you, if you'll stand by for a minute," Carr called by way of a preliminary.

"I imagine you have," Conrad responded. Then he asked if the flight controllers were ready to monitor a hot-fire test of the reaction control system. The RCS of the service module were in four identical 'quads' at 90-degree intervals around its body, each of which contained four bipropellant engines, two monomethyl hydrazine fuel tanks, two nitrogen tetroxide oxidiser tanks, and a helium pressurant sphere. These engines delivered a thrust of 100 pounds. In addition to attitude control, the system could perform minor velocity changes. Conrad's concern was that rain had collected in the upward-facing nozzles and, upon turning to ice in the vacuum of space, might prevent the thrusters from firing properly.

Carr said that they were still checking the memory dump, and would be ready to perform the test soon. But first he called for another computer self-test.

"I don't think they're firing, Houston," Conrad mused when the hot-fire test was performed.

"We saw all the events on telemetry," Carr assured.

"We didn't hear a thing," Conrad pointed out.

They performed the test again.

"We saw your manifold pressures change a little bit, so it looks like they fired," Carr confirmed. "Also, Neil's here, and he says he didn't hear his go, operating on Minimum Impulse, either."

After watching out the window during a third test, Conrad came back, "We can see them firing."

The next item on the list of things to verify was the ability of the SPS engine of the service module to gimbal in order to direct its thrust through the vehicle's centre of mass. Meanwhile, Carr reported that the memory dump had checked out.

Just before Apollo 12 lost contact heading across the Atlantic on its second orbit, Carr read up the details for the translunar injection manoeuvre. After he had read the information back to verify that he had written it correctly into the flight plan, Conrad said, "We're waiting for those golden words 'Go for TLI'."

"We'll give them to you at Carnarvon," Carr promised.

A few minutes later, relaying through Canary, Carr advised Conrad, "Your idea that it was probably lightning that did it, looks like the best theory right now."

Conrad offered the observation, "We might have just discharged ourselves."

"That's exactly the theory that people are thinking here," Carr agreed.

When out of communication, Conrad, mulling over what he had taken to calling the "big glitch" during the ascent, mused to his colleagues, "I bet those guys almost fell off their consoles."

"They weren't the only ones," Gordon observed.

After pondering further, Conrad said, "I'll bet my wife, your wife, and Al's wife fainted dead away."

"They don't know enough," Bean replied.

"I'll bet they did when you started calling out about 18 lights!" countered Gordon, prompting yet another round of laughter.

"The best part of all," Conrad added a little later, "was Al Bean's little voice kept saying, 'I have power on my buses'."

As they penetrated the Earth's shadow for the second time, Gordon returned to the optics. It was the judgment of the flight controllers that the platform had been lost due to the low voltage, and the inertial measurement unit was not damaged, but he ran another P52 to verify that the platform was not drifting. With this done, Conrad hit the GDC Align button to transfer this alignment to the stabilisation and control system – as would be done in the run up to every major manoeuvre.

Continuing to work through the checklist, Conrad, thinking about the reporters covering the launch, chuckled again, "That'll give them something to write about tonight."

Still peering at the stars in darkness, Gordon saw light flashes from the auxiliary propulsion system of the S-IVB working to maintain the orbital rate pitch rotation. The parameters of the parking orbit were changing due to the propulsive venting which the stage used to maintain its propellants settled in their tanks, but this was taken into account in calculating the translunar injection burn.

Resuming their reflection on the big glitch, Bean asked, "I wonder why it took the fuel cells?"

"It just disconnected them," Gordon replied. "That's all it did."

"But why nothing else?" Bean persisted.

"I don't understand that either," Gordon admitted. He had been surprised to hear that the erasable memory had not been corrupted.

Bean mused that both main buses must have been "spiked up" simultaneously.

"We must have gotten some horrendous spike," Conrad agreed. "I think we gave off a big static electrical discharge that drained the spacecraft for about 2 seconds' worth of power, you know, and then it just fell off."

"That must have been it," Gordon agreed.

"I imagine they can sort it out on their telemetry," Conrad concluded. In point of fact, there was no telemetry during the time that the signal conditioning equipment was inoperative.

Approaching Carnarvon's acquisition, Conrad said, "Let's hope nothing's wrong with the LM." There was no way to determine whether the lightning had damaged the lunar module – it was launched in a powered down state and was not providing telemetry.

After confirming that Mission Control would not wish further direct access to the computer, Gordon loaded into its erasable memory a program to count down to the start sequence for translunar injection. Although it was passive, the crew had found this useful in training.

It was decision time for Gerry Griffin. The spacecraft had not been designed to survive a lightning strike in flight. The 'unknown' factors primarily concerned the pyrotechnics of the parachute system, which there was no way to check because they did not provide telemetry. If the parachutes would not deploy, the astronauts were as good as dead anyway, so they might as well fly to the Moon! In '*Apollo*' by Charles Murray and Catherine Bly Cox, Griffin recalls Chris Kraft saying, "Don't forget that we don't have to go to the Moon today." As Griffin saw it, "We kept clicking off the checklist, and when we got to the end we all kind of said 'We don't know where all that stray electricity may have run around in the cabin, but everything we can check looks okay. Is there any reason not to go?' We looked at each other and said 'Hell no, let's go!'"

"Apollo 12, Houston," Carr relayed through Carnarvon. "The good word is you are Go for TLI."

"Whoop-ee-doo!" replied an exuberant Conrad. "We're ready! We didn't expect anything else."

"We didn't train for anything else, Pete," Carr observed.

"You better believe it," said Conrad.

Chris Kraft (foreground) confers with Flight Director Gerry Griffin.

TRANSLUNAR INJECTION

Nominally, TLI would take place on revolution 2. The maximum time available in parking orbit was three revolutions. After this, the Earth's axial rotation would not only carry the MSFN stations away from the space-fixed orbital plane, preventing them from monitoring this crucial propulsive manoeuvre, it might also mean that an emergency re-entry would land out of reach of the recovery force. Another issue was that progressive solar heating of the S-IVB would boil off so much cryogenic propellant as to leave the vehicle unable to achieve the delta-V of the window for a translunar trajectory. On revolution 2, the flight controllers discovered an error in the S-IVB's state vector. It would have been possible to correct this at the expense of postponing the burn to the third revolution, but it was decided to proceed since the divergence of the trajectory from the intended free-return would be minor, and because if the mission went to plan the spacecraft was in any case to manoeuvre off the free-return.

On this revolution, the ground track did not permit contact with Honeysuckle Creek after Carnarvon. Leaving Australia, Apollo 12 headed across the Pacific in the direction of Hawaii. An ARIA was a modified Boeing EC-135A aircraft fitted with a 7-foot parabolic antenna in its nose fairing. There were two aircraft on station along the route to relay voice communication with the spacecraft in real-time and to record its telemetry for subsequent replay to Mission Control. The flight controllers would not gain a real-time look at the telemetry until Hawaii gained acquisition, about half way through the burn. Ten minutes before the scheduled time of ignition, the S-IVB began its restart preparations. The instrument unit lit a lamp in the cabin to inform the crew, who would be mere observers unless required to intervene. As the vehicle refined its orientation, the adjustments in yaw were distinctly jerky. It maintained them in a heads-down attitude.

"There's an island down there," said Conrad peering out of the hatch window. "Two of them; three of them. We're probably passing over the Fiji's."

One of the ARIAs made contact.

"That poor guy's out in the middle of nowhere, isn't he?" Conrad mused about the pilot of the aircraft.

After several unsuccessful attempts, Carr managed to establish contact, but the link was poor.

To prepare the J-2 for reignition, the recirculation system chilled the feed lines, pumps and valves of the engine, the liquid oxygen tank was repressurised by helium at ambient temperature from two of the ten bottles carried on the thrust structure, and the hydrogen tank was repressurised with helium from another seven of the bottles. The aft-firing motors of the auxiliary propulsion system were then to fire in order to ensure that the propellants were settled in their tanks.

"Okay, S-IVB ullage starts right now," called out Conrad when the instrument unit lit another cue lamp.

"I don't feel any ullage," Bean pointed out.

"Don't worry about it," Gordon said. The acceleration was very low.

With the ullage complete, the instrument unit issued the reignition command, and a few seconds later the J-2 started up. The flow from the eight chilled helium bottles housed inside the hydrogen tank was regulated to sustain the liquid oxygen pressure, and (as previously) liquid hydrogen was siphoned from the engine feed line, heated, and the resulting gas fed back into the top of the tank as pressurant.

"Pressures look beautiful," noted Gordon, referring to the S-IVB. "Listen to that baby." As it accelerated, the vehicle began to rattle.

As on previous missions, the steering was discernible. It began with a swing to the right on a slow cycle. "Yaw should be zero at 1 minute," Gordon noted.

"Zero yaw," announced Conrad, watching the instruments.

"56 seconds," Gordon pointed out. Good enough.

"There's 1 minute, and it's sliding out to the left, now," Conrad said.

Twenty seconds later, having achieved a given change in velocity and monitored the propellant consumption rates, the propellant utilisation system of the instrument unit adjusted the mixture ratio. The valve started at its fully open setting, for a ratio of liquid oxygen to hydrogen by weight of 4.5:1, yielding a thrust of about 175,000 pounds at ignition. It now reverted to its null setting with a ratio of 5:1, increasing the thrust to about 206,000 pounds for the remainder of the burn.[11] If the translunar injection burn had been delayed to the third orbit, this shift would have been made almost immediately. The longer time spent at the initial setting in an injection burn on the second orbit would consume the hydrogen that was carried to compensate for the boil-off that would occur during the additional time spent in parking orbit in the event of the manoeuvre having to be postponed to the third orbit.

"Al Bean," chuckled Conrad, "you're on you way to the Moon!"

"Yeah," mused Bean. "Y'all can come along if you like."

"All right," agreed Gordon.

The S-IVB continued its slow yaw excursion. After verifying the attitude on the spacecraft's computer, Conrad asked, "All right, what are our shutdown rules?"

"Six seconds, and it's late," Gordon replied.

As the S-IVB burned off its propellants, its rate of acceleration increased, in turn increasing both the inertial load and the rattle.

At 3 minutes, Gordon compared the performance of the engine against figures on a cue card and reported, "We're right on velocity; we're right on H-dot; we're right on altitude."

When Hawaii made contact, this improved the voice link and provided real-time telemetry to Mission Control.

"Apollo 12, Houston through Hawaii," called Carr. "How do you read?"

"Loud and clear," replied Conrad. "It's steaming right down the pike."

"Right on the money," verified Gordon at 3 minutes 30 seconds.

"Your trajectory and the S-IVB both look good," Carr called.

"Everything's sticky-poo," agreed Conrad.

[11] A propellant mixture ratio of 5.5:1 would yield the J-2 engine's rated thrust of 230,000 pounds. Note that whereas the S-II *decreased* its thrust at the propellant utilisation shift, the S-IVB *increased* its thrust.

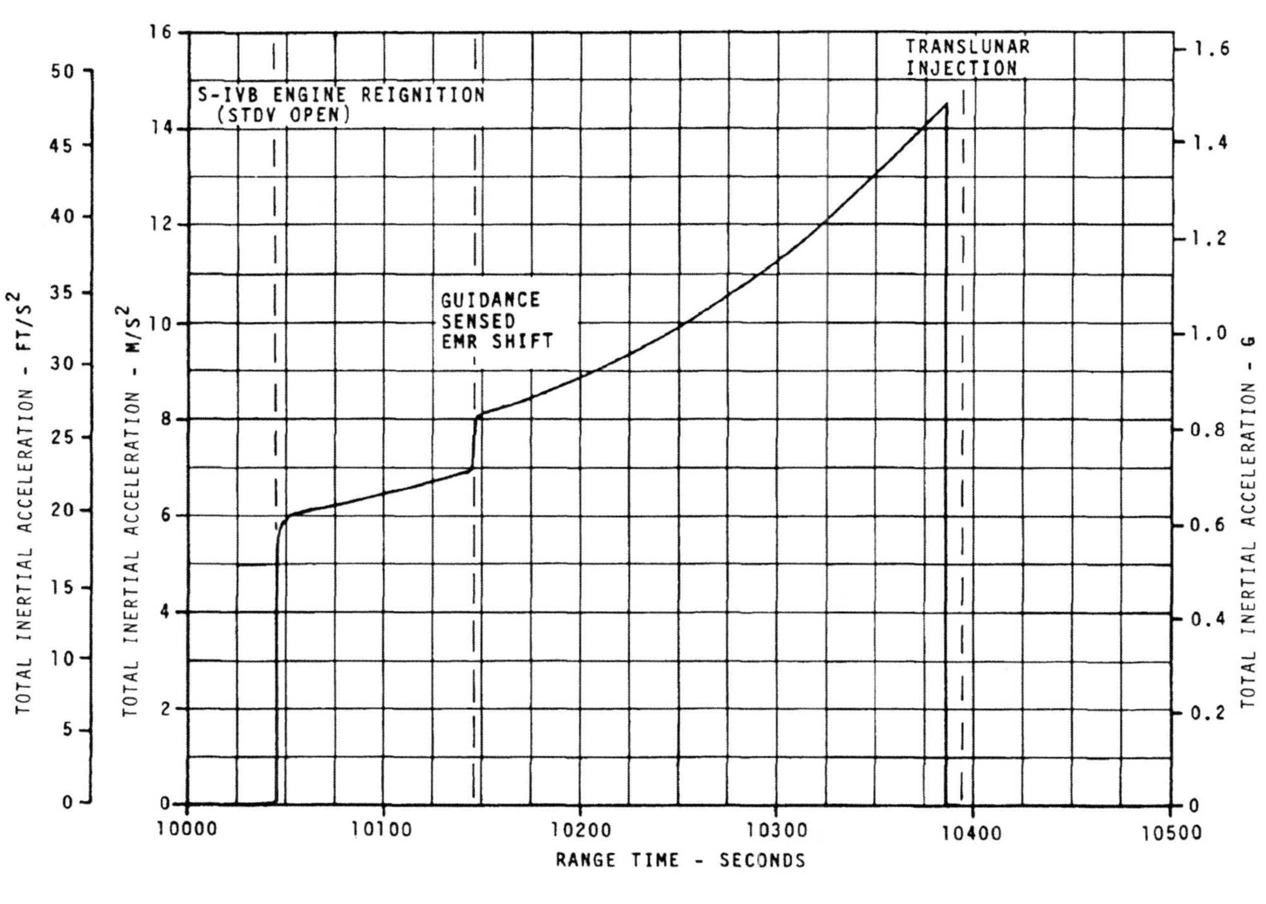

The total inertial acceleration during the translunar injection manoeuvre, with the propellant utilisation shift.

"Good show," said Carr.

At 5 minutes, the vehicle oscillated several degrees in pitch and then resumed its heading.

On determining that it had gained the desired velocity 1 second ahead of schedule, the instrument unit cut off the J-2 engine. The maximum acceleration on this stage of 1.48 g occurred at this time. The burn lasted 344 seconds and achieved a delta-V of 10,515 feet per second. Injection occurred at an altitude of 199 nautical miles, with a total space-fixed velocity of 35,389.8 feet per second and a flight path angle of 8.6 degrees relative to local horizontal; it had been zero at the start of the burn. At ignition, the vehicle had just crossed the equator at the International Date Line on a northeasterly track. At shutdown it was a few degrees due south of the Big Island of Hawaii. Translunar trajectories were viewed as either nominal or hybrid. A nominal trajectory would yield a low-pericynthion free-return passage behind the Moon that would return the spacecraft to Earth without further thrusting in the event of lunar orbit insertion not occurring. A hybrid trajectory was to yield a high-pericynthion free-return (with 'high' being defined as an altitude greater than 100 nautical miles), but at some point during the translunar coast the spacecraft would manoeuvre onto a non-free-return trajectory. In the event of lunar orbit insertion not occurring, this would require a major manoeuvre to re-establish a return to Earth. The pericynthion target was 1,850 nautical miles for launch on 14 November and 1,000 nautical miles for 16 November. Although the pericynthion would be 470 nautical miles owing to the error in the S-IVB's state vector, the path was still free-return. An SPS burn was already scheduled for T+31 hours to adopt the non-free-return trajectory required to reach the planned lunar landing site, so it was decided simply to factor the error inherited from the S-IVB into a revised calculation of that manoeuvre, in which the pericynthion would be lowered to 60 nautical miles.[12]

RETRIEVING INTREPID

For a translunar injection on the second revolution in parking orbit, the operational capability of the S-IVB allowed 2 hours of post-TLI activities. After safing itself, it adopted a local-horizontal attitude and initiated both continuous propulsive venting of the residual hydrogen boil-off and dumping the helium from the bottles inside the hydrogen tank. The oxygen vents were briefly opened, then closed to enable the tank pressure to build up in readiness for the forthcoming slingshot manoeuvre. After 15 minutes, this venting was terminated and the S-IVB adopted an attitude that would produce favourable illumination for separation by Yankee Clipper and the ensuing transposition, docking and extraction sequence. Conrad and Gordon swapped

[12] A free-return is a translunar trajectory that will achieve satisfactory Earth entry within the velocity correction capability of the service module RCS. The major advantage of the 'hybrid' non-free-return trajectory was increased flexibility in mission planning, but it was constrained so that in the event of a failure to enter lunar orbit it would be possible to set up a return to Earth using the descent propulsion system of the lunar module.

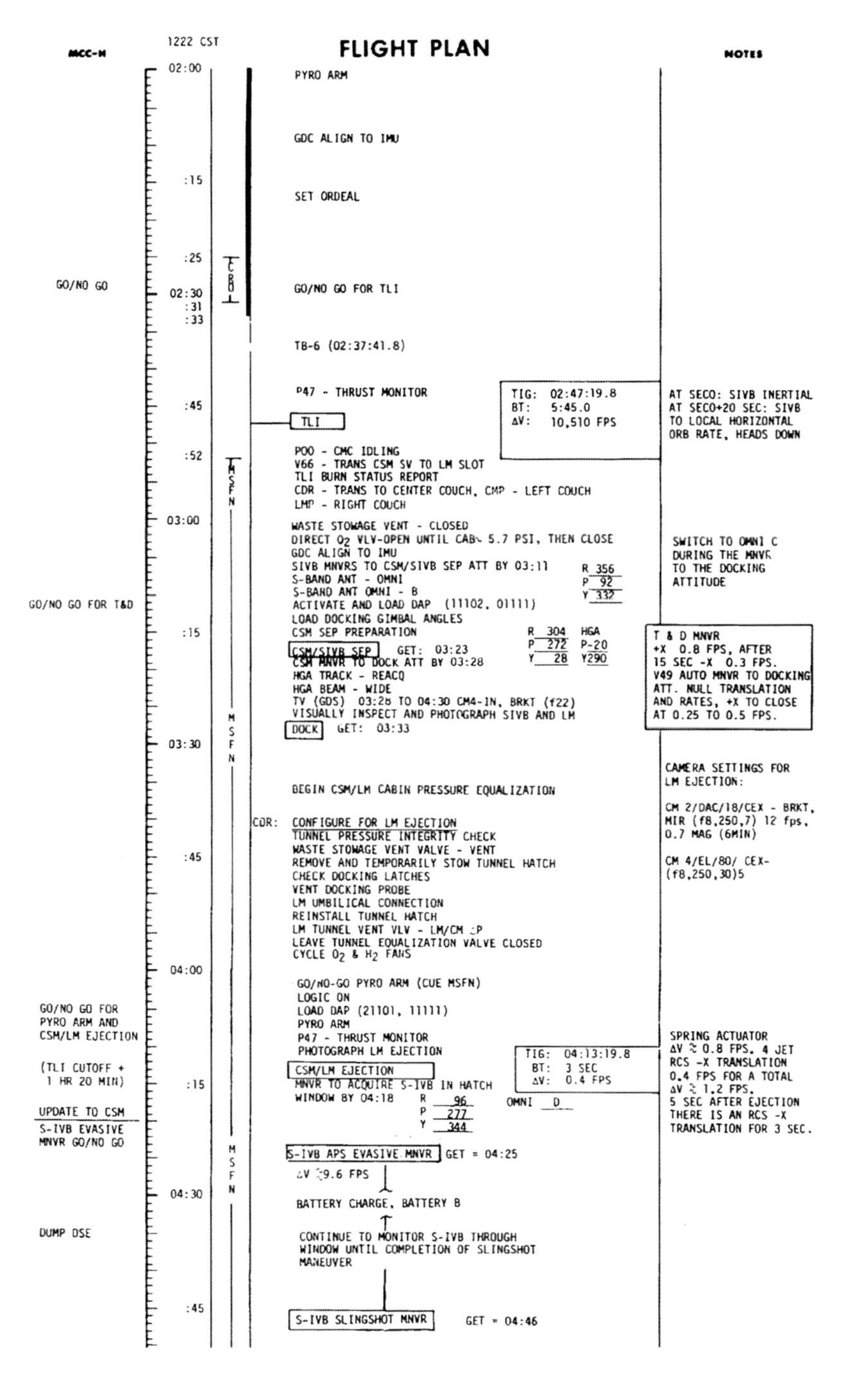

MCC-H

1222 CST

FLIGHT PLAN

NOTES

02:00 PYRO ARM

GDC ALIGN TO IMU

:15 SET ORDEAL

:25 CRO

GO/NO GO — 02:30 — GO/NO GO FOR TLI

:31

:33

TB-6 (02:37:41.8)

P47 - THRUST MONITOR

:45 TLI

TIG: 02:47:19.8
BT: 5:45.0
ΔV: 10,510 FPS

AT SECO: SIVB INERTIAL
AT SECO+20 SEC: SIVB
TO LOCAL HORIZONTAL
ORB RATE, HEADS DOWN

:52 MSFN

P00 - CMC IDLING
V66 - TRANS CSM SV TO LM SLOT
TLI BURN STATUS REPORT
CDR - TRANS TO CENTER COUCH, CMP - LEFT COUCH
LMP - RIGHT COUCH

03:00

WASTE STOWAGE VENT - CLOSED
DIRECT O_2 VLV-OPEN UNTIL CAB ~ 5.7 PSI, THEN CLOSE
GDC ALIGN TO IMU
SIVB MNVRS TO CSM/SIVB SEP ATT BY 03:11 R 356 P 92 Y 332
S-BAND ANT - OMNI
S-BAND ANT OMNI - B
ACTIVATE AND LOAD DAP (11102, 01111)
LOAD DOCKING GIMBAL ANGLES

SWITCH TO OMNI C
DURING THE MNVR
TO THE DOCKING
ATTITUDE

GO/NO GO FOR T&D

:15

CSM SEP PREPARATION R 304 P 272 Y 28 HGA P-20 Y290
CSM/SIVB SEP GET: 03:23
CSM MNVR TO DOCK ATT BY 03:28
HGA TRACK - REACQ
HGA BEAM - WIDE
TV (GDS) 03:28 TO 04:30 CM4-IN, BRKT (f22)
VISUALLY INSPECT AND PHOTOGRAPH SIVB AND LM
DOCK GET: 03:33

T & D MNVR
+X 0.8 FPS, AFTER
15 SEC -X 0.3 FPS.
V49 AUTO MNVR TO DOCKING
ATT. NULL TRANSLATION
AND RATES, +X TO CLOSE
AT 0.25 TO 0.5 FPS.

03:30 MSFN

BEGIN CSM/LM CABIN PRESSURE EQUALIZATION

CAMERA SETTINGS FOR
LM EJECTION:

CM 2/DAC/18/CEX - BRKT,
MIR (f8,250,7) 12 fps,
0.7 MAG (6MIN)

CM 4/EL/80/ CEX-
(f8,250,30)5

CDR: CONFIGURE FOR LM EJECTION
TUNNEL PRESSURE INTEGRITY CHECK
WASTE STOWAGE VENT VALVE - VENT
:45 REMOVE AND TEMPORARILY STOW TUNNEL HATCH
CHECK DOCKING LATCHES
VENT DOCKING PROBE
LM UMBILICAL CONNECTION
REINSTALL TUNNEL HATCH
LM TUNNEL VENT VLV - LM/CM ΔP
LEAVE TUNNEL EQUALIZATION VALVE CLOSED
CYCLE O_2 & H_2 FANS

04:00

GO/NO-GO PYRO ARM (CUE MSFN)
LOGIC ON
LOAD DAP (21101, 11111)
PYRO ARM
P47 - THRUST MONITOR
PHOTOGRAPH LM EJECTION

GO/NO GO FOR
PYRO ARM AND
CSM/LM EJECTION

(TLI CUTOFF +
1 HR 20 MIN)

:15

CSM/LM EJECTION
MNVR TO ACQUIRE S-IVB IN HATCH
WINDOW BY 04:18 R 96 P 277 Y 344 OMNI D

TIG: 04:13:19.8
BT: 3 SEC
ΔV: 0.4 FPS

SPRING ACTUATOR
ΔV ≈ 0.8 FPS. 4 JET
RCS -X TRANSLATION
0.4 FPS FOR A TOTAL
ΔV ≈ 1.2 FPS.
5 SEC AFTER EJECTION
THERE IS AN RCS -X
TRANSLATION FOR 3 SEC.

UPDATE TO CSM

S-IVB EVASIVE
MNVR GO/NO GO

MSFN

S-IVB APS EVASIVE MNVR GET = 04:25
ΔV ≈ 9.6 FPS

04:30

BATTERY CHARGE, BATTERY B

DUMP DSE

CONTINUE TO MONITOR S-IVB THROUGH
WINDOW UNTIL COMPLETION OF SLINGSHOT
MANEUVER

:45

S-IVB SLINGSHOT MNVR GET = 04:46

A section of the flight plan dealing with translunar injection, transposition, docking and extraction of the lunar module, and the evasive and slingshot manoeuvres performed by the S-IVB stage.

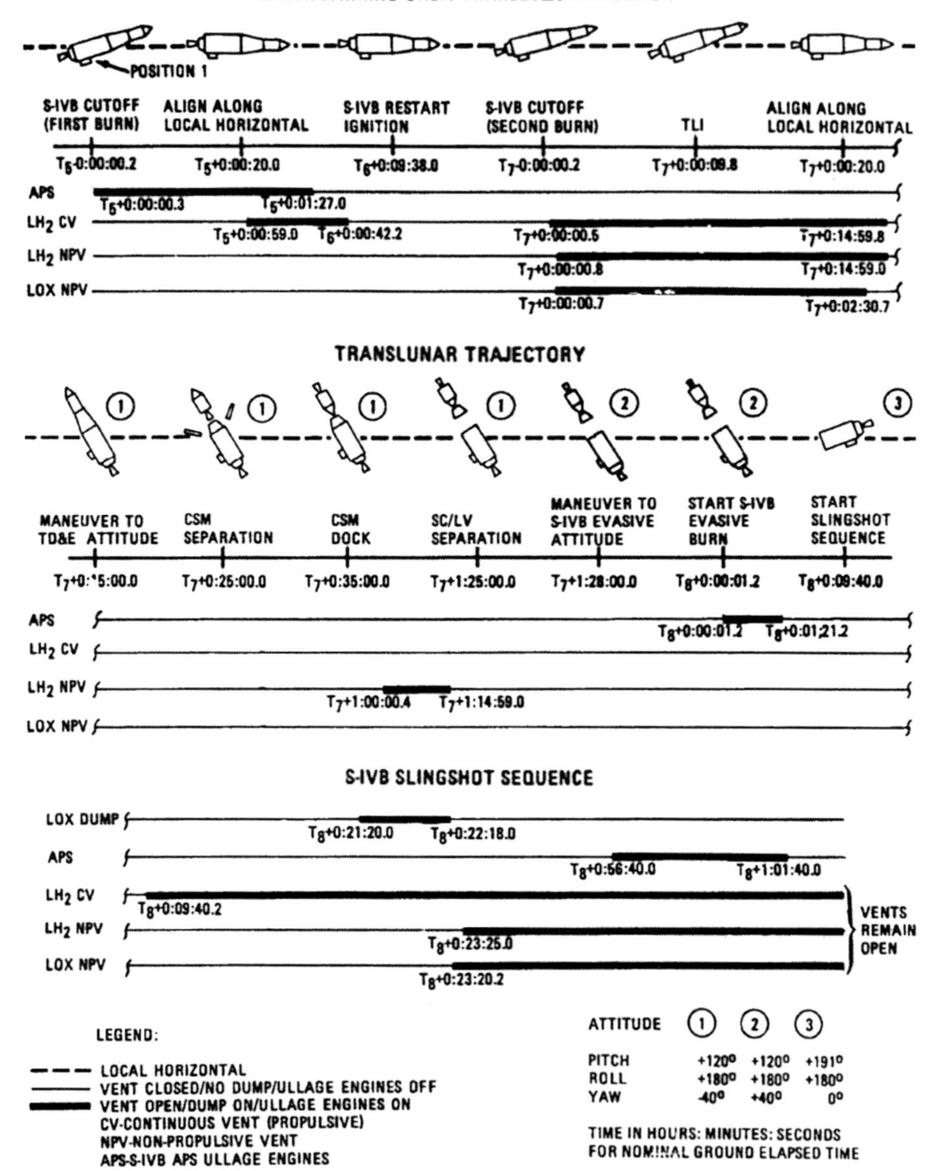

Further details of translunar injection, transposition, docking and extraction of the lunar module, and the evasive and slingshot manoeuvres performed by the S-IVB stage.

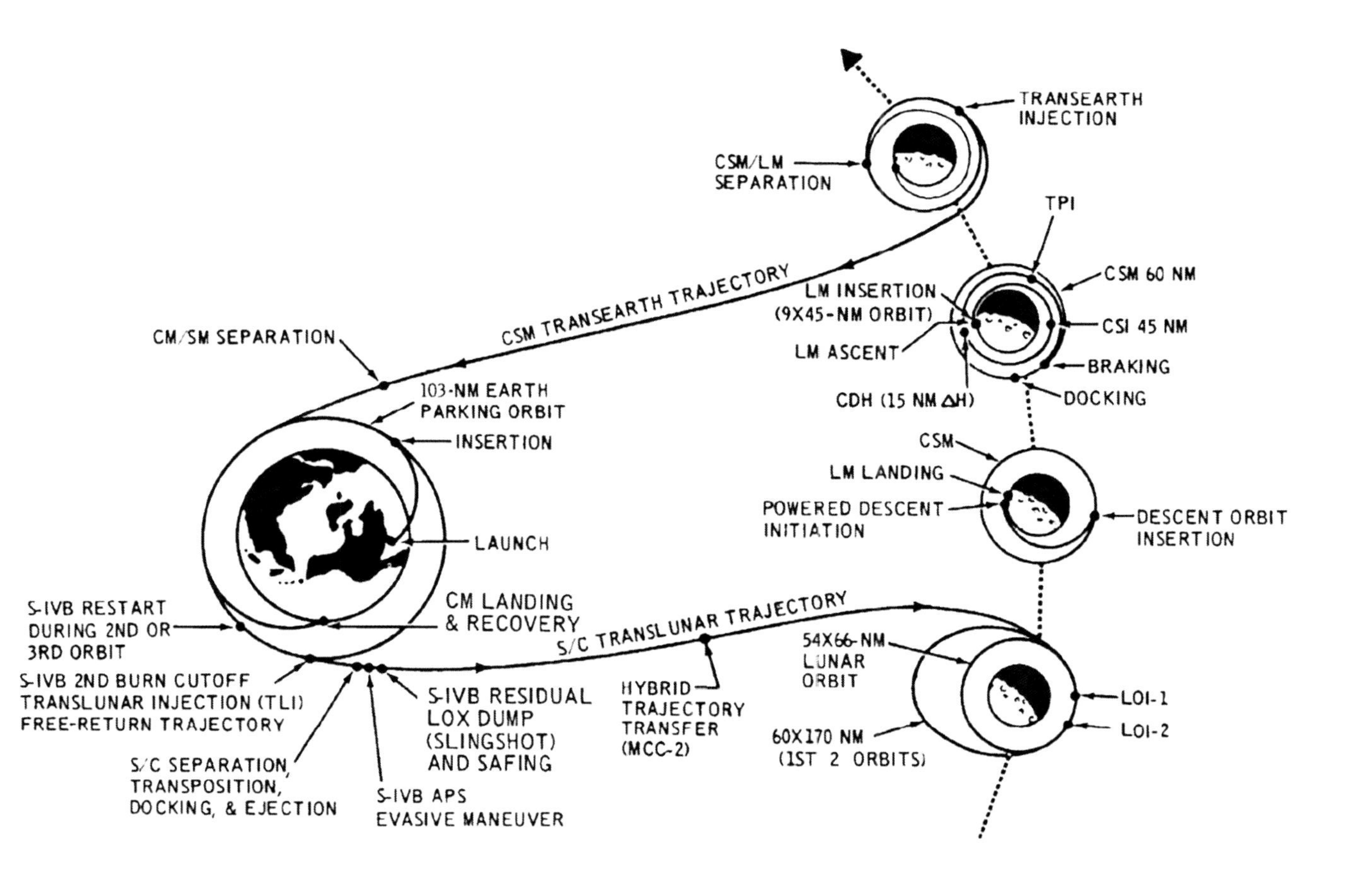

Details of various stages of the Apollo 12 mission, correlated with the motion of the Moon in its monthly orbit of Earth.

places so that the command module pilot could fly the spacecraft. Some 25 minutes after injection, Gordon fired the aft-facing thrusters and detonated the ring of pyrotechnic cord to separate Yankee Clipper from the S-IVB. The four panels of the adapter first slowly hinged open like the petals of a flower, then disconnected and drifted away.

"Okay, we separated," Conrad advised Houston. The shock of pyrotechnic separation had inadvertently closed two isolation valves in the RCS system, but he was alert to the possibility and promptly reset them.

The spacecraft used its thrusters to move directly away from the S-IVB, pitch up and rotate by 180 degrees. On gaining their first post-TLI view of Earth, they started to snap pictures. While in parking orbit, their motion had been eastward because the vehicle had been circling the globe more rapidly than Earth was rotating axially in that direction. At the time of translunar injection it was just south of Hawaii. During the first 7 hours of the translunar coast, the eastward motion of the ground track first slowed and then reversed as the rotation of the Earth overtook it. An injection over the Pacific had the advantage of placing the early part of the translunar coast over the Americas. At separation, the track was over northern Florida, heading southeast, and the altitude was sufficient for the 210-foot-diameter antenna at the Deep Space Network station at Goldstone in the Mojave Desert of California to establish a line of sight. "We can see the whole United States, Houston," announced Bean in awe.

As Gordon manoeuvred Yankee Clipper to close in, Conrad reported, "I've got an awfully pretty looking Intrepid sitting out the window here, gang."

"You're Go for docking," Carr relayed. "We are configured for television early if you want to punch it up."

The television camera for use in the Apollo 12 command module was the same one as carried on Apollo 11. It had been developed by the Aerospace Division of the Westinghouse Electric Corporation in Baltimore, Maryland, and had a zoom lens for both wide-angle and narrow fields of view. It had a 3-inch monitor which could be detached and mounted at a convenient location in the cabin so that its operator could adjust the zoom. Overall, it weighed 12 pounds. It produced a standard 525-line, 30 frame-per-second signal in colour by using a rotating filter wheel. The sequence of monochrome signals were carried on the S-band downlink, received by Goldstone and routed by landline to Mission Control, where they were processed into a colour image. The conversion process caused the image to lag about 12 seconds behind the audio link. As Yankee Clipper closed in, the television showed an ever diminishing section of the roof of Intrepid, finally looking into it through the overhead window, on which scribe marks were clearly visible.

"This Dick Gordon's smooth as silk," Bean praised the command module pilot's control of the spacecraft.

Gordon nudged the docking probe on the apex of the command module into the conical drogue of the lunar module. As the tip of the probe penetrated a hole at the centre of the drogue, three small capture latches engaged to achieve 'soft docking'. When he was sure that the alignment was stable, Gordon put the autopilot into Free mode and threw the switch to hydraulically retract the probe, draw the collars

A view of the S-IVB with the lunar module on top, during the transposition and docking sequence.

into contact and engage the 12 main latches that would rigidly lock the vehicles together.

"We've got a hard dock, Houston," Conrad announced.

In fact, Gordon had found the 9-minute transposition and docking exercise to be rather frustrating. The procedure required him to use the inertial sensors of the entry monitor system to measure his velocity relative to the S-IVB. In principle it was a simple procedure: thrust away from the S-IVB until achieving a velocity of 0.8 feet per second and then turn around and come back to dock. However, the entry monitor system was designed to measure large changes in velocity, not extremely small ones, and the reading jumped unaccountably. Unsure of his separation rate, he allowed extra time before initiating the rotation, with the result that by the time he halted he was almost 125 feet away from the S-IVB, twice his intended distance. The procedure had apparently been undermined by a slight accelerometer bias. Concluding that what was essentially "a very simple operation" had been made unnecessarily complicated by attempting to use the entry monitor system, Gordon recommended that future crews simply apply a thrust for a fixed interval and wait until they were sure they were clear. "I got myself in a box," he said in the post-mission debriefing, "and it's due to my own stupidity in not recognising these bad features. My whole philosophy [...] on the EMS [...] is that it should be used only two times: one is for entry and the other is for backing up an SPS burn. [...] The EMS was never designed for anything else other than those two functions."

By now, Apollo 12 was at an altitude of 3,819.3 nautical miles on a flight path angle which had increased to 45.1 degrees relative to local horizontal. Although the Earth's unrelenting gravity had slowed it to a total space-fixed velocity of 24,865.5 feet per second, it was rapidly transiting the fringe of the inner van Allen radiation belt. Each man wore a personal radiation dosimeter to measure his exposure.

As the crew worked through the post-docking checklist, Carr, who was watching the television view of a section of Intrepid's roof, saw some little white flecks adrift and enquired, "What just floated past the window?"

"We're in a great big cloud of ice balls up here, they're just all over everywhere," said Bean. "And there is a lot of stuff floating up out of the S-IVB itself that looks like ice or white paint chips."

"We can even see it here," Carr said.

"Just a second, and we'll move the camera over to the other side and give you a good Earth view," Bean said. By now their altitude had increased to 6,000 nautical miles.

"Hey, Gerry, this is Dick. How much fuel did I waste during that docking?" He could not check this himself because the system for measuring propellant quantity had been disabled by the lightning strikes.

"70 pounds," replied Carr after checking with the people who were trying to keep track of the situation.

"That's too much," Gordon chastised himself.

On the basis of training, he had expected to use only 20 pounds of propellant. But the EMS in the simulator had been driven by a computer rather than sensors, and so had been crisp. The extra propellant had been consumed in dealing with the quirk of

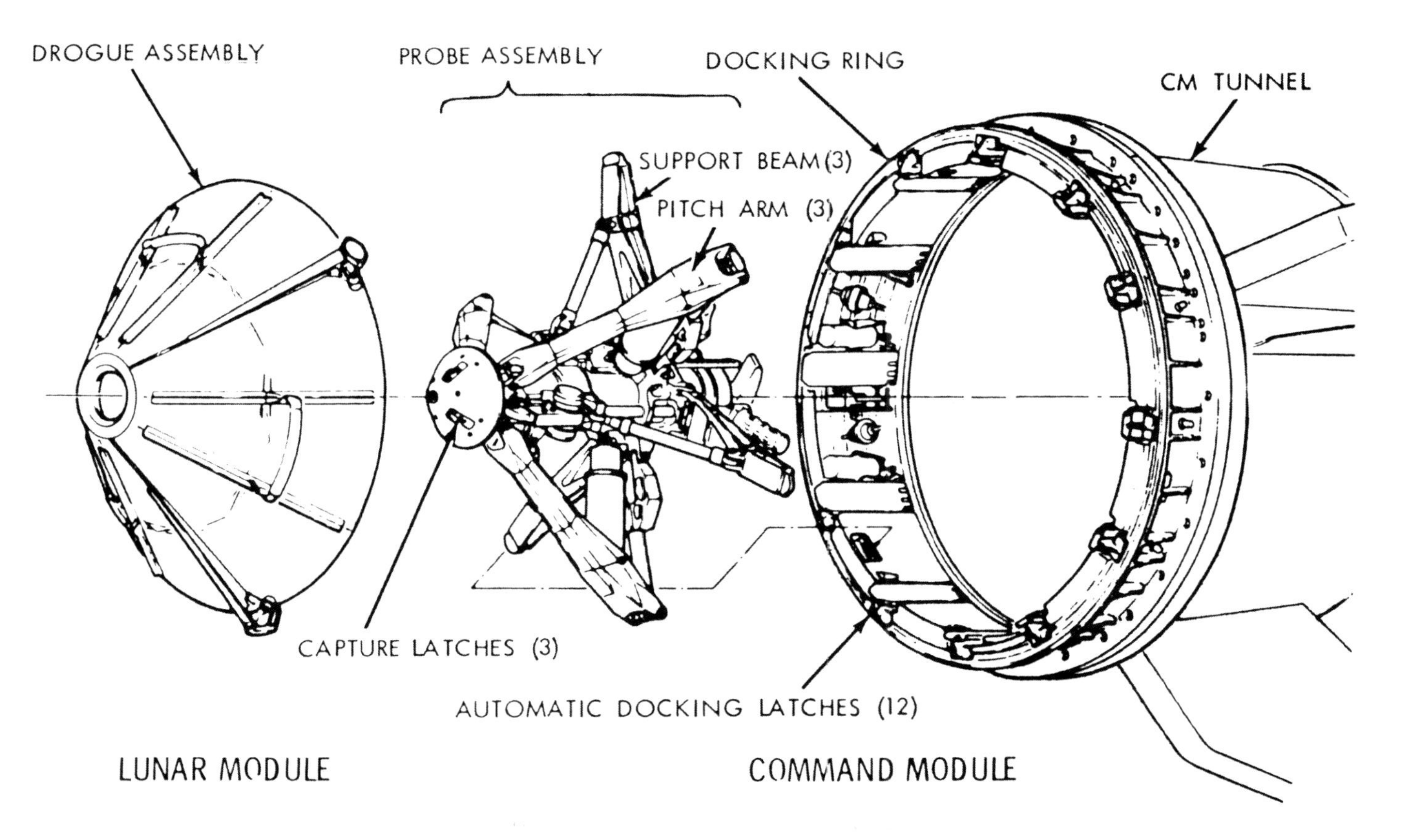

Details of the probe and drogue docking mechanism developed for Apollo.

the real device using a procedure that he now wished he had never adopted. For the pilot of a spacecraft, this waste of propellant was particularly frustrating.

The mood was lightened by a call from Carr, "Apollo 12, Houston. We're having a little trouble recognising things here. How about giving us a little commentary?"

"That's the Earth you're looking at, friend," offered Gordon.

"Oh," Carr mused. "I thought it was the Moon."

"Charlie isn't working again, is he?" Gordon asked mischievously. While serving as Capcom for Apollo 11, Charlie Duke, who was selected as an astronaut in April 1966 and had yet to fly, had embarrassingly mistaken a television shot of Earth for a view of the Moon.

"No," Carr laughed. "We've got him locked in a closet."

"You should be seeing the Yucatan Peninsula and Mexico, and Baja California is in plain sight. It's a pretty nice day down there," said Gordon. "The western Gulf of Mexico has a cloud coverage along the coast; it looks like it is almost up to Houston – it's south and west of it."

"Roger," Carr acknowledged.

"It looks like that garbage we came through down at the Cape is off the coast at this time." He was referring to the cold front that was moving southeast across the Cape at the time of launch. The weather at the Cape was now clearing.

"See," said Carr, "you could have waited and missed it, Dick."

"Oh, I wouldn't have missed that for the world," Gordon insisted, referring to the electrifying launch.

After asking for the camera to be zoomed out in order to show more of the globe, Carr observed, "Now we can see the Earth is indeed round."

"It's a fantastic sight," Gordon continued. "The Mississippi Valley has a little bit of cloud coverage coming down from Canada, and there's some in the northeast part of the country, up in the New England States – looks like they may be getting some snow over there in a day or two. Florida is cut in half by that front that went through this morning. The West Coast looks absolutely gorgeous. Baja California is clear. It looks a little cloud covered to the south and west of the San Diego/Los Angeles area. I won't say anything about smog. As I look up north, there's nothing but clouds."

Bean chipped in, "We see the Moon out the right window here, window number 5; it looks like about one quarter [phase]. We see the Earth out the left window."

Meanwhile, Conrad opened a valve to allow oxygen to pass from the command module into the 32-inch-diameter hermetic tunnel formed by the docking collars. A valve in the roof hatch of the lunar module had been left open for launch so that the air in the cabin would vent during the ascent. Pressurising the tunnel allowed oxygen to flow back through this valve into the cabin. As a preliminary to this operation, the pressure in the command module had been boosted. With the tunnel pressurised, Gordon left the television viewing Intrepid and went to assist in removing the apex hatch.

Conrad reported "a sort of funny smell" in the tunnel, but earlier crews had noted something similar and so he was not concerned. After discovering that the docking probe was hot, he made a slow 360-degree roll, inspecting each of the dozen latches in turn. On finding one which had not fully engaged, he banged it

A view of Earth shortly after translunar injection, featuring Central America and the Pacific coast up to the US border. The dark object is one of the jettisoned panels from the spacecraft adapter.

closed. Then he hooked up one umbilical that would provide power to the lunar module and another to relay the command to release the docked combination from the S-IVB. In making a second inspection of the latches he discovered another one that was not quite set, so he fixed it. Once Gordon had verified that Intrepid was drawing power, Conrad switched off the tunnel lights, reinstalled the apex hatch and closed the valve.

With a quarter of an hour remaining to the scheduled time for separating from the S-IVB, Bean asked Carr, "What's going on with the booster?" The reply was that it was venting. While their attention had been inside the command module, the S-IVB had dumped both its engine-start tank and the ambient helium tanks, and resumed dumping its cold helium. Hydrogen venting was also underway. Taking a look, Bean said it was a "pretty spectacular" sight. Conrad added that the sunlight had "made a rainbow" in the cloud. Bean was intrigued by the blobs of material that were floating around, "There's all kinds of things going on back there." They were 4 hours into the mission, and climbing through an altitude of 10,000 nautical miles.

"You're Go for ejection," Carr announced.

After a switch was thrown to detonate the pyrotechnics on the adapter ring atop the S-IVB that supported the 'knee' points of Intrepid's landing legs, small spring devices pushed the docked combination away at a separation rate of about 1 foot per second. Once clear, the autopilot reoriented the spacecraft to place the S-IVB in the centre of the hatch window.

"There's the S-IVB, and I can see it venting," Conrad reported when the stage came into view. "What does it keep venting, anyhow, Houston?"

"We're not supposed to be venting anything now," Carr replied.

"It's throwing stuff off the sides and out the back like crazy," Conrad told him. "Big radial clouds of it coming out the back." To ensure that the preprogrammed sequence of activities by the S-IVB did not interfere with a delayed LM extraction, it was to be initiated by radio command from Earth rather than by an onboard timer. Despite the protracted venting, the stage was stable but because that situation might change Conrad urged, "Maybe you ought to enable that manoeuvre right now."

Carr agreed, and 20 seconds later reported, "The manoeuvre is initiated."

"I can see it firing thrusters," Conrad reported. "And I can see it starting to yaw."

The S-IVB was using its auxiliary propulsion system to adopt the inertial attitude for the imminent evasive manoeuvre. This was the same as that for the transposition sequence, but with the yaw flipped from −40 to +40 degrees.

"I'll try to get it out of the centre hatch window now, Houston," Bean said, as he adjusted the reactivated television camera.

"We're copying your TV real clear now," Carr confirmed.

"Earth is about one and a half times the size of a basketball right now," Bean told Carr.

"Al, how far away is that basketball?" Carr asked.

"You've probably got a better idea than I do about that one," Bean replied, having misinterpreted. In comparing the size of the Earth in the void to a basketball, it was important to know how far in front of the eye that ball was being held, which is what Carr had asked. Bean interpreted the question to mean how far away the spacecraft was from Earth, and so answered by referring to the fact that the ground was aware of their altitude, which had just exceeded 14,000 nautical miles. It was an interesting difference of context.

"Did you see that thing throwing stuff out the back, Houston?" Conrad asked.

"We could a while ago," confirmed Carr, "and it looks like it's got a halo around it now." He offered a "hunch" about the unexpected venting: after J-2 shutdown, the plan had called for the ongoing oxygen boil-off to pressurise the tank, but one of the two non-propulsive valves had failed in its open position and the rising pressure was causing intense venting. A few seconds after Houston announced that the S-IVB had achieved the attitude for the evasive manoeuvre, it issued another large cloud.

"Houston, we're changing the scenery on you," Gordon advised, as he retrieved the camera from the hatch window. "We'll come back to that S-IVB just before it goes." He presented a view of Earth. "How does the homeland look to you?"

"It's beginning to look kind of small," Carr admitted.

Conrad was still intrigued by the S-IVB. "It's really weird, Houston. It reminds me of some guy standing back there with a water hose, just spraying it around in any old direction."

Gordon restored the television camera in the hatch window so that the engineers could watch the ullage motors of the auxiliary propulsion system fire for 80 seconds for an delta-V of 10 feet per second designed to displace the spent stage from the spacecraft. On Apollo 11 this evasive manoeuvre had been performed by the spacecraft using its service propulsion system.

"The burn was nominal," Carr reported.

"That thing did a fantastic job for us today," Gordon pointed out, referring to the superb performance of the S-IVB earlier in the mission.

"Sure did," Carr agreed.

"The sunlight is starting to come in the window," Gordon told Carr, "and we're a little concerned about the TV." If the Sun were to directly illuminate the camera's imaging sensor, this might be damaged. "So I guess you've seen the show for today on the S-IVB and we'll look at the Earth for a little bit for you." By now the ground track was just east of Puerto Rico, heading south and about to turn west.

"Roger, Dick," Carr replied. "And we'd sure like to see what you guys look like." He was suggesting they continue the transmission, switching the view to the interior of the cabin.

"We look just like we did this morning when we got out of bed," Gordon pointed out wryly.

"Now, there's a really reasonable guy for you," Conrad chortled.

Finished observing the S-IVB, Gordon reoriented the docked combination to the attitude suitable for the P52 star sightings he was soon to make in order to check the alignment of the inertial platform.

In retrospect, Gordon criticised the manoeuvring to watch the S-IVB making its evasive manoeuvre, calling it "a bunch of nonsense". He had been surprised to find that his digital autopilot had been programmed to undertake rotations at 0.5 degree per second. With the heavy lunar module attached, a turn at this rate required a lot of propellant to initiate and terminate. A rate of 0.2 degree per second would have been preferable.[13] Judging this to have essentially wasted 60 pounds of propellant, his recommendation to future crews was that once they had cleared their S-IVB they should achieve passive thermal control as soon and as efficiently as possible.

When the changing line of sight denied the television camera its view of Earth, Conrad switched it off. Carr reiterated his request for "an interior shot". This time he was successful. "There's Dick Gordon with his sunglasses on," Conrad announced as the television came to life. The signal faded out as the vehicle manoeuvred to the P52 attitude so Carr read up gimbal angles to aim the high-gain antenna on a boom projecting from the rear of the service module to re-establish the connection, then the camera was left running as a 'fly on the wall' while the crew performed the post-separation tasks.

"That's a nice-looking hat you're wearing, Al," Carr observed when the camera swung around to show Bean, who had donned a ball cap. Conrad had arranged for each man to have one – although as commander his had a propeller on top.[14]

"We're going to stow the TV," Gordon announced several minutes later. "We'll come back at you later with something."

At about this time, the S-IVB began to adopt the attitude for its final act, which was to execute a retrograde manoeuvre to deflect its trajectory to pass by the trailing hemisphere of the Moon in a manner which would 'slingshot' it into an independent solar orbit – as had been done by all previous Apollo lunar missions. The motivation was to minimise the probability of (in descending order of priority): impact between the S-IVB and the Apollo spacecraft in the translunar coast; impact of the S-IVB with Earth; and impact of the S-IVB with the Moon.

"Is it going to be in our window in the attitude we're in now?" Conrad asked.

"I kind of doubt it, Pete," Carr replied. "Do you want to watch it go?"

"No, no, we'll just stay put," Conrad decided. "We're getting hungry, and I think we're going to start getting out of these suits."

Ten minutes later, the S-IVB initiated its slingshot manoeuvre. The first act was the 'blow down' of the remaining oxygen through the J-2 engine, which lasted about 1 minute. This was followed over the next hour by venting the remaining hydrogen, dumping the remaining helium, and two burns by the ullage motors of the auxiliary propulsion system – the second one being an impromptu effort to compensate for the incorrect state vector. The delta-V requirement depended upon factors such as the launch time, flight azimuth and the translunar injection opportunity used, and in this case was 102 feet per second. The oxygen dump contributed 32.8 feet per second, the hydrogen venting added 9.2 feet per second, the programmed ullage burn lasting 300 seconds provided 38.7 feet per second and the ground-commanded 270-second ullage burn yielded 35.1 feet per second. By the time this activity was complete, the sub-Earth point was approaching Panama. Shortly thereafter, it was back over the Pacific and heading westward a few degrees north of the equator. The spent S-IVB adopted an attitude suitable for communication, so that it could be tracked until its battery drained. As the total delta-V of 115.8 feet per second significantly exceeded

[13] Later in the translunar coast, the autopilot was reset to use this slower rate.

[14] One can hardly imagine Neil Armstrong doing this.

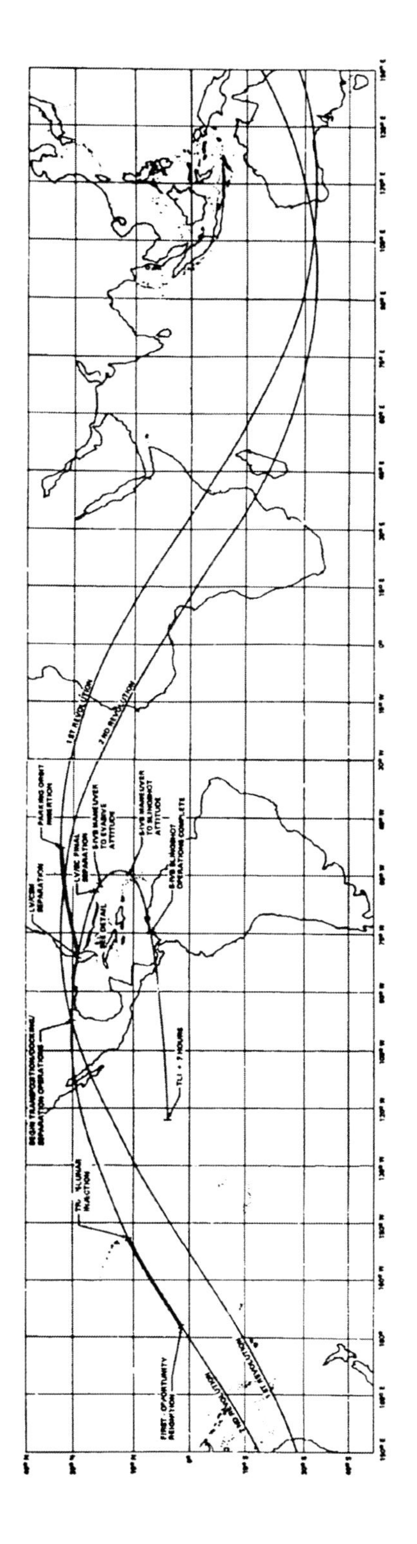

The track of Apollo 12 through to translunar injection on the second revolution, and the migration of the sub-Earth point over the first 7 hours of the translunar coast.

that desired, the altitude of the spent stage at the time of its closest point of approach to the Moon was 3,082 nautical miles, which was too great for it to be deflected into solar orbit.

CHECKING INTREPID

As the S-IVB was undertaking its slingshot manoeuvre unobserved, Carr called the spacecraft, "We would like you to consider the idea of getting into the LM tonight, before bedtime, and going through the housekeeping portion of your checklist." This would provide an opportunity to find out whether Intrepid had been impaired by the lightning strikes.

"That sounds like a good idea," replied Gordon.

"We're not going to do MCC-1, Dick. It looks like you won't need it, so you can do the inspection during that time when you'd normally be doing MCC-1."

The flight plan included options for four midcourse correction manoeuvres during the translunar coast, nominally scheduled at TLI + 9 and TLI + 31 hours, and LOI-22 and LOI-5 hours. The first one was meant to 'clean up' the injection dispersions and the second would adopt a non-free-return trajectory. Even although the state vector of the S-IVB had been flawed, the spacecraft's trajectory was satisfactory, and it had therefore been decided not to waste propellant in restoring a trajectory which would soon be abandoned.

"It sounds good," agreed Gordon. "We really don't have any place to go tonight, so we don't mind working late."

Carr estimated that the activity would occur at about 11 hours GET and promised, "We'll work up a good solid plan for you."

By 5 hours into the mission, Apollo 12 was 16,827 nautical miles from Earth. In response to a query by Conrad about a temporary interruption to communications, Carr explained that the MSFN was switching between ground station.[15]

While the crew ate ham sandwiches that had been prepared earlier that morning, and which Bean said were "delicious", Mission Control replaced the REFSMMAT in the spacecraft's computer. The initial reference had used local vertical at the pad at the time of launch. Now it was time to adopt one which was more appropriate for cislunar space.[16] The plane in which Earth travels around the Sun is the ecliptic. This is inclined at 5.15 degrees to the plane in which the Moon orbits around Earth. The REFSMMAT for cislunar space was therefore defined with its Z-axis perpendicular to the ecliptic and its other axes in the ecliptic such that the Y-axis paralleled the projection onto the ecliptic of the Earth-Moon line at the time of translunar injection and pointed at Earth.

[15] After several such interruptions, the crew asked that the Capcom provide alerts several minutes in advance of antenna switches.

[16] In Latin 'cislunar' means 'not beyond the Moon', but for Apollo it essentially meant the translunar and transearth coasting phases of a mission.

As things settled down, Carr called, "What do you say we break this simulation down and debrief it now, and the backup crew is ready to get in."

"You can tell Simsup that's a new one to work on," chuckled Conrad.

"It's a good thing we'd never seen it before," Bean chipped in, "because we sure didn't know what to do about it."

"Oh, you did pretty good," Carr observed.

"That's right. Absolutely nothing," Bean replied pointedly.

Six hours into the mission, with Apollo 12 at an altitude of 26,534 nautical miles and climbing at 11,602 feet per second, Pete Frank took over from Gerry Griffin for the second shift. M.P. Frank had been promoted from mission planning to Flight Director in 1967. Edward George Gibson was a PhD physicist selected as a scientist-astronaut in June 1965. Like Gerry Carr, he was a member of the support crew. He took over the communications console.

After taking star sightings for a P52 to check the alignment of the inertial platform using the new REFSMMAT, Gordon told the autopilot to manoeuvre to an attitude suitable for a P23 cislunar navigation check in which he would measure the angle of a star relative to Earth. The main source of error in making such an observation was the Earth's atmosphere, because the planetary limb is masked by layers of airglow. This had been factored into the procedure, and Gordon was to measure the elevation of the star relative to an airglow occurring at an altitude of 19 nautical miles. After inadvertently measuring against an airglow at over twice that altitude, he repeated the observation. Previous crews had found that the measurement became easier with increasing distance from Earth, as the angular difference represented by the altitude of the airglow diminished into insignificance. Gordon used the P23 results to update the state vector provided earlier by Mission Control. The object of the exercise was that if the vehicle were to irretrievably lose communication with Earth, he would be able to navigate a safe return. As such, cislunar navigation was merely one of many contingency procedures which crews routinely undertook and never actually needed. Similarly, much of the hardware was to guard against the loss of a primary system, and in critical cases the backup was not simply a copy, it functioned in an entirely different manner as a precaution against a generic failure.

"We've been thinking about the LM checkout procedures," Gibson called. "We'd like your thoughts on whether you want to go in there as soon as you finish up with the P23?"

"We can do that," Conrad concurred. "We could get into PTC and then go on in there. What have you got in mind?" Here he was referring to the passive thermal control regime that the spacecraft was to adopt during the translunar coast.

"We'd like to go ahead and check the position of several breakers, just to make sure that we get the heaters coming off and on for all of the systems as they should be," Gibson explained.

"We'll get ready to do that now," Conrad said. "I've gotten all the way out of my suit, but still have to stow it. Al is working out of his right now, while Dick is doing the P23. So we will want to get Dick unsuited, get everybody cleaned up and get this spacecraft stowed."

Pete Frank.

"It sounds good," Gibson agreed. "We'd like to get into the LM before 8 hours."

"What's your reason for wanting to get in so quick?" Conrad asked. The original suggestion had been to enter Intrepid at about 11 hours.

"Stand by," replied Gibson.

Conrad prompted: "Old curious Pete. You've got to give me a reason."

"The reason we'd like to get in there before 8 hours, is that the ASA heater may not be cycling." He was referring to the Abort Sensor Assembly associated with the Abort Guidance System. "Its thermal limit is about 8 hours, so we would like to get in before 8 hours and check the position of that circuit breaker to make sure it hasn't popped, and look at the status of the system." It was conceivable that the vibration of launch had caused the push-pull breaker to pop out.

"We'll hustle," Conrad promised. Five minutes later he was back, "We are done with the P23's, and I guess we're going to get ready to manoeuvre to PTC attitude. Or do you want us to go into the LM with the attitude we have right now?"

"If you'll hold your present attitude, we'll get an answer up to you in a couple of minutes," Gibson replied. The flight controllers were debating whether to establish passive thermal control before or after the LM inspection.

"Okay," Conrad replied. "We're not going anywhere."

It was 10 minutes before Gibson was able to relay the decision that they should perform the inspection in their current attitude. He then specified the list of checks they were to perform. In essence, they were to undertake portions of the activation checklist – in particular performing a self-check of the computer, providing a dump of its erasable memory so that this could be inspected for corruption, and a stream of telemetry to allow TELMU, the flight controller for the lunar module, to inspect his systems.

As the mission entered its seventh hour, Conrad opened the apex hatch and then extracted the probe and the drogue from the tunnel, swung the lunar module's roof hatch down into its cabin and made his entry. On emerging through the hatch head first, he had to perform a half-somersault to match the local vertical. Some earlier astronauts had found emerging from the hatch inverted momentarily disconcerting, but Conrad had no problem. He reported that the ASA circuit breaker that Mission Control was concerned about had not popped, so the heaters ought to be running as intended.

Bean joined Conrad several minutes later, after tying down the probe and drogue in the command module, and announced, "Things look real tidy up here, Houston."

Methodically checking the circuit breakers, Conrad reported that one which the checklist specified should be in was actually out, but it was only for a utility light. A few minutes later Gibson advised that the checklist was out of date, because it had been decided to launch with that circuit breaker out – it had not popped; they were to leave it out until they required the light. This would prove to be a rare error in the onboard documentation.

Intrepid was drawing the small amount of power that it required in its hibernation state from Yankee Clipper, but to perform the abbreviated activation would require internal power. Bean checked the voltages of the batteries, declared them healthy, and activated the communication system. As they waited for Houston to configure to receive the transmission, Conrad took a look through the telescope that was at eye level between the two crew stations and received a surprise.

"Hey, Houston, I was going to report that I had another person in sight, looking out the AOT, but it turns out that the left rear detent looks right in through Dick's rendezvous window and he's looking right back at us." And as Gordon would say later, from his perspective he saw a giant eyeball staring at him!

Levity aside, Conrad was irritated that 7 minutes had passed since activating the communication system. "How soon are you going to be reconfigured? We hate to use these batteries up."

"We've got about another minute or two," Gibson replied. "If we can't make it by then, we'll not go on with it."

After a total of 10 minutes, Gibson was finally able to report, "We're picking up some data from the LM now. Stand by. We're looking at it."

Meanwhile, Bean powered up the computer, performed a self-test and dumped the erasable memory – which the subsequent inspection confirmed had suffered no adverse effects. Instead of using P06 to shut down the computer, they were to pull the circuit breaker so that when the activation checklist was executed for real later in the mission the computer would be in the same configuration as at launch. It had been decided not to check the inertial measurement unit, so the batteries were taken off line.

"Have you got anything else for us before we close up the LM?" Conrad asked.

After a brief pause, Gibson said, "Go ahead and button it up."

As he made his way back through the tunnel, Conrad checked the marked scale which indicated the angular offset between the two docking collars, known as the index, and advised Mission Control that it was minus 0.3 degrees, which was a trivial misalignment.

After less than an hour in Intrepid, Conrad and Bean were back in Yankee Clipper with the probe and drogue restored. It was sensible to inspect the lunar module for obvious damage as early as possible – certainly before manoeuvring onto the non-free-return, since its propulsion systems would be required to restore a free-return to Earth if the lunar orbit insertion burn by the service propulsion system of the service module were to fail.

"We're looking at you down there, Houston," Bean said, "and now you're about the size of a volleyball."

"How far away is that volleyball?" Gibson asked.

"About 2 feet," estimated Bean, this time interpreting the query properly. At this time, the spacecraft's altitude was 40,000 nautical miles. "I can't see any landmass at all. All I can see is water with lots of clouds, and I can see sort of a glare point on the Earth – I think that must be the zero-phase point to us. Other than that, it's very, very bright. And another interesting thing is, on the dark side, I can't see where the Earth stops and space begins." The glare was the specular reflection of sunlight off the ocean.

When Gibson asked Bean why the "old heads" onboard were not commenting on the view, Gordon interjected, "He won't let us near the window!"

At this point, Gibson banished the levity by asking Conrad to confirm that, apart from the improperly documented utility light, they had left all the circuit breakers in the lunar module in the configuration specified by the activation checklist.

"That's affirmative," Conrad replied.

"Why'd you ask, Houston?" Bean asked.

Gibson explained that the current being drawn by Intrepid through the umbilical was about 1 ampere greater than prior to the inspection visit.

"What's that mean?" Conrad asked.

"Give us another long pause, Pete," Gibson replied, indicating that the matter was under discussion, "and we'll be back to you."

While waiting, Gordon suggested, "I think it's about time I went to PTC, don't you?" With the spacecraft holding a fixed attitude with one side baking in constant sunlight and the other freezing in the dark vacuum of space, the thermal stress was building up.

"That's affirmative, Dick," Gibson replied immediately.

"We're going to go ahead and manoeuvre to 090 and set up PTC," Gordon said, indicating that he intended to orient the X-axis of the command and service modules 90 degrees to the X-axis of the current REFSMMAT, thereby placing the long axis of the docked combination perpendicular to the Sun in order that imparting a slow roll would serve to smooth out the thermal stresses. This passive thermal control roll was known as the 'barbeque mode'. While it made no difference in terms of thermal control whether the spacecraft's axis pointed north or south of the ecliptic, Gordon later wished he had pitched to 270 degrees and pointed the other way, as that would have better positioned Earth and the Moon in the side windows for taking pictures. They were to photograph Earth several times per day on the way out, whenever their schedule allowed, in support of oceanographic and global weather research. As the apparent diameter of the Earth's disk shrank, they were to upgrade their Hasselblad from a 250-mm lens to a 500-mm lens. While Gordon was waiting for the digital autopilot to stabilise the attitude of the docked combination, preparatory to initiating the roll, he activated fans to stir the tanks of cryogenic reactants of the fuel cell system. This would be done regularly throughout the mission. The first time that any piece of equipment was operated, the monitoring flight controllers paid particular attention in order to gain a 'feel' for the system, since each spacecraft had its own quirks.

"You can go ahead and start the roll," Gibson announced.

"You say our pitch and yaw looks pretty good, huh?" Conrad asked.

"That's affirmative," Gibson replied.

"We've only been waiting about 10 minutes," Gordon noted. It was more usual to wait 20 minutes for the autopilot to fully stabilise the vehicle.

"Dick, it looks good down here."

Satisfied, Gordon initiated a roll that would complete three rotations per hour. A few minutes later, as the mission entered its ninth hour, they exchanged the lithium hydroxide canister that filtered carbon dioxide from the cabin atmosphere. This, too, would be done regularly during the flight. Since translunar injection, the mainframes in the Real-Time Computer Complex at Mission Control had been processing the radio-tracking data to produce an improved state vector, and this was now uplinked, superseding the one developed by Gordon from his P23 navigational sightings.

At this point, the crew had their first proper meal. The food was similar to that of the previous mission. In addition to balanced meals for 3 men available in man/day wraps, individual items had been packed in a 'snack pantry' so that each man could eat without having to raid a proper meal – this was known as 'smörgåsbord mode'. There were in excess of 70 items on the list of freeze-dried rehydratable, wet-pack, and spoon-bowl foods. Apollo 12 introduced both rehydratable scrambled eggs and wet-pack beef and gravy. Maximum use was made of the spoon-bowl packages for the rehydratable food items, and the spoon size was increased from one teaspoon to one tablespoon. Prior to the mission, each man had evaluated the available items and made his own menu. A dispenser in the command module issued a flow of drinking water for as long as the trigger was pressed. In addition, a spigot dispensed water in 1-ounce increments at temperatures ranging from 13°C to 68°C to rehydrate food. After water was injected, the bag was kneaded for several minutes, the neck cut off using scissors and the contents squeezed directly into the mouth. The water from the fuel cells tended to contain gaseous hydrogen and oxygen. The hydrogen had given the Apollo 11 crew severe flatulence. For Apollo 12, an inline separator had been installed to prevent hydrogen from entering the potable water tank. Although there continued to be bubbles of oxygen, these were not considered objectionable. Gas in the hot-water supply tended to inhibit complete rehydration of food. If this occurred in a spoon-bowl package, some of the gas was removed by opening the package and using a spoon to mix the food. A menu was designed to provide 2,300 kilocalories per man per day, and the men were to maintain a log of what they consumed. After a meal, a germicide pill affixed to the outside of a food bag was popped into the bag to prevent fermentation, and the bag was rolled up and stowed in a compartment in the lower equipment bay. Each meal pack included a 4 × 4-inch wet-wipe for personal hygiene.

For this first meal, Conrad had tuna salad in a rehydratable spoon-bowl, beef and gravy in a wet-pack, jellied candy in the form of intermediate-moisture bites, with rehydrated grape punch. As he would observe later in the mission, the food was "a lot better" than on the Gemini missions. Gordon, who had also flown Gemini, agreed that the food was better, but criticised the wet-packs because, as he put it, "there is just something about chunks of meat immersed in cold gravy that isn't nearly as appetising as it would be if it could be warmed".

Two hours after first raising the issue of the anomalous current flowing through the umbilical into Intrepid, Gibson came back, "We'd like to suggest that you go on back over to the LM and check the circuit breakers. The possibility here, is that you have got a system on-line which is not called for, and doesn't have proper cooling. I'd like to have your thoughts on that."

After saying that they were sure they had left the circuit breakers configured as specified, Conrad offered his own insight into the issue. As the hatch in the roof of the lunar module was opened, this triggered a microswitch that activated the cabin lights. When the hatch was closed, it would deactivate the lights. Before leaving, he had manually triggered the microswitch to verify that it doused the lights, but had not been able to verify that closing the hatch triggered the switch. If the lights were still on, this could explain the additional current. "That's about the only thing I can think," he offered.

"So those floodlights did go out when the hatch was closed?" Gibson asked.

"I don't know that they went out," Conrad replied. "I'm saying that if you push the switch, they'd go out."

There was only one way to find out, so they opened the apex hatch and removed the probe and drogue from the tunnel. When Gibson reported that telemetry showed no change in the current being drawn through the umbilical as they opened the lunar module hatch, this confirmed the hypothesis. It was concluded that the microswitch was slightly misaligned relative to the hatch.

"We show a drop in amps back to what it looked to be before you first went in," Gibson reported after Conrad had pulled out the circuit breaker for the floodlights. "We'd like you to leave it in that configuration when you leave the LM, and when you go back in you'll just have to punch it in."

With the mystery resolved and a solution implemented, they closed up the lunar module and restored the probe and drogue in the tunnel.

Shortly after translunar injection, Gordon had initiated a recharge of battery B. Gibson now called to say that due to the heavy load placed on the batteries when the fuel cells dropped off-line during the launch phase, the schedule had been revised to extend the first recharging of each battery involved. The water which was produced as a byproduct of the fuel cell reaction was used to cool the electronics and made available to the crew for food preparation, drinking and hygiene. Excess water was periodically dumped overboard through a valve in the side of the service module. Gibson requested that they dump the tank down to about 15 per cent of its capacity rather than the nominal 25 per cent, in order to ensure that they would not require to perform another dump for a day or so.

"Houston," Conrad called at the time of the cancelled first midcourse correction, "how far out from Earth are we?"

After checking, Gibson replied, "You're about 56,000 nautical miles out now and are smoking along at 7,600 feet per second."

"Thank-you," said Conrad.

"Al," Gibson called, "are you still at the window?"

"That's affirmative," Bean replied. In line with the slowing of their rate of climb, the rate at which the Earth's disk appeared to shrink had diminished. "The Earth is sort of funny. It just seems to hang out there. You can't see it rotate, you can't see it move or anything. It just sort of hangs out there in this black space. And the Moon just doesn't seem to be any bigger than it was when we left. However, it does look more like a sphere. It sort of looks like a ball that's being hung out there somehow. It's really crazy."

"Which way does it look like it is hanging from?" Gibson asked.

"The north pole, naturally," Bean explained. "Otherwise the string would get all tangled up!"

"Just scientists are supposed to know that," Gordon interjected sagely.

"You need some experimental proof," Gibson challenged.

Bean was discovering that travelling through cislunar space was psychologically different to orbiting a body. The distance involved was so great that in the absence of nearby objects to provide a sense of motion it seemed that the spacecraft was suspended between Earth and the Moon.

Having spent 8 days cooped up in Gemini 5 with Gordon Cooper in 1965, Conrad appreciated the need for translunar crews to have some form of entertainment, and their luxury item was a battery-operated tape player for music cassettes. One of the tracks that Conrad had selected was a version of 'San Antonio Rose' by Bob Wills and his Texas Playboys. When Bean used the monocular to obtain a better view of Earth, Conrad played the song. When this was remarked upon by Gibson, Conrad replied, "Al's getting homesick up here, and we're just trying to keep him happy."

Gordon had been using the 8-ball to monitor the passive thermal control roll. The spin axis had recently begun to wander, and was now nutating almost 30 degrees off the perpendicular to the ecliptic. When consulted, Gibson said that the water dump had perturbed the alignment. Continuing the housekeeping, Gordon terminated the recharge of battery B and started recharging battery A.

Meanwhile, Clifford E. Charlesworth took over as Flight Director. On graduating with a degree in physics from Mississippi College in 1958, Charlesworth joined the Navy as a civilian scientist and was hired by NASA in 1962. In Mission Control he was known as the 'Mississippi Gambler' on account of his laid-back personal style. This was to be his final mission as a Flight Director before taking on a management role. Don Leslie Lind was a reserve naval aviator with a physics doctorate who was selected as an astronaut in April 1966. He worked with Bendix to develop tools and procedures for deploying the ALSEP. Although not a member of the support crew, he had been assigned Capcom duty.

On his first call, Lind forwarded the recommendation that on terminating the PTC roll to perform a P52 platform alignment, in order to save propellant they should go to the attitude for a later P23 navigational sighting and perform the P52 using whichever stars were available in that attitude. Once this was underway, Lind observed, "You certainly had an exciting one this morning."

"Yes. It keeps recurring in our conversations throughout the day, today," Conrad laughed.

"Anybody see anything from the ground in all that business?" Gordon enquired as he terminated the PTC roll.

"That's affirmative," Lind replied. "We saw lightning coming right down your plume, right to the ground."

"Are you kidding me?" Gordon asked.

"That's some of the reports we've been getting back," Lind said. "I keep telling you, you don't fly through thunderstorms."

"I keep wondering why they write that in all the handbooks," Conrad laughed.

Gordon had to check the line of sight for a number of stars before finding a pair that were not masked by the lunar module, and then took his P52 sightings. When the results were analysed, Lind reported negligible platform drift since the previous P52, "It looks like you've got a pretty good platform up there."

Clifford Charlesworth.

Don Lind (left) and Paul Weitz.

Gordon, meanwhile, had measured the position of a star relative to Earth for the navigational check.

"How do those P23 observations look?" Gordon asked.

"They look great," Lind replied.

"They're real happy, huh?" Gordon asked, alluding to the trajectory specialists.

"They sure are," Lind assured.

"Okay," Gordon acknowledged. "If they're happy, I'm happy."

As the translunar coast would be some 8 hours longer than the free-return flown by Apollo 11, this difference was to be taken up by phasing the outbound sleep periods so that the activities in lunar orbit would occur on a familiar cycle. As a result, the first sleep period was timed to start later into the mission than previously. At the start of the sixteenth hour, Lind asked whether the crew intended to retire as planned.

"We may go a little bit earlier," Conrad replied. "We're all pretty tired."

"That's fine with us," Lind said. "We can understand that you might want a little extra sleep."

While his colleagues prepared the evening meal, Gordon reinstated the PTC roll, dumped the erasable memory to be verified by Mission Control, changed the lithium hydroxide canister, and suspended battery recharging overnight.

With the pre-sleep checklist complete, Lind wished the crew "pleasant dreams". They had been awakened 5 hours prior to launch and were now 17.5 hours into the mission, 78,000 nautical miles from Earth and travelling at 6,000 feet per second. In addition to being fatigued by the long first day, they were undergoing an adaptation to weightlessness that manifested itself as a sensation of fullness in the head. Conrad and Bean snuggled into sleeping bags which were held in place by restraints beneath the side couches, and Gordon strapped himself onto a couch.

They had been assigned a 10-hour sleep period. Gordon slept well, but awakened several times, inspected the state of the spacecraft and then went back to sleep. On awakening after 5 hours, Bean took a sleeping pill. Conrad awoke after 8 hours and spent the remaining time quietly studying the 'light flashes' that previous crews had reported while in cislunar space and could be seen irrespective of whether the eyes were open or closed.[17] As Conrad said afterwards, "By paying a little attention, you could pin down that it was happening to one eye at a time. The discharges appeared in two manners. They appeared as either a bright round flash, or a particle streaking rapidly across your eyeball in a long thin illuminated line."

HYBRID MANOEUVRE

The 'graveyard shift' was worked by Gerry Griffin's team. Paul Joseph Weitz was a naval aviator selected as an astronaut in April 1966. As a member of the support crew, he was on the Capcom roster. As the mission approached its 28th hour Weitz was snoozing, awaiting the time to place the wake-up call, but Conrad beat him to it: "Hello, Houston, Apollo 12."

"Good morning, Pete."

"Morning," Conrad replied.

"You have everybody up and about there?" Weitz asked.

"Everybody's up," Conrad confirmed.

After giving the post-sleep report that included reading the personal radiation dosimeters, the astronauts set about preparing breakfast and Weitz updated them on the morning news: "World attention is on the flight of Apollo 12. The Soviet Union held the crew as 'courageous', and *Tass*, the official Soviet news agency, reported the start of the mission in a brief factual report in both of its Russian and foreign language reports. Czechoslovak television carried live coverage of the liftoff, complete with an explanation of technical details. In West Germany, all radio and television networks carried the launch live. So too, did the Japanese Broadcasting Company. The launch is being described by such adjectives as 'spooky' and 'cliff hanging'. Even President Nixon, a one-time Navy man himself, admitted he had some anxious moments, but added, 'I'm really proud of those three men up there.' Weather is a news item in Houston, where temperatures are expected to dip into the 20's tonight. Automobile owners are being advised to put anti-freeze in their car radiators. Today is a voting day here in Houston, as Houston picks a mayor, eight

[17] The Mission Report stated that "efforts are continuing to explain this phenomenon", and it was eventually concluded that the flashes were caused by high-energy cosmic rays passing through the optic nerves of the crewmen.

councilmen, four school-board members, and decides upon a number of special issues. In sports, Houston Oiler, Woody Campbell, ended rumors and speculation yesterday by strolling into the Oiler training room and putting on his uniform. He says he's in good shape after 10 months as an MP [Military Policeman] with the First Infantry Division in Vietnam, and hopes to be in action very soon."

After thanking Weitz for the update, Conrad pointed out that it was particularly nice to have hot coffee with breakfast.

An hour into the day, Weitz advised, "For your information, the burn attitude for MCC-2 will also be a good attitude for P52 and for all your star checks if you want to come out of PTC and just go right to that attitude." Then he read up the details of the manoeuvre that would depart the free-return trajectory.

At 029:16:53 Apollo 12 passed the halfway point between Earth and the Moon in terms of distance, being 112,899 nautical miles from both bodies. However, it was still within the Earth's gravitational sphere of influence and continuing to slow as it climbed, and therefore was not at the halfway point in terms of time. The crew were preparing for a television transmission which would include the MCC-2 burn. After Weitz gave a Go to perform the burn, Gordon waited until the angle of the PTC roll was convenient and then halted the rotation and adopted the desired orientation. As Gordon performed the sextant sightings for the P52 platform alignment that would precede the manoeuvre, Bean swapped the lithium hydroxide canister. In Mission Control, Pete Frank took over as Flight Director.

As the crew were preparing the service propulsion system for the burn, Weitz was providing Conrad with a running commentary on a car race in Phoenix, "Al Unser is leading after 59 laps. Mario Andretti is second. Bobby Unser is third. A.J. Foyt was forced out of the race. My information doesn't indicate when, but he was running with the leaders and his car was damaged in a collision; there were no injuries."

"Roger-roger," Conrad acknowledged. It is difficult to imagine Neil Armstrong engaging in a sports discussion while preparing for a major manoeuvre, but Conrad wasn't Armstrong and the tone of this mission was lighter than its predecessor.

A few minutes later Weitz was back, "A hot flash on the Phoenix 200. Al Unser and Mario Andretti collided with each other, and Bobby Unser is now in the lead."

"Okay. Very good," Conrad replied.

And then again, "A correction: that collision was between Mario Andretti and Bobby Unser. Al Unser still has the lead."

"Do you know how far along they are in the race?" Conrad asked.

"No, we don't know right now, Pete," Weitz replied. "We'll get it for you."

"Okay," said Conrad.

Once the television camera was ready, Conrad prompted, "Houston, you want the TV now?"

"Stand by, Pete," Weitz replied. He then concluded his commentary on the race. "They stopped the race in Phoenix after 84 laps because of rain, if you can believe it; and they're going to wait awhile and then restart when it slacks off." With that, Weitz handed over to Gerry Carr.

"We're ready for the TV whenever you are, Pete," Carr called.

The transmission began. "We're just going through the pre-burn checklist at this time," Conrad announced. "Al's down in the bilge stowing some gear for the burn. Dick's in the left-hand couch, and I'm in the centre couch right now. We have the TV mounted so that you can watch the instrument panel and the switch-throwing." They had donned their ball caps for the show. Conrad had arranged for Carr to be provided with a similar cap, labelled 'CAPCOM'.

Normally, the astronauts had their microphones configured such that they had to depress a push-to-talk switch to speak to Mission Control, but for the burn they used the Vox voice-activated system.

As a water dump was due, it was decided to perform it prior to rather than after the manoeuvre so as not to induce a perturbation that would complicate the post-burn tracking. Gordon aligned the stabilisation and control system to the attitude indicated by the inertial measurement unit, to ensure that it was ready to serve as a backup to the primary system. Then Conrad read the checklist and he and Gordon performed the preliminary switch-throwing.

"Al, let's bring on the bus ties," said Conrad, with 6 minutes remaining.

"Okay-doke," replied Bean. He tied the two main buses together so that if one bus lost its power supply the equipment hanging off it would still receive power from the other one.

The service propulsion system provided the thrust for large spacecraft velocity changes. The 20,500-pound-thrust bipropellant engine employed a 50:50 mixture of hydrazine and unsymmetrical dimethyl hydrazine (a combination which was known as aerozine-50) as fuel and nitrogen tetroxide as oxidiser. Milled aluminium radial beams divided the interior of the service module into six sectors around the central cylinder that housed the SPS engine and two helium pressurant spheres. Four of these sectors held fuel and oxidiser tankage; another held the fuel cells and their cryogenic reactants; and (by design) the sixth sector was essentially empty. The engine could respond either to firing commands from the guidance and navigation system or to manual commands specified by a hand-controller. The thrust was not throttleable, but the propellant mixture ratio could be adjusted manually for optimal performance. The engine was automatically gimballed to direct its thrust through the centre of mass. As the tanks were full, on this occasion the plan did not include firing the thrusters to settle the propellants.

As Conrad counted down the final 10 seconds, the computer flashed Verb 99 on the DSKY for authority to execute the burn, and he pushed Proceed in consent. At the appointed time, nitrogen forced open the valves to allow the helium-pressurised hypergolic propellants to enter the combustion chamber.

"Ignition," called Conrad.

"Thrust," added Gordon.

Bean verified that both sets of ball valves had opened to deliver propellants to the engine – having two 'banks' of valves ensured that the burn would continue even if one valve were to close. The engine remained firing until the computer decided that the specified change in velocity had been achieved, which occurred after a duration of 9.2 seconds.

The manoeuvre occurred at an altitude of 116,930 nautical miles from Earth, and the delta-V of 61.8 feet per second was within 0.1 foot per second of that desired. It reduced the pericynthion of 470 nautical miles inherited from the S-IVB to just 60 nautical miles, at which point the lunar orbit insertion manoeuvre would be made. The accuracy of this new trajectory was such that both of the remaining optional corrections were cancelled. Although the burn was brief, Bean had time to optimise the propellant mixture ratio by monitoring their individual flow rates. This firing of the SPS served to calibrate the engine's performance in readiness for calculating the lunar orbit insertion manoeuvre. Just as Apollo 11 had flown the trail blazed by its predecessors, this mission would further expand the capability of the Apollo system. In particular, the more efficient lunar orbit insertion manoeuvre allowed Apollo 12 to be the heaviest payload yet sent to the Moon.

After the post-burn checklist had been completed, Conrad dimmed the floodlights and Bean aimed the television camera out of the hatch window to let the viewers see the small particles that were accompanying them in space.

"It looks like a snow storm," said Carr.

"I think what that is," Conrad ventured, "is ice from these water and urine dumps and everything that has collected on the vehicle, and then when we make a burn we shake them loose again."

After the television equipment had been stowed away, Gordon re-established the passive thermal control attitude and initiated the roll. He resumed the recharging of battery A, to fully restore it from the load it carried after the lightning strikes. When the master alarm sounded, Carr explained that this was due to a higher than nominal oxygen flow to maintain the cabin at the desired pressure of 5.0 psia. The increased consumption was in part due to the redesigned urine receptacle, which tended not to drain very well, prompting the crew to leave its nozzle slightly open to vacuum. The activities for the remainder of the day were light. As Conrad told Carr, "We're going to clean up the spacecraft, and I think the three of us will shave. Then we want to get some exercise." Other than using the struts and the flat areas in the lower equipment bay for doing push-ups and arm-pulls, there was a 'gym' which each crewman used several times per day. As Conrad reported after the mission, "I would exercise until I was just getting warm. I didn't want to exercise heavily enough to really perspire, because I wanted to keep my clothes as clean as possible. I think we exercised for periods longer than half an hour, but at a slow, steady pace."

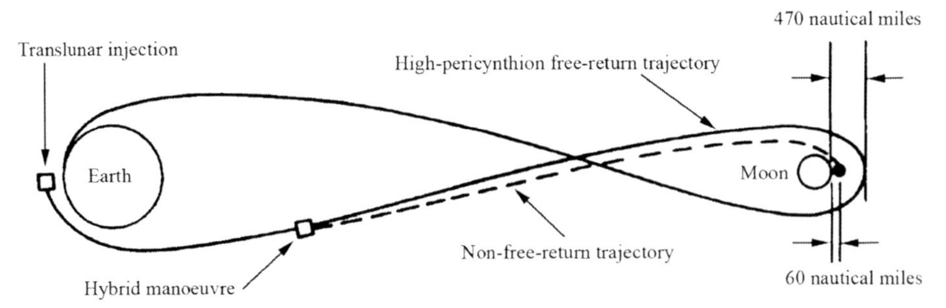

How the high-pericynthion free-return trajectory of the translunar injection was transformed into a non-free-return trajectory with a more efficient lunar orbit insertion.

With music playing in the background, Conrad reported, "We are really having a ball up here."

"All dressed up and no place to go," Carr replied.

"Oh, we're going someplace," Conrad assured. "We can see it getting bigger and bigger all the time." Whereas the crew of Apollo 11 had been unable to observe the Moon on their trip out due to their landing site being east of the lunar meridian and the Moon being close to the Sun in the sky, in this case the western target increased the angular separation of the two bodies.

After receiving some sports news, including the fact that the race in Phoenix was to be rerun tomorrow, Conrad asked, "How are our families doing, Gerry?"

"I haven't talked to them," Carr answered, "but I'll make a few calls and give you some answers."

"Every day, if you could," Conrad suggested.

They were now 120,000 nautical miles from Earth, which appeared no larger than the size of a golf ball at arm's length. "Hey, Gerry," Gordon called, "somebody has probably already said this before, but that place looks like an oasis down there."

An hour after Conrad's enquiry about their families, Carr was able to report, "Al, I checked home for you. Sue got back this afternoon from the Cape. She said to say that yesterday was a pretty exciting day for her; kind of wet, too. And she wanted to let you know that she attended a birthday dinner party for Vice President Agnew and had a real fine time. She says her Mom and Dad are with her at home now. They just finished dinner and the kids are headed for bed, and she wishes you Godspeed."

"Thanks, Gerry," Bean replied.

"I guess Jane and the boys are out to dinner," Carr told Conrad. "I haven't been able to get ahold of them yet." As for Gordon's family, "Barbara says that Barb and Norman and your mother are all there now, everybody's doing fine, and they've had enough excitement for a while, thanks. And if you don't mind, they'd just as soon you lay off the spectaculars."

"Why? What in the world has happened?" Gordon asked innocently.

Carr laughed, "She said Aunt Dorothy is passing through Houston now, and she called in to say 'hello'. She said also to let you know that Bill Der Bing is there with them and is doing a great job." Bill Der Bing was a Public Affairs staffer.

"Very good, thank-you," said Gordon.

When Conrad reported that they had an object in sight that was holding a fixed position relative to the spacecraft at a particular point in the passive thermal control roll and appeared to be tumbling, Carr said it could not be the S-IVB because that was on the opposite side of the sky and in any case was 2,500 nautical miles away – too far to be seen by the unaided eye. It wasn't clear what the object was, but similar sightings had been made by earlier crews.

"We'll assume it's friendly anyway," Gordon announced.

At Mission Control's request, Gordon ended the recharge of battery A. Now that all the spacecraft's batteries were fully replenished, the original topping up schedule would be reinstated.

Cliff Charlesworth took over as Flight Director and Ed Gibson as Capcom. After a lengthy period of silence from the spacecraft, Gibson asked, "You folks have been pretty quiet. What's up?"

"We're just exercising, listening to the tape recorder and looking at Earth and the Moon," Conrad replied.

"How's the cloud cover over the Pacific?"

"Australia is real clear again," Conrad replied.

"What's the smallest piece of land you can pick out?" Gibson asked. "Can you see any of the Pacific islands?"

"No, we can't see any of them," Conrad said. "Dick says he can see Borneo right now. He's looking through the monocular."

"You can't see it with your naked eye, though," Gordon pointed out.

"We're not looking at that much of an Earth," Conrad said, referring to the phase of the globe, "and the region of the Pacific that is close to the terminator has a very shallow Sun angle on it and is fairly washed out."

"You still getting a pretty good glint off the surface then?" Gibson asked.

"Yes. Dick says the subsolar point is just west of Australia. By the way, how far out are we now?"

"You're 137,720 nautical miles out and going along at 3,580 feet per second," Gibson advised. Several minutes later he was back, "Al, is that Moon beginning to look a little bigger to you now?"

"It sure is," Bean replied. "We were watching it through the sextant a while ago, and the features on the Moon are much more pronounced than we ever saw before we left Earth." The terminator was a few degrees east of the meridian, cutting across the dark plain of the Sea of Serenity. "It's very stark and beautiful from this point of view. I imagine tomorrow it's going to be even more impressive."

"It'll probably be very impressive from a distance of around 3 or 4 feet," Gibson teased, alluding to moonwalking.

"That's true!" Bean laughed.

After initiating a water dump and a fuel cell purge at Mission Control's request, Gordon reported, "I don't think that water dump did the PTC any good at all." The roll had developed a pronounced wobble.

"In looking at your angles here," Gibson replied, "it looks as though you might be touch-and-go during the sleep period if you don't reinitialise the PTC. When you do that, we would like you to reinitialise the DAP to 0.2 degrees per second." With the lunar module on the nose, making fast turns was consuming an unwarranted amount of propellant, so they were to reset the digital autopilot to employ a slower and more efficient rate of rotation. While the dumps continued, Mission Control uploaded a new state vector and inspected the erasable memory of the computer, then Gordon cancelled the unstable roll, restored the desired attitude perpendicular to the ecliptic and restarted the roll for passive thermal control just in time for supper.

The second flight day had been fairly relaxed because the astronauts did not want to arrive in lunar orbit in an exhausted state. This time, Gordon and Bean rigged the sleeping bags beneath the couches and Conrad slept strapped into a couch. To please the Flight Surgeon, Conrad and Bean provided biomedical telemetry overnight.

INTREPID AGAIN

To get flight day 3 off to a good start, Paul Weitz piped up a rendition of 'Reveille' played on the bugle.

"We're just stirring," Conrad replied. "We'll be with you in a few minutes."

After a sleep report from the crew and a consumables update from Earth, Weitz played 'First Call To Formation', prompting Conrad to assure him, "Everybody is at attention in here."

"Let me know when you're settled in the breakfast nook," Weitz said, "and I'll give you your morning news." A few minutes later, he continued, "The news reports on the flight of Apollo 12 are highlighting yesterday's midcourse correction and the fact that the flight is moving along smoothly. One wire-service story mentions the improvements in the menu on this flight. Both local and network stations featured video-taped highlights of your television show yesterday. Prayers for the mission's success were said in churches everywhere. In Houston, Mayor Louie Welch won his fourth consecutive term by defeating five candidates. He got 53 per cent of the vote. A minimum housing code and a freedom-of-choice integration plan were strongly supported. That's about it for news."

As Pete Frank took over as Flight Director, Weitz relayed the sports scores and then handed over the communications console to Don Lind, who updated Conrad on the race that had been rerun, "The Phoenix 200 is over. We still have only unofficial results, but the unofficial results show Al Unser as the winner. Number 2 is Ruby and number 3 is Dallenbach."

"Thank-you very much," Conrad replied.

"Everybody has had breakfast, brushed their teeth, combed their hair, and we're even thinking about shaving today," Gordon reported.

"You're all cleaned up and nowhere to go," Lind said.

"We're going somewhere," Conrad replied, adding, "We're just not sure where."

"We are," Lind assured.

With the spacecraft 165,000 nautical miles from Earth, Gordon performed a P52 platform alignment.

The main item on the flight plan for the day was to be opening the lunar module, which was to be televised. Conrad called with a request, "If we could get a high-gain antenna angle that puts the Sun in our centre hatch, so we could get as much light as possible into the command module, we would like to show the removal of the hatch, probe and drogue and then take the TV over into the LM."

After promising to provide the attitude information, Lind pointed out, "We'll get the first 29 minutes of that presentation live through Goldstone; the rest of it will be recorded at Honeysuckle and shipped back to us. We'll get it in several days."

"We'll move it up," Conrad decided, to avoid losing the live feed to Houston. As scheduled, the transmission would have switched part way through from Goldstone to Honeysuckle Creek, which did not currently have a satellite relay. Advancing it by half an hour would also enable the American commercial television networks to carry the entire transmission, although it would not start until well after midnight.

At this point, Bean called, "We had a can of tunafish spread salad last night, and there's about a half a can left today. That stuff's still good to eat, isn't it?"

"The Flight Surgeon suggests that you try a new can," Lind advised.

"We just opened it last night. Are you sure it isn't all right?" Bean persisted.

Several minutes later Lind was back, "We're checking with some people whether there is any problem with the tunafish. Why don't you hold off eating it until we get a better answer for you?"

"Okay," Bean said.

After 15 minutes Lind called again, "You can't imagine what consternation your tunafish question has raised. We have a wide diversity of opinion. The majority says throw it away."

The Public Affairs Officer duly reported to his listening audience, "We've closed the tunafish question."

Half an hour later Bean asked, "Do you have a report on our families today?"

"Negative," replied Lind. "We'll see if we can find out what's been going on."

"I'll tell you what *we've* been doing since we got up this morning, just for your information," Conrad reported, speaking for his all-Navy crew. "We've cleaned the spacecraft fore and aft and all lower decks and ladders, and cleaned up the garbage and re-stowed everything. Everybody has had a bath and shaved. And Al's studying the Moon. I'm studying the lunar descent. Dick has been fitting the S-158 camera package, making sure it fits in the hatch window and works. All that has occupied us for about the last 3 hours."

"You sound incredibly neat," replied Lind.

"We haven't got much else to do, pal," Conrad observed.

After half an hour of telephoning around, Lind announced, "Pete, Jane says they have had a very quiet Sunday afternoon there. Everybody is home, and everybody is well. There just really isn't much excitement going on. It's just been a very quiet afternoon over at your house."

"Okay," acknowledged Conrad.

"Barbara reports to you, Dick, that Sharon and Lynn Diamond are over for the evening. They're expecting the Irwin's over momentarily, and Jim McDivitt has just left. She says the boys have gone back to school and she thought things were going to be pretty quiet, but between Barbara and Karen she's got so many giggling girls around there that it's more noisy than she thought. The other thing she pointed out, was that Father Connolly had been over in the afternoon and had noticed Barbara having a nap. So coming out of church this evening, in front of some of the members of the congregation, he said that the last time he had seen her she was asleep. And this was very embarrassing to her at the moment."

"Well, that's better than what he could have said," interjected Gordon.

Lind laughed. "Also, Barbara had the comment for the whole crew that she is a little disappointed in how much you're talking – she expected more conversation out of you than she's been hearing lately."

"We're talking," Gordon said. "She's just not hearing."

"She also requests that when you talk, you should try to be a little funnier," Lind said. Moving on, "Al, your wife reported that the family are missing you, and they are extremely proud of how the flight is going along. They'll be watching tomorrow. She concurs with the decision on the day-old tunafish. That's about all the family has to report."

As his colleagues prepared lunch, Bean placed a dark glass on the monocular and inspected the Sun. He said that he could see several dark sunspots in the equatorial zone about 35 degrees from the western limb of the solar disk. Lind pointed out that spots in that position would pose no threat because a spot near the western limb was receding from view as the Sun rotated on its axis and any flare which it might issue would not be directed Earthward.[18] When Lind congratulated them on their interest in solar astronomy, Bean responded, "We've been studying astronomy, geography, geology and a few other things up here." A minute later he explained, "Really there isn't a lot to do on the way out to the Moon. You have got your systems to monitor, and you eat, keep yourself clean and get some sleep. And except for that, you're free to look out the window and to study the checklists and maps and things that you're going to be using when you get into lunar orbit."

Tom Stafford came on the line, announcing himself as "your old Social Director speaking".

"What are you doing up so late, Social Director?" Conrad asked.

"Oh, I just thought I'd keep check on you, Pete," Stafford replied. "It looks like everything is just going great."

"There's not too many places we can go," Conrad mused.

"It's quite a view out there, isn't it?" Stafford said. As commander of Apollo 10, he was familiar with cislunar space.

"Sure is," Conrad agreed.

Although travelling very rapidly at the moment of translunar injection, the Earth's gravity had been progressively retarding the climbing spacecraft. But of course it had sufficient energy to reach lunar distance. In fact, translunar injection was not a true 'escape' burn, and the trajectory was not aimed at the Moon as much as at where the Moon would be as the spacecraft approached lunar distance. If not for the presence of the Moon at that time, the trajectory would continue out and peak at a point some distance beyond the radius of the Moon's orbit, then fall back in a highly elliptical orbit. The trajectory was designed to make Apollo 12 pass just ahead of the Moon, in order that lunar gravity would draw the spacecraft in, to pass around the far-side. Ironically, therefore, although the terminator had just crossed the meridian into the western hemisphere, the fact that the leading hemisphere was in darkness meant that as they approached the Moon the astronauts were able to see less and less of it.

When Stafford handed the microphone back to Lind, Gordon reported, "Pete and I just polished off four frankfurters and some applesauce. Al Bean hasn't quite eaten his yet. This food up here has been really good."

[18] By tradition, the western limb of the Sun is the one facing west when the Sun is viewed from the Earth's surface. This used to be the case for the Moon, but in 1961 the International Astronomical Union changed it so that an astronaut on the lunar surface would see the Sun rise in the east.

"I'm sitting down here without anything to eat," Lind mused. "Just remember to do your mild exercise today, so you don't gain weight."

After lunch, a waste water dump perturbed the passive thermal control roll. Dave Scott, the commander of the backup crew, was sitting in as Capcom and relayed the decision to ignore the instability since they were soon to halt the roll and adopt the attitude for the television transmission. Meanwhile, Cliff Charlesworth took over as Flight Director and Gerry Carr as Capcom.

"When do you want us to start this show?" Conrad asked, once they had adopted the required attitude.

"We're ready whenever you are," Carr said.

In preparation for the intravehicular transfer, the command module was pumped up so that the valve in the apex hatch could be opened to allow air to flow through the tunnel to restore the pressure of the lunar module cabin to the desired value.

Approaching 63 hours into the mission, Carr called, "Lights! Camera! Action!"

Conrad provided the commentary, "As you can see, Dick is up there in the tunnel. He is opening the vent valve now, to pressurise the LM." The lighting was excellent. "Out our number 1 window we have the Earth. Out the number 3 window we have the Sun shining in. And out the number 5 window we have the Moon." In the current attitude, the forward-facing rendezvous windows viewed open space.

"And the hatch is open," Bean commentated a minute later.

"We have a good picture of the hatch," Carr confirmed.

"This big old hatch wrestles pretty easily here in zero-g," Bean observed, as they stowed the hatch in a protective bag.

"It beats one-g all to heck, doesn't it?" Carr said.

"Sure does, Gerry," Bean agreed.

Gordon stowed the hatch under the left-hand couch, then went up into the tunnel to remove the probe. On the flight plan this would have been the first time that they did this, but as he wryly pointed out, "we've had it in and out of there several times already".

"The lighting in the tunnel is just a tad dim," Carr remarked, "but we can make out what you're doing."

"That's wide open on the f-stop now," Bean said. "I think part of the problem is there is a lot of contrast between that dark tunnel and the white garment of Dick's. It's giving the camera a little fit."

While Gordon stowed the probe under the right-hand couch, Conrad went into the tunnel to fetch the drogue.

"You got any late-night watchers?" Gordon asked. It was 2 a.m. in Houston.

"That's affirmative," Carr confirmed. There were people in the viewing gallery at the rear of the MOCR.

"It sounds like you're handling empty milk cans up there," Carr said, referring to the clanking noise as Conrad passed the drogue down to Gordon.

Conrad opened the lunar module's roof hatch, dragging hoses with him. "We're putting these CSM hoses down inside the LM, since there's no ventilation in there now. We'll just lay them around down there where we're going to be working and it will make it real nice and cool and gives us clean air down in there."

Bean took up position in the tunnel, pointing the camera through the hatch into the lunar module's cabin. "We're getting a good view of the main control panels, upside down," Carr reported.

"The way we stopped," said Bean, referring to the attitude of the vehicle, "has the Sun coming in through one window, and it's heating it up pretty good in here." Bean gave the camera to Conrad and entered the cabin to start stowing the equipment that Gordon was passing through the tunnel.

After a tour of the cramped lunar module, Conrad aimed the camera back through the tunnel, showing Gordon in the command module hatch. "Pete," Carr called, "are you going to let Dick get any LM time?"

"He's been in and out with us," Conrad replied.

"They have got to let me get LM time, Gerry," Gordon quipped, "since they get command module time."

"That's right," Carr agreed. "Equal time."

When Bean announced that he was finished, Conrad asked Carr if anyone there had anything else for them to do, and there was not, so they sealed the lunar module. Back in the command module, Conrad pointed the camera at the Moon. Although to a terrestrial observer the Moon was currently a 'half' phase, from the vantage point of the spacecraft it was a crescent.

With ventilation hoses, the television cable and communications umbilicals in a tangle, Bean said that the cabin looked "like a snake pit".

Switching the camera over to Earth, on which Australia was centrally located and Antarctica was prominent, Conrad said, "I know it's been said before, but this is a really spectacular sight." At this point, Apollo 12 was almost 178,000 nautical miles from Earth.

To round out the transmission they showed Gordon reinstalling the hardware into the tunnel, leaving the probe loose, just for stowage. "After all the wrestling we have done with that probe in practice, it's really amazing to see that big thing float around in here," Bean pointed out.

As Apollo 12 was about to pass below the horizon at Goldstone, terminating the television transmission, Gordon spoke to the general audience, "Gerry, I have some comments to the folks at home: We've enjoyed doing this for them. All three of us are in good spirits, we're feeling great, we've exercised and slept well, the food's been good, we've lots of nice cold water to drink, and we are enjoying the scenery. However, I'll tell you one thing: we do miss the good people back home." With the television stowed, he resumed the PTC roll.

The day's major activities behind them, it was time for a little exercise and then supper. As Conrad was doing his workout, he ripped loose the EKG sensor. Flight Surgeon John Zieglschmid pointed out the loss of data to Carr, who asked Conrad to attend to the sensor. Three days prior to launch, each crewman had taken a laxative to empty his bowel, and the space food was designed to maximise the time between bowel movements. Almost on cue, each man made his first bowel movement on the third flight day.

At 68 hours 30 minutes, Apollo 12 reached the point at which the gravitational attractions of Earth and the Moon were matched, and pulling in opposite directions.

By Isaac Newton's law of gravity and the fact that the mass of the Moon is just 1.2 per cent of the Earth, the neutral point is 83 per cent of the way to the Moon in terms of distance. However, although the ratio is fixed, the actual distances vary because the orbit of the Moon is elliptical. On crossing this point, the spacecraft passed from the terrestrial sphere of influence into the lunar sphere of influence. Consequently, instead of continuing to slow as it 'climbed' from Earth it began to accelerate as it 'fell' towards the Moon.

After the final housekeeping activities, including halting the deteriorating passive thermal control roll and restarting it, the crew retired for the night, this time without providing biomedical telemetry.

Gerry Griffin took over as Flight Director for the graveyard shift, and he and his team of flight controllers watched a replay of the recent television transmission. The flight plan included the option of a midcourse correction several hours prior to lunar orbit insertion, but the trajectory was satisfactory and it had been decided to extend the sleep period by 2 hours to allow the crew the best possible rest in preparation for what promised to be a very busy day.

INTO LUNAR ORBIT

With the clock approaching 78 hours, Ed Gibson piped up 'Reveille' as the wake-up call and Gordon reported, "All present and accounted for, sir,"

As the crew had their breakfast, Gibson read up the morning news. "The flight of Apollo 12 continues to maintain world-wide interest, and your television broadcasts are getting priority preference on the local and network newscasts. There are a lot of foreign press here at Houston Press Center. Incidentally, a baby boy was born to a mother in Baltimore, Maryland, at the precise time of your liftoff. His name is Charles Richard Alan Wilson." Charles, Richard and Alan were the first names of the crew. Gibson wrapped up with a comprehensive sports report.

At this point, Glynn S. Lunney took over as Flight Director. While working at the Lewis Laboratory of the National Advisory Committee for Aeronautics he gained a degree in aeronautical engineering from the University of Detroit. A month after he graduated, Lewis became part of NASA and he became the youngest member of the newly established Space Task Group. After serving as a Flight Dynamics Officer for the Mercury program he was promoted to Flight Director.

"Boy, that Moon looks big today, Houston," Conrad called. "It's about the size of a baseball held at arm's length. And you can see all the mountains and craters. It's really a beautiful sight. We're starting to move on the far side of it from the Sun, so we only see about an eighth of it illuminated by the Sun. But that eighth of it is really stark. Particularly up near the poles, on the limb, you can start to see that it's not a nice smooth ball anymore. It's got some little ridges and bumps that would be mountains or craters if you could see them right head on."

"It's a good sign if it's looking bigger," replied Paul Weitz, having taken over as Capcom. At that time, Apollo 12 was 11,000 nautical miles from the Moon and had accelerated to 4,000 feet per second.

Glynn Lunney.

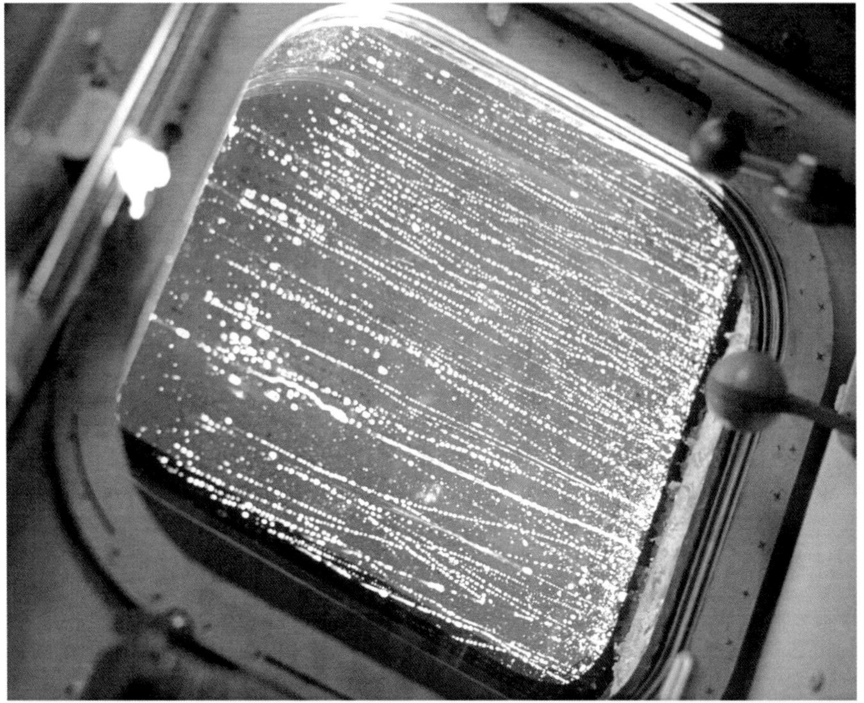

The launch phase contamination coating the round hatch window and the commander's square window.

"On the other hand, the Earth looks like about the size of a quarter held at arm's length, which is pretty small," Conrad added.

As they worked through housekeeping chores, Weitz asked them to power up the transponder in the command module for the lunar module's rendezvous radar. The self-test on the transponder confirmed that it had not been damaged by the lightning at launch.

"Houston," Bean called, "we're closing to the Moon fast enough now that every time we do a 360 degree roll and pick it back up in the windows again you can see it grow quite considerably." They were 7,500 nautical miles from the Moon, which was currently about the size of a volleyball at arm's length.

"The next trip past 300 degrees, we'll stop the roll," Conrad reported. As this was being done, Lunney polled his team about lunar orbit insertion, then ordered a new REFSMMAT be uploaded into the spacecraft's computer, this one defined relative to the intended lunar landing site. Gordon then took star sightings for a platform alignment in readiness for the lunar orbit insertion manoeuvre.

The flight plan called for a television transmission showing the approach to the Moon, but from the spacecraft's perspective the Sun was so near the Moon in the sky that it was highlighting the contamination on the windows. After switching on the camera and inspecting the monitor screen, Conrad reported, "All that the TV is doing, is an excellent job of picking up all the droplets and glare and rivulets on the window. It's pretty hopeless." They didn't make the transmission. "This is quite a sight," Conrad enthused. "Our motion to the left is not as apparent as our motion towards the Moon, giving us the decided impression that we're going right into the centre of that baby!" The trajectory required precise navigation, because the Moon is 1,877 nautical miles in diameter and its mean velocity as it pursues its orbit around Earth is 3,380 feet per second.

"We'll check it out for you," Weitz offered.

"No, I trust you," Conrad chuckled. "It's really a shame we can't show you this, because we're dropping behind it in a hurry with respect to the Sun and have only about 2 degrees of a crescent Moon right now. We can see the illuminated portion get smaller and smaller. And of course, it's filling more and more of the window."

As the range reduced to 4,000 nautical miles, it was time to adopt the attitude for lunar orbit insertion. Although it would have been possible to enter lunar orbit by executing a single manoeuvre, if the service propulsion system engine were to fire for too long then the spacecraft might thereafter dip so low as to crash. Because the insertion could only occur on the far side of the Moon, out of communication, in planning Apollo 8 it had been decided to guard against this outcome by splitting the manoeuvre. The initial burn (LOI-1) would produce an orbit with a high apolune on the near-side, and the Manned Space Flight Network would track the spacecraft during the first two revolutions to make a precise calculation of the delta-V for the follow-on burn (LOI-2) that would circularise the orbit at the desired altitude. This strategy had been retained by the later missions. Throughout the morning, Houston had been providing revised data for the lunar orbit insertion manoeuvre and, since Apollo 12 had departed from the free-return trajectory, data on the various abort options in the event of a failed insertion.

With an hour remaining to the time that Apollo 12 would pass behind the Moon, Chris Kraft, George Low, Jim McDivitt and Rocco Petrone settled into the rearmost consoles of the Mission Operations Control Room.

Once in the Moon's shadow, Gordon verified the burn attitude by taking a sextant sighting on a star.

"You're Go for LOI," Weitz confirmed.

Conrad acknowledged, and said they were holding at the 6-minute point on the burn checklist. All loose items had been stowed safely away and the urine vent valve closed in order to prevent the distraction of excessive oxygen flow triggering a master alarm.

Loss of signal occurred right on schedule. At that moment, the spacecraft was at an altitude of 472 nautical miles and travelling at 7,188 feet per second. If the burn were not made, then Apollo 12 would reappear around the trailing limb after about 24 minutes 50 seconds, but if it braked into the desired orbit then the signal would resume after 32 minutes 11 seconds. If the SPS engine failed to ignite, or if its burn was interrupted, the docked combination would emerge from behind the Moon on a non-free-return and the abort action would depend upon the timing. If the LOI-1 manoeuvre failed within 90 seconds of its scheduled start time, Conrad and Bean were to scramble into the lunar module, power it up and fire its descent engine 30 minutes after the scheduled time of LOI-1 in order to adopt a trajectory that would approximate a free-return. If the delta-V capacity of that engine was insufficient to establish the desired transearth trajectory, then ideally an SPS burn would finish the task 2.5 hours later; but if the SPS remained unavailable the descent stage of the lunar module would be jettisoned to use the ascent engine. If LOI-1 was interrupted between 90 and 170 seconds after igniting the SPS, the vehicle would end up in a highly elliptical capture orbit and the abort would be achieved by firing the descent engine twice on successive revolutions: once to stretch the orbit and again to break away and head for home. If the SPS had fired for longer than 170 seconds in the LOI-1 manoeuvre, the abort would be achieved by a single firing of the descent engine after spending only one revolution in the capture orbit.

As Conrad reflected later, when it was decided to adopt the hybrid trajectory the abort scenario, albeit limited, was fairly simple, but as the options were explored "it got exercised into a very complicated thing". It was good training though, "because we had a lot of confidence before we went that we could burn both the DPS and the APS to get ourselves out of trouble".

With P40 in the computer ready to ignite the SPS, the stabilisation and control system was aligned to the inertial measurement unit so that if the computer were to fail during the manoeuvre then this backup system would be able to hold the vehicle stable while the burn was completed manually.

Moments before Apollo 12 passed into sunlight, Conrad and Bean donned their ball caps in order not to be dazzled by sunlight reflecting off the roof of the LM and gleaming in the rendezvous windows. Bean noted that wearing the hats had the additional benefit of blocking their view out of the hatch window, meaning that they would not be distracted by the sight of the lunar surface passing by.

As previously, there was no need for an ullage burn to settle the SPS propellants

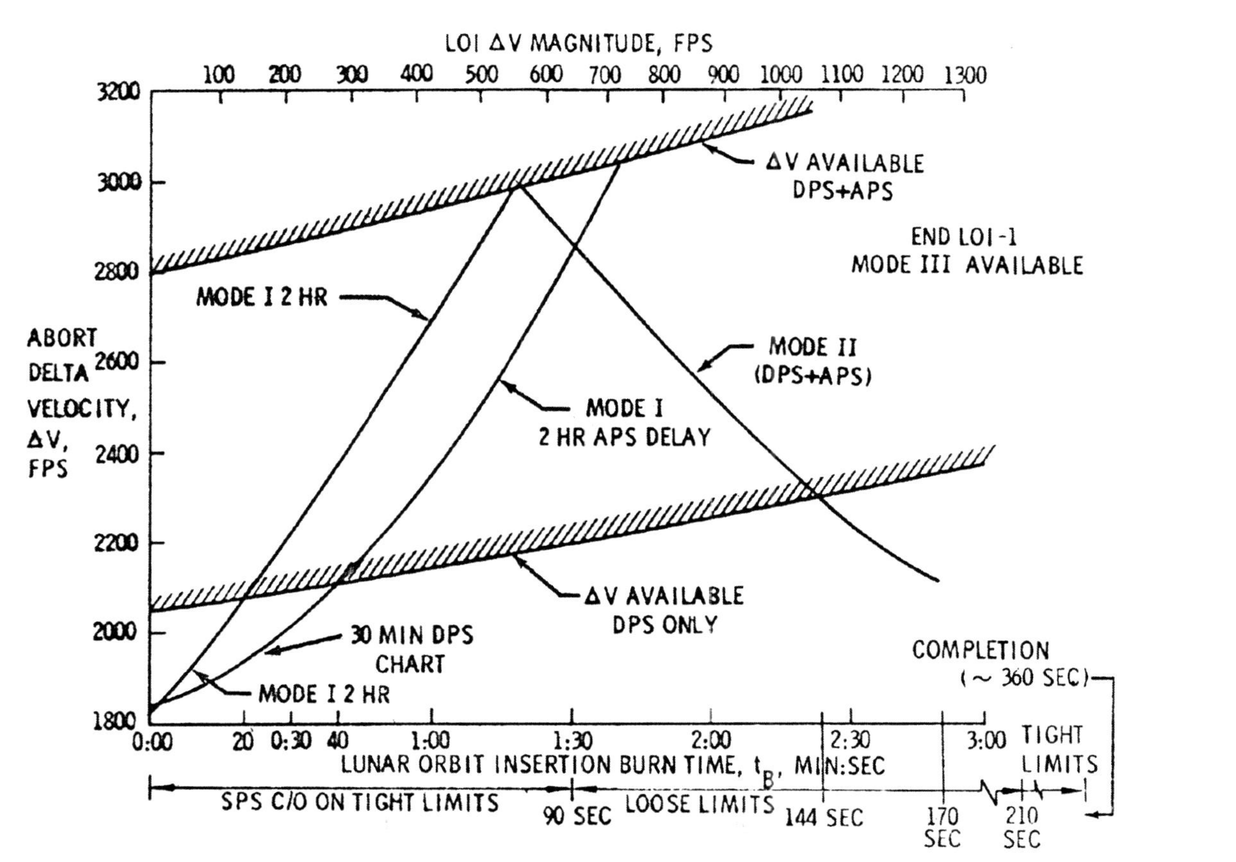

The complicated rules for recovering from a failed LOI-1 burn, with the actions to be undertaken being dependent on how far that manoeuvre progressed.

in their tanks, and at the calculated moment the computer ignited the engine, thereby concluding the translunar coast which had lasted 80 hours 38 minutes 1.67 seconds.

"Ignition," confirmed Conrad.

Bean reported that the second set of propellant feed valves had opened to back up the primary valves.

"Nice and smooth," Conrad observed.

"Twenty seconds," called Bean. "You're right on the money."

"Everything looks good over here," Conrad added.

"Roll's good," Gordon noted.

"Don't worry about roll," Bean said. So long as the engine gimbals were directing the thrust through the vehicle's centre of mass a small amount of roll did not matter.

"Okay, gang," Conrad announced. "We're out of Mode 1 and into Mode 1-A."

The propellant pressures and the combustion chamber pressure all looked good, and the gimbals were maintaining the thrust aligned.

"Mode 2," called Conrad.

"Loose rules now," Gordon noted.

As Conrad explained after the flight, "there were too many abort modes to call out in the cockpit, so we went on the tight/loose scheme: whenever we were in the tight rules we'd call 'tight', and when we were in loose rules we'd call 'loose', and in that way we let the modes fall out the way they would".

The engine continued to perform as it should.

"We're in Mode 3," Conrad called. "We're still in the loose rules." Half a minute later, "Tight rules."

"That balance looks good," said Bean, referring to the propellant mix indicator. The nitrogen that was holding open the feed valves was stable.

"It looks to me like we got a hot engine," Conrad said, having checked the actual rate of change in velocity against the values listed on a cue chart, "and we're going to shut down early."

"Okay," acknowledged Gordon.

"About 5 seconds early," Conrad estimated.

"100 per cent chamber pressure," Gordon noted.

"It's going to shut down 6 seconds early," Conrad decided. He started counting down, and the computer cut off the engine precisely as he called the event.

With a duration of 5 minutes 52 seconds, it slowed the vehicle by 2,889.5 feet per second with residuals of less than 0.1 foot per second in each of the three axes. The altitude was 82.5 nautical miles at the start of the burn, and 61.7 nautical miles at its end. The intended capture orbit was 62.3 × 169.3 nautical miles, with its high point on the near side of the Moon. The onboard solution for the orbit achieved was 61.8 × 170 nautical miles. The selenographical coordinates were 5.7°N, 175.6°E, at the start of the burn and 1.6°S, 154.0°E, at its conclusion. This marked the start of the mission's first revolution of the Moon. Whereas the spin axis of the Earth is inclined 23.5 degrees to the ecliptic, the Moon's axis is within 1.6 degrees of this plane. The spacecraft's orbit was inclined at 14.5 degrees to the lunar equator, with the latitude of its most southerly point on the near-side in the eastern hemisphere. The weight of the docked combination was 96,076 pounds prior to the burn; on completion it was 72,225 pounds, the difference being the propellant consumed.

As they worked through the post-burn checklist, Bean was first to peer out of the window, "Look at that Moon!"

"Son of a gun," said Conrad.

"Look at the size of some of those craters," Bean enthused. Although he resumed the checklist, he kept sneaking glimpses of the landscape passing by. "Man! Look, at that place."

"That's a God-forsaken place!" exclaimed Conrad. "But it's beautiful, isn't it?"

"It's good to be here, is all I can say," Gordon said.

Conrad noted that because the Moon was not very large, the horizon had "a nice arch to it".

Gordon rolled 180 degrees, pitched the nose down and initiated Orb Rate so that the docked combination would rotate more or less in time with its travel around the Moon, thereby maintaining the surface in the command module hatch. At this point they were 10 minutes from crossing the trailing limb of the Moon. Several minutes later there was a collective sigh of relief in Mission Control when the time passed at which the signal would have been reacquired if the SPS had failed to ignite.

"It doesn't look like we're at 60 miles, does it?" Conrad mused.

"That's because the craters are so much bigger than anything you've ever seen," Bean suggested.

Although they had a map of the planned ground track, in these first minutes they had a poor sense of situational awareness. "We're right around here," Conrad said. "We're looking backwards, so we ought to turn this map around onto the way we're going." To confirm that they had the map the right way around, he said, "Out your side you ought to see a huge old crater."

"Where?" Bean asked.

"Right out there," Conrad indicated.

"I do!" Bean confirmed. "A monster."

"The secret," Conrad observed, "is to point the map down towards the LEB." The lower equipment bay was on the opposite side of the cabin from the hatch.

"There's a huge monster with a central peak," Gordon called.

Bean recognised it because part of its floor was covered by a patch of mare, but had trouble with the Russian name, "That's the one you said was called Tsi - Tsi - Oh, yes. Tsiolkovsky." He scrambled to fetch a Hasselblad.

"What's the big hurry, Beano?" Gordon asked.

"Take pictures," Bean replied.

"Take pictures next rev," Conrad said.

As the spacecraft crossed further onto the illuminated hemisphere, the absence of shadows made it difficult to discern topographic detail and the light-toned terrain became dazzlingly bright. It was a stark contrast with the black sky. "I wonder if we ought to have sunglasses for this," Bean mused, pondering the extreme contrast. As Conrad later reported, "I used to wear sunglasses all the time in Gemini, orbiting the Earth and looking at the ground. I'm used to the changes in colour of the ground on

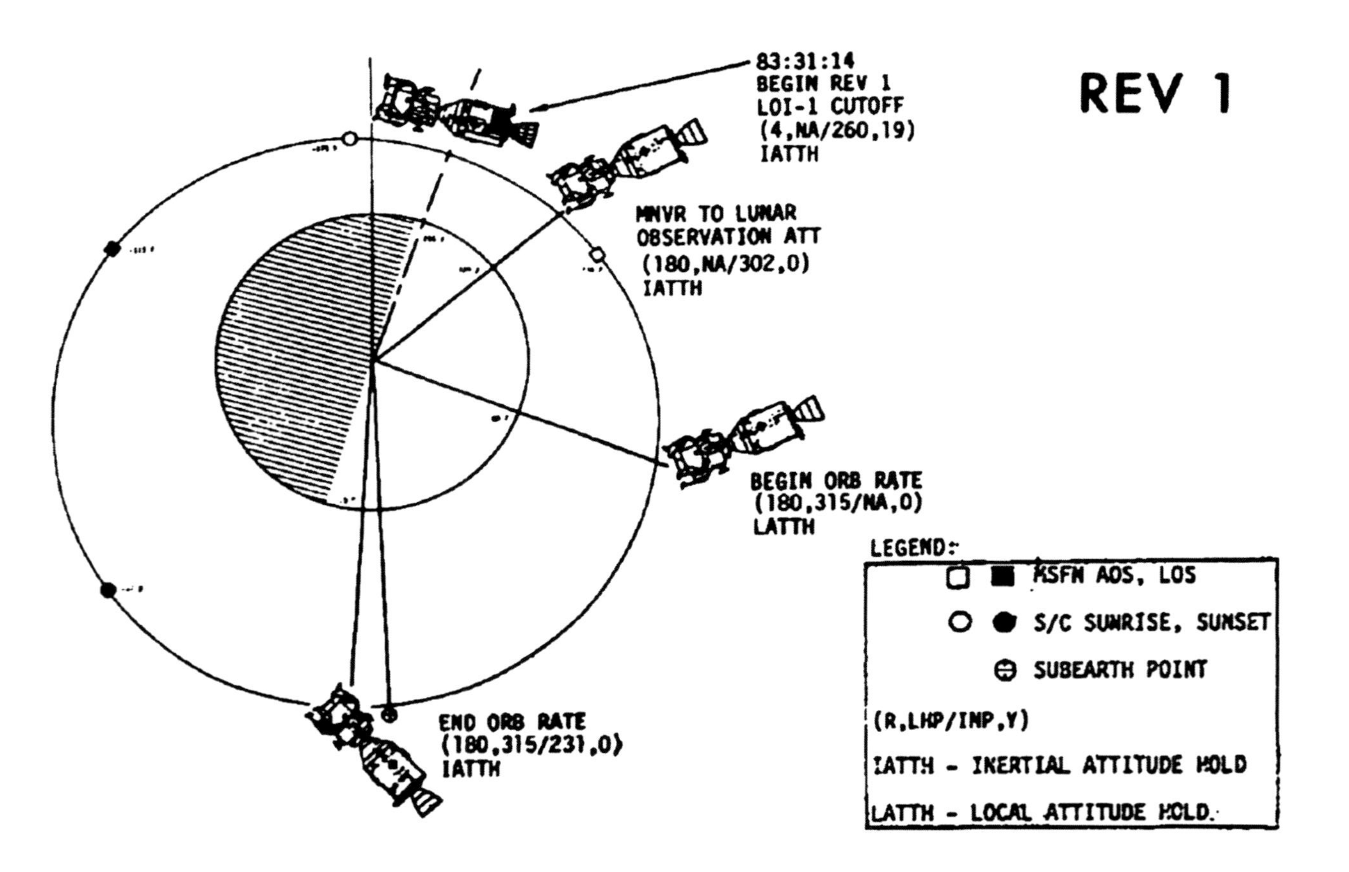

A page from the flight plan showing the manoeuvres to be performed during the first lunar revolution.

the Earth from flying aircraft and using sunglasses. I never used them in lunar orbit. I put them on a couple of times, but I didn't like the colour that it made the Moon. I felt it degraded my observations of the lunar surface, so I never wore them. I don't think that any of us felt it was so bright that we needed them." Wearing sunglasses would actually degrade scientific observations.

"We're climbing," Conrad pointed out just as communications were restored with Earth.

"Apollo 12, Houston," Weitz called.

"Yankee Clipper with Intrepid in tow has arrived on time," Conrad replied. After giving the burn status report, he added, "I guess like everybody else that just arrived in lunar orbit, all three of us are plastered to the windows, looking."

"Understand," said Weitz.

"To Navy troops, it doesn't look like a very good place to pull liberty, though," Bean chipped in.

Several minutes later, Conrad complimented the cislunar navigation, "That was an excellent long-range rifle shot you guys gave us."

As the television camera was being set up, Weitz asked about their impressions of the Moon from orbit. Conrad said that they were having difficulty gauging the sizes of the craters, "We're just sitting here discussing various sizes and getting ourselves oriented on the map."

On schedule at 84 hours, they began the television transmission about 60 degrees east of the meridian. Because the landing site was so far west, more of the near-side was illuminated than on previous missions; in fact, it was illuminated for fully 60 of the 85 minutes that it took to pass through the apolune of the capture orbit and cross the near-side.

While Gordon operated the television camera, Conrad inspected surface features using the monocular. Their ground track provided excellent views of the very large craters Langrenus and Petavius. Conrad reported that although the central peak of Langrenus looked smooth to the naked eye, the monocular showed tiny black dots which were evidently very large boulders.

They passed the camera to Bean for a view through another window. Weitz had asked about the colour of the surface, so Bean addressed this point, "It's changing slightly as we travel around it. At first, it had a very, very light grey-white concrete appearance. Now, it's still a light-grey concrete, but has a touch of brown in it. At least, that's the way it appears to me." After further thought, he added, "In fact, if I wanted to look at something that I thought was about the same colour as the Moon, I'd go out and look at my driveway!"

Bean passed the camera back to Gordon, and Conrad described a crater that was "the first one I've seen with fractures in the bottom of it [...]. The fractures run right through the middle of the crater including the rim, perpendicular to it, all the way across the crater, which gives me the feeling that the fracture pattern doesn't have anything to do with the crater."

As they passed over the Sea of Fertility, Conrad said it was only a little bit darker than the surrounding terrain, and compared it to "the beach sand down at Galveston whenever that is wet".

Weitz quipped, "We had a team of geologists checking out the concrete in Al's driveway. We'll send them to Galveston now."

"Okay," laughed Conrad. And then, "Looking down into a real fresh impact crater in the Sea of Fertility with the monocular, I can see some pretty large boulders. So I guess, as high as we are, if I can see those boulders, they must be pretty darn big!"

As they flew on, the large crater Theophilus came into view and Conrad reported that yet again the monocular revealed what appeared to be enormous boulders on its central peak.

Further on, Conrad noted something familiar about a trench-like feature, "One of the things we saw at the Cape when we talked with a geologist was an experiment he did by blowing air through sand, and sure enough I've got some examples of that right here in these trench-like structures." Some scientists believed that many of the lunar features were volcanic, but others disagreed; it was a long-running debate with each side considering its own ideas to be self-evidently true.

Gordon swung the camera north to show another one, "What you're seeing is a rille with a whole bunch of what looks like vent holes running along it."

Over the central highlands, Conrad offered a scientific speculation, "There's a whole bunch of areas in here that give you the feeling that, as we've talked about with the geologists, some of this is volcanic action."

The shadows lengthened as they crossed the meridian. To the south, the Straight Wall was prominently displayed. In making their approach to the Moon, with the illuminated portion diminishing to a thin crescent, Bean had noted that the peaks of isolated mountains on the mare were reminiscent of cumulus clouds over the ocean. Now viewing them from orbit and from the sunlit side, this impression persisted. "Do you see that very high mountain on the horizon, Houston? All you can see is reflected light." In reply, Weitz asked if Bean could open the aperture to improve the image, which he did, but once they were beyond the terminator at 16°W he had to end the transmission. The dark portion of the near side was visible to the astronauts in Earthlight, but the camera could not pick it up.

As the crew stowed the television camera, Weitz asked if they had any idea why the signal strength from the boom-mounted steerable S-band antenna had begun to oscillate, but they could offer no insight.[19] As they flew around the far-side several minutes later they were preparing lunch. While behind the Moon, they undertook housekeeping chores such as purging the fuel cells and dumping waste water. On emerging back into sunlight they did some more sightseeing of the rugged terrain on the far-side.

Conrad pointed to a row of craters. "You know what that is?"

"Vent tubes?" Gordon mused.

"Yes," Conrad said. "I haven't seen anything like that before."

Gordon grabbed a Hasselblad and snapped a photograph for the scientists. This was a remarkably audacious bit of geologising, as vent holes were of volcanic origin and, as such, contentious.

[19] This would prove to be a persistent problem.

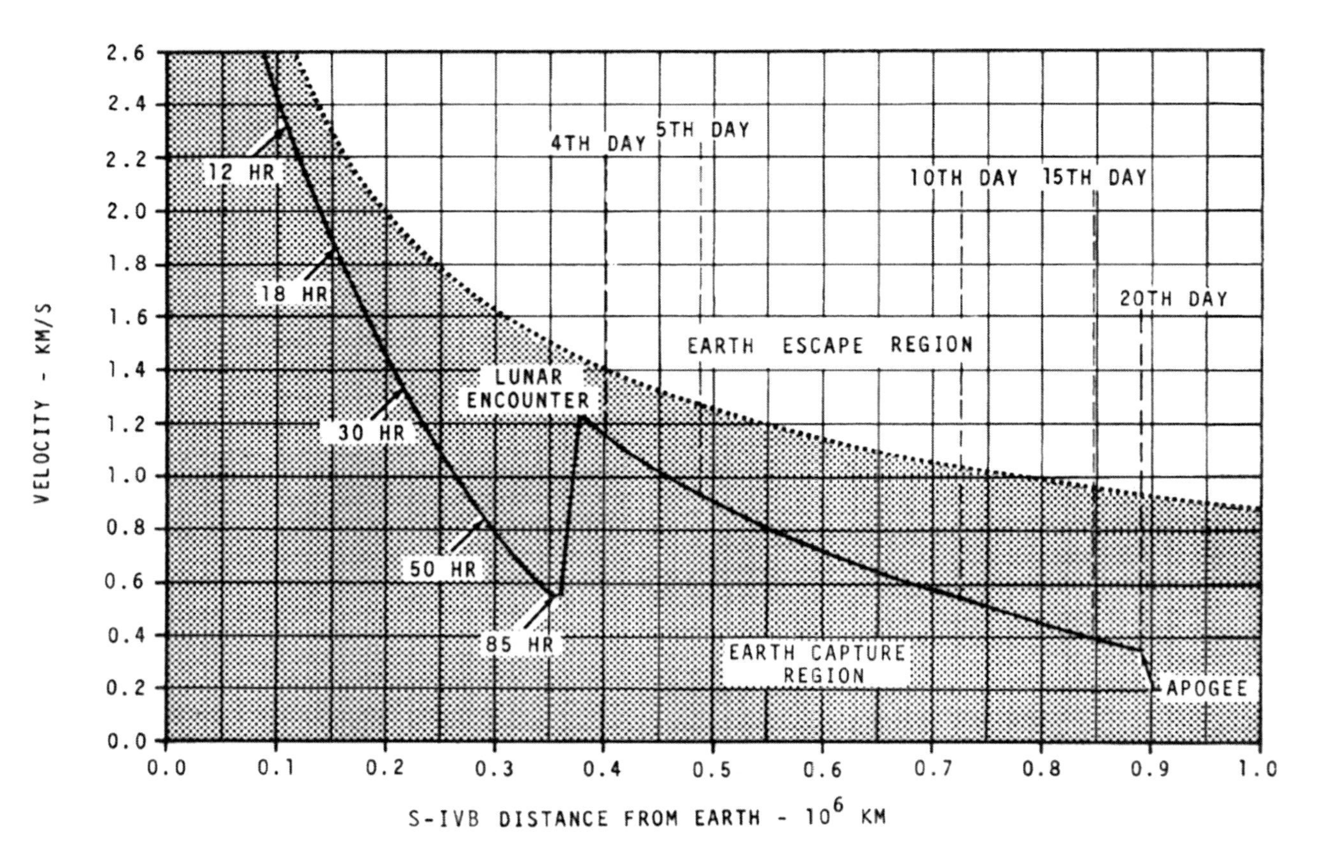

Owing to the off-nominal translunar injection, when the S-IVB flew by the Moon the gravity assist was insufficient for a slingshot into solar orbit.

At about this time, the spent S-IVB made its closest approach to the Moon at an altitude of 3,082 nautical miles. Due to the trajectory error, the lunar slingshot did not attain the desired heliocentric orbit and instead left the vehicle in a geocentric orbit which was almost coplanar with the orbit of the Moon and had a perigee at about 65,400 nautical miles, an apogee at about 483,500 nautical miles and a period of about 42 days – it was difficult to be precise because the S-IVB did not provide long-term radio-tracking, but the perigee was well inside the radius of the Moon's orbit and the apogee was far beyond it. It was inevitable that the S-IVB would be significantly perturbed by subsequent encounters with the Moon. When Canadian amateur astronomer Bill Yeung discovered a near-Earth object as a 16th magnitude speck of light in the constellation of Pisces on 3 September 2002, it was designated J002E3. Interestingly, its trajectory suggested that it had been captured by Earth from a heliocentric orbit. Spectroscopy indicated a painted surface similar to that of an S-IVB. Calculations suggested that the S-IVB of Apollo 12 had made nine or ten orbits of Earth and then passed through the sunward Lagrange point (L1) and escaped into solar orbit in March 1971. After 33 solar orbits, it passed through this region in April 2002 and was recaptured by Earth. After six orbits, it escaped again in June 2003.

As Apollo 12 emerged around the limb on revolution 2, Weitz resumed efforts to troubleshoot the S-band antenna problem. Having analysed the radio-tracking from the first near-side pass and measured the orbit to be 61.6 × 169.5 nautical miles, Mission Control uploaded a revised state vector and Weitz read up the data for the forthcoming LOI-2 manoeuvre.

There was no tourist commentary while passing over the illuminated near-side on this revolution – it was a working pass as the astronauts prepared for the SPS burn. In fact, some of the wonder of being in lunar orbit was already wearing off. Shortly after crossing the terminator into darkness, Gordon took star sightings to verify the alignment of the inertial platform in preparation for the manoeuvre.

After Glynn Lunney polled his flight controllers, Weitz told the crew they were "Go for LOI-2." After loss of signal, Pete Frank took over as Flight Director. Weitz would remain as Capcom.

Onboard, the final preparations were made to the accompaniment of music on the cassette player.

"If they made a Hollywood movie just like this, you wouldn't believe it," Bean volunteered.

"What do you mean?" Conrad asked.

"Listening to this music on the back side of the Moon," Bean explained.

"That's right," laughed Conrad.

"You got something against music?" Gordon asked innocently.

"Nobody would buy it," Bean insisted. "This is corny; you've got to be 'hard' out here."

"The biggest thing I missed on Gemini 5 was not having any music," reflected Conrad. "Every once in a while they'd pipe some up over HF, but I'd have given my right arm for our old nickel-dime tape recorder there."

Nevertheless, several minutes later Conrad ordered the music off, "Why don't you rewind it right now, then we're in good shape."

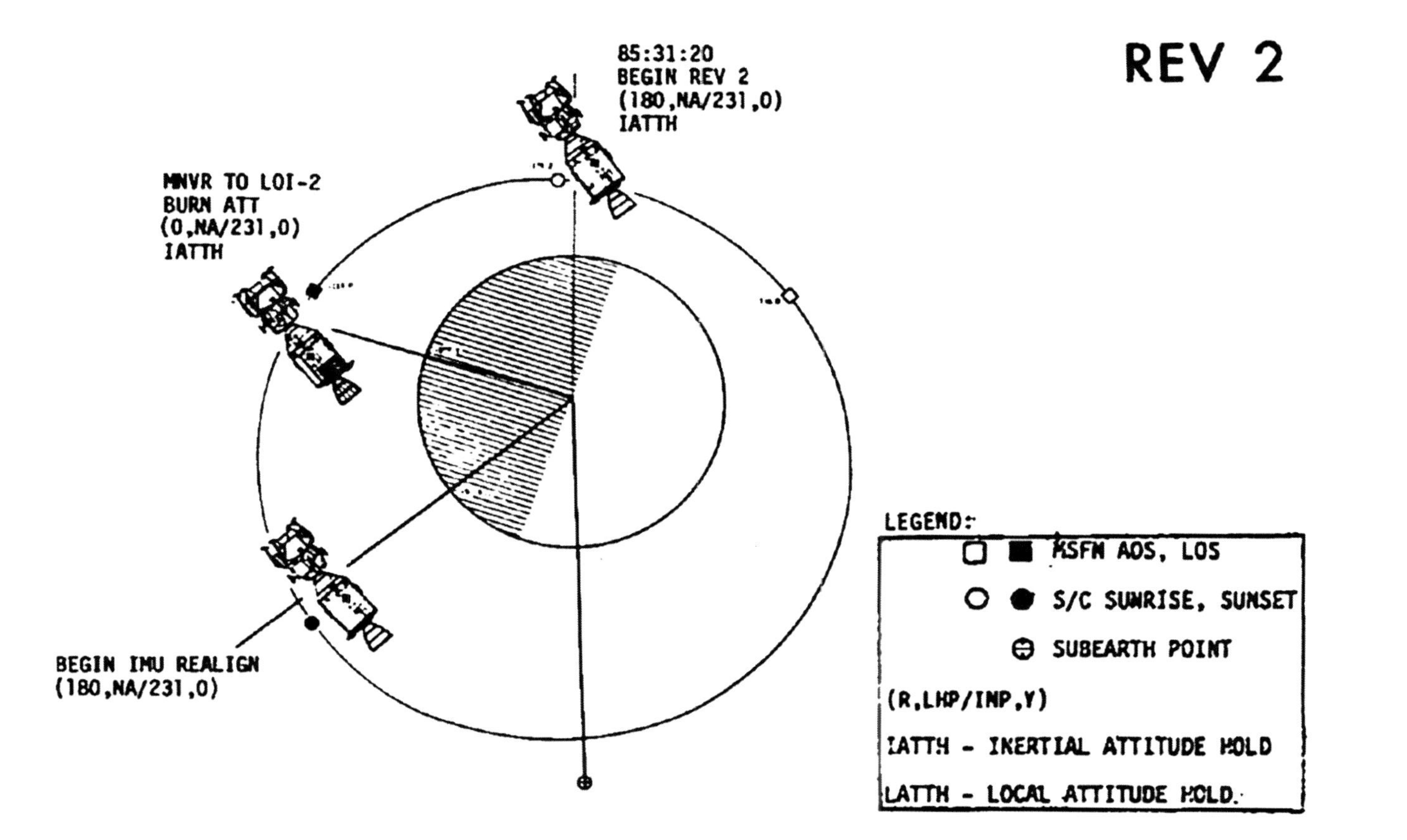

A page from the flight plan showing the manoeuvres to be performed during the second lunar revolution.

"After the burn's over, we can have a little more music," Bean said.

"I like the kind of ship you run," Gordon announced.

"Keep the sailors happy," Conrad laughed. "Give them exactly what they want."

The levity continued. "I'll be glad when you shave that moustache off your face," Bean teased Conrad.

"Yes," Gordon agreed. "It's terrible."

"I might keep it, now that I hear that comment," Conrad countered.

"You wouldn't like it if you saw yourself," Bean replied.

"If it's that repulsive, I ought to wear it!" Conrad insisted. "God, I hate to shave! Oh! It's going to hurt."

"It'll feel good," Bean assured. "Take a bath."

"Yes, I'm going to take a bath tonight," Conrad decided.

"Tonight's the night," Bean agreed.

"Take a bath, shave, get all cleaned up, and a good night's sleep," Conrad said. Then to Bean he continued, "You got anything else to do tomorrow? All right, that's what we'll do then. We will go for a little lunar landing. How's that? Unless you've got something better in mind; like maybe a little surfing at the beach or something."

"Hell, yes!" Bean replied.

Gordon had a bright idea, "How about you go play in the sand on the back side?"

"Take the LM down and land on the back side?" Conrad laughed, imagining the reaction in Mission Control if they were to do such a thing. "Wouldn't that shake them up!"

"We could do our DOI burn an hour later, telling them we'd seen something on the back side that was a little more interesting than the front side," Bean reasoned. If they were to postpone descent orbit insertion by 1 hour in a 2-hour orbit, this would yield a landing on the far-side rather than the near-side.

What a difference to the all-business Apollo 11 crew!

For this SPS burn, two of the thrusters were to fire for 19 seconds to settle the main engine's propellants. By now the tanks contained a lot of gaseous helium pressurant. There was no way of knowing where this was in the tanks in the weightless state, but the act of applying a small thrust against the direction of motion served to displace the lower density gas away from the valves that would feed the fluids to the engine. Although the burn was to last only 17 seconds, this would reduce the apolune on the near side of the Moon from 170 nautical miles to 54 nautical miles. It was essential that the burn not overrun, lest the altitude be reduced so far as to cause the vehicle to crash. In the event, the burn lasted 16.91 seconds and yielded a delta-V of 165.2 feet per second. It was made at an altitude of 61.6 nautical miles over selenographical coordinates centred on 1.75°S, 151.2°E. The onboard solution for the resulting orbit was 54.7 × 66.3 nautical miles, which was within 0.3 nautical miles of that intended. The eccentricity was deliberate, the expectation being that by the time of rendezvous after the lunar surface portion of the mission the gravitational perturbations of the mascons would have circularised Yankee Clipper's orbit at 60 nautical miles.

During the post-burn checklist, Bean's attention strayed to the landscape below, but Gordon drew him back inside. Conrad decided there was an omission from the

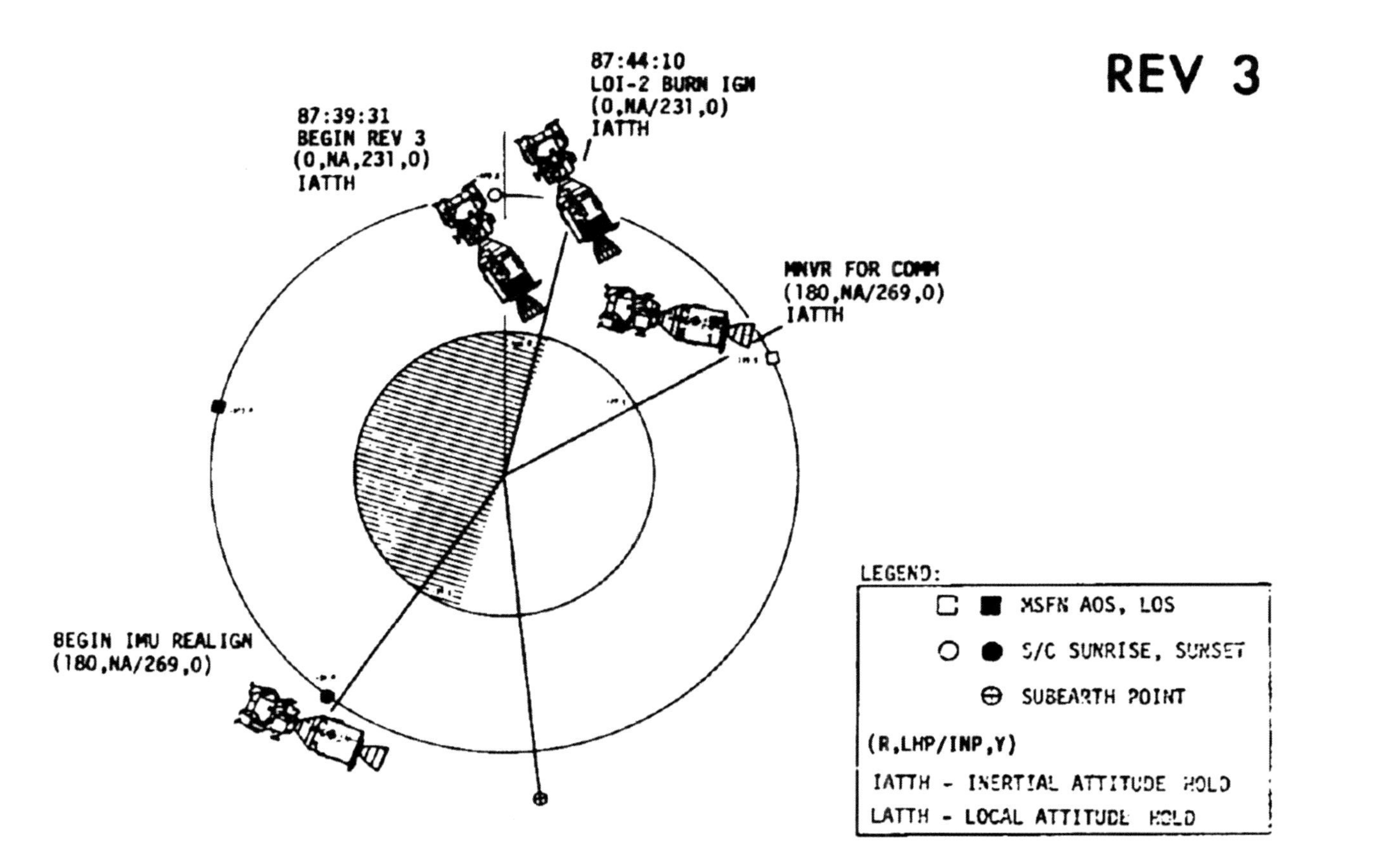

A page from the flight plan showing the manoeuvres to be performed during the third lunar revolution.

checklist to lock the translational hand-controller, so he wrote it in for future use. On coming around the limb on revolution 3, he informed Mission Control that the burn had been successful.

Conrad and Bean set about reopening the tunnel, and upon passing into darkness Gordon took star sightings for another platform alignment. Ten minutes before loss of signal, Conrad informed Mission Control that he and Bean were ready to perform housekeeping chores and communications checks in preparation for the next day's lunar landing. Intrepid had two S-band transmitter-receivers, two VHF transmitter-receivers, a signal-processor, the 26-inch-diameter steerable S-band antenna on the roof, a pair of fixed S-band antennas, two VHF antennas for use in flight and a lunar surface antenna that would relay between the vehicle and the astronauts while they were outside. In addition to the voice links, the system could transmit and receive tracking and ranging data, and could transmit telemetry to Earth and television from the lunar surface. Voice communication between the lunar module and Earth would be by S-band (as in the case of Yankee Clipper), but voice between the two vehicles would use VHF and would be feasible only when there was a direct line of sight. For these preparations, however, Conrad and Bean communicated with Gordon using the internal umbilicals.

At sunrise, Conrad glanced up through the small overhead window, "Oh, man, the command module looks pretty."

Bean floated over to see for himself, "Would you look at all the water! The whole outside of that spacecraft is covered with that crap." Actually, both spacecraft were coated with frozen fluids from the waste water and urine dumps.

As if on cue, Conrad said, "I've got to go back to the command module and take a leak." Once he was in place, he called, "Hey, Al, you ought to look out the overhead window there and watch this urine dump."

"Let me know when you go," Bean replied.

"I am right now," Conrad said.

"It's coming right out in all directions," Bean reported.

The urination system was proving to be an annoyance. Because the nozzle in the cabin was not draining properly the crew were leaving the vent valve open longer than recommended, and thus losing oxygen to space. They were also going through towels at a prodigious rate, using them to clean up the urine that was always on the urination facility and found its way onto clothing and the couch areas. The situation was exacerbated by the fact that each man had only one set of clothing. It consisted of long-john underwear, pants and a jacket. There were moves afoot to introduce a one-piece garment, but Conrad was against this because a separate jacket permitted each man to wear either the jacket or pants, or both, as per individual comfort. The pants were good because they had large zippered pockets in which to stow personal items such as dosimeters, tooth brushes and spoons. The jackets were worn only if the temperature dipped. The long-johns, however, soon became filthy and a regular change would have been welcome.

At acquisition of signal on revolution 4, Conrad returned to the lunar module and Bean called Mission Control using Intrepid's call sign.

"Go, Intrepid," replied Don Lind, who had taken over as Capcom.

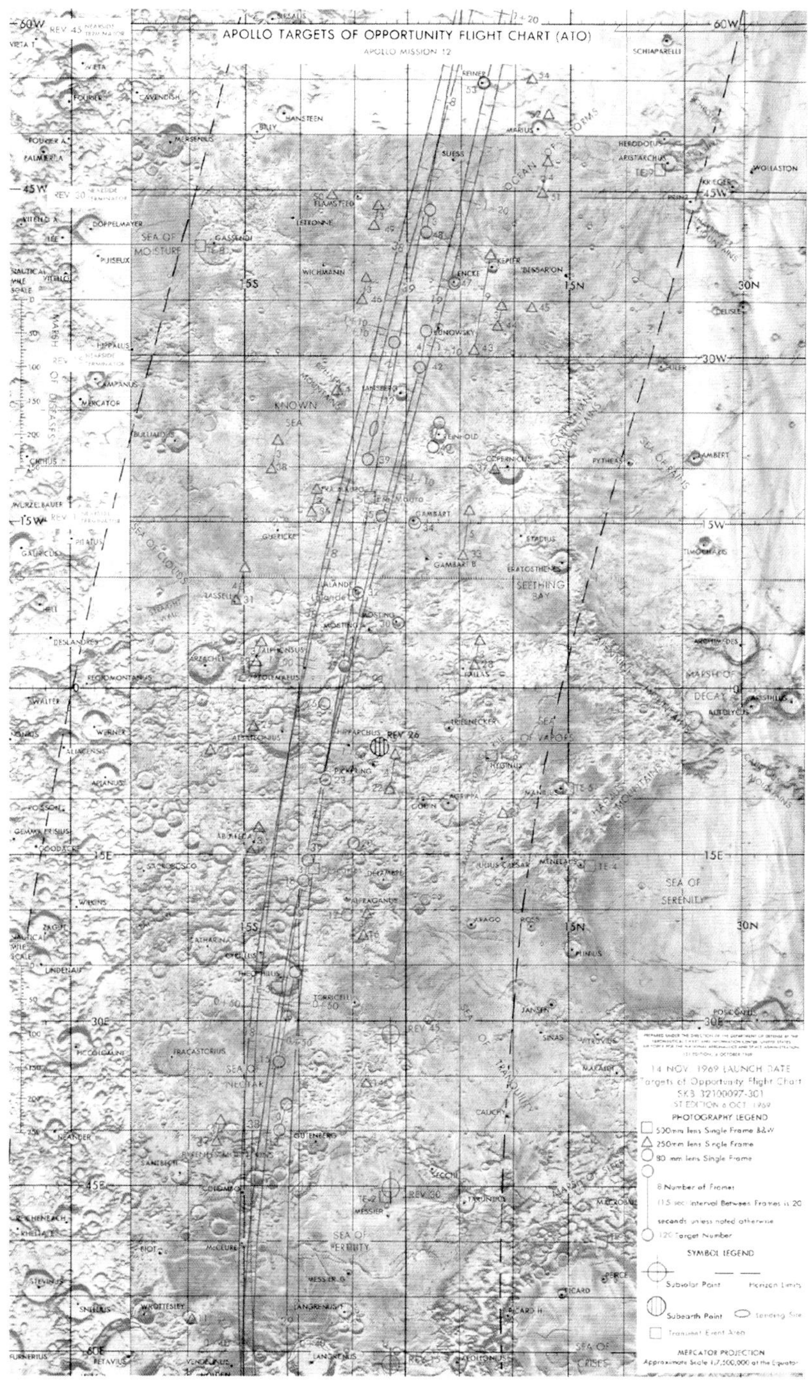

The ground track of Apollo 12 in lunar orbit for a portion of the near-side, annotated with photographic assignments.

"We're just standing by until 090:30, where we do our comm activation," Bean reported. "Everything looks good in the lunar module. We've checked all the things we're supposed to and they're all ship-shape and ready to go."

When Lind said Mission Control was ready to do the communications checks at any time, it was decided to go straight ahead. As a result, the lunar module's power system was brought on line and, following the checkout, the spacecraft returned to the umbilical from the command module at the time that the flight plan called for it to have gone onto internal power. On now finding themselves 20 minutes ahead of schedule, Conrad and Bean decided to return to the command module early, taking care to pull the circuit breaker to extinguish the cabin floodlights. They decided to leave the tunnel clear for a faster ingress the following morning.

"Don't bug me," Gordon warned his colleagues when they rejoined him. He was running a P22 landmark tracking exercise. This involved adopting an orientation in which the command module optics pointed at the looming target, then imparting a pitch rate to hold this in view as the vehicle flew overhead. After the computer had aimed the optics at the target, Gordon refined the pointing and took sextant sightings at predefined viewing angles. In this case, the target, designated H-1, was a circular crater about 2,500 feet in diameter that was inside a rille about 10 degrees east of the landing site. "The attitudes and the 0.3-degree-per-second pitch rate were easy to perform," he recalled in the post-mission debriefing. "The Auto Optics feature of the P22 puts you so close to the actual landmark that there is no doubt about it; but it is reassuring to have a good photograph of the landmark and so know exactly what you are going after." Landmark tracking enabled the orbital parameters to be measured extremely precisely.

Several minutes before Apollo 12 passed over the limb, Conrad announced, "We have you out our window for the first time since we've been in lunar orbit. We've been too busy to take a look at you."

"We're all smiling for you," Lind replied.

"We're happy up here," Conrad assured. "Soon we're going to have movies on the fantail."

"So long," Lind said. "We'll see you coming around the other side."

Amid housekeeping on the far-side, Conrad suddenly said, "I'm hungry enough to eat the ass out of a porcupine."

"So am I," Bean agreed

In preparing day 4, meal C, they speculated on the RCS propellant being used in attitude manoeuvres with the heavy lunar module on the nose. With the propellant monitoring sensors disabled by the lightning at launch, they were reliant on Mission Control's calculation of the consumption rate. "I'll tell you," Conrad said, "the guy that could really milk this son of a bitch was John Young. I think that he figured out every thrust before he [left Earth]."

On acquisition of signal, Lind picked up on the earlier remark, "We liked your idea of movies on the fantail so much that while you were running around behind the Moon there this time we replayed your last television coverage for the boys down in the hangar deck. It was very nice."

"Very good," said Conrad as he finished his supper.

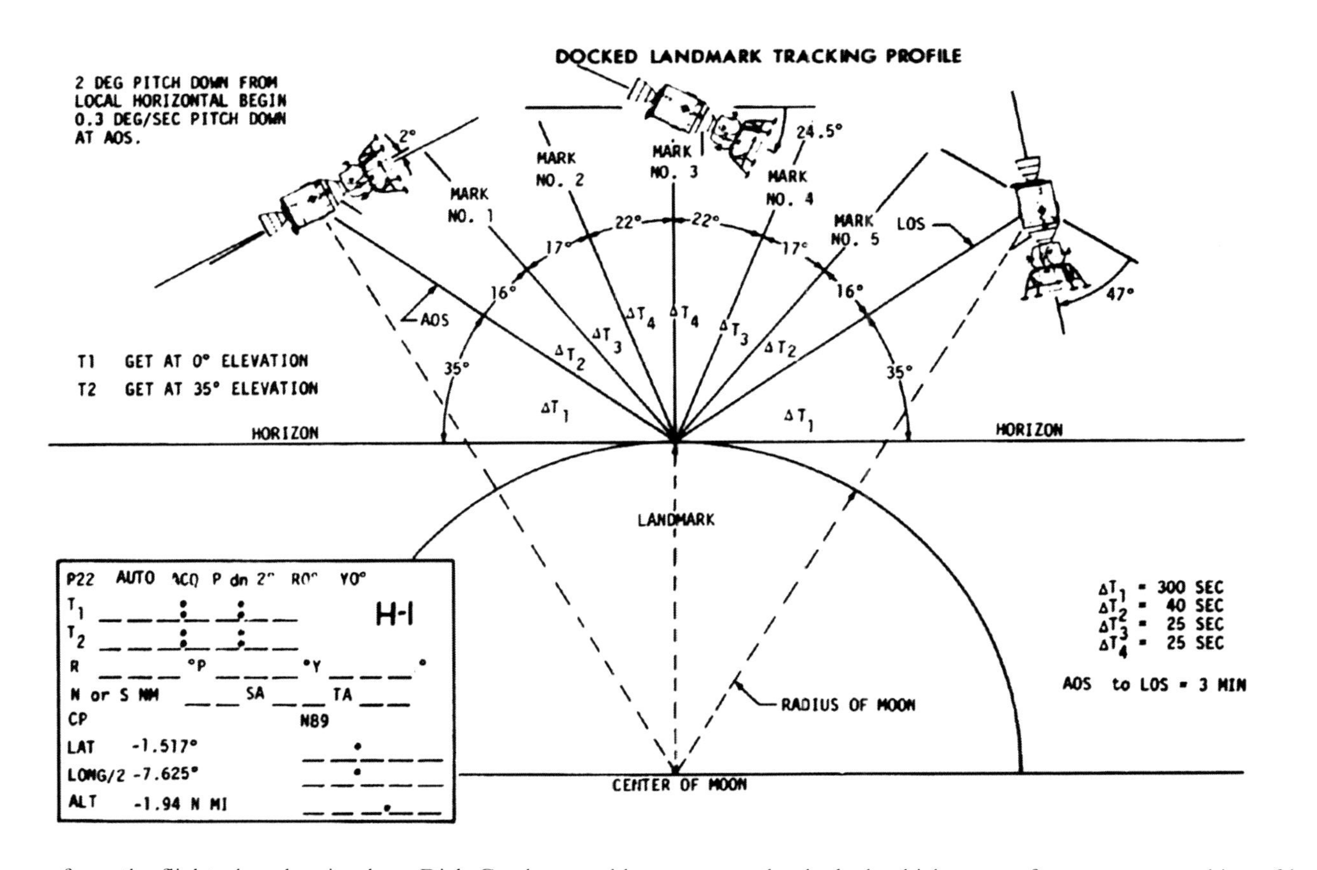

A page from the flight plan showing how Dick Gordon would manoeuvre the docked vehicles to perform sextant tracking of lunar landmark H-1.

"It looks a lot better on the big screen than it does on my home television set," Lind observed.

"It looks a lot better from right here with the old eyeball," Conrad said. "I wish we could give you television from the back side of the Moon. I'm more impressed with the back side of the Moon, it's prettier than the front side."

"You're making us feel envious," said Lind. And when Conrad warned that his crewmates disputed his opinion, Lind said, "I guess everyone has his favorite spot some place."

Then something occurred to Lind, "Al, on your TV broadcast, when you were commenting about mountains looking like clouds, were you aware the mountains were in the Sea of Clouds?"

"No," Bean replied. Then he added, "I guess that's where it got its name, huh?"

"We thought you had planned a pun for us," Lind said.

"I wasn't smart enough for that," Bean said.

The sleep period was to start during the far-side pass of revolution 5. The Flight Surgeon wanted overnight biomedical telemetry from Conrad and Bean. One of the electrocardiogram sensors had been causing Conrad some irritation, so he removed and reapplied it and asked Mission Control to confirm that it was functioning. The data would be monitored in real-time when on the near side of the Moon and taped along with all the other telemetry during the far-side pass, then dumped to Earth at high speed upon acquisition of signal.

Shortly before Apollo 12 appeared on revolution 7, Glynn Lunney took over as Flight Director for the graveyard shift. News came in from the Solar Particles Alert Network that a moderate solar flare had been detected, but this was one which had erupted on 2 November and whose active site had rotated right around the far side of the Sun and was coming back into view, so was not expected to present a problem to the mission.

At acquisition of signal on revolution 8 Bean came on the line and reported, "I've been getting a little stuffy in the head. In fact, I have been that way since launch. I don't have a cold or anything. My ears are sometimes clear, and sometimes they're not. I took a decongestant pill several hours ago. To be sure that my ears are clear tomorrow for all the LM activity and the EVA, how often should I take one of those pills to get the maximum effectiveness?" After consulting, Lind advised him to take one every 8 hours. Bean signed off and snuggled back into his sleeping bag, but it was difficult to drift off to sleep knowing that in a few hours he would attempt a lunar landing.

PINPOINT LANDING

As the clock reached 101 hours, with Apollo 12 a few minutes from loss of signal on revolution 9, Don Lind piped up 'Reveille' and called, "Good morning, gentlemen. Today's the big day. Hit the deck!"

"We've been there for a while," Pete Conrad assured him. They had been awake for some time but remained radio silent.

The timing of the formal start of flight day 5 allowed the crew to catch updates to the mission prior to going over the hill, then have their breakfast in peace during the far-side pass. After telling them that there was a solar flare that posed no threat, Lind asked about the biomedical sensor which Conrad had repositioned. Conrad replied, "Yes. I want to talk to you about that." In fact, he was "developing a little bit of a problem with all of them". The blue gel used to affix the sensors had antagonised his skin and after several days had produced blisters that were leaking. As he had never had this problem before, he reasoned that the gel had changed and he was allergic to it. "So, what I propose is, I'm going to leave them just the way they are until we get all the way through the EVAs and I get back up here; then, I want to take them off." On receiving no reply, he asked onboard if they still had a line of sight, which they did. A moment later, Lind reappeared and said that the steerable S-band antenna had temporarily lost lock – that issue was getting worse. Gordon reconfigured to an omni antenna that was sufficient for voice but could not relay much telemetry from lunar distance. It turned out that most of Conrad's message had been missed, so he tried again, "For some reason, it's making me break out in a skin rash. It looks like I've got poison ivy under those things, and they're weeping. Now, I do not want to take the rest of them off, because I'm afraid of what I'm going to find underneath." He reiterated that he wished to retain them until after the lunar surface part of the mission and then get rid of them. "They're driving me buggy," he concluded.

"We're going to talk that one over," Lind replied, but Apollo 12 flew around the corner before the Flight Surgeon could respond.

"We're on the other side," Conrad observed onboard. On checking the flight plan, he noted that they had to purge the fuel cells, dump waste water overboard, begin a battery recharge and generally clean up the command module.

Bean broke out the medical kit and extracted nasal sprays and nasal emollient for transfer to the lunar module, whose medical kit did not contain such items. Conrad decided to apply antibiotic to his skin rash. He asked Bean to take some aspirin with them. Observing the raid on the medical kit, Gordon quipped, "Are you guys leaving anything for me?" In fact, the 5 × 5 × 8-inch medical kit started out with three pain suppression injectors, three motion-sickness injectors, two 1-ounce bottles of eye drop, three nasal sprays, one 2-ounce bottle of ointment, two compress bandages, a dozen adhesive bandages and an oral thermometer. It also provided 60 antibiotic, 12 nausea, 12 stimulant, 18 pain-killing, 60 decongestant, 24 diarrhea, 72 aspirin and 21 sleeping pills. And there were several spare biomedical harnesses and the gel to affix them. Eight hours having passed since he took his first decongestant pill, Bean took a second to see him through the descent to the Moon.

Each man had defecated on the third flight day. Gordon decided he needed to go again. "I wish I could shit," Conrad declared. "I'd feel a lot better about it. I don't have the slightest inclination, but I just know what's going to happen: it's going to be the first shit on the lunar surface." In the command module they used Gemini-type plastic solid-waste bags. These contained a germicide to prevent bacteria and gas formation, and were generally referred to as 'blue bags' due to the colour of the

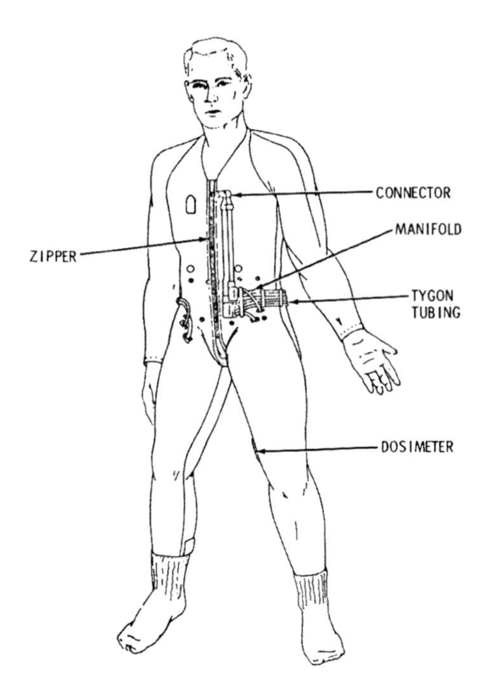

Details of the liquid-cooled garment that Intrepid's astronauts would wear throughout their time away from Yankee Clipper.

germicide. While in the lunar module, they would have to remain suited and hence defecate into diapers.[20]

Having finished attending to his skin rash, Conrad warned, "We ought to get as much done as we can in the next hour, because as soon as we come back out from behind the Moon it's going to be yak, yak, yak, yak." He meant that once back in communication, Mission Control would be feeding them information for the lunar landing and constantly checking on their progress.

Immediately after breakfast, Conrad and Bean were to remove their long-johns and don the liquid-cooled garments which they would wear under their suits for the lunar surface activities. As the meal was being prepared, Bean pointed out that the flight plan called for him to enter the lunar module, power it up at 104 hours, and make a start on the activation checklist while Conrad suited up, then, shortly after Conrad entered the lunar module at 104:30, Bean was to go back to the command module and suit up. Bean instead suggested that he simply suit up prior to ingress and remain there; Conrad said it was a good idea.

Several minutes later, munching his breakfast, Bean put a music cassette on and admitted, "I'm about as jumpy as I can be, this morning." And then, to provide a scale, he added, "Jumpier than I was on launch day."

[20] In fact, they wore diapers at launch as a precaution against having to perform a post-TLI abort in a situation in which they would be obliged to remain suited throughout the flight.

"I kind of have the same feeling," Conrad admitted. "It's a bigger day."

"It is," Bean agreed. "You've got more things under your command today. On launch day, you're just kind of along for the ride." Today would be an exercise in *flying*. "Stick to the checklist, and be careful not to throw the wrong switch. Don't get in a hurry. Don't get too fancy."

"No fancy shit, you're right," Conrad agreed.

"It'll be good to get down on the lunar surface and do some physical work. You know that?" Bean said.

"Speak for yourself," Conrad replied. "I'm a lazy son of a bitch." After chasing a bacon bite that had floated away, he mused, "I just hope we find the old Snowman! Then, I hope we find a place to land! Then, I hope I can set it down all right!"

After a discussion of the illumination conditions at the landing site, with the Sun just 5 degrees above the horizon and casting lengthy shadows, Bean laughed, "If we pull this off, we'll be slicker than owl shit."

Flight Dynamics Officer Dave Reed had promised to program the computer to aim right for the crater in which Surveyor 3 was standing, but Conrad was sceptical. "It's driving me buggy. I just don't know what I'm going to see when I pitch over. You know, I'm going to say, 'Ahhh!! There it is!' or I'm going to say, 'Fuck it! I don't recognise anything' and then I'm in deep yogurt. If I don't recognise anything, I'll just let it keep going. That's the best I can do."

"If you don't recognise a thing, just tell me and I'll look out my side," Bean said, "and you look at the computer for a few seconds to let me see if I see anything out there." He peered at the far-side passing below, and was moved to say in awe, "Isn't that fantastic!" On reflection, he wondered, "How did we ever get here anyway?" prompting laughter all round. "Why would a guy want to put his ass on the line like this?"

By the time they emerged around the limb on revolution 10, Gerry Carr had taken over as Capcom. He picked up on the matter of Conrad's biomedical sensors. The Flight Surgeon's recommendation was to remove all the sensors, put healing cream on the irritation, and then apply Band-Aids. He should then retrieve a spare harness from the medical kit and apply its sensors to fresh skin. "Of course," Carr added, "if you don't like this particular proposal, they have no objection to you sticking to the way you wanted to do it."

If this had been suggested the previous evening, Conrad would almost certainly have accepted, but the start of preparations for a lunar landing was no time to mess around with the harness. "I want to stick with it the way it is," he decided. "You're getting good data now, aren't you?"

Carr confirmed that the biomedical telemetry was good, and then asked whether Bean was still stuffy. Bean said he had been feeling stuffy since Earth orbit and had assumed that it was an adaptation to weightlessness which would ease off during the way out but it had persisted. But he was upbeat, "I don't sneeze, cough, or have any other symptoms. It just seems to be a fullness in the ears and nose. I took a couple of decongestants, which worked well. Right now my ears are clear and my nose feels real well." Carr relayed the revised recommendation that Bean take decongestants at 6-hourly rather than 8-hourly intervals. Having started applying a nasal spray, Bean asked how often he should use this, and was told every 3 hours. Carr asked whether anyone else was suffering similarly; the answer was no. Conrad suggested that Bean might be particularly sensitive to something in the cabin atmosphere.[21]

After the crew provided their post-sleep report and Carr read up some technical data, Conrad and Bean unpacked and inspected their liquid-cooled garments.

"What's going on down there in the world today?" Conrad enquired.

"I don't know, Pete," Carr said. "I just got here myself." Several minutes later he was back. "News coverage on your flight is beginning to pick up as the touchdown gets closer. I guess most of the reports right now are about your medical ailments: your 'code in de dose' and all that stuff."

Conrad was irritated that the news of the mission was focusing on the health of the crew. "I'd like to square something away down there. Al does not have a cold. All I have is a 1-inch itch, and I don't consider that any major medical problem. As a matter of fact, we're in pretty damn good shape."

Meanwhile, Gordon was preparing to use a Hasselblad with an 80-mm lens and black-and-white film to photograph the target assigned to Apollo 13, which was in the hummocky terrain north of the crater Fra Mauro. Jim Lovell would command that mission, and back in their Navy days Conrad had given Lovell the nickname of 'Shaky' owing to his nervous energy. With the job done, Conrad called down, "You can tell good Captain Shaky that he can relax. We've got his pictures."

"He'll be tickled to hear that," Carr replied.

Conrad added ominously that Lovell might be alarmed when he saw the pictures, because the target for that mission was in much rougher terrain than the Snowman of craters on the Ocean of Storms.

"Dick," Carr called several minutes later, "I just talked to Barbara. She has been resting up, taking naps and everything, getting all set for spending the night up with you. And she said to tell you that she thinks you're just great."

"Well, tell her thank-you," Gordon replied. "After 16 years, it's about time!"

Once Conrad and Bean had donned their liquid-cooled garments, Carr noted that Bean's electrocardiogram was giving poor data, probably because the gel had dried out, and asked him to attend to it. This reinforced the lesson that in future the sensor harness should be replaced the evening prior to the lunar landing in order to ensure a clean setup.

Meanwhile, Gordon took star sightings. "Gerry, this platform has done real well, in spite of that glitch we gave it at launch."

"Do you recommend that we glitch them all like that?" Carr asked.

"No, sir. Not at all," Gordon retorted.

On passing around the corner a few minutes later, the astronauts put on a tape of the bubble-gum hit 'Sugar, Sugar'. Usually, they would hold onto cabin struts and shuffle their weightless bodies in time to the beat, but this time it was played while Conrad helped Bean to wriggle into his pressure suit. When Conrad retrieved his

[21] In fact, Bean had contracted a cold, and Conrad would start to show the symptoms while away in the LM.

own suit, he got a surprise: water vapour had condensed on the bulkhead behind where his suit had been stowed. "God damn! Oh, man, that's bad news! My suit is sopping wet all over the legs." After further inspection he complained, "It's all down the back of the legs." As there was no quick way to dry it out, he would have to live with the inconvenience.

Since the probe and drogue had not been reinstalled, all that was required to gain access to Intrepid was to remove the apex hatch and pass through the tunnel, but this was a tight squeeze for a man wearing a pressure garment assembly. Once Bean was in, Conrad passed through his helmet and gloves. No sooner had Gordon mounted the television and 16-mm movie cameras on brackets in the rendezvous windows than the spacecraft reappeared around the limb on revolution 11. During the far-side pass Cliff Charlesworth had relieved Glynn Lunney as Flight Director. When Carr established contact, Gordon informed him that Bean had entered the lunar module on time and that Conrad was suiting up; however, he did not explain that Bean had opted to suit up early.

Bean called Carr using the lunar module's call sign and reported that he had just opened the shades and discovered both windows to be frosted over, but he expected they would clear after he powered up and activated the heaters. The power system of the lunar module comprised four silver-zinc oxide storage batteries in the descent stage for the descent and lunar surface period, and two more in the ascent stage for the rendezvous. They supplied 28 V_{DC}, and a pair of inverters provided 117 V_{AC} at 400 hertz for those systems which required alternating current. He switched over to internal power on schedule at 104 hours and began the lengthy activation checklist, starting with the communications system. Having suited up, Conrad joined Bean as per the flight plan. Once the computer had passed its self-check, its erasable memory was downloaded to Mission Control to be checked for corruption.

Because the lunar module had been inert, its clock had to be set manually. Conrad asked Gordon what time it was and so Gordon told him. "No, no, no. Give me one for the future," Conrad explained. Gordon duly gave a 15 second countdown to the next whole minute and Conrad started his clock on the mark. A check showed that there was an error of 0.8 seconds, so they did it again. With the clocks synchronised, the next task was to initialise the inertial platform. This involved Gordon reading the gimbal angles of his platform and Conrad converting them to the reference axes of his own vehicle, taking into account the docking tunnel index. The procedure called for Mission Control to make the same calculations for verification. "Hey, Houston," Conrad called. "Are you doing your mathematics on the ground?"

"We're working," Carr assured, and a moment later read up the results.

Given the go-ahead, Conrad entered the data. This procedure coarsely aligned the platform. It would be fine-aligned later by taking star sightings.

At this point the plan envisaged Bean returning to Yankee Clipper, but since he had already suited up he remained in place to continue the activation.

Conrad's approach to a checklist was 'get ahead and stay ahead'. The flight plan did not call for the landing gear to be deployed for another hour, but he was well ahead and saw no reason to wait. "Have you got any objections if I blow the gear down?" When Carr said to proceed, Gordon warned that because he was viewing the lunar module against the dark sky he would not be able to provide visual verification of the deployment. They went ahead anyway. "Any doubt about that?" Conrad asked Carr, who reported that the telemetry showed the legs had been explosively splayed out from their stowed configuration in tight against the base of the descent stage. As Conrad would say later, "Deployment of the landing gear left no doubt in our minds that the pyros fired. Dick was able to see three of the four gears from the command module, but we got a grey talkback which indicated they were all down and locked."

To remain ahead, Conrad decided to close the tunnel early. Andrew Chaikin, in '*A Man on the Moon*', poignantly tells of Gordon looking down the length of the tunnel at his friends. It was a time for good-byes, and yet none were said. Gordon dearly wished it had been possible for all three of them to go down to the lunar surface. For his part, Bean wondered whether he would ever see the man again. Gordon passed the drogue through the tunnel for Conrad to install, and then he himself installed the probe so that the capture latches at its tip mated with the drogue. With both hatches closed, the tunnel could be depressurised – a process that would take some time.

The lunar module was now operating on its own environmental control system. This comprised the atmosphere revitalisation section, the oxygen supply and cabin pressure control section, the water management section, the heat transport section, and outlets for servicing the portable life-support systems with oxygen and water for the moonwalks. The components of the atmospheric revitalisation section were the suit circuit assembly that cooled and ventilated the pressure garments and removed carbon dioxide, odors, noxious gases and excess moisture; the cabin recirculation assembly that ventilated the cabin and controlled its air temperature; and a duct that vented steam from the suit circuit water evaporator to space. The oxygen supply and cabin pressure section supplied gaseous oxygen to maintain cabin and suit pressures. The descent stage supplied oxygen for the descent and lunar surface phases, and the ascent stage carried oxygen for the rendezvous phase. Water for drinking, preparing food, the liquid-cooled garments and refilling the coolant of the portable life-support systems was supplied by the water management section. The water was held in three nitrogen-pressurised-bladder tanks, a large one in the descent stage and two smaller ones in the ascent stage. The heat transport section utilised primary and secondary water-glycol-solution coolant loops. The primary loop was for temperature control of cabin and suit circuit oxygen, and for thermal control of batteries and electrical components on cold plates and rails. If this became inoperative, the secondary loop would supply the cold plates and rails only, and suit circuit cooling would be by the suit loop water boiler. The waste heat from both loops was vented overboard using water evaporation or sublimation systems. Operating the environmental systems fell within lunar module pilot's remit.

Moving on, Bean activated the guidance system which would be used to achieve a rendezvous with Yankee Clipper in the event of a failure in the primary system, then Mission Control uploaded a REFSMMAT into the computer. The primary guidance and navigation system (PGNS) comprised an inertial measurement unit that could be updated by star sightings by the Alignment Optical Telescope, and the rendezvous and landing radars. It supplied inertial reference data for computations, provided an inertial alignment reference by feeding optical sighting data into the LM computer, displayed position and velocity data, controlled both attitude and thrust in order to maintain a specific trajectory, controlled descent engine gimballing and throttling, and processed radar inputs to compute rendezvous data. The abort guidance system (AGS) was an independent backup system which had its own computer, but it used body-mounted gyroscopes instead of a gimballed inertial platform. The radar section comprised the landing radar which supplied the lunar module's main computer with altitude and velocity data during a lunar landing, and the rendezvous radar which measured the range and range-rate of the CSM and provided line-of-sight angles for computing manoeuvres. The ranging tone transfer assembly was a passive responder for the VHF ranging device carried by the CSM, and as such it served as a backup to the active rendezvous radar. The control electronics section controlled the vehicle's attitude and translation in all three Cartesian axes, controlled (by PGNS command) the automatic operation of the ascent engine, the descent engine, and the reaction control thrusters, and also handled commands provided by the manual attitude and thrust-translation controllers. And the orbital rate system displayed the computed local vertical on the pitch axis of the 8-ball during circular orbits.[22]

Just before Apollo 12 disappeared around the far-side, Conrad, determined to remain ahead on the checklist, told Carr they would proceed with everything that did not require the participation of Mission Control. However, although they were able to pressurise the propellant tanks of the reaction control system with helium, they were so far ahead of Gordon that they then had to wait for him. But Conrad was content to be idle. As he explained in the post-mission debriefing, "Had we run into a problem somewhere along the line, we would have had more than adequate time to cope with it."

While waiting, Conrad watched the landscape passing by. "Man, that's a mighty impressive territory down there." He knew the altitude was 60 nautical miles, but he found the motion somewhat misleading, "I keep estimating that if I was flying over the desert, I would be about, at the most, 30,000 feet." And then he mused, "I hope I can get myself oriented when we get down lower."

As they approached the point at which they would reacquire Earth on revolution 12, Gordon used the VHF link between the vehicles to report, "Things are looking awful good over here, Pete."

"Looks good here, too," Conrad replied. "Have you checked out your TV yet?"

"No, I haven't," Gordon said.

Once Carr had established contact, he confirmed that the telemetry from the two vehicles indicated they were in good condition and then he read up the latest data. After checking the drift rates of the inertial platforms in the two vehicles, Mission Control uploaded revised state vectors.

After Conrad and Bean had each exercised their hand-controllers and the systems had been verified by telemetry, Conrad called, "Houston, I can step ahead here and

[22] This latter system was known by the contrived acronym of Orbital Rate Drive Earth and Lunar (ORDEAL).

do RCS checkout, if you want, right now." When Carr said he should wait several minutes, Conrad asked Gordon how long it would be until Gordon would require to manoeuvre the docked combination. On being told that he had at most 20 minutes, Conrad said he thought he should be able to complete the firing tests by then. When Carr issued the go-ahead, Conrad started with a 'cold-fire' in which the system was exercised right to the point of actually igniting the thrusters, while Mission Control monitored the telemetry. Yankee Clipper had been maintaining the attitude of the docked combination, so Gordon placed his digital autopilot in Free so that it would not try to counteract the motions imparted by the lunar module during the ensuing 'hot-fire' tests. As soon as Conrad was done, Gordon reactivated his digital autopilot and manoeuvred to the attitude for P22 tracking of a landmark known simply as point 193. It was an elliptical crater about 1,000 feet in size, situated about 6 nautical miles south and 3 nautical miles east of the crater in which Surveyor 3 stood. When the target came into the field of his sextant, he initiated the slow pitch rotation that would enable the optics to lock on.

"Pete, my boy, I gave you five of my best ones," Gordon reported, referring to the five time-spaced sextant marks that he had taken while passing over the landmark.

"Good show, Richard," Conrad replied.

After analysing the data, Mission Control determined that Apollo 12's landing site was about 4,700 feet south of the pre-flight estimate. It was also about 2,400 feet higher in relation to the idealised spherical Moon. These revisions, along with other navigational errors, would be able to be taken into account in planning the powered descent.

"How's the Snowman look, Dick?" Carr asked. By now, the dawn terminator had advanced beyond the landing site.

"I didn't even have a chance to really look at it," Gordon replied, as he began the manoeuvre to the attitude in which the lunar module was to calibrate its AGS. With this achieved, he once again set his digital autopilot to Free.

"What'd you just do, Clipper?" Conrad demanded.

"I didn't do anything," Gordon replied.

"I thought I heard a big clunk somewhere," Conrad explained.

"Well I've been accused of a lot of things, but that's the worst," mused Gordon, "I'm in here minding my own business, Pete."

A self-test of the rendezvous radar of the lunar module showed an anomaly in its range indication. This irritated Conrad because, as he explained later, "We didn't get 500 feet as advertised on our tapemeter, we got 493 feet. The system specification was 500, and 493 was specified for our particular radar. The checklist didn't reflect our specific number. This is something that we argued about before flight. We didn't want the specification number in there, we wanted the *right* number. Somebody chose to ignore us." The transmitter power output was about 1 volt below nominal, and would limit the maximum range at which the radar could operate, but he was not concerned about this. Meanwhile, the AGS had been calibrating itself. As a check, Bean cycled the input to the 8-ball back and forth between the PGNS and the AGS. Although it jumped about a quarter of a degree in both roll and pitch, he deemed this to be an acceptable divergence.

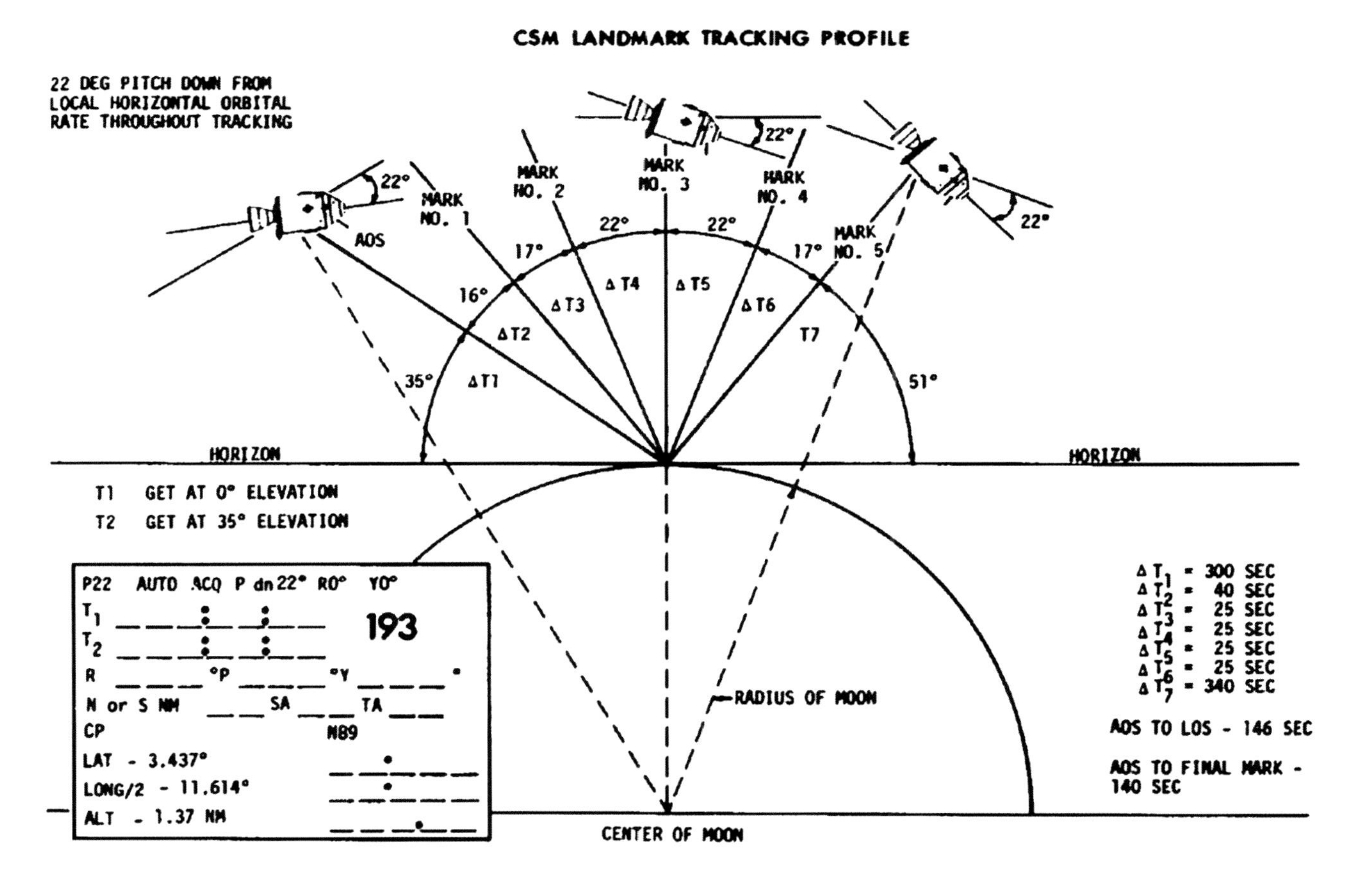

A page from the flight plan showing how Dick Gordon would use an orbital rate pitch rotation to stabilise Yankee Clipper for sextant tracking of lunar landmark 193.

"Al," Gordon called, "are you through with the AGS Cal and everything?"

"That's affirmative," Bean replied.

Several minutes later, Apollo 12 passed around the limb onto the far-side.

As Apollo 11 approached the point of initiating its powered descent the PGNS thought it was flying the intended trajectory, but by timing the passage of landmarks Neil Armstrong had realised he was coming in 'long'. Several factors contributed to this error, one being the manoeuvre that the lunar module had executed shortly after entering the descent orbit, where a sighting of the Sun was used to measure the pitch drift rate. On Apollo 12 this telescopic observation had been eliminated. Instead, a technique was employed in which the platforms of the two vehicles were compared after each of the various attitude manoeuvres performed during the recent near-side pass. As the drift of Yankee Clipper's platform had been established, any differences served to measure the drift of Intrepid's platform.

"I'm going to manoeuvre to the undocking attitude," Gordon announced. Several minutes later, he asked, "What's all that oil-canning about over there, Pete?"

"We're doing our pressure integrity check, Dick," Conrad explained. The skin of the cabin flexed as the pressure changed. The tunnel had been vented, but the sound was conducting through the docking collars.

"You're sure making all kinds of noise," Gordon said.

"We're going to pressurise the DPS now," Conrad announced. "You may hear something." He was referring to the descent propulsion system which would execute the powered descent.

After 12 minutes of radio silence, Gordon asked, "How're you guys doing over there?"

"We're just completing the prep for undocking," Bean replied. "How's it going over there? All buttoned up?"

"Lots of space," Gordon observed. "I've got the seat down and the struts off. I'm about to take this suit off." The plan called for him to remain in his suit whilst alone, but he had changed his mind. "I figure I can get back into it." By hooking a lanyard to the zipper, he would be able to suit up without assistance.

On re-establishing contact with Earth on revolution 13, Gordon alerted Carr that he was going to activate the television.

"We're ready for undocking," Conrad announced.

After Cliff Charlesworth had polled his flight controllers, Carr relayed, "Intrepid, Houston. You're Go for undocking."

"Roger-roger," Conrad acknowledged.

If the LM had been judged No-Go for undocking owing to some fault that would not affect the descent propulsion system, the two vehicles would remain docked and that engine would perform a plane change to improve the photographic coverage of candidate landing sites. There would be no need for the SPS to make a plane change for a rendezvous, but this would later perform the scheduled post-rendezvous plane change. The revised mission would be to achieve the planned photography whilst awaiting the scheduled time to head home.

When Carr reported that the television was operating and featuring the overhead window of Intrepid, Conrad asked, "Can you see me waving at you, Gerry?"

"You better have Dick focus it," Carr suggested.

"It's focused," Gordon advised. "It's just dark where he is."

While the lunar landing procedures and profile were generally similar to those of Apollo 11, this time the landing was to be a precision operation and several changes had been introduced in an effort to reduce the landing point dispersions. Previously, the docking probe had been retracted prior to unlatching the collars, and residual air in the tunnel had pushed the vehicles apart and imparted an unwanted motion to the lunar module. Although Neil Armstrong had nulled this out, when the perturbation was propagated through the descent orbit insertion manoeuvre it contributed to the downrange error. To ensure that the state vector for the descent orbit insertion was well defined, it had been decided to extend the probe, hold Intrepid in a soft-docked state so that any residual air in the tunnel could escape harmlessly, and then, when everything was stable, disengage the capture latches and retract the probe.

After Bean had initialised the AGS to match the more accurate PGNS, Conrad told Gordon, "I'm going to dial up a Verb 77 and stand by on the Enter until you release me." By this, he meant he would initiate Attitude Hold once the probe had released his vehicle.

Fifteen minutes prior to the scheduled time of undocking, Gordon had oriented the docked combination in an inertial attitude in which it would be vertical relative to the Moon at undocking, with Yankee Clipper on top.[23] At the appointed time, he hydraulically extended the probe. It did so rather more rapidly than he had expected and induced a distinct rebound, but the spring-loaded mechanism rapidly damped out the oscillation. Satisfied, he released the latches and retracted the probe.

"Watch him, watch him," Bean urged Conrad. "Don't let him bump into any of our equipment." A lateral motion between the vehicles could all too easily cause the probe to nudge one of the antennas on the roof of the lunar module.

"Back off, Dick," Conrad said.

Gordon fired his thrusters for 2 seconds to initiate a slow withdrawal.

"There he goes," Conrad assured Bean.

As Conrad said in the post-mission debriefing, "The soft-undocking worked very well. When Dick undocked, we hit the end of the probe and there were some slight longitudinal oscillations. The probe damped these out very well. When he undid the capture latches, the two spacecraft were completely null to each other and did not separate. Dick physically had to back off. I had no indications from either the AGS or the PGNS that we got any velocities."

Undocking occurred at selenographical coordinates 13.5°S, 87.0°E, at an altitude of 63.0 nautical miles.[24]

As he drifted clear, Gordon snapped Hasselblad pictures of Intrepid through the left-hand rendezvous window, which he was sharing with the 16-mm movie camera. The television was installed in the right-hand window. After checking the monitor,

[23] This was the inverse of the situation on Apollo 11.

[24] It was 11:16 p.m. EST on 18 November.

which he had alongside the left couch, he adjusted the focus and asked, "How's the picture, Gerry?"

"The picture's beautiful," Carr replied.

"Quite a sight, isn't it?"

"It sure is, Dick," Carr said.

On Apollo 11, Eagle had manoeuvred to perform a 360-degree yaw 'pirouette' in order to enable Mike Collins to visually verify that all four landing legs were down and locked. This time it had been decided to make this manoeuvre only if there was an indication that the legs had not deployed properly, but, as the television pictures clearly showed, such a manoeuvre was unnecessary. To further minimise perturbing Intrepid's trajectory, the active station-keeping activities that Eagle had undertaken had also been deleted. All Conrad did was yaw left 60 degrees and pitch up 90 to an attitude in which he and Bean could view Yankee Clipper through their forward windows.

"Get his picture," Conrad told Bean.

"I can't. He's not in my window," Bean replied.

"The hell he isn't; look up!"

"Now, he is," Bean acknowledged. He started to take pictures of Yankee Clipper nose on, but they were midway across the illuminated hemisphere of the Moon and with Yankee Clipper above Intrepid the Sun was in the field of view. "You've got to keep him like that for SEP," Bean said, meaning that Conrad was to maintain visual contact until Gordon performed the separation manoeuvre. "How far would you say he is out?"

"I haven't any idea," Conrad admitted.

"I'd say 50 feet," Bean estimated.

"Look at his antenna wobbling around," Conrad said to Bean. A few moments earlier, Carr had told Gordon that his signal was cutting out. The steerable S-band antenna ought to be locked on Earth, but it was oscillating. Evidently, this was the cause of the intermittent communications difficulties they had been suffering since entering lunar orbit. The problem was getting worse: initially the signal had been merely fluctuating in strength but now it was cutting out. Conrad attempted to call Gordon, without any response. Bean suggested that the sound of the drive motor of the movie camera that Gordon was running might be being picked up by his voice-activated microphone, which, being in transmit mode, would in turn block reception. Conrad asked Carr to call Gordon and tell him to deactivate his Vox. This restored VHF communication between the vehicles. When Carr again said the S-band link with Yankee Clipper was intermittent, Conrad passed on his observation that the antenna was "wandering" in two dimensions, as if performing a search algorithm.

Resuming their checklist, Bean switched on Intrepid's tracking light briefly and Gordon confirmed that it was operating. Then Carr read up the data for the descent orbit insertion, for the powered descent and for the contingency manoeuvre referred to as 'No-PDI + 12'. In the event of the powered descent being ruled out by a fault not involving the descent propulsion system, this engine would be fired 12 minutes after the time that the descent would otherwise have started.

As Carr switched his attention back and forth between vehicles, supplying data, Conrad wryly observed to Bean, "That's one busy mother."

A shot of Yankee Clipper taken by Al Bean in Intrepid shortly after undocking, with the Sun close to the line of sight.

Meanwhile, the landing radar successfully passed its self-test.

After some tests, Gordon decided that the S-band antenna had lost its Auto Track mode, and he resorted to steering it manually. In this mode, the antenna was steady. When Conrad offered some advice, Carr retorted, "All right, Conrad. You're stealing our thunder now."

"Just trying to help out," Conrad laughed. When Carr said Conrad had spoken just as he, himself, was preparing to speak, Conrad responded, "The only reason I'm beating you, Gerry, is I'm a quarter of a million miles closer to him than you are."

"A good answer!" Bean observed onboard.

Meanwhile, as the two vehicles flew in formation, Bean decided to activate the movie camera mounted in his window to record several minutes of Yankee Clipper slowly drifting away.

Tracking by the Manned Space Flight Network had revealed that the translunar navigation and two-stage lunar orbit insertion had produced a ground track which, it was calculated, would be 4 or 5 nautical miles north of that desired at the time of powered descent initiation. However, Intrepid would be able to correct this cross-range error in the braking phase of its powered descent. When Carr informed him of this, Conrad joked, "As long as the PGNS knows it, that's okay with me."

In essence, everyone was waiting for the time at which Gordon would make the separation burn. There was some confusion, because the time written down in the lunar module's flight plan was 20 seconds early.

At undocking, Yankee Clipper had been oriented with its apex pointing straight downward. But because its attitude was inertial it was horizontal by the time of the separation manoeuvre, one-quarter of a revolution later. This manoeuvre was made at an altitude of 59.2 nautical miles at selenographical coordinates 6.5°S, 7.7°W. It lasted 14.4 seconds, and was done by firing the thrusters in the plus-Z direction so as to descend. The radial delta-V of 2.4 feet per second nudged Yankee Clipper into an orbit of 56.3 × 63.5 nautical miles that had the same period as the original orbit and would yield a separation from Intrepid of 2.2 nautical miles half a revolution later, at the time of that vehicle's descent orbit insertion manoeuvre.

"Bye-bye," Conrad called.

"See you, troops," Gordon replied.

As the line of sight to Yankee Clipper began to include the Moon, Bean restarted the movie camera, "That's going to be a beautiful shot, because he is coming down below us."

"You look neat, down there against the Moon, Dick," Conrad reported.

As Bean continued to fuss with the camera, Conrad advised, "Al, I think you've had enough." It was time to resume the checklist. Meanwhile, Gordon had curtailed his television transmission.

Yankee Clipper repositioned itself below Intrepid, oriented in order to perform a sextant tracking check and test the VHF ranging equipment. Meanwhile, the ability of the rendezvous radar on the lunar module to operate with the transponder on the command module was established and the data fed to Intrepid's computer to verify that it could process rendezvous parameters. Next, Conrad keyed P40, the program

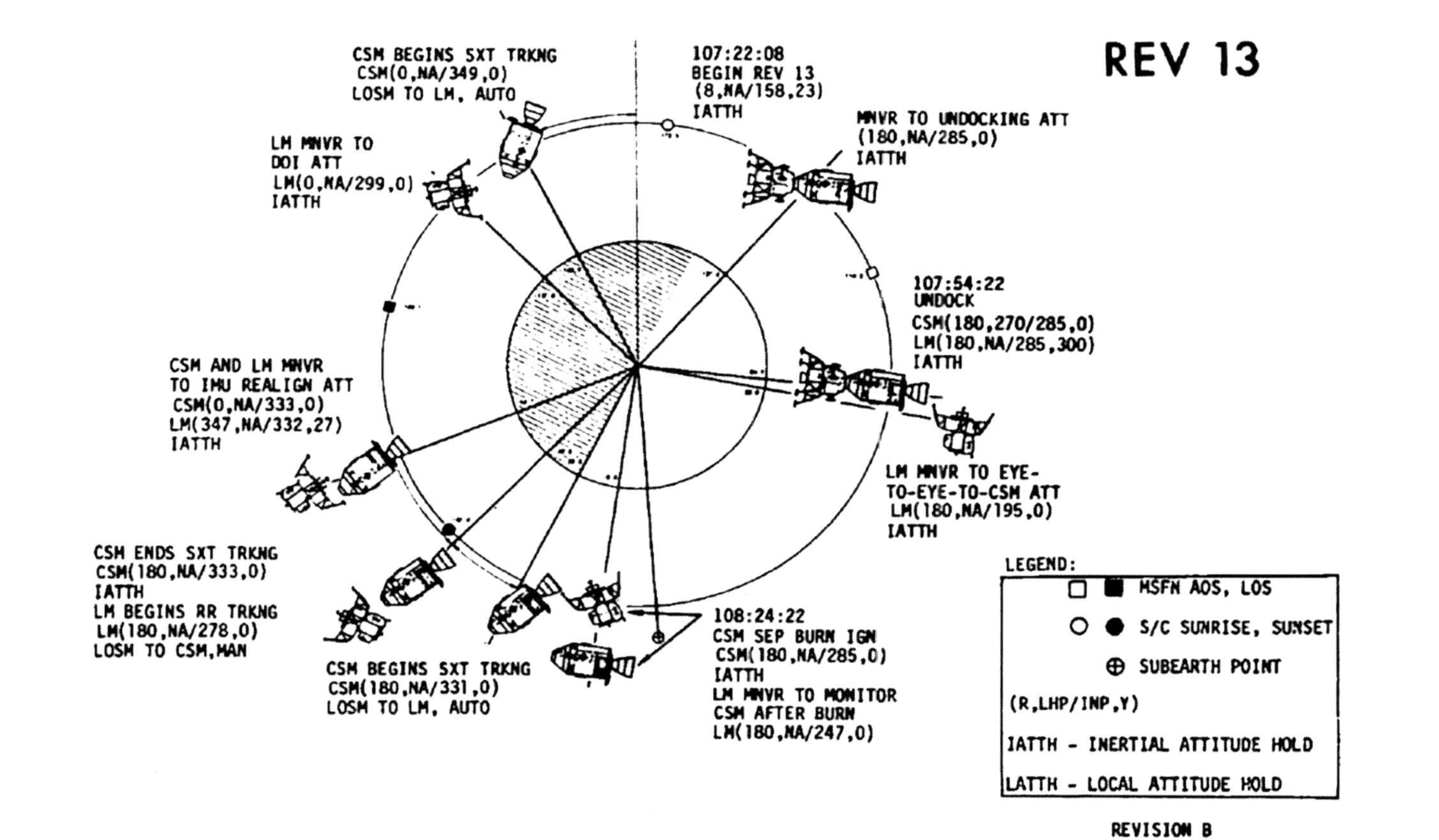

A page from the flight plan showing (clockwise from the top) how Dick Gordon would orient Apollo 12 in preparation for undocking, and the preparation for the descent orbit insertion manoeuvre by Intrepid.

A view of Intrepid taken by Dick Gordon shortly after undocking. The smooth area in the left-foreground is the floor of the crater Ptolemaeus, and the largest crater to the right is Herschel.

that would fire the descent propulsion system for descent orbit insertion, in order to let Mission Control inspect the resulting telemetry.

While waiting for an AGS update to complete, Conrad complained, "Goddamn, we ought to have this alignment done."

"Heck, Pete," Bean pointed out, "we're not even supposed to be there yet."

"I know," Conrad admitted. "I just want to stay ahead."

"Shit," Bean complained, "you make me think we're behind all the time."

"Gotta hustle, boy," Conrad insisted. He regarded losing their lead time as being equivalent to falling behind. When the update completed he told Bean, "Let's don't bother checking it."

Bean, more cautious, said, "Let me ask Houston about the AGS and PGNS, and see what they think."

While Conrad and Bean advanced through the checklist the flight controllers in Houston compared the primary and abort guidance systems, then Carr relayed their verdict, "You're PGNS and AGS look real good."

With this confirmed, Conrad initiated P52 to perform a star sighting to check the drift of the inertial platform, preparatory to descent orbit insertion. "Hang on, you're going for a ride," Conrad told Bean as the digital autopilot began the manoeuvre.

"Man, this is a fast manoeuvre rate," Bean observed. "What've you got in for this thing?"

"Two degrees a second," Conrad replied.

"You're kidding."

"No, I'm not kidding," Conrad insisted.

"That's not 2 degrees a second."

"Look at the 8-ball; look at your rate needles," Conrad said as the stars raced by the window in the yaw rotation.

"I hope the Moon isn't in the way," Bean said, thinking that perhaps by making the sighting ahead of schedule the Moon might block the line of sight to the assigned star.

"The Moon's not in the way," Conrad said. "There's nothing but stars out there, buddy."

Once in the specified attitude, they dimmed the cabin lights and Bean donned an eye patch and peered through the telescope mounted above the forward instrument panel. As with the telescope in Yankee Clipper, it provided no magnification with a field of view spanning 60 degrees. The star Capella was not quite centred, so Conrad refined the vehicle's attitude in response to a series of verbal cues from Bean based on the position of the star in his field of view. Then the autopilot manoeuvred to the star Rigel, and the process was repeated. As Conrad reflected in the post-mission debriefing, "Al and I had practiced doing a two-man alignment. Al looked at the stars. I flew the spacecraft and ran the computer, and we got the alignment done in a pretty snappy order." Bean agreed that this was a good way to do the alignment, "It lets the fellow who's looking out at the stars keep his night vision." The gyroscopic drifts were well within allowable limits. With the platform tweaked, it was time to perform a final star sighting to check the angular scale incorporated into Conrad's window. In the visual phase of the powered descent, this landing point designator would show him where the computer was aiming for. "What star is it I'm looking for?"

"Aldebaran in Taurus the Bull," Bean replied.

"Okay," said Conrad. "That one I know." Once Intrepid had pitched to the angle corresponding to 40 degrees on the window scale, he peered through a small optical sight. There had been some doubt about whether such a sighting would be feasible, but he had no difficulty. "Hell, I can see the stars in broad daylight with the cockpit lights up." After carefully checking the star, he announced, "Old Aldebaran is half a degree off in yaw and half a degree off in pitch, which is below the noise level." The misalignment revealed by this calibration was negligible.

"Do you want to mark it?" Bean asked, offering a pen to annotate the scale on the window.

"Hell no; I know where it is," Conrad retorted.

Just before they passed around the far-side, Carr gave the go-ahead to undertake the descent orbit insertion manoeuvre.

On resuming their preparations Bean said, "We've got to don helmets and gloves in a minute."

"Why?" Conrad asked.

"Because it says so," Bean said, pointing to the checklist.

"Okay," laughed Conrad.

Although they sealed their suits, they did not pressurise them. Upon growing hot, Conrad asked, "Hey, why don't you turn up a little of that LCG cooling." A minute later, he whistled and exclaimed, "Hey, that's cold enough. Goddamn."

"I just barely moved that son of a gun," Bean pointed out. The flow setting for the liquid-cooled garment was rather coarse. "Let me turn her down."

In addition to specifying switch settings, the checklist contained a reminder that the lunar module pilot should refresh the commander's memory of the abort rules.

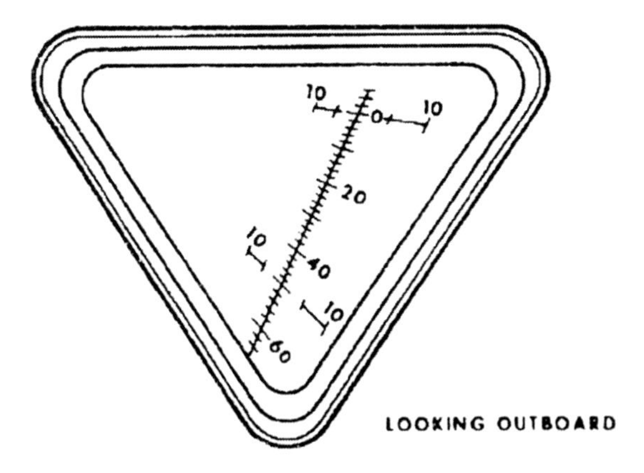

Details of the landing point designator scale incorporated into the commander's window of the lunar module.

Conrad undoubtedly *knew* the rules, but Bean read him the list of contingencies and their requisite actions.

"Yankee Clipper, Intrepid," Conrad called, "Is it all right with you if we turn our tracking light off?" The flashing beacon was to enable Gordon to track them while in darkness, and Conrad wanted to save its limited life for the rendezvous. Gordon said to go ahead.

"Clipper," Conrad called, "we're going to the DOI burn attitude at this time." The digital autopilot manoeuvred to the attitude required for the descent orbit insertion. It was an inertial attitude which, when the time came to make the burn, would point the descent engine directly down the velocity vector so that its thrust would serve as a brake. Flying on instruments, Conrad remarked to Bean, "I don't know whether I'm right-side-up or upside-down in this damned inertial world."

With the manoeuvre imminent, they ran through the final setup. With just over a minute remaining, Conrad announced, "I'm ready to burn."

The lunar module used hypergolic aerozine-50 and nitrogen tetroxide pressurised by helium. The ascent stage had four quads of 100-pound-thrust engines. In addition to providing attitude control, these engines were to deliver pre-manoeuvre ullage to settle the propellants of the main engine. The descent orbit insertion manoeuvre was preceded by a two-jet ullage firing.

"Ignition," announced Conrad, with Gordon listening in. "It's burning; it looks good."

For the first 15 seconds of the 29-second burn, the descent propulsion system ran with its throttle at 10 per cent.[25] The engine could be gimballed by 6 degrees in any direction, and in this initial period the computer aligned the thrust precisely through the centre of mass. Then the throttle was opened to 40 per cent for the remainder of the burn. As soon as the desired change in velocity had been achieved, the computer cut off the engine.

"Shutdown," Conrad called.

The manoeuvre occurred at an altitude of 61.0 nautical miles at selenographical coordinates centred on 6.45°N, 171.5°E. It lasted 28.97 seconds, and the onboard solution said that the 72.4-foot-per-second retrograde delta-V had placed Intrepid in an orbit with an apolune of 60.5 nautical miles and a perilune of 8.0 nautical miles. By design, the low point was 16 degrees (260 nautical miles) uprange of the ALS-7 target.

Having performed the separation manoeuvre radially downward, Gordon was currently below and ahead of Intrepid at a range of 2.2 nautical miles. He had his vehicle oriented to point the optics at the lunar module, whose engine was aimed against the direction of travel. As he explained in the post-mission debriefing, his view of the burn "was almost like looking straight up the tailpipe of an airplane".

As they worked through the post-burn checklist, Conrad, observing the residuals, remarked, "Oh, that's a good burn."

[25] The nominal 'full thrust' of the descent propulsion system was 9,870 pounds. The engine could be throttled, but its minimum setting was 10 per cent.

"Do not trim," Bean reminded him. The velocity residuals indicated by the PGNS were very low, and in close agreement with those displayed by the AGS. Normally the commander would fire the thrusters after a major manoeuvre in order to zero the residuals in each axis, but on this occasion, to minimise the effect of accelerometer bias errors, the residuals were written down and after acquisition of signal would be reported to Mission Control for incorporation into a state vector update.

After finishing the post-burn checklist, Conrad turned Intrepid to the inertial attitude appropriate for the initiation of the powered descent. Since at *that* time they were to be flying with their windows facing away from the Moon, this meant that they now faced towards the lunar surface and on emerging around the limb would be 'upright' and facing Earth.

Gordon had intended to track Intrepid in its descent orbit right down to where it initiated the powered descent. His technique involved taking sextant sightings and VHF ranging measurements to enable his computer to estimate the lunar module's state vector in readiness for a possible rescue. Unfortunately, he was preoccupied with the sextant when the computer displayed the first VHF ranging measurement and he accepted it without thinking. Only as further data came in did he realise that the first result was wide of the mark. As he said in the post-mission debriefing, this "kind of blew the state vector. I looked out the sextant and saw the LM again, but by then I really wasn't very concerned about keeping its state vector up to date. In fact, by accepting that bad VHF, the LM state vector was essentially lost."

Meanwhile, Bean was looking ahead to the final phase of the powered descent, "Fly this thing in there."

"I'll do my best," Conrad promised.

"I know it," Bean said.

"I just hope it's enough," Conrad added.

A moment later, with everything going to plan, Bean praised their vehicle, "What a machine!"

When Bean began to review the No-PDI + 12 contingency manoeuvre, noting that it would put them into an orbit ranging between 13 and 147 nautical miles, Conrad, sure that they were about to initiate a landing, not a rendezvous, cut him off, saying, "Forget it. We ain't doing that one."

"Hope not," Bean agreed.

As they continued to advance the checklist, Conrad again felt hot, "I need another shot of cold."

"Another shot of cold water coming up," said Bean, adjusting the temperature of the flow to their liquid-cooled garments.

During a break in the checklist shortly before acquisition of signal, Conrad looked out at the Moon and called Gordon, "You know, it doesn't look any closer to me down here than it did at 60 miles. I frankly can't tell the difference."

"Well, you're not that close yet, are you?" Gordon pointed out.

"Maybe that's it," Conrad said. "Heck, I don't know."

"I don't imagine we've come down more than 6 miles," Bean ventured. He asked the computer to display their current altitude, which was 50 nautical miles.

The glassed-in gallery of the Mission Operations Control Room was packed. The

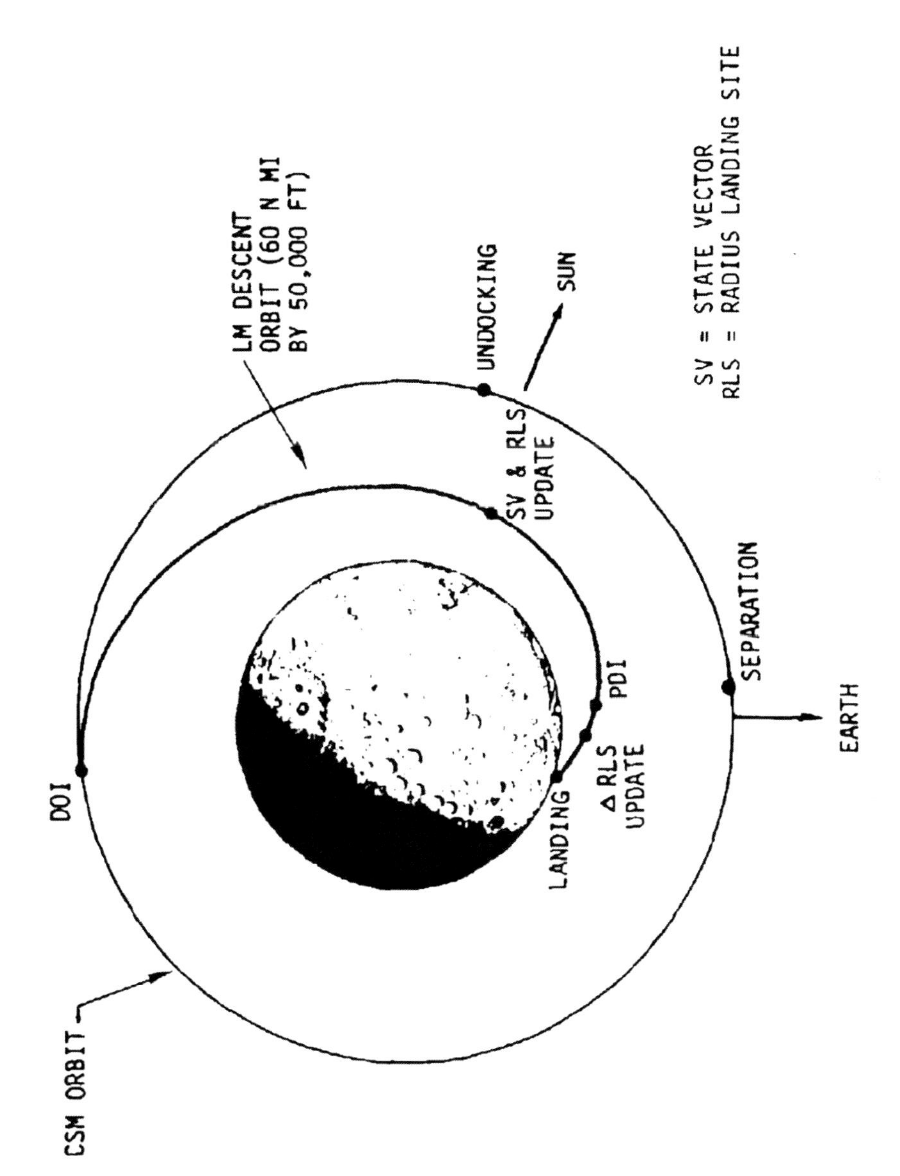

The manoeuvres to be made by Intrepid from descent orbit insertion to lunar landing.

VIPs included NASA Administrator Thomas Paine and his wife; Deputy Administrator George Low; Associate Administrator Homer Newell; Associate Administrator for Manned Space Flight George Mueller; Associate Administrator for Space Science and Applications John Naugle; Wernher von Braun, Director of the Marshall Space Flight Center; and Stark Draper, Director of the Instrumentation Laboratory at the Massachusetts Institute of Technology that developed the Apollo guidance system. Also present were astronauts Frank Borman, Neil Armstrong and Buzz Aldrin.

"We're going to get Earthrise in a second," Conrad noted. "You ought to have Earthrise by now, Dick."

Yankee Clipper had been slightly below and ahead of Intrepid at the time of the descent orbit insertion manoeuvre, but as the lunar module descended Gordon was maintaining a pitch which would hold his optics facing it. "I'm looking backwards," he pointed out.

"Oh, that's right," Conrad acknowledged. "I forgot."

Eager to snap pictures of Earthrise, Bean noted that since the vehicle was flying with its windows in-plane, the fact that the orbit was inclined to the lunar equator by 15 degrees meant the view should be better through Conrad's window than through his own. "There!" he exclaimed.

"No," said Conrad.

"I thought I saw it coming up," Bean said.

"There it is!" Conrad called several seconds later, as Earth began to rise over the lunar horizon and Bean started to snap pictures.

"We just watched our first Earthrise, which was fantastic!" Conrad replied when Carr established contact on revolution 14. "And we had a great DOI burn." He then provided the PGNS residuals from the burn and Bean supplied the corresponding values from the AGS.

The fact that the landing site was further west than that for the previous mission provided sufficient time between the appearance around the limb and the initiation of the powered descent for Mission Control to calculate a revised state vector based on tracking obtained during the previous near-side pass and the residuals from the descent orbit insertion burn. In addition, as soon Intrepid appeared, the deep-space navigation expert, Emil Schiesser, seated with the trajectory specialists in the row of consoles known as the Trench, started to compare the predicted and actual Doppler data in order to calculate the 'range correction' that would be the key to attempting a precision landing.

Onboard, Bean stowed the Hasselblad camera with which he had been snapping pictures of Earth and declared, "Time to get down to business."

"Yes," Conrad agreed.

One item on the checklist was to configure the biomedical telemetry. Although it was possible to monitor all three men simultaneously in the command module, the lunar module's system could transmit only one data stream and there was a selector switch. The Flight Surgeon wanted to monitor Conrad during the descent.

Bean told Carr that Earth was about 30 degrees above the horizon, concluding, "you really are beautiful".

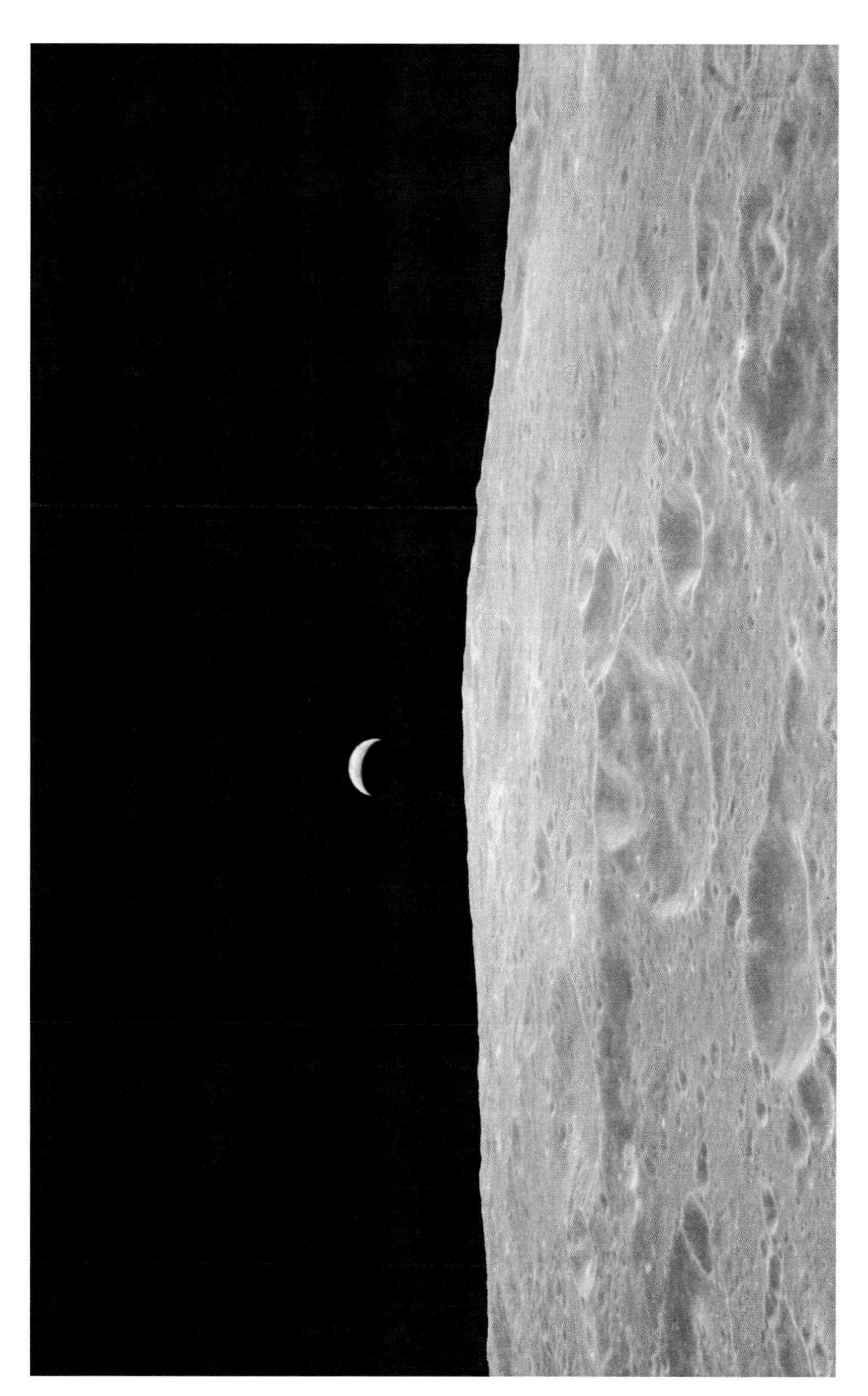

A shot of Earthrise taken by Al Bean as Intrepid emerged around the limb in the descent orbit.

"We put on our Sunday best for you," Carr replied, although it was not actually Sunday.

"We're pretty well suited-out up here ourselves," Bean laughed.

Conrad asked Carr if they were indeed coming in 5 nautical miles north of their intended line of approach. When this was confirmed, he asked whether this would change the azimuth of the approach to the Snowman, but the cross-range correction would be made during the braking phase of the powered descent and they would be essentially on-track by the time the lunar module pitched over and gave Conrad his first view of the terrain ahead.

Bean was confident, "It'll be just like landing anytime on Earth, Pete."

"What?" Conrad asked, somewhat puzzled.

"The closer I get, the more I feel like it will be just like landing on Earth," Bean explained. "Just let it fly down and you park it somewhere, then we'll bound out and grab a rock!"

Moving down the checklist, Bean refreshed Conrad's memory of the abort rules for the powered descent. Meanwhile, Mission Control updated Intrepid's computer with a revised state vector and radius-to-landing-site, the latter being an estimate of how far the target was from the centre of the Moon, and thus its elevation relative to the spherical globe assumed by the navigation system.

Conrad had decided not to employ the multiple methods used by Neil Armstrong and Buzz Aldrin to verify the line and timing of their ground track during the run in to the powered descent. In particular, he did not want to fly 'windows down' to use the landing point designator to time the passage of landmarks approaching perilune as a means of judging the altitude at perilune and then perform a 180-degree yaw to 'windows up' in the braking phase. Armstrong had used these 'manual' checks in order to assess the accuracy of the PGNS prior to committing himself. But Conrad did intend to use rendezvous radar tracking of Yankee Clipper to measure Intrepid's position relative to that vehicle's orbit as a means of calculating what his altitude would be at perilune – if this was satisfactory, he was willing to put his trust in the spacecraft and the MSFN tracking and approach PDI essentially blind. The nominal perilune altitude was 50,000 feet. The first estimate was 57,500 feet. A minute later, the second estimate was 59,000 feet. As the estimates increased to 64,000 feet it became evident that something was wrong. However, Mission Control assured them that the radio tracking was still showing a good perilune and recommended that they ignore the rendezvous radar procedure. The point at issue was that if they were to initiate the descent from too high an altitude, Intrepid would run out of fuel before it could finish the descent. If the perilune proved too high, they would have to forgo the powered descent on this pass and manoeuvre at apolune to lower the perilune in order to try again on the next pass. It was subsequently realised that the method was based on the mothership being in a particular orbit, and it was not in that orbit. As Conrad wryly recommended in the post-mission debriefing, "I'd like to recommend that this check be eliminated. It is just busy work. As you can see, it didn't provide any intelligence that let us make a determination."

With the decision to proceed made, Conrad started P63, the program that would undertake the braking phase of the powered descent, in order to enable Houston to examine the telemetry, and then he deactivated it.

In crossing the eastern longitudes of the near-side in the attitude for the initiation of the powered descent, the lunar surface was still visible in the lower corner of the triangular windows. When Conrad observed that they were getting pretty low, Bean said that it helped not to think about it! Carr offered sensible advice, "All you've got to do, is not fly through any clouds." This was a reference to the comment by Bean the previous day that bright mountains on the dark plain looked like cumulus clouds over an ocean. After another look outside, Conrad called Carr, "I sure hope you have us lined up right, Houston, because there sure are some big mountains in front of us right now." In fact, they were still east of the meridian and well short of the perilune point.

As Conrad continued to comment on the terrain ahead, Carr, an aviator with the Marine Corps, suggested, "Call when you want us to turn on the mirror." This was a reference to the mirror on an aircraft carrier that indicated to an approaching aircraft when it was on the correct glide slope.

In that spirit, Conrad said, "You can turn the mirror on. Give me the fox-corpen. And we've got the hook down." A naval aircraft had a hook on its tail to engage the retarding cable strung across the carrier's deck.

Gordon promptly chipped in, "Fox-corpen, 285. Clear deck." This indicated the azimuth. As due west was 270 degrees and their orbit was inclined at 15 degrees to the equator, this meant they would make their final approach on an azimuth of 285 degrees.

Shortly thereafter, Conrad reported that he could no longer see the surface unless he leant fully forward against his restraint harness. At this point, Apollo 11 had been 'windows down' and S-band communications had been impaired by the structure of the spacecraft masking the line of sight of the antennas. By flying in a 'windows up' attitude, communications with Intrepid remained excellent throughout.

After starting P63, Conrad had the computer repeat the manoeuvre to the attitude appropriate for initiating the powered descent, to cancel any drift that had occurred since originally adopting this attitude. Then they reinitialised the AGS to the PGNS and activated the landing radar. The radar was on the underside of the descent stage on a two-position hinge whose initial configuration faced to the rear. With Intrepid oriented 'windows up', this aimed the radar at the ground. In the cabin, the ALT and VEL lights came on – they would go *out* when the radar was successfully measuring the vehicle's altitude and horizontal velocity respectively. The rate of descent, referred to as H-dot, would be calculated from the change in altitude.

In Mission Control, Cliff Charlesworth polled his flight control team, then Carr relayed the result, "Intrepid, Houston. Go for PDI."

Some 12 minutes later, Conrad, satisfied that everything was ready, armed the descent engine. The computer fired a pair of thrusters for ullage, and then, with 5 seconds remaining, it flashed Verb 99 seeking final authority to initiate the descent. Conrad pressed Proceed.

"I have ignition," Conrad announced, and 5 seconds later continued, "Descent Engine, Command Override, On." This would prevent a spurious command from switching off the engine.

At the start of the burn, the spacecraft was at selenographical coordinates 7.76°S, 7.82°W, at an altitude of 7.96 nautical miles and travelling at 5,560 feet per second. The target was 260 nautical miles further on. The Press Kit scheduled the burn for 110:20, and it was precisely on time. One factor in designing the 'hybrid' translunar trajectory was to arrange the timing to enable the 210-foot-diameter antenna at Goldstone to provide radio tracking throughout the descent. As previously, the engine began at 10 per cent thrust and was gimballed to align the thrust vector precisely through the centre of mass, because an offset thrust would induce rotations which the thrusters would require to correct and (more seriously) cause 'wasted' thrust which would not contribute to the braking. This time, the preliminary phase lasted 26 seconds.

"Throttle up!" Conrad called when the thrust was increased to the 'fixed-throttle' position of 93 per cent. It would remain at this thrust for most of the braking phase.

"DPS is looking good, Pete," Carr reported, relaying Mission Control's view of the telemetry.

"Alrighty," confirmed Conrad.

As the landing site was approximately 5 nautical miles south of the orbital plane, the computer commanded a roll angle of 4 degrees to correct this cross-range error during the braking phase.

"There's a little RCS activity, not too much," Bean noted. The thrusters sounded like someone hammering on the wall of the cabin. It was more thrusting than in the simulator, but Neil Armstrong had told them to expect this.

Flying blind, Conrad keyed the computer at 30-second intervals and checked the trajectory parameters measured by the PGNS against predictions on a cue card. As part of his routine, Bean compared the AGS against the PGNS to ensure that it was accurate in case they had to abort. One minute into the burn, they were descending through 48,000 feet with a horizontal velocity of 5,208 feet per second and a rate of descent of 20 feet per second.

Since Intrepid appeared around the limb, Emil Schiesser had been monitoring the Doppler shift on its radio signal, comparing how it varied against the prediction in order to compute the 'range correction'. This calculation found that the PGNS was aiming for a point 4,200 feet short of the target, and the chosen means of telling it to extend by that distance was to imply that the crew had decided to change the target. This downrange redesignation could have been postponed to the approach phase, when Conrad could see the surface ahead, but it would have consumed more propellant than doing it in the braking phase.

The correction was now relayed by Carr to the crew in a typically prosaic form, "Intrepid, Houston. Noun 69, plus 04200."

"Plus 04200," Bean echoed.

"That's affirmative," Carr said.

After Bean had entered this value into the input register, he waited for Mission Control to verify it. Because this update would significantly affect their trajectory, it was essential that it be made properly – as otherwise they might crash.

"Intrepid, Houston. Go for Enter," Carr confirmed.

Bean told the computer to accept the update, then advised Conrad, "It's in, babe."

Everything was still going well at the 2-minute mark.

Half a minute later, Bean observed, "It feels good to be standing up in a g-field again." Unlike in the command module, in which they occupied their couches for thrusting, in the lunar module they were standing upright and the thrust imposed a welcome sense of gravity.

Dropping through 44,700 feet with a horizontal velocity of 4,276 feet per second and a rate of descent of 53 feet per second, Conrad ventured, "We're smoking right down there."

"Intrepid, Houston," Carr called, "You're looking good at 3 minutes."

Landing radar acquisition in altitude occurred at 41,438 feet, and acquisition in velocity occurred several seconds later at an altitude of 40,100 feet. The two lights went out to indicate that the radar was yielding data. The computer would not use the radar data in its guidance calculations until instructed to do so. At this point, the computer was flying on the basis of the recently updated state vector. If the radar was functioning properly, then its data would be better and the computer would need to incorporate this into its navigation calculations. As the first step, the computer compared the altitude estimated by the PGNS against the radar value. The delta-H value initially showed the actual trajectory to be 1,000 feet low. It was fluctuating with the undulations in the terrain, but never exceeded 2,000 feet.[26]

"It looks good. Recommend you incorporate it," Carr advised.

"No sooner said than done," replied Conrad.

The computer's strategy was to split the difference and iterate the process in order to smooth out the correction. After 30 seconds the delta-H had converged to 400 feet, and 2 minutes later was a mere 100 feet.

The thruster activity picked up as they descended through 35,000 feet, caused by the computer compensating for the change in Intrepid's centre of mass as the main engine's propellants were burned off and the remaining propellants sloshed in their tanks. Throughout the burn, the computer monitored how it was required to fire the thrusters to maintain the vehicle in the desired attitude and used this information to gimbal the engine. Ideally, this feedback loop would both align the engine's thrust vector through the centre of mass and minimise thruster activity, but the process was perturbed by propellant sloshing. Conrad asked Carr how the gimbal looked in the telemetry and was assured that it was nominal.[27] At the 5-minute mark the lateral velocity component required to correct for the cross-range error reached its peak of 78 feet per second. With the PGNS now incorporating radar data into its navigation, Bean reinitialised the AGS. It was at about this point on Apollo 11 that the first of several program alarms occurred owing to the computer being overloaded, but that problem had been eliminated and Intrepid flew on without such distractions.

Although the PGNS had been commanding the descent propulsion system thrust

[26] If they had found themselves in excess of 10,000 feet *higher* than the PGNS estimated, they would have been obliged to abort, since to have continued would have resulted in Intrepid running out of fuel before it reached the surface.

[27] The increased awareness of the thrusters in flight merely reflected an inadequacy of the simulator, which was not affected by sloshing.

as a decreasing function, the need to limit erosion of the engine nozzle meant it was throttleable only between 10 and 60 per cent and, as a result, the thrust was to be held at the 'fixed-throttle' position until the commanded thrust decreased below 60 per cent, whereupon the throttle would start to operate as commanded. In fact, the throat erosion was substantially greater than predicted, but this was interpreted as merely a deficiency in the reference model, which was duly revised to better reflect actuality. On a nominal descent, this stratagem would allow about 2 minutes of true throttling towards the end of the braking phase in order to control dispersions in thrust and trajectory.

"Throttle down will be at 6 plus 22," Carr announced, a minute in advance of the time estimated by the flight controllers based upon the vehicle's performance.

"After the engine throttles down, I'm going to put on the camera," Bean said to Conrad, referring to the 16-mm movie camera he had mounted in his window.

"You're looking good at 6 minutes," Carr advised.

Bean again updated the AGS.

"Throttle down!" Conrad called, when the event occurred within 1 second of the predicted moment. They were at about 25,000 feet. After leaning forward against his harness to peer through the bottom corner of his triangular window he told Bean, "I can just barely see the horizon."

"Why don't I go ahead and put that camera on now?" Bean suggested.

"All right," Conrad agreed.

"It's running," Bean confirmed.

At the 7-minute mark, Carr advised, "Monitor descent fuel 2." After studying the two redundant propellant gauging systems, each of which had 'low level' sensors for fuel and oxidiser, Mission Control's recommendation was to set the Quantity Light in the cabin to monitor gauging system number 2.

"We're out of 19,000 feet," Conrad said, half a minute later. "I've got some kind of a horizon out there. I've got some craters, too, but I don't know where I am yet."

Bean reconfirmed that the movie camera was working.

"You're looking good at 8 minutes," Carr advised.

Bean pointed out that they were descending through 12,000 feet, and Conrad told him, "When we get to 10,000 feet, hook up your lanyard." This was another aviation term. A low-altitude lanyard would automatically deploy the pilot's parachute as he separated from his seat after ejecting from an aircraft, but it was hooked up only when flying below 10,000 feet. Of course, in this case he was joking.

The braking phase was to end at the 'High Gate', at an altitude of about 7,000 feet and some 5.2 nautical miles uprange of the computer's aim point, whereupon the vehicle pitches over to transition from firing its engine entirely against the velocity vector to directing an increasing portion of its thrust downward. This transition also faces the windows forward for the approach phase of the descent. In performing this pitch manoeuvre, the computer automatically switches from P63 to P64.

Conrad announced, "Standing by for P64."

"Okay," replied Bean.

Conrad stood on his toes for a better view through the corner of his window. "I'm trying to cheat and look out there," he explained, but was unable to get his bearings.

As Bean reported they were at 7,000 feet, the computer switched programs and he announced, "P64, Pete."

"P64," Conrad acknowledged.

"Pitching over," Bean confirmed.

As the vehicle pitched, the radar on its underside hinged to its second position in order to continue to face the ground.

On initiating the approach phase, the computer commanded a ramping decrease in throttle from 54 per cent down to 33 per cent over a period of 90 seconds.

As they passed through 6,000 feet, Bean updated the AGS for the third and final time in the powered descent.

During the braking phase, the PGNS had corrected the 5-nautical-mile cross-range error and performed the 'redesignation' to extend the trajectory by 4,200 feet. If all was well, then, as Dave Reed had promised, Intrepid should be heading directly for Surveyor crater. As more of the terrain head became visible, Conrad aligned himself with the landing point designator scale on his window and looked for the distinctive crescent of craters beyond the Snowman formation. With the Sun barely above the horizon behind him, craters were readily discerned by the shadows in their interiors. Suddenly he saw the arc of craters near the horizon. And, remarkably, the projected trajectory passed through the belly of the Snowman.

"Hey, there it is!" he excitedly called. "There it is! Son-of-a-gun! Right down the middle of the road!!!"

"Outstanding!" Bean congratulated. Although they were clearly on track, they had yet to determine where on that line the computer was aiming. In P64, the computer supplied angles for the landing point designator. "42 degrees, Pete."

Conrad sighted at that angle on the scale and announced in amazement, "Hey, it's targeted right for the centre of the crater!" All the refinements to achieve a pinpoint landing were working out precisely as intended.

Although the commander would cycle his attention between his instruments and the view outside during the visual phase of the descent, the lunar module pilot's role was to remain 'inside' and recite information. But Conrad prompted Bean to take a look for himself. Bean did not have the advantage of a scale on his window, but the Snowman was clearly visible. Returning to the computer, Bean reported that their altitude was 3,500 feet and the rate of descent was 99 feet per second. On checking the propellant gauge, he pointed out that they had 15 per cent fuel remaining.

"Intrepid, Houston. Go for landing," Carr announced.

"Roger," Conrad acknowledged.

To redesignate the target, Conrad could click his rotational hand-controller left or right, forward or back, for a measured displacement. The computer appeared to be aiming right for Surveyor crater, but he did not actually want to land inside it. Also, they had been asked to remain at least 500 feet from the Surveyor 3 lander. His plan was to set down on a flat-looking patch, dubbed Pete's Parking Lot, northeast of the crater. "I'm going to flip over one," he decided, and clicked to displace the aim point 2 degrees to the right which, for their approach azimuth of 15 degrees north of west, was essentially a displacement to the north.

"40 degrees, Pete," Bean continued.

"That's so fantastic, I can't believe it!" Conrad said.

"You're at 2,000 feet," Bean reported their altitude.

As they got closer, the 'footprint' of the LPD shrank, improving Conrad's ability to judge the computer's aim. Initially, they had appeared to be heading for a point inside Surveyor crater, but the 4,200-foot correction matched a point just outside the crater's southwestern rim. The redesignation right was meant to move the aim point onto the ground north of the crater, into the general area in which Conrad intended to seek a safe site. But he now realised that his redesignation had actually moved the aim point *into* the crater, so he moved it downrange. On seeing that this put the aim point in the crater that formed the Snowman's head, he promptly added another such adjustment, relocating the aim point beyond the head and even further from Pete's Parking Lot. This was not necessarily a problem, as in planning the mission they had picked provisional landing spots at four places near the Snowman and sketched out a geological traverse for each. But Conrad wanted to guarantee that they would be able to visit Surveyor 3, and that required achieving a landing within about 2,000 feet of it. He would therefore have to draw the aim point nearer to the crater in which the lander resided.

Bean maintained his recitation, "38 degrees. 36 degrees; you're 1,200 feet, Pete."

"Okay," Conrad acknowledged.

On passing through 1,000 feet with a rate of descent of 30 feet per second, Bean pointed out that they had 14 per cent fuel remaining. On stealing a glance out of his window, he was astonished by the speed of their approach. Nevertheless, it was clear that they were in the right ball park. "It looks good out there, babe!"

Conrad redesignated the aim point north.

"You're at 800 feet," Bean said. "33 degrees."

Rather than attempt to manoeuvre for Pete's Parking Lot, Conrad opted to try for a spot somewhere northwest of Surveyor crater, and he redesignated short twice in rapid succession to draw the aim point back around the Snowman's head. As he did so, Intrepid descended through 500 feet about 2,000 feet uprange, a position known as the 'Low Gate'. Several seconds later, Conrad moved the aim point north, closer to his chosen area.

When Bean reported their altitude to be 400 feet, Conrad announced, "I've got it." By selecting the rate-of-descent switch, he caused the computer to change to P66 to execute the landing phase in a semi-automatic mode in which he specified the rate of descent and the computer controlled the engine in a manner that would produce that rate.

"You're in P66, Pete," Bean confirmed.

On initiating this program, Conrad had slowed the rate of descent from 9 feet per second to half that rate.

Bean reported that they had 11 per cent of their fuel remaining. This put them in a much better situation than their predecessors at this point. Stealing a look outside, he remarked, "Oh! Look at that crater. It's right where it's supposed to be!" Not having followed Conrad's redesignations Bean did not know where the aim point was, but to him the line of approach looked excellent.

Their track was approaching the Snowman's northern 'foot' and the aim point

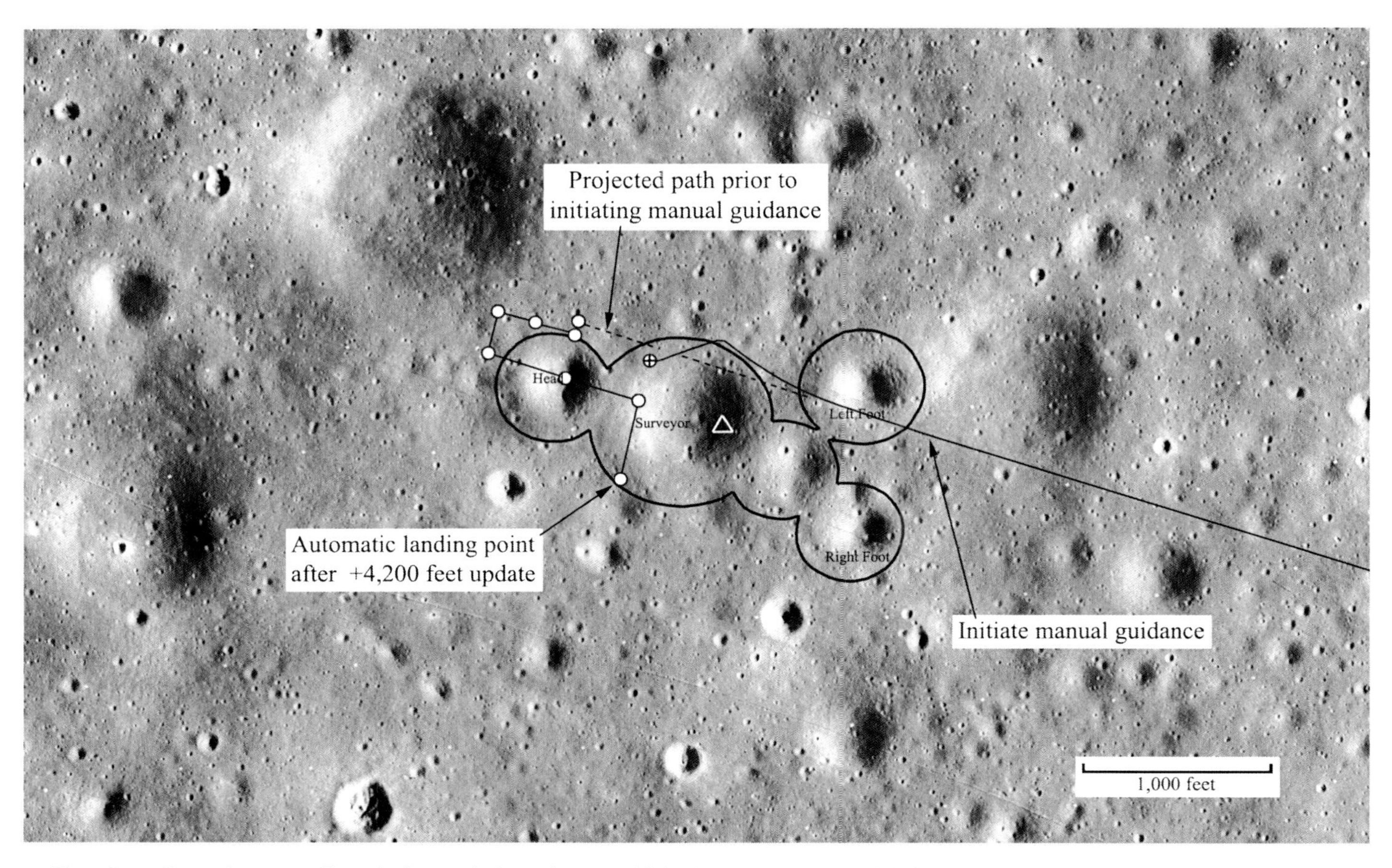

How Pete Conrad repeatedly redesignated the point to which the computer should aim, prior to initiating manual guidance and manoeuvring Intrepid around the north side of Surveyor crater.

was north of its head. Conrad would have to fly a detour around the northern side of Surveyor crater in order to reach the area northwest of that crater. Since P66 did not employ the landing point designator, he would have to steer manually. With Intrepid supported by the pillar of thrust from the main engine, this meant flying in the same sluggish manner as the LLTV on which he had trained. To move horizontally in any given direction he had first to tilt Intrepid so that a portion of its thrust initiated a suitable drift rate, and then before he had strayed too far he had to tilt in the opposite direction in order to cancel this motion. Meanwhile, the computer would throttle the main engine to hold the rate of descent during these attitude changes. Conrad began his detour by rolling right, to move off-track in that direction. Not having flown the LLTV, Bean was unfamiliar with the exaggerated rotations required to start and stop lateral displacements. As he was thrown about in his harness and the 8-ball gyrated, he said, "Hey, you're really manoeuvring around."

"Yep," Conrad replied, preoccupied.

The landing radar data dropped out for 6 seconds during this manoeuvring, owing to the manner in which the relative velocities caused a zero Doppler in one or more of the beams.

"Come on down, Pete," Bean urged, because to make the detour around the crater Conrad had reduced the rate of descent to just 3 feet per second and the engine was guzzling fuel in supporting them against the 5.3-foot-per-second-squared acceleration of lunar gravity.

"Okay," Conrad replied.

Having skirted around the northern rim of Surveyor crater, Conrad had now to swing left to land northwest of that crater. He initiated this leftward translation at an altitude of about 200 feet. During the gyrations to perform this detour, the computer commanded approximately 100 throttle changes in the 19 to 44 per cent range. The time between successive commands was generally less that 0.20 second, which was only slightly longer than the engine's response time. As would be realised later, the engine was on the verge of throttle-instability, which, if it had occurred, would have resulted in wild variations in thrust that would have mandated an abort. (It was also belatedly realised that the same situation had occurred on Apollo 11. The problem was eliminated by a software modification for later missions.)

As they descended through 175 feet, the engine's exhaust plume impinged on the surface and began to stir up dust. Bean said they had 10 per cent fuel remaining and again urged Conrad to increase the rate of descent, which was still 3 feet per second. The landing radar dropped out again, this time for 10 seconds. "130 feet; 124 feet, Pete." Bean reported. Conrad doubled the rate of descent to 6 feet per second. "8 per cent. You're looking okay." At 98 feet, with Conrad maintaining this rate of descent, Bean warned, "Slow the descent rate!" Conrad did so. At 80 feet the rate of descent was 4 feet per second, and at 60 feet it was 3 feet per second once again.

At this point, the low-level fuel sensor of propellant gauging system 2 illuminated the Quantity Light. This was meant to occur when there was 5.6 ($\pm$0.25) per cent of the propellant remaining. In hovering flight with the throttle operating at 32 per cent this would cause the engine to cut off in 116 seconds. With 20 seconds reserved for the preliminary action of an abort, during which the engine would be throttled up to

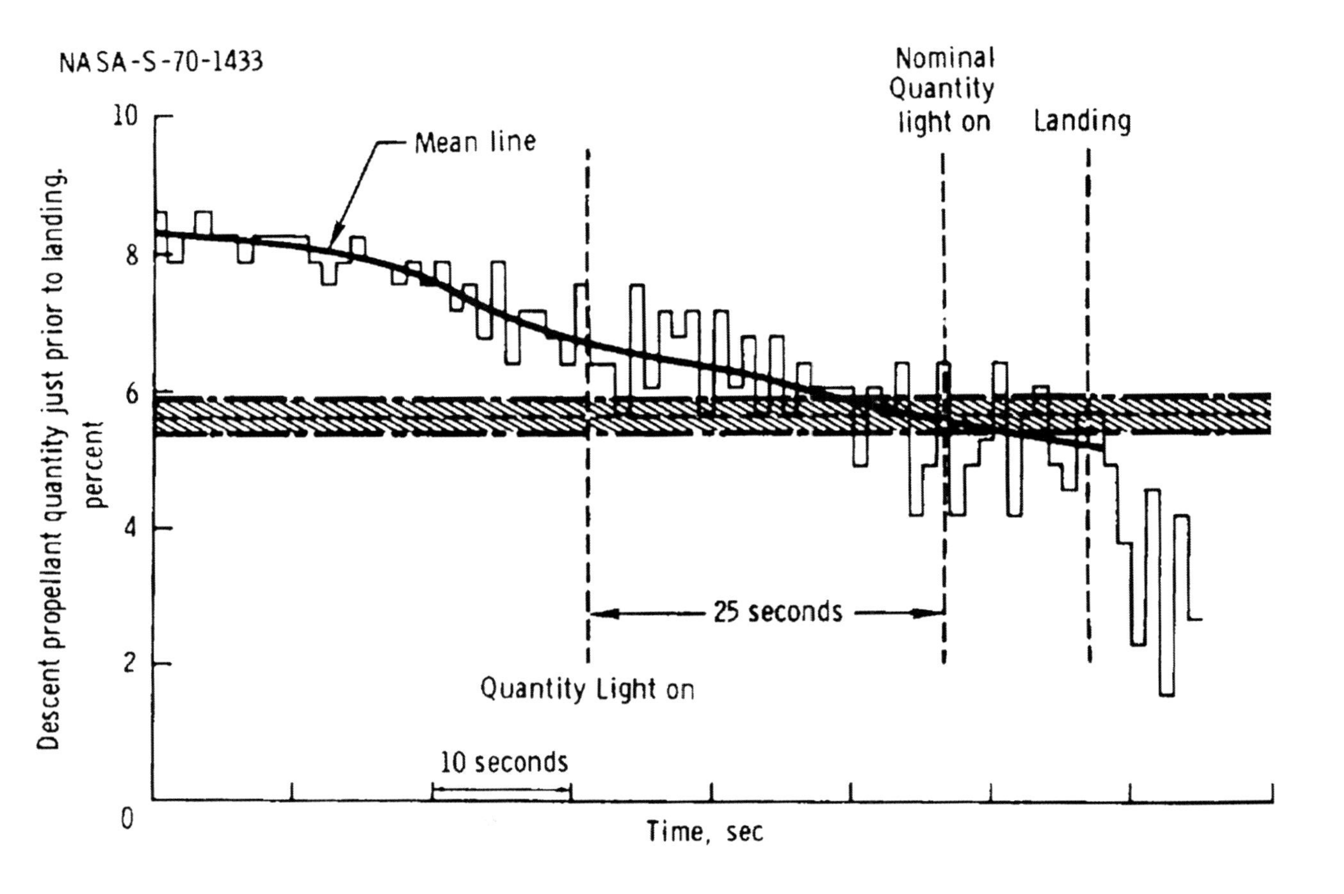

How propellant sloshing caused the Quantity Light to illuminate prematurely during the final phase of Intrepid's powered descent.

cancel the rate of descent and impart a positive climb prior to abort-staging, the low-level signal meant that in 96 seconds Conrad would have either to abort or forgo the option of aborting and commit himself to touching down within the next 20 seconds. To aviators, this decision point was known as the 'bingo' call. However, subsequent analysis of the telemetry revealed that the signal was about 25 seconds premature. In fact, the mean fuel quantity was 6.7 per cent and the sloshing induced by Conrad's manoeuvring had produced a peak-to-peak oscillation of 2.3 per cent which briefly uncovered the sensor. Once triggered, the signal was maintained because the sensor was designed to remain latched. The averaged fuel quantity did not reach 5.6 per cent until almost half a minute later. This early signal did not pose a significant issue, however, as Conrad was already very close to setting Intrepid down.

At 50 feet, still descending at 3 feet per second, the dust displaced by the engine plume was impairing the ability of the radar to measure the horizontal velocities and side-lobes off the irregular surface caused the altimeter to intermittently break lock. Worse, the cross-pointer instrument that was to display the horizontal rates appeared to have failed. The obscuration was due to the entrainment of particles as the plume of exhaust gas flowed across the surface. The result was an ever-thickening sheet of dust which radiated out on shallow trajectories just above ground level. It was much worse than on Apollo 11. As Conrad endeavoured to ease Intrepid down, he cycled his attention between the 8-ball to check his attitude and the window to estimate his rates relative to the boulders that poked up through the sheet of dust. He also yawed 10 degrees to the right of the approach path in order to improve his view through the left-hand window.

"42 feet, coming down at 3," Bean recited. And then, "30 feet, coming down at 2, Pete. You've got plenty of gas, plenty of gas, babe. Hang in there."

By now, the dust was completely obscuring Conrad's view of the surface. He had the 8-ball for attitude control, but his concern was that he might drift backwards. He did not know that the cross-pointer was actually working and was centred because he had completely nulled his horizontal rates while hovering.[28]

Carr announced that they had 30 seconds of flying time remaining, as indicated by the low-level signal.

"He's got it made!" Bean assured Carr.

Several seconds later, as Conrad was easing Intrepid down at 2 feet per second, one of the 68-inch-long probes that projected from the lateral and rear legs touched the surface.

Bean called out, "Contact Light!"

"Roger. Copy contact," acknowledged Carr matter-of-factly.[29]

Having noted the lamp himself, Conrad immediately hit the Engine Stop button, as planned, in order to prevent a blow-back of pressure should the engine bell come

[28] In reviewing the dust situation during the final phase of the descent in the post-flight debriefing, Conrad recommended that future commanders level off prior to disturbing the dust, select a spot, station themselves above it, and then descend with a slight forward motion in order not to creep backwards.

[29] It was 01:54:36 a.m. EST on 19 November.

into contact with the surface. At the moment he pushed this button, the vertical rate was approximately 0.4 feet per second downward. As the engine thrust decayed, Intrepid fell freely for 1.2 seconds before the first of its four 37-inch-diameter foot pads made contact with the surface, and from his perspective it was the right-hand pad. By that time, the vertical rate had increased to about 3.4 feet per second. Intrepid was also drifting forward at about 1.7 feet per second and leftward at 0.4 feet per second. The pitch and roll attitudes when the first foot pad made contact were roughly 3 degrees down and 1.4 degrees left. The at-rest attitude, from the perspective of the crew, was 3 degrees up in pitch and a 3.8-degree roll left, indicating a surface slope of several degrees down to the left/rear. Hence, the act of touching down involved a rotation from the 3-degree pitch-down attitude to the final 3-degree pitch-up attitude, with a maximum angular rate of 19.5 degrees per second, accompanied by a slight left roll and right yaw with maximum rates of 7.8 and 4.2 degrees per second respectively. The legs were designed so that the shock of the impact would drive the lower struts up inside the main struts, which contained crushable aluminium honeycomb. In fact, the landing was so smooth that it barely stroked the shock absorbers. The duration of the powered descent was 11 minutes 57 seconds, and it was later calculated that on touchdown 3.8 per cent of the descent propulsion system propellants remained onboard, which was sufficient for approximately another 103 seconds of firing time.

There was no 'Eagle has landed' announcement this time, just the crew reciting the checklist to shut down the descent stage and ready the ascent stage for a surface abort should this become necessary, although unlike their predecessors they did not actually run a simulated countdown.

The descent stage was eight-sided, alternating long and short sides. It was 14 feet 1 inch in diameter and stood 10 feet 7 inches tall. Its exposed surfaces were covered in a mylar and aluminium alloy to serve as a thermal and micrometeoroid shield. The square central section held the descent engine, and the propellant tanks were housed in the four square peripheral bays: two for fuel and two for oxidiser. The procedure for venting the pressure in the propellant tanks was changed after Apollo 11. On that occasion the tank of supercritical helium pressurant was vented. Unfortunately, this caused the stagnant fuel in the helium/fuel heat-exchanger to freeze, and then when the still-hot engine caused the fuel to expand against the plugged heat exchanger this resulted in a pressure increase in the feed lines which threatened a burst that could pose a risk of contamination for an astronaut working near the vehicle on a moonwalk. This time, the helium tank was isolated prior to releasing the propellant tank pressures.[30]

"Good landing, Pete! Outstanding, man! Beautiful!" Bean praised.

"We're in real good shape!" Conrad reported.

"Roger, Pete," Carr replied.

After just over 1 minute of hectic activity, Conrad and Bean reached the point of waiting for Houston to examine the telemetry, which offered a much more detailed view of Intrepid's systems than was available to the crew. Because four of Intrepid's

[30] The supercritical helium tank itself was vented just before liftoff.

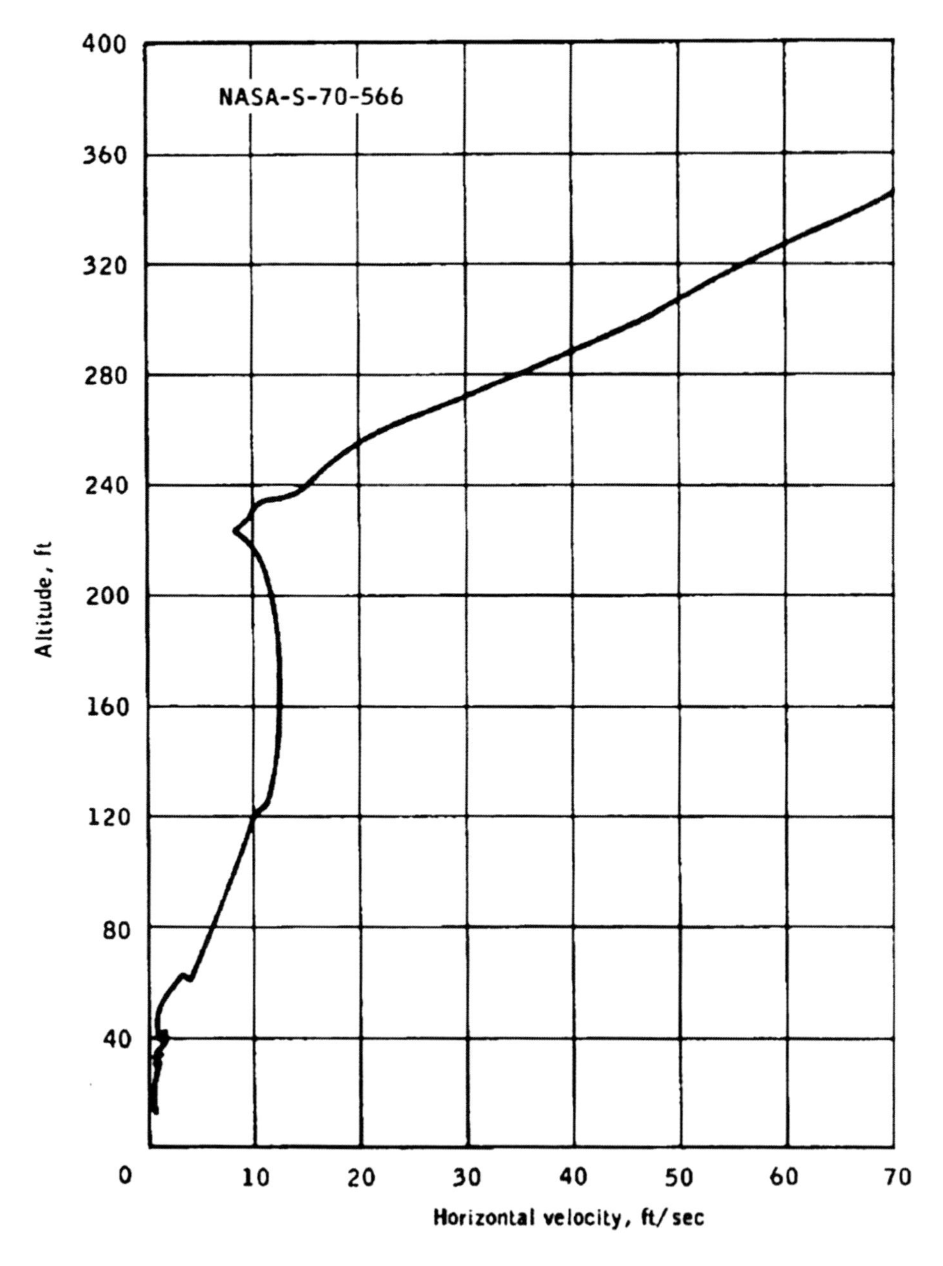

A plot of altitude and velocity calculated onboard during the final phase of Intrepid's powered descent.

six batteries were in the descent stage, the ascent stage had power only for the time required to make a rendezvous. This could begin only when Yankee Clipper was conveniently positioned, and there was a window once per revolution. A rendezvous would be feasible if Intrepid were to lift off within the first 12 minutes of arrival. In planning, two decision points had been defined: T1 and T2. The first decision, 3 minutes after landing, would be on the basis of first impressions. The second, several minutes later, would be made after a more thorough inspection of the telemetry. By tradition, decisions were expressed as Go/No-Go, but Bill Tindall of the mission planning and analysis division had drawn attention during training for Apollo 11 to scope for confusion: "Once we get to the Moon, does 'Go' mean stay on the surface, and does 'No-Go' mean abort from the surface? I think the decision should be changed to 'Stay/No-Stay' or something like that." His advice had been accepted.

While awaiting the T1 decision, Conrad called, "Man, oh man, Houston. I'll tell you, I think we're in a place that's a lot dustier than Neil's. It's a good thing we had a simulator, because that was an IFR landing." He meant it had been an 'instrument landing' because in the final descent he had lost visibility.

"Roger, Pete," Carr replied.

Bean exploited the pause to take his first look outside, "Holy cran, it's beautiful out there!"

"It sure is," Conrad agreed.

At this point, Gordon, knowing that his colleagues were not busy, called through heavy VHF static, "Hello, Intrepid. Congratulations from Yankee Clipper."

"Thank-you, sir," Conrad replied. "We'll see you in 32 hours."

"Okay. Have a ball," Gordon said.

Meanwhile, Bean deactivated the 16-mm movie camera that had been filming the descent through his window.

A few seconds later, Carr gave the welcome news, "You're Stay for T1."

"Roger. Stay for T1," Conrad confirmed.

When Conrad said they had landed beyond Surveyor crater, but he did not think they would have any difficulty in walking back to it, Carr asked, "Where did you put down, Pete? Over on Site 4?"

"About halfway between Site 4 and Site 3," Conrad reckoned. "I flew by the right side of the crater and then had to fly over to the left and land." The sites referred to were two of four used in planning the surface activities. The intended landing point at Pete's Parking Lot northeast of Surveyor crater was Site 1. His initial guesstimate that they were just west of Head crater was based simply on the fact that he could not recognise this crater in the view out of his window. "You guys did outstanding targeting!" he praised Mission Control.

Meanwhile, to be ready in case they were to receive a No-Stay at the T2 decision, Conrad initiated P12, the computer program that would perform the liftoff.

One of Bean's early tasks after landing had been to have the AGS store its current attitude, to preclude its measurement drifting while the vehicle was stationary. "How does the AGS look?" he asked Houston.

"PGNS and AGS both looking great," Carr assured.

"Man. I can't wait to get outside," Conrad declared. Harking back to the descent,

"The Snowman stood out so clear, I couldn't believe it. I'm sorry I flew by [it], but I was just going too fast. It's a good thing that we levelled off high and came down, because I sure couldn't see what was underneath us once I got into that dust."

"Look at those boulders out there on the horizon, Pete," Bean exclaimed. "Jiminy! This is a pretty good place." The geologists wanted samples of boulders in the crater ejecta blankets, but distances were very difficult to gauge on the lunar surface and it would transpire that these boulders were probably associated with a large crater that was several miles away and hence well beyond their reach.

"Are we Stay for T2?" Conrad prompted Mission Control 7 minutes after touchdown.

"You're Stay," Carr replied. This period for Intrepid ran until Yankee Clipper was again in position for a rendezvous. "Would you like to recycle and try it again?" In training, descents were flown repeatedly, one after the other, and Carr was jokingly asking if they would like to have the simulator reset.

"No," Conrad laughed. "Not this time." Still high with adrenalin, he explained, "I think I did something I had bet I'd never do. I believe I shut that beauty off in the air before touchdown. I was on the gauges! That was the only way I could see where I was going. I saw that blue contact light and shut that baby down, and we zipped in from about six feet." In fact, the rule book called for the engine to be shut down in this manner, but Neil Armstrong had waited several seconds and landed with it still running and Conrad had decided to do likewise. But upon seeing the contact light he had followed his training and pushed the button without thinking.

"Pete, the Air Force guys here say that was a typical Navy landing!" relayed Carr, alluding to the rough manner in which naval aviators treated their aircraft.

"As long as the hook was down and we didn't bolt, I'm happy," Conrad replied. A naval aircraft maintains its engines at full thrust as it touches the deck of a carrier, so that if the hook fails to catch the arresting wire it can run down the deck and take to the air again in an action known as a bolter.

Meanwhile, after crossing the terminator into darkness in Yankee Clipper, Gordon had taken star sightings to realign his inertial platform and then disappeared around the corner.

Pursuing the checklist, Bean initiated a calibration of the gyroscopes of the AGS. Conrad instructed the computer to run P57 to make a lunar surface alignment which involved relating the vector of lunar gravity to the landing site REFSMMAT. This measured the orientation of the vehicle as it rested on the lunar surface relative to the REFSMMAT, thus providing 'attitude' data that would be required in processing star sightings. Meanwhile, they doffed their helmets and gloves and installed shades in both windows to darken the cabin to enable Bean to use the telescope. The optics could be swivelled through six detents. On Apollo 11 the locations of the Earth and Sun in the sky were such that Buzz Aldrin was able to use only two of the detents, in which there were no bright stars near the centre of the fixed fields of view. Bean was more fortunate because at the westerly landing site the Earth was east of rather than west of the zenith, giving him a clear view of more of the sky.

"Boy," Bean exclaimed, "you can sure see the stars out of this AOT. I'm in detent 1 looking at Sirius and I can see the whole constellation." From his perspective, the star was southwest and about 34 degrees in elevation. It is not only the brightest star in Canis Major, but also the brightest star in the heavens. After sighting on Sirius, Bean sought a second star to complete the observation and decided to use Pollux, a star in the constellation of Gemini that was conveniently located for viewing but was not known to the computer, so he instructed the computer to treat it as a planet, then consulted a data book for the celestial coordinates and keyed them in so that the computer could proceed. In the post-mission debriefing, Conrad recommended that as the computer knew only several dozen stars, instead of seeking a listed star that was well positioned in the field of view of the telescope, future crews should simply sight on a convenient star and supply its coordinates.

As the astronauts adapted to lunar gravity and the fluids began to drain from their heads, their voices lost their stuffiness.

"Hey, Houston," Bean called. "It's fun inside this spacecraft at one-sixth g."

"Don't break anything, Al," Carr warned.

"As soon as we landed, we started to handle the books like we did in the simulator at one g," Bean explained.

The computer processed the star sightings so that at liftoff it would know its starting point and, in particular, the azimuth to fly to attain the same orbital plane as the target. Then the inertial platform was checked by sightings of Procyon, alpha Canis Minoris, and Regor, one of the stars that had been mischievously named by Gus Grissom but was officially listed by the International Astronomical Union as gamma Velorum. After Mission Control had updated the computer, Bean aligned the AGS to the PGNS.

At this point, precisely 1 hour after the landing, Carr announced, "Stay for T3." In effect, this signified that Mission Control was satisfied that Intrepid would be able to remain on the surface for the full intended duration.

"Okay, we'll start our partial power down," replied Conrad. Because they were to remain on the surface for some 10 hours longer than their predecessors, most of the systems would have to be powered down to preserve the batteries. They took down the window shades and set to work.

Meanwhile, Gordon in Yankee Clipper reappeared on revolution 15 and Carr gave him the P22 data for tracking landmark 193. Gordon had tracked this shortly prior to undocking and was to do so again on his first post-landing overflight.

"I'll see if I can plot it in the same place I did last time," Gordon promised.

"You've got something new to look at down there, too," Carr pointed out.

"Let's save them for the next revolution," Gordon replied. With the optics aimed at landmark 193, he would not have time to look for Intrepid during this pass. "How are things going on the surface?"

"They're doing great, Dick," Carr replied.

3

A visit to the Snowman

WHERE ARE WE?

The "Stay for T3" decision that was made 1 hour after Intrepid landed on the Moon concluded Cliff Charlesworth's shift. He handed the Flight Director's chair over to Gerry Griffin. At the same time, Ed Gibson took the communications console. One of the small group of scientists hired as astronauts in June 1965, Gibson had worked closely with Jack Schmitt, a geologist from the same group of scientist-astronauts, in planning the moonwalks.

"Well done, Intrepid," Gibson congratulated, referring to the fact that the landing had been made very near the Snowman. "You've got a bunch of happy geologists in the back room waiting to go. We're standing by for your description." At this point, the flight plan called for Pete Conrad and Al Bean to describe what they could see of the landscape.

"We were just working on that," Conrad replied.

Rather than have to rely on his memory for the names of features in the vicinity of the Snowman, shortly prior to the mission Conrad asked Al Chidester, who managed the participation of the US Geological Survey in their training, to provide coloured maps which they could carry on their traverse.[1] These were annotated Lunar Orbiter pictures, and Conrad had carried them aboard ship as part of his personal kit. With the assistance of one of these maps they had been trying to identify terrain features to figure out precisely where Intrepid stood. There was a trade-off: they would gain a broader perspective once outside, but they could see much further from inside the elevated cabin. In addition, they took Hasselblad photographs to enable the scientists to interpret the scene for themselves after the mission.

Looking directly down-Sun, the surface detail was washed out by the 'zero-phase' glare. In part this was because with the Sun only 5.5 degrees above the horizon the protuberances hid their own shadows, but mainly it was due to coherent backscatter in which the tiny lithic fragments that comprised the surface reflected incident light

[1] The maps were hand coloured by Ray Sabala and James Vandivier at the Cape.

D.M. Harland, *Apollo 12 – On the Ocean of Storms*, Springer Praxis Books,
DOI 10.1007/978-1-4419-7607-9_3, © Springer Science+Business Media, LLC 2011

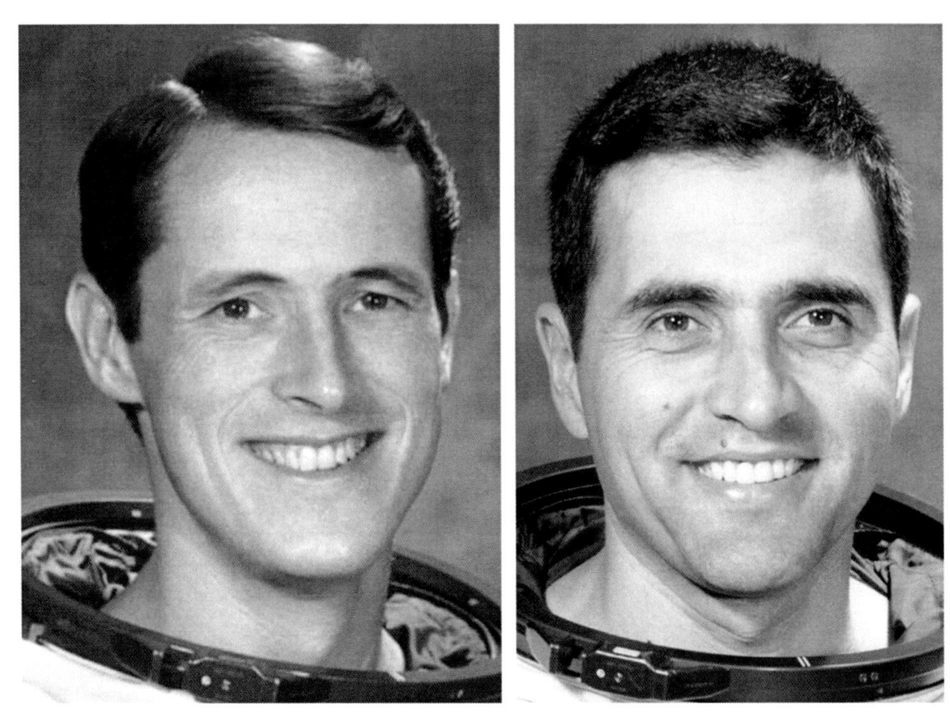

Ed Gibson (left) and Jack Schmitt.

Al Chidester of the US Geological Survey's Branch of Astrogeology, Flagstaff, Arizona.

directly back towards its source. The only shadow visible down-Sun was that cast by Intrepid. Consequently, they obtained their bearings primarily by leaning forward to place their heads against the windows in order to look out to either side, where some shadow detail was visible.

As Conrad noted in the post-mission debriefing, "The thing that confused both of us in the beginning, was the fact that distant objects looked much closer. It took us a while to realise that we were seeing many more craters than were on our map. They also looked smaller on the lunar surface than they did on the map." And Bean made the additional point, "When you're sitting on the ground, none of the shadows in the bottoms of the craters are visible, so you end up always seeing the bright part of the landscape, and it's difficult to find the craters. You say, 'There's a crater over there', and it's difficult for the other guy to see it for a while, until you learn to look for the edges."

Conrad's initial guess that they stood west of Head crater had been based on the fact that he could not see this crater. However, as they learned to read the landscape they recognised narrow ellipses off to the left as the rims of craters and he began to wonder whether one of these might be Head. Referring to the fact that the geologists could not finalise the route of the second moonwalk until they knew precisely where Intrepid was, he offered map coordinates which put them a little to the northeast of Head. But a few minutes later he pointed out, "We're having a little trouble judging distance. How long is my shadow?"

After consulting, Gibson replied, "Your shadow length on a level surface is 250 feet."

"You've got to be kidding me!" Conrad exclaimed in astonishment. Intrepid stood 23 feet tall, the Sun was at a very low elevation and he could rationalise this figure, but due to the absence of familiar objects to provide a sense of scale he would never have guessed the shadow was that long.

"We could shorten that to 230 feet," Gibson updated, and then asked whether they were judging their shadow to be longer or shorter than it was.

"Well, if my shadow's 230 feet long, we're really misjudging distances," Conrad replied.

"Are you short or long?" Gibson persisted.

"I would have said my shadow was much shorter than that," Conrad replied. They had been greatly underestimating distances.

As Yankee Clipper flew overhead, it was tracked by Intrepid's rendezvous radar in order to help to refine the selenographical coordinates of the landing site. Gordon was optically tracking landmark 193. On each such pass, the vehicle was above the horizon at the landing site for barely 12 minutes and direct communication had to be by VHF. Not having established this link, Gordon called Earth by S-band to inform Intrepid that his transponder indicated that it was locked on. Peering up through his overhead window, Conrad saw the spacecraft pass over.

In an effort to help to refine where Intrepid was standing, Gibson asked whether a certain landscape feature was recognisable.

"As soon as we get done eating, we'll get with it," Conrad replied. "We're pretty hungry."

"Okay, you deserve it," Gibson replied.

Hearing Gibson's praise, Gordon interjected, "No, they don't. Tell them to get to work. For heaven's sakes, Ed, they're down there having all the fun and you and I are doing all the work."

"They'll get with it soon," Gibson pointed out, referring to the moonwalk.

Four meal periods on the lunar surface were scheduled. The food was the same as in Yankee Clipper, except that rehydration was cold since Intrepid did not have any hot water. As Conrad said in the post-mission debriefing, speaking generally of their time on Moon, "I ate everything in sight! Al didn't eat quite as much as I did. As a matter of fact, I ate a lot of Al's food. I ate the whole ham paste tube and a couple of other things of his. I really wanted to keep the old energy level up, and I enjoyed the food."

Resuming the window report, Conrad announced, "There's a large crater smack in front of us." There was the narrow ellipse of a crater seen from an oblique angle that subtended about 60 degrees of arc. "Unless we look very carefully, it's not obvious that there is a big crater out there. I suspect it could be the head of the Snowman." Using the length of their shadow as a measure, this put them about 350 feet from the northeast rim of Head, which in turn meant Intrepid was just northwest of Surveyor crater. "I think really, the best thing for us to do is to get out and look around. The sooner we do that, the quicker we'll figure out where we are."

"Roger, Intrepid," Gibson replied. "We concur with that."

Whilst a major goal of Apollo 12 was to demonstrate a pinpoint landing, what this actually meant was that they land within walking distance of Surveyor 3. As Conrad wryly admitted later, it was obvious they were within the 2,000-foot limit. "We were trying to pinpoint where we were to within 10 or 20 feet, which was ridiculous."

Before Conrad and Bean could begin to prepare for the moonwalk, the flight plan required them to offer a scientific assessment of the landing site. The geology team were in the science support room to monitor activities and provide advice, and were represented in the MOCR by the Experiments Officer who relayed their input to the scientist-astronaut Capcom.[2]

While Bean finished his meal, Conrad inspected the landscape using a monocular which provided a magnification factor of ten. "My general impression is that we're in country where I see mostly angular rocks," he began. The appearance varied with viewing angle. "At first glance out of the spacecraft, I can distinguish absolutely no colour difference in anything. About the only difference is looking cross-Sun versus down-Sun. I am sure some of these rocks must have different colours and different

[2] The leader of the field geology investigation (S-059) was Eugene M. Shoemaker. He established the Branch of Astrogeology of the US Geological Survey at Flagstaff, Arizona, and had recently moved to the California Institute of Technology. His co-investigators were Gordon Swann, M.H. Hait, E.H. Holt and R.M. Batson, all of the USGS, E.M. Goddard of the University of Michigan, Aaron Waters of the University of California at Santa Cruz, J.J. Rennilson of the Jet Propulsion Laboratory, T.H. Foss of NASA, and scientist-astronaut Jack Schmitt.

textures, but from here they don't appear that way. Looking at all the materials on the horizon and the blocks on the horizon, they all appear to be of the same material and they all appear to be pure white."

Noting that they were falling behind schedule, Gibson nevertheless asked them to estimate the azimuths to some of the large boulders on the horizon to enable Mission Control to triangulate Intrepid's position on the maps drawn using pictures taken by the Lunar Orbiter missions.

Bean, also aware that they were running late, countered, "I'll tell you what we're going to do. I'll give you a good description here, then we're going to get ready, and when we get out we'll take the TV and show you the craters. I think that'll give you a pretty good handle on it." He meant Mission Control could find the landing site by making their own observations. As to the landscape, "Generally, it's just sort of like an undulating plain. You can see quite far in all directions. There doesn't seem to be any particularly high objects to interfere with the view. There are blocky-rim craters in almost every direction. Some of them are quite close and some are far away." He described some of the largest craters, underestimating their distances and hence also their diameters. "There are many rocks scattered around. Most of them are partially buried and you can see that there are little fillets of dirt built up around almost all of them." This last point was of particular interest to the geologists. Filleting of rocks close to Intrepid had no doubt been enhanced by the dust which had been disturbed by the engine plume. On level surfaces further away, filleting was ejecta from small impacts splashing against the sides of rocks and accumulating at their base. In that impacts of various sizes occur at random distances and azimuths, a fillet will tend to be symmetric, but a rock that has been exposed for longer will grow a taller fillet. A large impact will tend to build up the fillets on the facing sides of nearby rocks. On the walls of craters, the progressive downslope movement of loose material will tend to accumulate on the upslope sides of rocks. Continuing his report, Bean said, "One interesting feature that is directly at our 12 o'clock, about 20 feet, is a whole surface area that is a bit different from the rest in the fact that it has got sort of parallel lines or trenches, perhaps an eighth of an inch deep, which run north to south from about my 2 or 3 o'clock position all the way over to Pete's window." He ventured that this pattern was not an artefact of the engine plume, since such an effect would be radial to the vehicle. The scientists were also hoping that the astronauts would be able to discern transitions between different types of terrain. However, as Bean reported, "There doesn't seem to be any possibility here of seeing anything like a contact between different-colored surfaces. There may be a chance to notice the contacts by looking at the texture [...] but other than that, it just looks like one uniform surface with many, many craters in it."

One feature of this site that had prompted the geologists to favour in it preference to Surveyor 1 was that the edge of a bright ray from Copernicus was just west of the Snowman and it was hoped that even if this contact was not visible, recent impacts might have excavated material which produced distinctive bright rims. But as Bean reported, "There's no immediately apparent white-rim craters near us." But the news was not all bad. One of their tasks was to sample rocks from crater rims in the hope that these would have been excavated from the lava believed to be present beneath the blanket of fragmental debris known as the regolith. "I think you're going to like this place, though, Houston, because we can see, in the not-too-far distance, some nicely sized rocks that are on the edge of craters that we suspect could be bedrock." And, contrary to concerns expressed by some planners that the ground between the craters of the Snowman would be so rough as to impede extravehicular activities, he was enthusiastic, "This is a lot better surface, I think, than Pete or I had imagined before we got here. It looks like we're going to be able to move around pretty well, and it looks like there's going to be a lot of different types of samples lying about. So I think, probably, with that we'll go ahead and start rigging out."

When Gibson asked the distance to a large crater that they had earlier reported at their 1 o'clock azimuth, Bean replied, "I'd say about 500 feet. It runs from about my 12:30 to my 2 o'clock position. It doesn't look like it has any particular blocks on its rim." Although Gibson did not say so, the scientists were considering tacking a visit to this crater onto the end of the first moonwalk, if there should prove to be the time available. But the scientists could not know the crew were seriously underestimating distances and the near rim of this crater was over 1,000 feet away. Bean continued, "I think we'll be able to pinpoint ourselves pretty well when get out and look behind us a little." Having leaned his head right up against his window to extend his view as far as possible around to the left, Conrad found that, about 20 feet away, the ground sloped dramatically down at about 10 degrees. "We landed right past a fairly large crater," he reported.

"Roger," Gibson replied. "That'll give us a lot to work with while you're in your EVA prep."

With that, Conrad and Bean resumed powering down Intrepid's systems. Unlike their predecessors, they shut down the platform of the inertial measurement unit and both the primary and abort guidance systems. The plan for Apollo 11 had scheduled a rest period before the extravehicular activity, but upon arrival Neil Armstrong and Buzz Aldrin had decided to forgo this. Conrad and Bean's plan was to start the first moonwalk as soon as possible. However, due to the protracted window-observations they were half an hour late in starting the preparations. Reviewing these activities in the post-mission debriefing, Bean said, "I'd recommend on the next trip, you make a quick evaluation knowing that you may not be precisely right. Then make a quick judgment of the general geological features. Don't spend more than 5 minutes on it, because the minute you get outside all those guesses you made through the window will be either right or wrong. [...] Then, maybe in the time between the two periods of extravehicular activity, when you know exactly where you are, you can provide a better geological description, if that is what's in order."

After 10 minutes of reconfiguring and stowing loose items, Conrad asked, "What time are we scheduled to go out?"

After Gibson said that the timeline had them starting to depressurise the cabin in about an hour, Conrad and Bean set to work in radio silence.

When Yankee Clipper reappeared on revolution 16, Gibson read up the latest data for his part of the mission.

"Hey, Ed, how come you ain't giving me a TEI?" Gordon enquired cheekily.

"I think we'll work out the rendezvous solution first," Gibson replied deadpan.

"Oh, okay," Gordon concurred.

EVA PREPARATIONS

The checklist for donning and verifying the accoutrements required to venture out onto the surface was lengthy.

First, each man emptied the urine collection bag worn inside his suit and then he drank a healthy measure of water to stave off dehydration whilst outside. Next they retrieved their backpacks. The commander's had been stowed against the side wall, and the lunar module pilot's against the front hatch. The portable life-support system (PLSS) made by Hamilton Standard of Windsor Locks, Connecticut, was 26 inches high, 18 inches wide and 10 inches deep. It contained: (a) a primary oxygen system to regulate the suit at 3.7 psi; (b) a ventilator to circulate oxygen, both for breathing and to cool, dehumidify, and cleanse the suit of carbon dioxide and miscellaneous other contaminants; (c) a loop to circulate 4 pounds of water per minute through the liquid-coolant garment; (d) a sublimation unit to shed excess heat to vacuum; and (e) a communications system for primary and backup voice relay via the LM. The entire pack was covered with aluminised kapton to minimise heat transfer, and fibreglass to protect against incidental damage. It had sufficient water and oxygen for at least 4 hours of nominal operation. Then they applied an anti-fogging agent to the inside of their bubble helmets, which they placed temporarily on the engine cover at the rear of the cabin. Finally, they retrieved the helmet augmentation, extravehicular gloves, chest-mounted remote control units for the backpacks, and the rubber overshoes which would provide traction outside in the weak lunar gravity. Each man also strapped an Omega Speedmaster Professional wristwatch onto one arm of his suit.

Once Bean had helped Conrad to don his overshoes, Conrad assisted Bean. Next, they retrieved the OPS packages which were to be affixed atop the backpacks. In the event of the PLSS oxygen supply failing during the extravehicular activity, the OPS had sufficient oxygen for a 20-minute run back to Intrepid and 10 minutes to ingress and revert to the cabin's supply. After Bean had added his OPS, Conrad helped him to don the backpack and fasten the waist and shoulder harnesses, which clipped onto rings on the suit. Then they attached the chest unit to the shoulder harness, and ran oxygen hoses from the PLSS to the sockets on the front of the suit. Bean then helped Conrad do likewise. After almost an hour of radio silence, and now only 10 minutes behind, they disconnected their cabin communications umbilicals and replaced them with the ones from their backpacks, then each man started the checklist to verify the settings of his portable communications system.

Gordon's task when overflying the Snowman on this revolution was to try to spot Intrepid on the surface through his sextant. On Apollo 11, Mike Collins had not been able to locate Eagle because no-one knew where it had landed and because all that could realistically be achieved in the brief time available on each pass was to inspect a single field of view; it was impractical to scan the sextant and search. In contrast, Gordon knew more or less where Intrepid was. Once he had identified the Snowman in the telescope, he centred this in the field of view. The computer held the optics on target while he transferred his attention to the boresighted sextant. His objective was readily identifiable as a bright object which cast a distinctive pencil-

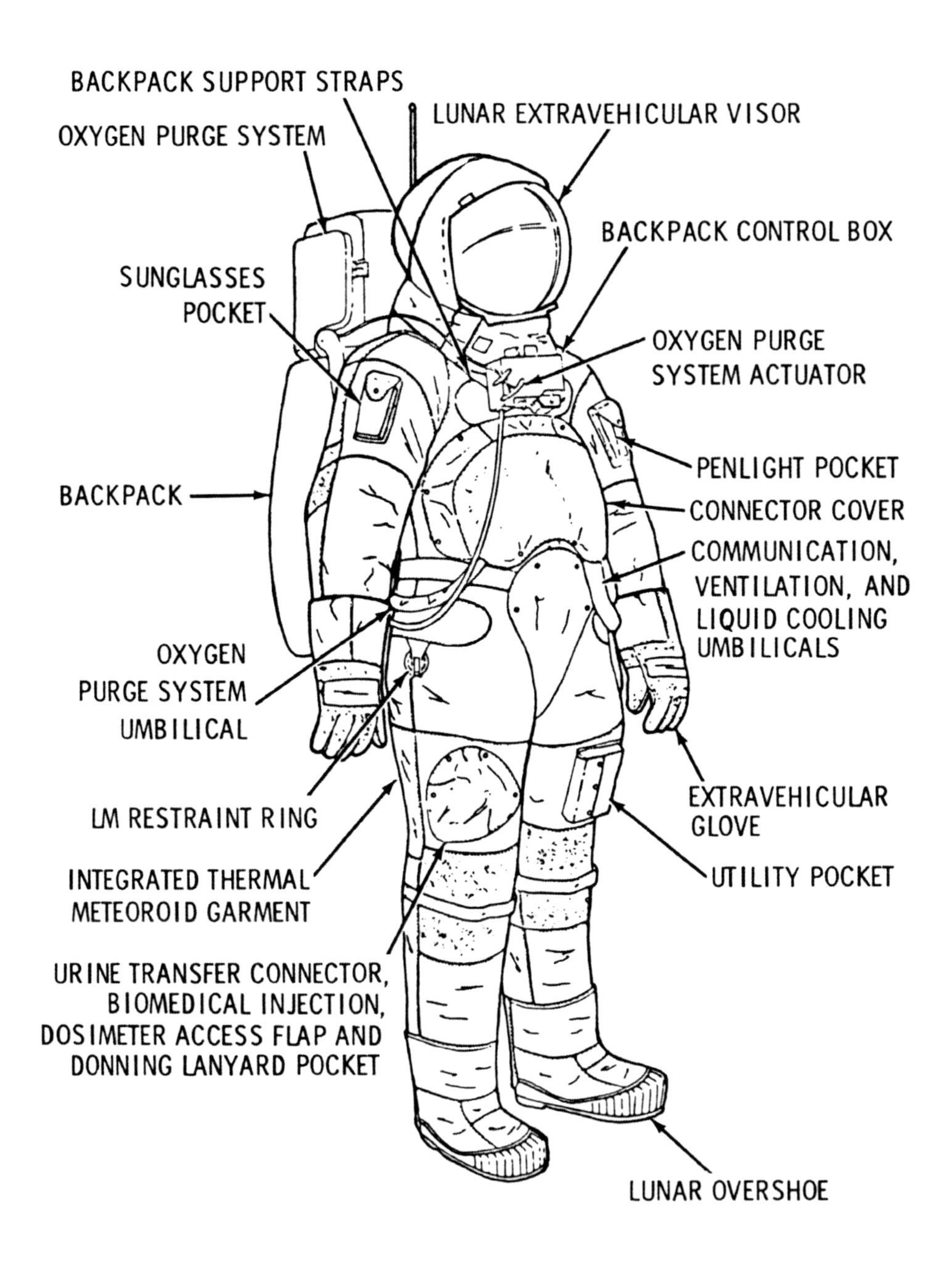

Details of the extravehicular mobility unit, including the portable life-support system and oxygen purge system.

thin shadow. "I have Intrepid! I have Intrepid!" he called to Houston. "He's on the Surveyor crater, about one-quarter of a Surveyor crater diameter to the northwest."

"Well done," Gibson congratulated.

"He's got a fairly good-sized crater just to the north and slightly east of him. But, directly behind him – he is on the Surveyor crater." As his line of sight became less oblique, the situation was easier to read. "Now I'm directly overhead. He's a third of the way between the Surveyor crater and Head." A moment later, he called, "I see the Surveyor! I see the Surveyor!"

"Good eyeball," Gibson replied.

"Ed, that's almost as good as being there," Gordon enthused.

In the post-mission debriefing Gordon reported, "It may be my imagination, but I thought I could see details of the descent stage and the landing gear extending from it."

By this time the elevation of the Sun had increased to 6.6 degrees, but the eastern interior wall of Surveyor crater, on which the old lander stood, was steeper than this and hence still in shadow. However, the mast which held the solar panel and planar antenna was tall enough to catch the Sun.

Gordon was delighted, as he explained in the debriefing, "This excited me quite a bit. I was pretty surprised that I was able to see that."

Meanwhile, in their effort to make up time, Conrad and Bean had fouled up their communications checks. Working from memory instead of methodically reciting the checklist, both men found themselves with inoperable headset microphones because the push-to-talk switches of the remote control units had *not* been moved from Off to Main. As Conrad admitted in the post-mission debriefing, "We were behind, and we started to hustle a little bit faster than we should have. We made several mistakes because I allowed us to get off the checklist. That cost us another 10 or 15 minutes, figuring out goofs that we'd made."

Once the communications check was finished, Gibson advised them that Gordon had had visually spotted both Intrepid and Surveyor. The latter was excellent news. After Surveyor 3 landed in 1967, Ewen A. Whitaker of the University of Arizona's Lunar and Planetary Laboratory identified the crater in which it resided. As Conrad wryly told Eric Jones in an interview in 1991, if Dave Reed had successfully steered them to the specified crater and it had turned out that the Surveyor wasn't in it, "We would have been pissed."

Bean connected his OPS hose and transferred the oxygen loops from Intrepid over to his PLSS. When Conrad did likewise, no oxygen flowed. It transpired that they had earlier omitted a step in the checklist by failing to plug in one of Conrad's hoses!

Bean had earlier prepared the 16-mm movie camera in his window to document the extravehicular activities close alongside the LM. With the cabin's environmental control system shut off, the water separator was no longer removing exhaled water vapour from the cabin. Noticing that the windows were fogging, Conrad decided to run the window heaters for a while in order to clear the right-hand window, lest the condensation crystallise when the cabin was vented to vacuum and thereby obscure the camera's view. Pushing on, they donned their bubble helmets and the protective cover with its thermal shield, visors, and sunshades. After donning their gloves and gauntlets, they chilled themselves with a surge of cold water from the environmental control system to their liquid-cooled garments before unplugging from Intrepid and switching to the water supply from their backpacks. In fact, this chill down was an improvisation, because the PLSS sublimation units that were to shed excess heat by dumping water to the exterior vacuum environment would not work efficiently in an atmosphere. On Apollo 11 they had been activated as soon as Neil Armstrong and Buzz Aldrin disconnected from the cabin's environmental control system, with the result that the vapour they produced had impaired the depressurisation of the cabin. This time the units were not to be switched on until after the hatch had been opened, and the astronauts chilled themselves down to prevent overheating in the final phase of the preparation process.[3]

"Those rocks have been waiting four-and-a-half billion years for us to come grab them," Bean mused while they were standing idle, monitoring their suits for pressure leaks. "Let's go grab a few."

"ALSEP first," Conrad noted, referring to the Apollo Lunar Surface Experiments Package that they were to deploy.

Satisfied with their suits, they were ready to begin the cabin depressurisation; now some 45 minutes behind the flight plan.

"Houston, are we Go for EVA?" Conrad called.

"Stand by," Gibson replied.

"*Stand by?*" Conrad echoed in surprise. "You guys ought to be spring-loaded!"

Several seconds later, having consulted, Gibson was back, "You're Go for EVA."

There were two valves available for dumping the cabin pressure. In addition to the main value on the forward hatch there was the option of using the one on the upper hatch which was really to equalise the pressure with the docking tunnel. Neither was particularly easy to reach when kitted out for extravehicular activity, one requiring bending down and the other stretching up. Bean opted for the forward hatch valve. The hatch was a large square below the central instrument console, with the valve on the right. After shuffling tight into his side of the cabin, he bent to reach blindly for the actuator and missed it. As Bean tried again, Conrad pushed his colleague's arm towards the valve. After venting down from 4.8 psi to 3.5 psi, Bean closed the valve. Once they were satisfied that their suit pressures were stable, the venting resumed. On Apollo 11 the valve included a bacterial filter to protect the lunar environment from terrestrial biota, lest this be sampled, returned to Earth and misinterpreted as evidence of lunar life. The presence of the filter had significantly slowed the rate of depressurisation. With the filter deleted from Intrepid, the venting proceeded much more rapidly.

With the pressure at 0.1 psi, Bean tried the egress handle, which was on the left side of the hatch. When the 32-inch-square hatch refused to open, he peeled back the ribbing on the upper corner to relieve the residual pressure, a trick he had discovered in altitude chamber training. As he explained in the post-mission debriefing, "When you're pulling the handle, that makes you open all of the surface area against 0.1 psi.

[3] This proved to be such a good idea that it was added to the checklist for future missions.

But when you're pulling back the corner, you're only moving a little part of it. And the thing was so flexible. It was one-piece machined titanium or something, and so thin that it was ridiculous." It took about 5 seconds for the cabin pressure to decline sufficiently for him to be able to hinge open the hatch.

With the cabin in vacuum Conrad and Bean turned on the pumps to circulate cold water through their liquid-cooled garments and started the sublimation units. As the principal limitation of the PLSS was not oxygen but the supply of coolant water, this marked the formal start of the extravehicular activity. Having made up a little time, they were now only about half an hour behind schedule.[4]

FIRST MOONWALK

Because the hatch hinged in towards Bean's side of the cabin, Conrad was to egress first. To fully open the hatch, Bean had to ease himself against the wall. Projecting about 10 inches to the rear, the backpack greatly reduced his scope for movement in the cramped cabin. Although in the weak lunar gravity this weighed one-sixth of its terrestrial 120 pounds, it retained its inertia. Buzz Aldrin had unwittingly nudged a panel of circuit breakers and snapped off the breaker that would later be required to supply power to the ascent engine! To preclude a recurrence, these panels were now protected by guard rails. Conrad turned his back and, with his left hand on the ascent engine cover, eased himself down onto his knees with his feet projecting through the open hatch. Bean then provided verbal cues to assist Conrad, now on his hands and knees, to reverse out onto the porch, which was the platform above the outrigger of the landing gear. In the process, the lower corner of his PLSS tore a 6-inch-long rip in the thermal and micrometeoroid shield of the hatch. Whilst this did not compromise the integrity of the cabin, the sharp exposed edge of the material did pose a potential risk of damaging a suit. There was a raised handrail above each side of the porch. As his boots projected beyond the end of the porch, Conrad hoisted himself up using the handrails and lowered his feet, one by one, onto the topmost of the nine rungs of the ladder affixed to the strut of the gear.

On the left-hand side of the porch, from Conrad's point of view, was a D-ring. He pulled a small pin to release the ring from its bracket, pulled the ring without effect, and exclaimed, "Man, that's a heck of a tug with that handle." He took another step down the ladder to improve his leverage. "Good Godfrey. That handle's in there like something I never saw before." Renowned for his profanity, Conrad was moderating his expressions for the listening audience. Giving up on the D-ring, he grasped ahold of the lanyard to which it was connected and tugged on that, thereby achieving the desired effect of releasing the modular equipment stowage assembly (MESA) that was contained in the wedge-shaped part of descent stage beneath Bean's window. "I just pulled the cable," he explained.[5] This was hinged at the base and swung down

[4] It was 6:32 a.m. EST on 19 November.

[5] In view of this problem, the D-ring was eliminated and in future the commander simply tugged on a loop added to the end of the lanyard.

about 120 degrees. "The MESA's down." This action exposed the television camera, whose signal would be received by the station at Honeysuckle Creek in Australia. Meanwhile, Bean had closed the circuit breaker to power the television and started the 16-mm movie camera in his window.

At this point Conrad looked to his right, to the southeast, and confirmed what he had glimpsed through the window. "Hey, I'll tell you what we're parked next to."

"What?" Bean asked.

Conrad laughed, "We're about 25 feet in front of the Surveyor crater!"

"That's good!" Bean replied. "That's where we wanted to be."

"I bet you that when I get down to the bottom of the ladder, I'll be able to see the Surveyor."

At this news, Gibson, who had been more or less silent since the start of the cabin depressurisation, chipped in, "Sounds good, Pete."

Bean passed out the lunar equipment conveyor. This was a clothesline-like device for transferring items to and from the surface. On Apollo 11 it had been a loop that was operated in the manner of a 'Brooklyn clothesline'. It had since been redesigned as two separate lines. Once a package had been connected between the lines, it would be drawn in the desired direction. The interior end was attached to a fixture on the roof of the cabin and the other end would remain outside.

"Do you have any TV, Houston?" Conrad asked.

"We've got a TV," Gibson confirmed. The camera was fixed in position to look at the ladder. With Intrepid facing essentially west, the ladder was in shadow. All that could be seen at this point was some of the vehicle structure, the base of the ladder, and a tilted horizon between the bright lunar surface and the black sky. The novelty over the inaugural lunar landing was that the picture was in colour. After being used in the Apollo 10 command module, this particular camera, the first colour television camera to fly on an Apollo spacecraft, had been modified for the harsh lunar surface environment by painting its exterior white for thermal control, substituting coated metal gears for plastic ones in the colour-wheel mechanism, adding a special bearing lubricant, and installing internal conduction paths to the outer shell to radiate heat to vacuum. "No Pete Conrad as yet though," Gibson pointed out.

"No," Conrad explained. "I'm at the top of the ladder."

"Don't go down yet," Bean reminded. "I've got to get my camera on you, babe." Earlier, Conrad had asked for a picture to be taken through the hatch showing him at the top of the ladder.

"I can't go down yet, anyhow," Conrad pointed out. He was still in the process of preparing the equipment conveyor. As he fiddled with it, he explained, "It came out of the bag in three pieces and, as you might well imagine, I picked the wrong piece."

"Do you want me to pull it back in and throw you the end?" Bean offered.

Conrad declined and persevered. The conveyor was designed to enable the outside man to stand well away from the ladder, so the line was a considerable length. As he worked to configure it some of the line draped down onto the surface, and when he pulled it back up it was coated with black dust. "Man, they aren't kidding when they say things get dusty." Finally finished, he exclaimed, "Whew! I'm headed down the ladder."

"Okay, wait," Bean said. "Let me get the old camera on you, babe." He took one of the two Hasselblad cameras that were ready to be transferred down to the surface, bent to aim it through the hatch and blindly snapped off several frames.[6]

As Conrad headed down the ladder his feet appeared in the television image. The audience watched him moving down from one rung to the next, leading with his left foot and following with his right; both hands gripping the outer rails. The 16-mm movie camera was recording his progress from a higher perspective. On reaching the final rung he pushed off backwards and, letting his hands slide on the rails, dropped in the weak lunar gravity to the foot pad. As his boots made contact, he exclaimed in delight, "Whoopie! Man, that may have been a small one for Neil, but it was a long one for me."

In the summer of 1969 Conrad and his wife Jane had hosted the Italian journalist Oriana Fallaci, who was firmly of the view that Neil Armstrong's famous 'one small step' must have been scripted for him by NASA. Conrad was unable to convince her otherwise. Finally, he said, "Look, I'll prove it to you. I'll make up my first words on the Moon right here and now." When he told her what he had decided to say, she was sceptical. He bet $500. She never paid the debt! His rationale was that at 5 foot 6 inches in height he was almost 6 inches shorter than Armstrong, and therefore the 30-inch jump from the bottom rung of the ladder to the pad was more intimidating to him. In fact, the reference was misleading, because Armstrong had made his remark when he lifted his left boot off the pad and made an imprint in the lunar dust close alongside. There was a plaque on the leg of Intrepid but, unlike its Apollo 11 counterpart, it did not bear a memorable inscription, so Conrad did not bother mentioning it.

"I'm going to step off the pad," Conrad reported. As Armstrong had done, he held onto the ladder with his right hand and lifted his left boot over the rim of the pad. He called "Mark!" as it touched the surface.[7] "Oooh, is that soft and queasy." Retaining his grip on the ladder, he stepped off the pad. "Hey, that's neat. I don't sink in too far." He released the ladder and stepped out of Intrepid's shadow. The angle of the Sun was lower than it had been at the time of the Apollo 11 moonwalk, and the glare was intense. The polycarbonate shell that augmented his bubble helmet incorporated an outer visor with a gold coating to reflect sunlight which, in the vacuum of space, was not only bright but also full-spectrum. Even 'squinting' his eyes was of little help. "Boy, that Sun is bright. It's just like somebody shining a spotlight."

The first minute or so of the moonwalk was to be spent evaluating his balance and mobility. "Well, I can walk pretty well, Al, but I have got to take it easy and watch what I'm doing." He walked slowly off-camera to the right, and then returned to the shadow where he peered around the other side of Intrepid, across the crater that was immediately behind it.

[6] What a shame that Neil Armstrong hadn't asked Buzz Aldrin to take such a view of his egress; that would have been the *Life* cover shot!

[7] It was 6:44 a.m. EST on 19 November.

As Pete Conrad egressed, Al Bean snapped a picture of him standing on top of the ladder and holding the porch rails. Note the contingency sampler projecting from Conrad's thigh pocket.

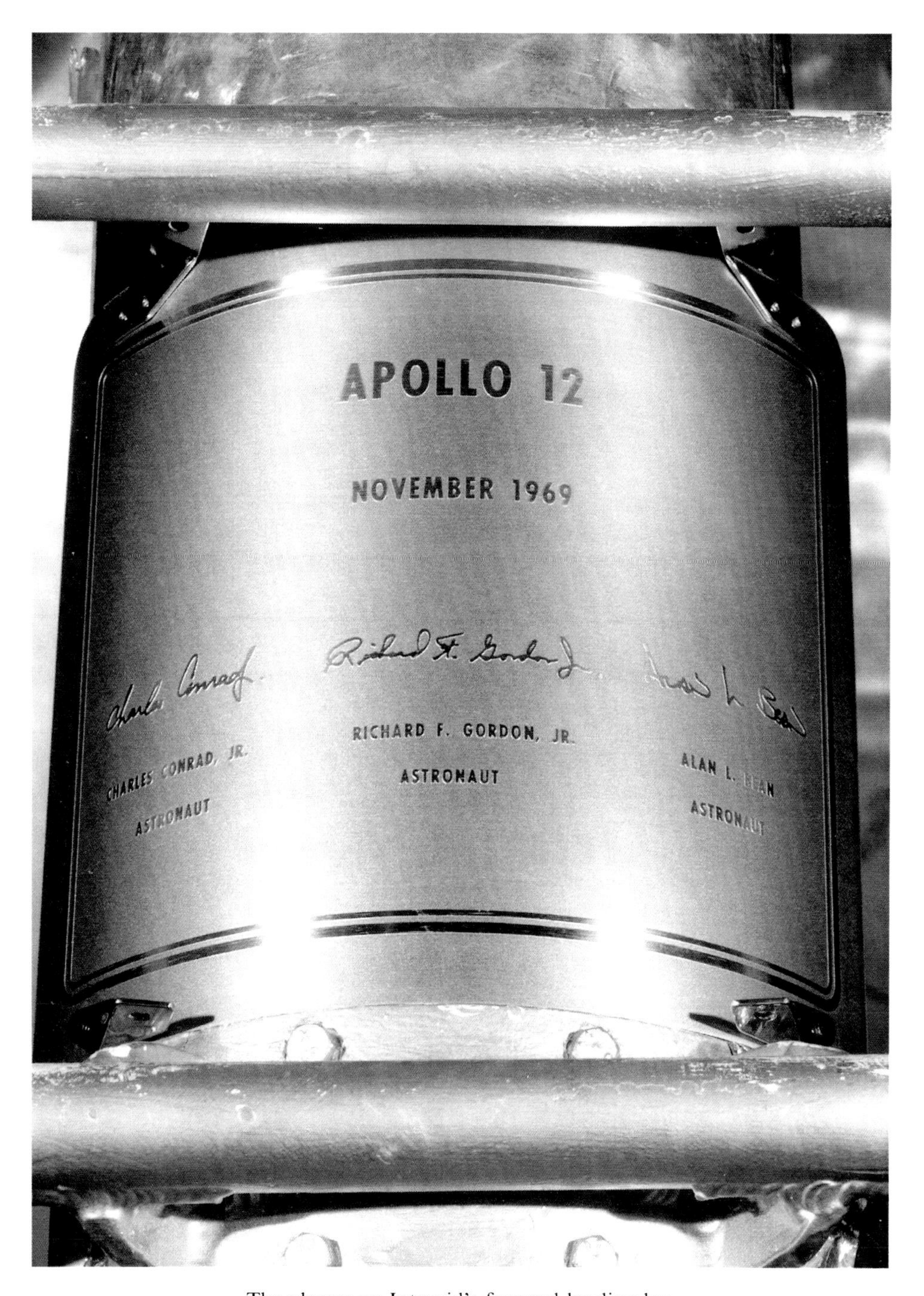

The plaque on Intrepid's forward landing leg.

"Boy, you'll never believe it!" he called delightedly. "Guess what I see sitting on the side of the crater!"

Bean ventured, "The old Surveyor, right?"

"The old Surveyor. Yes, sir. Does that look neat! It can't be any further than 600 feet from here. How about that?"

Over loud applause in the Control Room, Gibson gave his congratulations, "Well planned, Pete."

Conrad knew that the slope on which Surveyor 3 stood was about 12 degrees, but his experience with the manner in which air scatters sunlight and softens terrestrial shadows tricked the mind into interpreting any slope in dark shadow as being steep, and this one looked to be about 40 degrees. They had been provided with a tether to assist in descending into craters, but the goal of visiting the unmanned lander as the highlight of their second moonwalk looked pretty daunting.

Although they had begun the moonwalk later than planned, Conrad's motto was 'get ahead and stay ahead', so he was eager to start work. Neil Armstrong and Buzz Aldrin had each had a patch sewn onto the gauntlet of one glove listing the tasks to be conducted during their moonwalk. On Apollo 12 each man wore a 'book' on his left cuff. This had stiffened pages about 3 inches square on a spring-spine that would remain open at any given position due to the curvature. There was material on both sides, and index tabs projecting out. In essence, every action of the extravehicular activity had been sequenced and scheduled. The listing had been prepared by Bean but, unbeknownst to the prime crew, several days prior to launch their backups had added some annotations. In addition to cartoons which depicted Conrad and Bean as Snoopy-astronauts, drawn by Ernie Reyes, a member of the crew operations team, there were four photographs taken from the 1970 Playboy Playmate Calendar,[8] one with an annotation inspired by the repertoire of their field geology trainers, "Don't forget: Describe the protuberances."

Conrad's first task was the contingency sample, as a precaution against his being obliged to curtail the moonwalk before he and Bean could do some proper sampling. An early proposal for this tool was a scoop that had a long, detachable handle. After a sample had been obtained by scraping the loose material of the surface, the handle would be discarded. The concept was refined to a hoop that held a detachable teflon bag in the style of a butterfly net. The hoop was about 6 inches in diameter and the handle was about 2 feet long, and both were aluminium. The handle was collapsible, being made of several sections that were strung along a loose cable that ended with a T-shaped grip. The astronaut would retrieve the tool from his thigh pocket and pull the grip to draw the handle straight, and then the T would be inserted into a slot to maintain its rigidity. The bag had an aperture of about 4 inches and was about 1 foot long. Conrad moved some 50 feet northwest of Intrepid to obtain the sample from the rim of a small crater where the surface appeared relatively undisturbed by the

[8] Conrad had one picture each of Angela Dorian and Reagan Wilson (September and October 1967 respectively) and Bean had Cynthia Myers and Leslie Bianchini (December 1968 and January 1969).

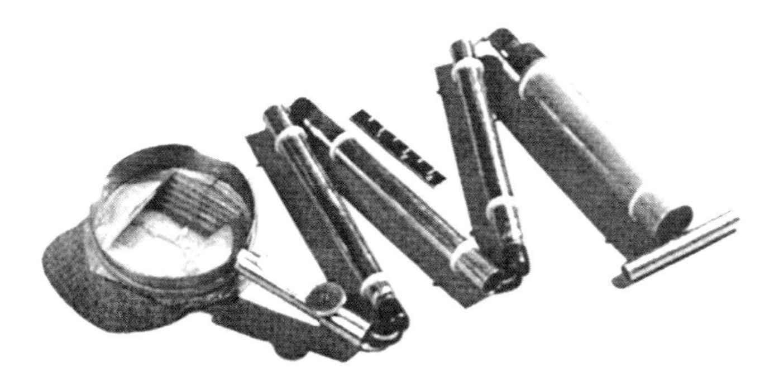

The contingency sampler in its stowed configuration. Drawing the cable tight using the T-handle would straighten out the tool. The circular ring had a bag in which to collect the lunar surface sample.

engine's plume. Although out of view of the television camera, he was, by design, visible to the 16-mm movie camera.

"Say, Houston," Conrad called, "one of the first things that I can see, by golly, is little glass beads. I've got a piece about a quarter of an inch in sight and I'm going to get that in the contingency sample bag, if I can get it. I got it!" At this stage of the program, 'lunar glass' was fascinating to both the astronauts and the scientists. He scraped the surface half a dozen times in order to include several rock fragments in the sample.

"Boy, you sure lean forward, Pete," Bean pointed out. "On Earth, you'd fall over leaning that far forward." On the Moon, an astronaut must lean forward about 12 degrees to balance the PLSS on his back, an angle which looks perilous but in lunar gravity isn't.

As Conrad walked back to the MESA, Bean offered another observation, "I'll tell you, your boots are digging in the soil quite a bit. If you don't pick up your feet, you really kick a load of dirt ahead of you. Your left foot's got a big mound ahead of it right now that it's just pushing along." In the airless environment, the displaced fine-grained material shot out on a low parabola and fell in an arc.

Back in view of the television camera, Conrad placed the contingency sampler on a horizontal portion of the landing leg (just as in training) to have both hands free to remove the thermal blanket from the MESA. "Would you believe it? The MESA is too low, for once." He raised it on its hinge to a more amenable position. "How long have I been out, Houston?"

"You've been out 25 minutes," Gibson replied, adding the good news that he was about 4 minutes ahead of the timeline.

On unstowing from the MESA the transfer bag designed for the lunar equipment conveyor, he placed it on a fold-out platform known as the table. After detaching the contingency sample bag from the tool, he closed the bag by crimping aluminium strips. Neil Armstrong had stowed the sample in his pocket, but this time it was to be transferred up into the cabin using the conveyor. The tool was now junk. Although Conrad was one of the shortest astronauts, his recommendation was that this sample could have been obtained more efficiently if the tool had been a few inches longer.[9] To complete the load, he added to the transfer bag the replacement PLSS batteries and lithium hydroxide canisters that would be required for the second moonwalk.

Bean now realised that he had inadvertently knocked the hatch closed against the line of the equipment conveyor. The efficiency of the sublimation unit of his PLSS had fallen as the accumulating vapour built up cabin pressure. When he reopened the hatch, the vapour escaped as a cloud of ice crystals, much to the surprise of Conrad below.

On retrieving the conveyor, which was hanging down between the landing leg and the MESA, Conrad muttered tunelessly as he hooked the transfer bag to the line. The dust on the section of the line which had come into contact with the surface made his gloves filthy. "I tell you one thing, we're going to be a couple of dirty boogers."

In preparation for the transfer, Bean began to pull in some of the slack on the line. As he did so, the retaining pin slipped out of the overhead fixture and the line fell to the floor and almost disappeared out of the hatch. He had to reset it. The verdict was that the pin was not large enough to prevent it from slipping out.

At this point, having flown around the far-side while Conrad and Bean were suiting up, Yankee Clipper reappeared on revolution 17. Gibson informed Gordon that Conrad was 32 minutes into the extravehicular activity and things were going well. Mission Control then activated an S-band relay to enable him to eavesdrop on the surface activities.

Meanwhile, Conrad carried the transfer bag, hooked to the conveyor, out in front of Intrepid. In training he stood directly in front of the hatch, but with the vehicle yawed to the right this put him in sunlight. To avoid the glare he stepped sideways into the shadow. Realising that in an offset position the passage of the bag would be obstructed by one of the handrails of the porch, he moved back out into sunlight. "I can't see a thing, looking into the Sun. Pull." As Bean hauled on the line Conrad let the free end run through his hands, and because the line had been on the ground he exclaimed, "Man, am I going to get dirty!" At first the bag danced about wildly, but as it approached the hatch and the line shortened it settled down and passed through without difficulty.

As a result of this experience, Conrad realised that they would have trouble when walking up-Sun, particularly during the long traverse of the second moonwalk. The helmet augmentation included small blinders that could be extended to block lateral glare from either side, but an additional one that could be pulled down would have improved visibility. As he pointed out in the post-mission debriefing, "We had a low enough Sun angle that anytime you put your hand up, looked directly up-Sun, and just blocked the Sun out, you could see perfectly up-Sun." With a shield that could be pulled down, it would have been possible to turn 360 degrees and see perfectly in any direction.

[9] An automated lander could have collected such a sample (and indeed the Soviet Union had such a project in an advanced stage of development) but for Apollo this was merely a preliminary task, not the objective of the mission.

"I'll have this stuff right back out to you in a flash," Bean called, as he unloaded the bag and reloaded it with two electric Hasselblads modified for the lunar surface environment.[10] Their 60-mm Zeiss Biogon lenses had apertures in the range f/5.6 to f/45 and a focal range from 3 feet to infinity. To assist operation using gloves, there were detents at the 5-foot, 15-foot and 74-foot settings. One lens was equipped with a filter that could be rotated in 45-degree increments for light polarisation studies. The magazines could be exchanged. For the first moonwalk they held 160 frames of colour film and for the second each would have 200 frames of thin-base black-and-white film. The cameras were mounted on chest brackets, whose assembly included an oversized trigger with which to fire the shutter and electrically advance the film. As there was no viewfinder, the astronauts learned to aim the camera in training by taking pictures and observing the results. The camera system incorporated a rigidly installed glass plate bearing a reference grid immediately in front of the image plane to impose the 'crosses' that characterise Apollo lunar surface photography.[11]

Eager not to waste time by standing idle, Conrad decided to tend to the television camera, which was to be removed from the MESA, mounted on a tripod, and placed about 20 feet from Intrepid to provide the audience with a view of activities at the MESA. "How long have we been out, Houston?"

"Pete, you're 34 minutes into the EVA," Gibson replied, "and you're right on the nominal timeline."

"That contingency sample is black!" Bean observed.

"You'd better believe it," Conrad replied. "I may have filled the bag too full."

After some difficulty removing the locking pins that held the television camera in place, Conrad flipped it upside down. At this point Bean announced that the transfer bag was ready to deploy. "Ah, all right," replied Conrad. "I've got to stop what I'm doing." Leaving the audience viewing an inverted image, he stepped in front of the ladder, retrieved the conveyor and started to haul on it in a hand-over-hand manner to lower the equipment transfer bag while Bean played out his end of the line. As soon as the bag was in reach, Conrad extracted the cameras and placed them on the MESA.

"I'll be out in a minute," Bean announced. Prior to egressing, he had to adjust the 16-mm camera and perform a final status check of the vehicle.

"Let me know, so I can photograph you," Conrad said. As he would soon need the Hasselblad anyway, he decided not to resume work on the television camera but to make a start on the early photographic tasks. On going to the area where he collected the contingency sample, he first shot a down-Sun picture and then moved around to view the area cross-Sun, snapped a picture, took one step sideways and shot another one to provide a baseline for stereoscopic analysis.

At this point Gibson informed Bean, "You're Go for egress."

"Okay, Pete, here I come."

"Wait; wait; wait; wait."

[10] Unlike on Apollo 11, Conrad and Bean were not to share a single camera.

[11] And set the challenge to authors to remove them!

"You ready now?"

"Here I come," Conrad said, as he moved in line with the hatch to provide verbal cues to help Bean egress.

Once he was safely on the porch, Bean announced, "I'm pulling the hatch closed here." The exterior of the spacecraft was protected with thermal shielding. To have left the hatch open would have exposed the interior to the backscattered sunlight of the zero-phase glare. It was standard practice to swing the hatch almost closed, with the valve in its automatic setting so that even if the ongoing outgassing by materials inside were to cause the hatch to close, the pressure could not build up to prevent the crew's ingress. Meanwhile, Conrad stepped around to the north to take pictures of Bean descending the ladder. The television audience was still watching an inverted view. On reaching the lowest rung, Bean pushed off and confidently dropped down onto the foot pad without sliding his hands on the handrails.[12]

"Turn around and give me a big smile," said Conrad. Grasping the ladder with his right hand, Bean turned to his left for the formal portrait. "Attaboy! You look great. Welcome aboard."

As Bean took his first tentative steps to evaluate his balance and mobility, Conrad reviewed his cuff checklist and announced, "I'm off for S-band antenna." Currently, the television was being transmitted by the steerable high-gain dish on Intrepid. The image quality would be improved by the deployable antenna that was carried on the side of the descent stage. After being erected close alongside Intrepid on a tripod, it would unfurl like an umbrella. Apollo 11 had carried one, but when Mission Control told Neil Armstrong that it was satisfied with the black-and-white transmission from the lunar module he had opted not to waste precious time on the larger antenna. One of the Apollo 12 goals was to evaluate a man's ability to perform useful work in the lunar environment, and erecting the antenna was being treated as an experiment. The length of its cable enabled it to be placed anywhere in the northwest quadrant which was within 20 feet of the MESA. When unfurled, the dish would be some 10 feet in diameter.

"If you turn and walk over to your right a little bit and look over that crater, you are going to see our pal sitting there," Conrad urged Bean, referring to Surveyor 3.

"Look at that!" exclaimed Bean.

"Will you look how close we almost landed to that crater!" Conrad laughed.

"Beautiful, Pete."

In '*A Man on the Moon*', Andrew Chaikin relates how at this point Bean reached into his pocket, extracted his silver astronaut pin and threw it into the crater, safe in the knowledge that he would be issued a gold replacement upon returning to Earth.[13]

When Gibson asked for "some comments on your boot penetration", Conrad said, "I think it's pretty much the same as Neil and Buzz found, don't you, Al?"

[12] It was 7:14 a.m. EST on 19 November.

[13] The 'astronaut pin' was designed by the Air Force to honour Gus Grissom after his Mercury flight. It was such a good idea that the Astronaut Office decided to give each astronaut a silver pin on being selected, and a gold replacement after returning from his first mission. Bean therefore had no further use for his silver pin.

Al Bean at the top of the ladder, holding the porch rails. Note the proximity of Intrepid to the rim of Surveyor crater.

Al Bean descends the ladder. Note that he has left the hatch partially open.

"I do," Bean agreed. "One thing I have noticed, is that it seems to compact into a very shiny surface. I guess the particles are very small and very cohesive, so every boot print, as you look at it, looks almost like it's a piece of rubber itself. It's so well defined you can't see any grains in it or anything." A moment later, he called, "Hey, you can see some little shiny glass in these rocks."

"Yeah, I reported that," Conrad pointed out. "You can also see some pure glass, if you look around."

Continuing his familiarisation with the lunar environment, Bean announced, "You can jump up in the air."

But Conrad, aware that time was passing, urged, "Hustle, boy, hustle. We've got a lot of work to do."

"I've got to do my fam for 5 minutes here," Bean pointed out. He had been on the surface for only 3.5 minutes.

"Fam some useful work," Conrad insisted, "like getting the TV camera going."

As Conrad wrestled with the antenna, Bean pulled a length of television cable out of its bin on the MESA and warned Conrad that he was going to run it in front of thc porch.

"Houston, I'm going to move the TV camera now," Bean advised.

Whereas the black-and-white camera of Apollo 11 required a manual exchange of the narrow lens for a wider one when installing the camera on the tripod, the colour camera had a zoom lens. The black-and-white camera had 320 lines per frame and scanned at 10 frames per second. The colour camera produced a national television network standard 525 lines at 30 frames per second. As previously, the framing and focus would be performed by verbal cues from Mission Control. After deploying the television camera, Bean was to sweep out a panorama, rotating the camera through less than the width of the field of view and pausing for several seconds to enable the scientists in Mission Control to snap a Polaroid off their screen each time. Because the camera was vulnerable to bright light, he was to omit the up-Sun direction from his panorama. He was later to relocate the camera in order to show the offloading of the ALSEP. The camera's view varied as he lifted it off the MESA and attached it to the tripod and, unfortunately, for several seconds the intense sunlight shone directly into the optics.

On Apollo 11, the camera had been placed at the 2 o'clock position relative to the lunar module. The plan this time proved problematic. As Bean explained in the post-mission debriefing, "I moved the tripod and the TV over to the deployment place. This was in front of the commander's window, which would be about 10 o'clock at about 20 or 30 feet. The only problem was that when I got over there I realised that because the LM had landed in about a 10- to 15-degree right yaw, the MESA was now in the Sun and that if I put the camera where it could view the MESA it'd be looking directly into sunlight. If I had put the camera in the shadow of the LM, as we planned to do originally when the MESA was in the shadow, you wouldn't have been able to see the MESA. So, I said, 'Well, I think I'll take it over and put it on the opposite side, at about 2 or 3 o'clock.' I carried the camera over to the opposite side, stuck it there, pointed it at the LM and called the ground."

"Al, we have a pretty bright image on the TV," Gibson advised. "Could you either move or stop it down?"

Bean stopped the aperture down. "That's as far as it goes, Houston. How does that look to you?"

"It still looks the same, Al. Why don't you try shifting the scene?"

Not realising that he had damaged the vidicon tube, Bean, who was trying to show Intrepid, presumed the camera to be dazzled by reflected sunlight. "The problem is the LM is very reflective. Well, I've got two choices. Let me go over here further to the side, and you check and see if it reflects too much. And if it does, I'll have to go stick it in the shade."

"And also, you might try setting the automatic light control to Outside," Gibson suggested.

"How does that look, Houston?" Bean asked, having complied.

"Still looks the same, Al. We have a very bright image at the top and blacked out for about 80 per cent of the bottom."

On successfully unfurling the S-band antenna, Conrad laughed heartily, "Man, oh, man; did that thing deploy!"

Continuing to troubleshoot the television camera, Bean aimed it down-Sun. "Al," Gibson called, "we haven't seen any change at all. Why don't you put your glove in front of the lens, but not over it, to see whether we can get any change at all."

"What do you see now?

"Still the same, Al."

Bean examined the plug for the extension cable. "All the connections look good."

"Al, why don't you take a good close look at that lens and make sure it is in the right configuration."

"I've got it focused at infinity. I've got the zoom at 30 or 40 or 50 feet; I'll put it 75. And I've got the f-stop at 22."

At this point Conrad plugged the S-band antenna into the lunar module and asked Bean to assist in pointing the antenna at Earth. It was no easy task to look up whilst obliged to stoop forward to balance the backpack. As the axial rotation of the Moon is synchronised with the period of its orbit around Earth, an astronaut on the lunar equator at zero longitude would see Earth more or less stationary at the zenith, with a slight oscillation arising from the eccentricity and inclination of the Moon's orbit. The fact that the landing site was at 23°W meant that Earth was displaced east of the meridian. It would rotate axially once per day, and over the duration of a month wax and wane in the opposite phase to the Moon as viewed by a terrestrial observer. At this time, it was only a thin crescent. The Moon spans about half a degree of arc to a terrestrial observer, but because Earth's physical diameter is four times that of the Moon it spans 2 degrees in the lunar sky. The antenna had a narrow beam, so Conrad was required to aim it almost directly at Earth using an optical sight and a cranking system for azimuth and elevation. As Conrad explained in the debriefing, "We had anticipated that it was going to be very difficult to align the antenna with the Earth. One, the sight doesn't allow any latitude. If the Earth is in the sight, the antenna is perfectly aligned. If it's not in the sight, you don't know where it is. Al came over and helped because the antenna had a tendency to tip over, especially when moving the crank. The crank was stiff." Bean gripped the antenna, pressed it into the surface, and held it steady while Conrad operated the cranks in accordance with the cues that Bean provided as he looked up through the fine mesh.

"The old Earth's just hanging up there," Bean observed. "It's amazing." He would later say he was particularly impressed by the sight of the crescent Earth in the black lunar sky.

After Conrad finally "picked up a corner of the Earth", it was just a matter of fine tuning.

While Mission Control considered the problem with the television camera, Bean reviewed his checklist and announced, "I'll put out the solar wind collector."

"Al," Gibson prompted, "when you finish up the solar wind, would you give one last try on that camera? Try opening the f-stop all the way and exercising the zoom."

"I sure will," Bean promised. He retrieved one of the Hasselblads from the MESA and mounted it on the bracket on his chest.

Conrad, finally finished with the S-band antenna, asked Gibson, "How long have we been out?"

"You've been out 1 hour and 2 minutes, and you're both running about 2 minutes off nominal," Gibson replied. Then he added for clarification, "Behind."

The solar wind composition experiment (S-080) was a 1×4.6-foot sheet of ultra-pure aluminium foil that was to be exposed to the Sun to collect ions carried by the solar wind. The sheet was thick enough to trap solar wind particles, but thin enough for high-energy cosmic rays to pass through. Erecting it on the Moon would enable it to sample the solar wind outside the Earth's magnetosphere. On Apollo 11, it had been exposed for just over an hour. The much longer exposure this time would yield a better sample. Also, because on Apollo 11 it was erected near the MESA the foil had been polluted by the dust kicked up as the astronauts went about their business. This time it was to be erected at least 60 feet from the lunar module. The hardware comprised a staff to be driven into the ground, with the sheet on a roller at the top and an attachment clip near the bottom.

Bean moved northwest of Intrepid to a position that they were otherwise unlikely to venture, unrolled the foil and clipped it into place so that it ran most of the length of the staff. At the Apollo 11 site the surface rapidly consolidated with depth, with the result that Neil Armstrong and Buzz Aldrin found it difficult to drive tubes into the ground. Bean was more fortunate and readily pushed the staff in to a depth of about 5 inches, which was more than sufficient for it to remain upright. However, as he explained in the debriefing, "As I started trotting back to the LM, I looked back at it. We were caught in the same predicament of not being able to estimate distances. It didn't look like I had gone out 60 feet, so I walked out, picked it up, carried it out another 20 or 30 feet, and stuck it in the ground quite quickly. Then I stood there, looked at it, and said again, 'Well, it looked like 60 feet, but now it doesn't.' So I pulled it out of the ground again, went another 20 feet and stuck it in." In fact, the final position was about 130 feet from Intrepid.

The experiment was designed by J. Geiss and P. Eberhardt of the University of Berne, and P. Signer of the Swiss Federal Institute of Technology – prompting its

The solar wind collector sheet with Intrepid on the rim of Surveyor crater, and Pete Conrad in the process of taking his 12 o'clock panorama.

nickname of the 'Swiss flag' experiment. On return to Earth, the foil would be baked in an ultra-high vacuum and the liberated noble gas atoms would be separated and analysed to measure their absolute quantities and isotopic ratios. The scientists were particularly interested in determining the average modern value of the He^3/He^4 ratio, for possible comparison with ancient ratios derived from gases extracted from lunar rocks and from recovered meteorites. In addition, it would facilitate a search for the isotope tritium, which is a hydrogen atom with a proton and a pair of neutrons in its nucleus.

No sooner had Bean completed deploying the solar wind composition experiment than Conrad asked him to assist in erecting the Stars and Stripes, which was carried to the Moon in a thermal shroud beneath the hand rail on the left side of the ladder. On Apollo 11, Armstrong and Aldrin had assembled the flag and then tried to push it into the ground, achieving a depth that was barely sufficient to hold the flag upright. This time, the top of the lower section of the staff had been hardened so that it could be hammered. Bean fetched the hammer from the MESA, pounded the staff into the ground to a satisfactory depth and then, taking the hammer with him, returned to the television camera. "Okay, Houston, I'm going to move the focus a bit and see what happens."

"Don't spend too much time on it. You're running a tad behind," Gibson advised.

Bean placed his hand on the camera to test the hypothesis that the filter wheel had jammed and was blocking the optical path, but he could feel it running. After he had varied the lens settings without producing any effect on the image, he announced, "I can try something else."

"Why don't you press on?" Gibson urged, having accepted that the problem was irrecoverable.

"How's that?" Bean asked.

"What change did you make?"

"I hit it on the top with the hammer. I figured we didn't have a thing to lose."

"Skillful fix, Al," Gibson laughed.

Bean gave the camera another tap, stopped it down to f/22 and zoomed out.

"Al, we're still not getting a good picture. Why don't you press on, and we'll try to get back to it later if we have time," Gibson prompted.

"I'll just leave it pointed toward the LM so that, if you do get a picture, you'll see something," Bean explained and then set off back towards Intrepid.

In fact, Mission Control wrote the television off, and later asked for the camera to be returned to Earth for failure analysis. It was decided that the intense sunlight had burned the light-sensitive coating off the vidicon tube. In '*Tracking Apollo to the Moon*', Hamish Lindsay related that Nevil Eyre, a video technician at Honeysuckle Creek, was watching his screen at the time. "I could see that Alan Bean was starting to point the TV camera at the Sun, because it was getting very bright up in the top left corner of the screen, then I could see it starting to peel away from the left. It was like somebody holding a sheet of paper and putting a match to it – no flames, just burning, rolling back in a boomerang shape."

The astronauts had not been able to train with a real camera, they had employed a

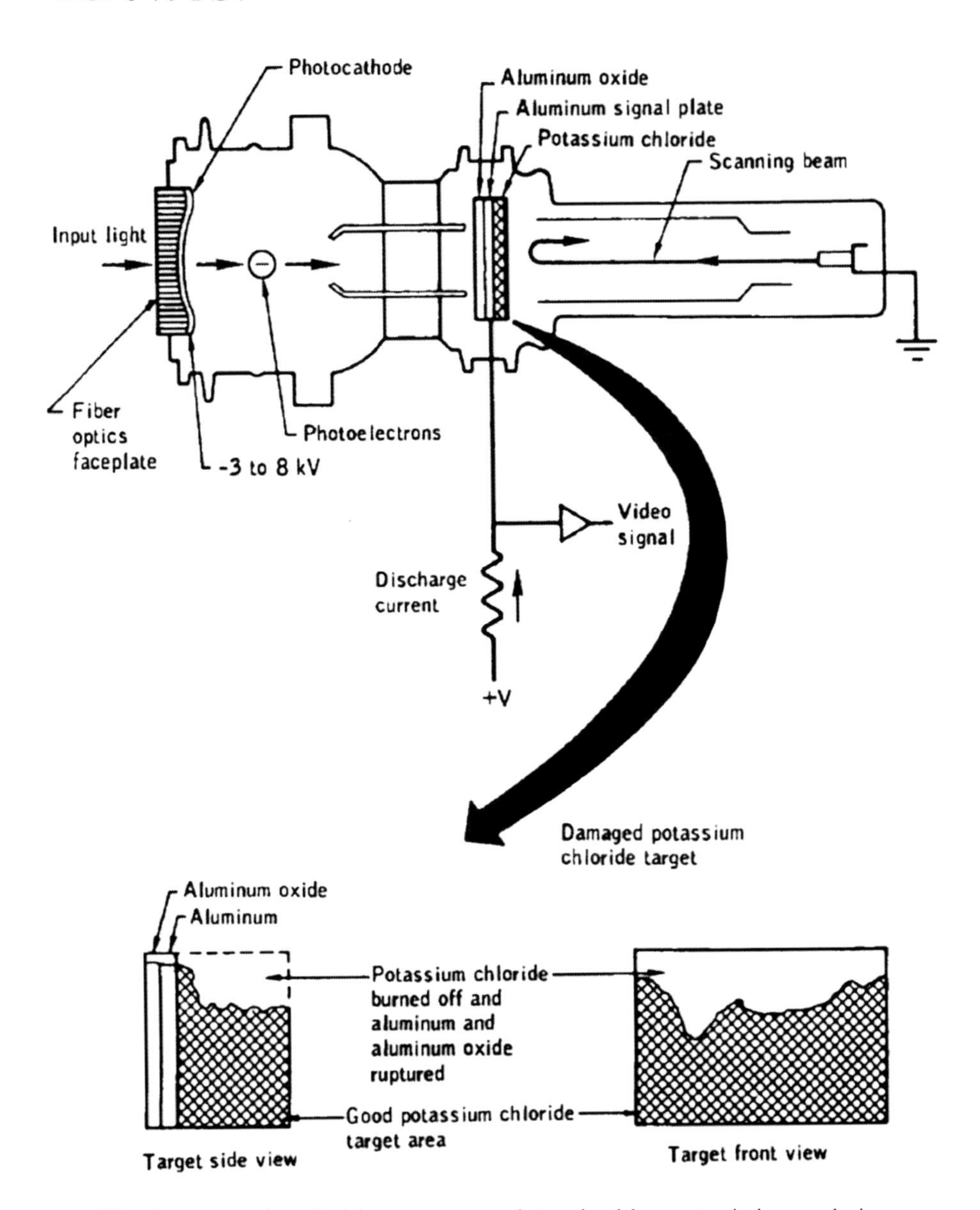

The damage to the television camera, as determined by post-mission analysis.

Pete Conrad is holding the Stars & Stripes because the locking hinge that was supposed to hold the crossbar out from the staff would not latch.

A portion of the 12 o'clock panorama taken by Pete Conrad west of Intrepid. The zero-phase glare washes out surface detail in the down-Sun direction, and only the ascent stage's shadow is in-frame.

A portion of the 4 o'clock panorama taken by Pete Conrad, showing the flag, solar wind collector sheet and television camera.

The 4 o'clock panorama taken by Pete Conrad included Al Bean working at the MESA.

To take his 8 o'clock panorama Pete Conrad moved over the lip of Surveyor crater, producing a low perspective. This portion the pan includes Al Bean in the process of taking pictures to document the state of Intrepid.

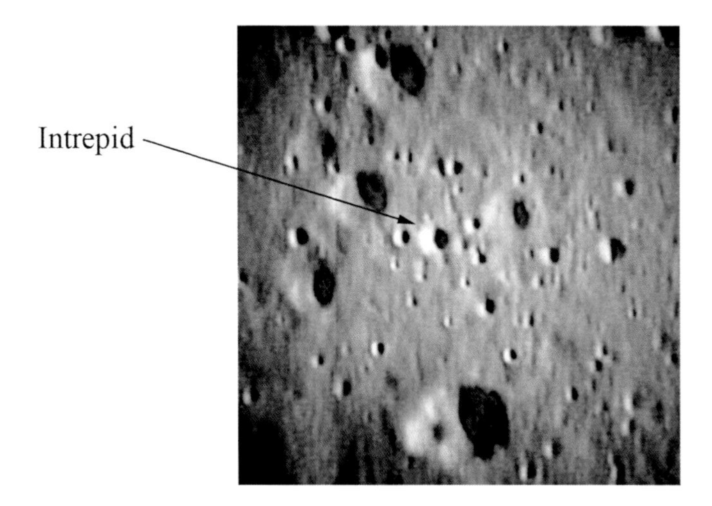

A frame from the 16-mm movie camera viewing the Snowman through the sextant on Yankee Clipper's third overflight of Intrepid.

wooden block as a substitute. In fact, the first time they saw their camera was on the Moon. As Bean explained to Eric Jones in a 1991 interview, television was regarded as communications equipment. Whereas the scientists had attended training sessions to offer advice on the deployment of their equipment, the television people did not. If they had, then perhaps someone would have suggested that a cap be placed on the lens while the camera was being handled. For later missions, circuitry was added to shut off the vidicon whenever the field of view was above a safe brightness.

With the television crippled, NBC, CBS and ABC hastily replaced their planned coverage of the moonwalks with other shows. Only PBS continued, improvising for the lack of video. In effect, the moonwalks were now a radio show! In '*All We Did Was Fly To The Moon*', Dick Lattimer argues that there was an up-side, "There was nothing stuffy about [Conrad and Bean], they laughed and chuckled, and just plain enjoyed their time on the Moon, and so did we back here on Earth who listened in. It was like listening to suppertime radio serials back in the 1940s; we had to imagine what it looked like, and that made it all the more fun."

Meanwhile, Conrad had discovered that the hinged crossbar that was to hold the flag out would not latch into place. As Bean had a Hasselblad, he snapped a tourist shot of Conrad holding the flag out by hand, then Conrad left the flag draped around its staff. His next task was to fetch his camera from the MESA and take panoramic sequences, each with a dozen overlapping frames at 30-degree intervals around the horizon. They were to be taken 20 feet from Intrepid in the 12 o'clock, 4 o'clock and 8 o'clock directions. In order to take the third panorama at the planned distance from Intrepid, he crossed over the rim of Surveyor crater and exclaimed "Whoops!" as his boots slipped in the dust on the slope.

"Watch yourself," warned Bean. "It's easy to slide." Having documented the solar wind composition experiment, he was inspecting Intrepid for damage in landing and taking pictures of how the descent engine had eroded the lunar surface. As far as he could see, the vehicle was in perfect condition.

Seeing the boot prints from when Bean threw his astronaut's pin into the crater, Conrad said, "I notice you have been over here."[14]

"Al," Gibson called, "do you have any comments on the LM foot pad interaction with the surface?"

"Actually, these pads went in a little bit further than did Neil's. I'd say most of the pads are in about an inch and a half to two inches. And it sort of looks like we were moving slightly forward, but had pretty well killed off our left/right velocity when we touched down. The right-hand pad seems to have bounced; the others don't. So it must have hit there first, and rocked back and forth or something." Later analysis of the telemetry showed that this was indeed what happened.

"Do you see anything on the surface?" Gibson asked, referring to erosion by the engine plume.

Bean replied, "It's kind of interesting, the surface under there is clean. It doesn't have the loose dust particles, as does the rest of the lunar surface about here. It also has a number of small round dirt clods, if you want, that seem to be rolling off in a radial direction from underneath the skirt of the engine [which is] about 8 inches or so off the ground."

"Roger, Al. Good description," Gibson acknowledged.

As Conrad pointed out in the post-mission debriefing, "There was a rock about 4 by 3 by 2 inches lying under the engine. I can't figure how it was lying right there at the skirt's edge. After seeing all that dust and stuff flying on landing, I was surprised it didn't blow a rock of that size away."

During this time, Gordon made his third overflight. Knowing that there would be scepticism that Intrepid and Surveyor 3 were visible, this time he maintained the landing site centred in the telescope while the 16-mm movie camera viewed through the sextant.

Bean announced, "I'm ready to start the ALSEP off-loading."

"I have something for you first," Conrad replied. On revisiting the MESA, he retrieved a pair of teflon 'saddlebags' that had been carried to the Moon in one of the sample return containers. Each man hooked a bag to the other's left hip, affixing it to one of the rings for the tethers which held him upright while the spacecraft was manoeuvring.

When Gibson announced they were running about 6 minutes late, with 2 hours 30 minutes nominally remaining, Bean reiterated that they were ready to begin work on the ALSEP, which was to be the main task of the first day. "We'll start catching up now. We've gotten over the initial checkout on how to walk and move around. And maybe we won't have any problem with this hardware – like we did with the TV and the antenna."

Since the inception of the space program many scientists had opposed the plan to

[14] Later suspecting that he had shot these panoramas at the wrong focus, Conrad asked Bean to retake them. As he admitted in the post-mission debriefing, "I got in a hurry, got off the checklist, and I took my three pans at 15 feet focus."

send astronauts to the Moon, arguing instead that robotic vehicles could perform the necessary tasks more cheaply. NASA therefore decided that astronauts on the Moon should be given work which specifically required the involvement of a human being. Offloading and deploying the ALSEP was to demonstrate definitively not only that a man could do useful work in the lunar environment, but also that he could undertake tasks that were beyond the capability of an unmanned lunar lander. As yet, the most complex activity by a robot was in January 1968, when Surveyor 7 used its remotely controlled arm to force to the ground an alpha-scattering analyser whose cable had snagged. This was considered a great achievement, but it paled into insignificance compared to deploying a suite of scientific instruments.

The ALSEP was in the scientific equipment (SEQ) bay, the wedge-shaped part of the descent stage diagonally opposite the MESA. It had two doors: a small one on the left that simply hinged open, and a larger one that folded upward. Inside, left and right, were the two subassemblies that comprised the ALSEP. As the suit did not allow an astronaut to raise his hands much above head height and the base of the bay was at chest height, he was to hold a lanyard and walk backwards, thereby extending a small boom that in turn hoisted the subassembly and dangled it outside the bay. Walking forwards then enabled the lanyard to lower the subassembly to the ground. So as not to clutter the cuff checklists, the tasks to offload and prepare the ALSEP for deployment were printed on the rear wall of the bay.

Conrad offloaded subassembly 1 and deposited it on its side, well away from the working area. Once its scientific payload of a seismometer, magnetometer and solar wind spectrometer were removed, it would become the central station of the ALSEP instrument suite. Bean offloaded the second subassembly and deposited it, too, on its side. This held the remaining scientific instruments, a variety of tools and the power system. Their conclusion was that the SEQ lanyards were excessively long and too readily entangled. Conrad closed the doors to protect the interior of the descent stage from the intense sunlight.

As the subassemblies were bulky and difficult to carry individually, one man was to carry both to the deployment site. On Apollo 11, Buzz Aldrin had carried the two instruments, one in each hand. But for the full ALSEP, which was not only heavier but also had to be carried further, making the task more strenuous in and of itself, it had been decided to use a carrying tool. After studying the possibility of dragging a sledge on which the two subassemblies would be mounted, the developer, Bendix of Ann Arbor, Michigan, had decided that the subassemblies should be attached to the ends of a single bar, which Bean would grasp in both hands and carry at waist level. This would leave his view of the ground immediately ahead clear, minimising the chance of stumbling. In order not to waste mass, the carrying bar would also act as the antenna staff for the central station. Conrad retrieved the two sections of the bar from Bean's subassembly, linked them together, and inserted one end of the bar into the socket on the base of his subassembly, leaving it projecting out horizontally.

The scientific experiments left by Apollo 11 were a mirror to reflect lasers back to Earth for ranging measurements and a passive seismometer. As the seismometer was powered by solar cells, it incorporated two small 'nuclear heaters', each of which had 1.2 ounces of plutonium-238 and generated 15 watts of heat to prevent sensitive parts

Pete Conrad uses a lanyard to offload one of the ALSEP subassemblies from the SEQ bay.

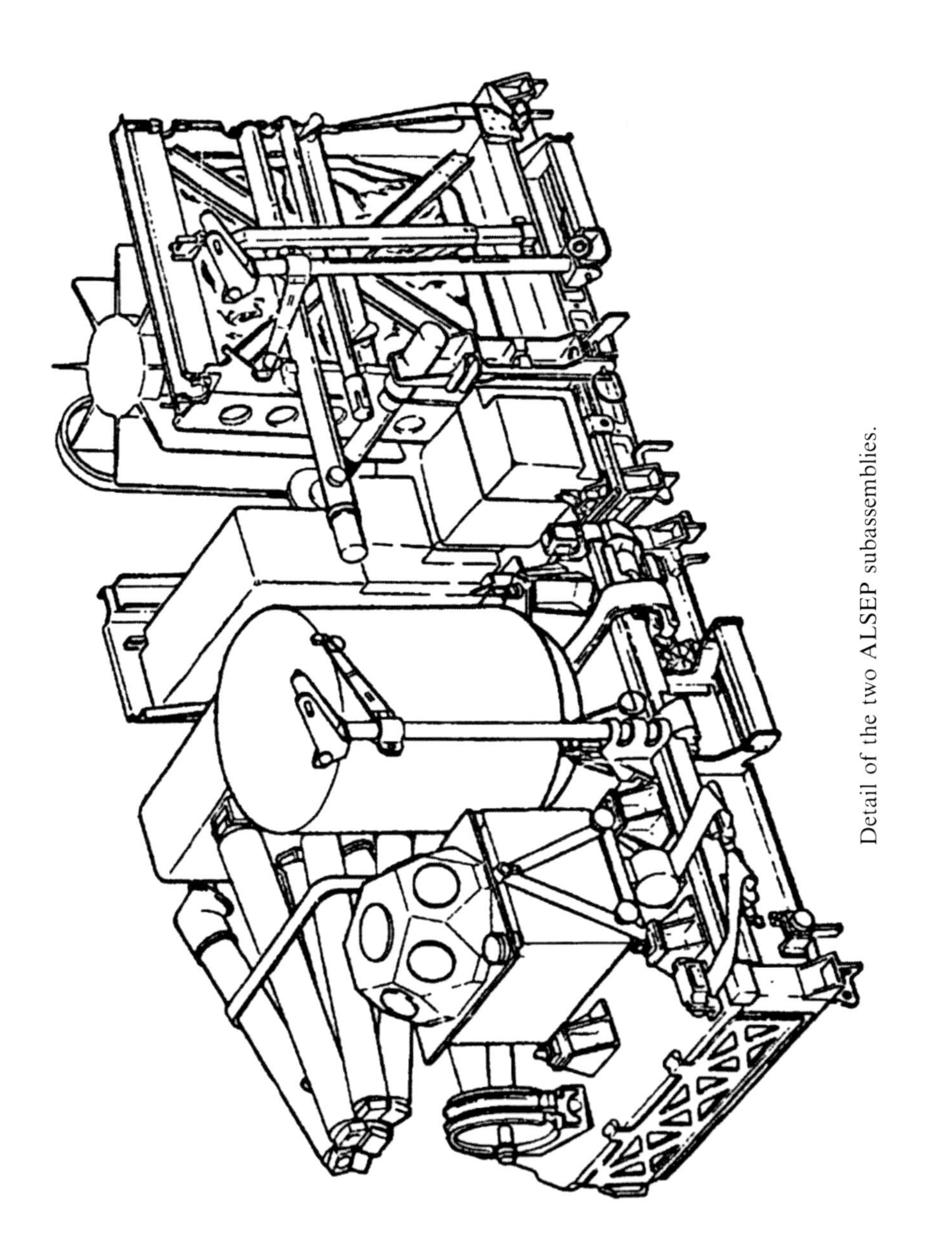

Detail of the two ALSEP subassemblies.

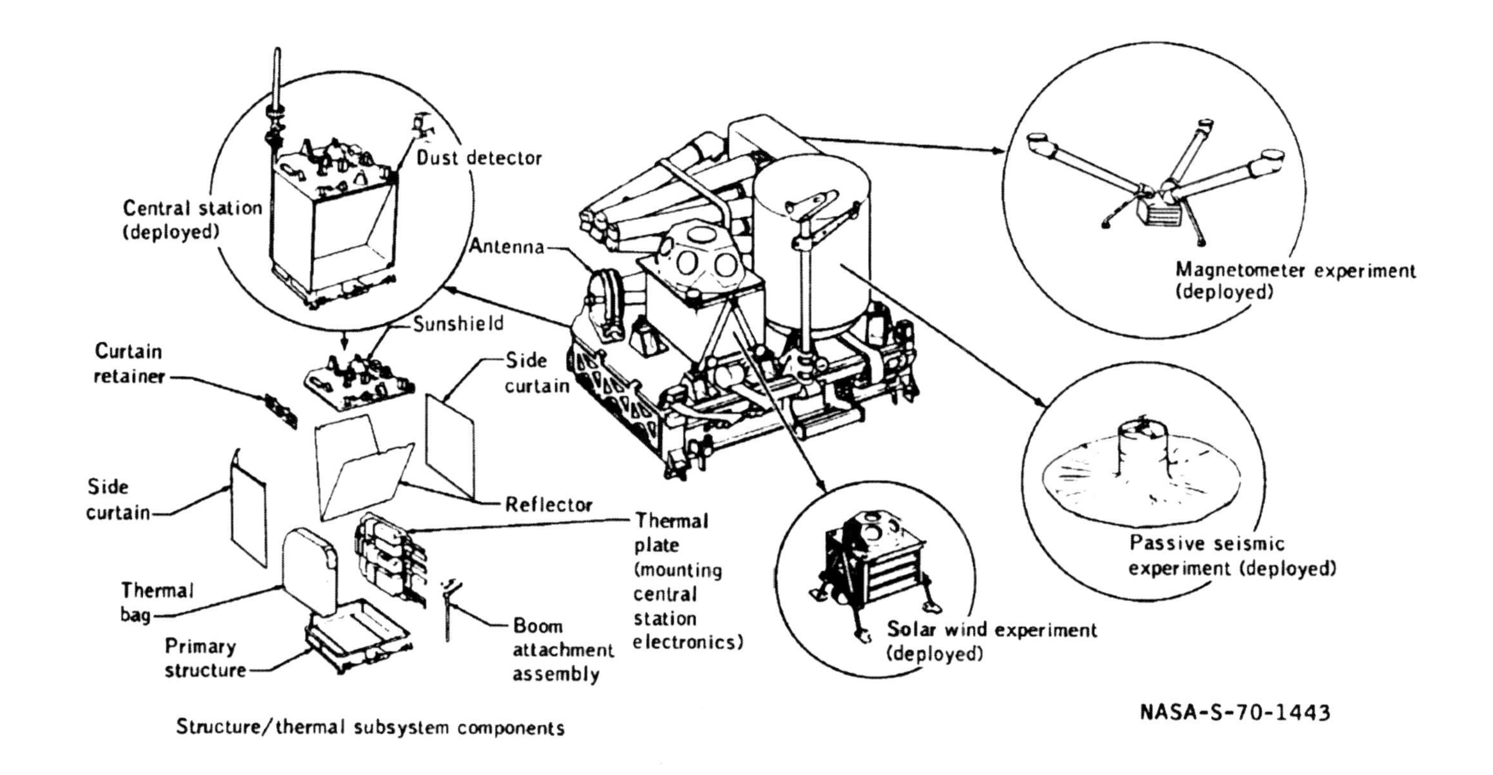

The central station subassembly of the ALSEP, with the magnetometer, solar wind spectrometer and seismometer bolted to its top deck.

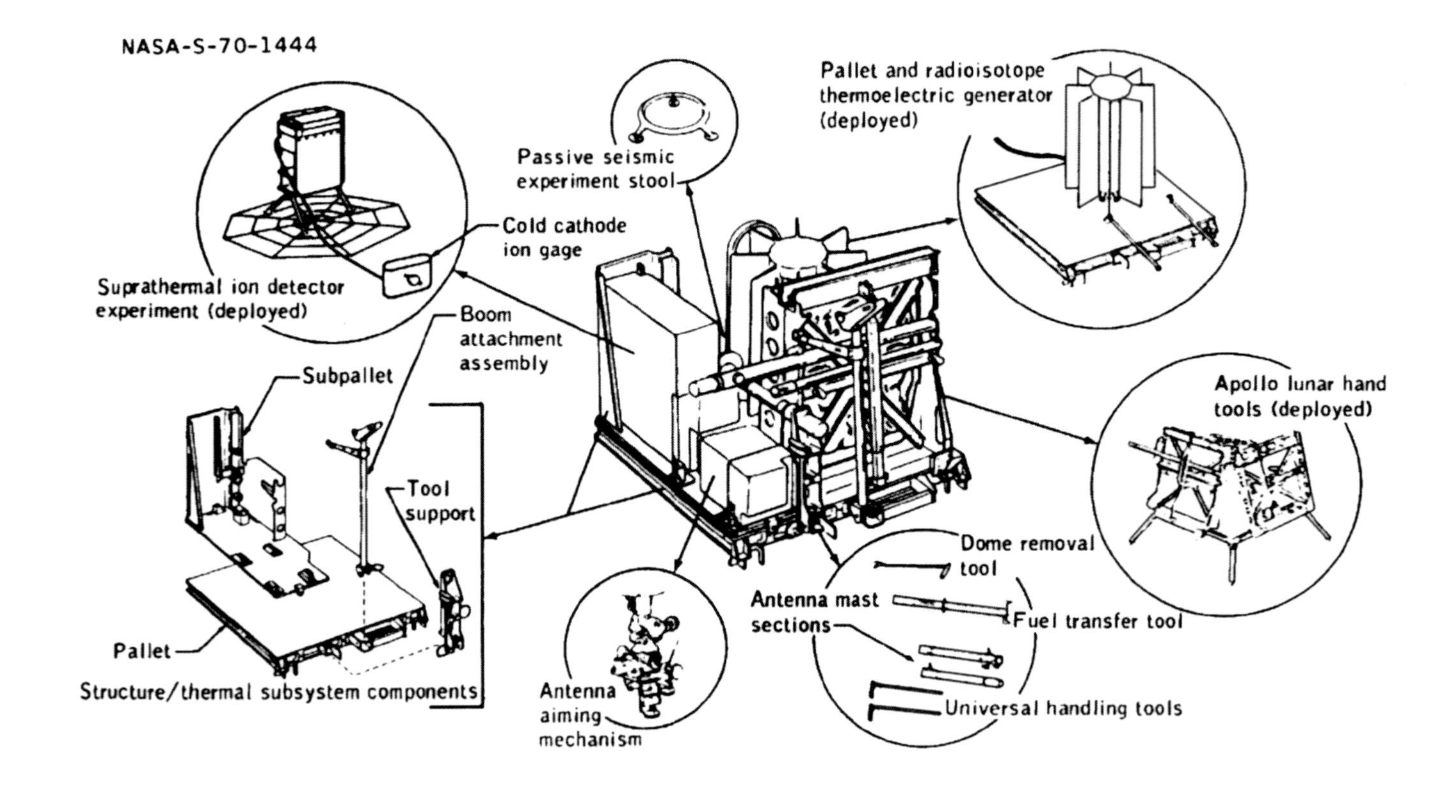

The RTG subassembly of the ALSEP, with the hand-tool carrier and separate subpallet.

The offloaded central station subassembly of the ALSEP.

Al Bean has retrieved the RTG subassembly of the ALSEP from the SEQ bay.

from cooling below −54°C during the long night, when the surface temperature plunged to −173°C, but it suffered from the chill and malfunctioned when awakened at sunrise. The ALSEP was to run day and night, drawing power from a radioisotope thermoelectric generator (RTG) which utilised thermocouples to convert the energy released by the fission of plutonium-238 nuclei into electricity.

The Atomic Energy Commission developed a series of 'atomic batteries' under its Systems for Nuclear Auxiliary Power on behalf of the Department of Defense as a replacement for short-lifetime chemical batteries rechargeable by solar power. The first system to be used in space, a SNAP 3A, was launched on 29 June 1961 onboard the Transit 4A navigational satellite. Unfortunately, on 21 April 1964 Transit 5BN-3 failed to achieve orbit and instead flew a ballistic arc with its apogee high above the south pole and, since the cell of its SNAP 9A was incapable of surviving entry, this released its plutonium in the upper atmosphere. NASA's first use of such a system was to supplement the solar arrays of its Nimbus weather satellites, but in view of the Transit loss the agency ordered that that its SNAP-19 be designed to survive re-entry. On 18 May 1968 a launch from Vandenberg Air Force Base in California had to be destroyed by ground command, dumping the wreckage of Nimbus B into the Santa Barbara Channel. After the Navy retrieved the two SNAP units, these were refurbished, installed on Nimbus 3 and launched on 14 April 1969. The SNAP-27 of the ALSEP was the first use of 'nuclear-electrical' power on the lunar surface.

The cylindrical RTG was about 18 inches tall and 16 inches across its eight radial fin-like heat radiators. The *thermoelectric unit* was made of lightweight beryllium and had an empty mass of 28 pounds. The *fuel capsule* was 16.5 inches long and 2.5 inches in diameter. Made of a rugged super-alloy, it had a mass of 15.5 pounds, of which 8.4 pounds was fuel. The plutonium-238 was fully oxidised in order to make it chemically and biologically inert. If the fuel capsule had been installed into the thermoelectric unit prior to launch, the system would have overheated in the closed SEQ compartment. In order to radiate the heat to space, it was carried in a *fuel cask* on the exterior of the lunar module. The cylindrical cask, which had hemispherical ends, was designed to withstand inadvertent entry into the Earth's atmosphere. It had a primary graphite heat shield, a secondary beryllium thermal shield, and an internal structure of titanium and inconel. It was 23 inches long and 8 inches in diameter, was attached to the descent stage by titanium struts, and weighed 24.5 pounds (about 40 pounds with the fuel capsule installed). If the cask were to come down on land or in shallow water, it ought to be recoverable; in deep water it would pose no risk and would be abandoned.[15] The power conditioning unit had the direct current voltage converters, shunt regulators, filters and amplifiers needed to convert and regulate the power. The requirement was for the assembly of 442 lead telluride thermocouples to convert the 1,480 thermal watts into at least 63 watts of electricity

[15] On Apollo 13, the lunar module was used to get the crew and their crippled command module back to Earth and was abandoned shortly before re-entry on a trajectory designed to dump the fuel cask of the RTG into the Tonga Trench in the western Pacific.

at a constant 16 volts for continuous operation of the ALSEP over a period of at least 1 year.[16]

One of the most dangerous tasks assigned to an astronaut on the lunar surface was to extract the fuel capsule from the cask and place it into the thermoelectric unit. The radiation from the spontaneous decay of plutonium-238 was not a problem, as it was mostly in the form of alpha particles, a very mild, short-range emission. The danger was the 500°C temperature of the capsule, and thus of the cask which was radiating this heat to space. As the fabric of the suit would melt at a temperature considerably lower than this, it was essential to take measures to ensure he did not bump against the cask while accessing the SEQ. Originally, the bay had a single door that opened upward. After a simulation performed by Don Lind for the human-factors engineers, it was decided that *part* of the door should be made to open to the left, towards the fuel cask, in order to act as a physical shield. This is why the doors were configured as they were. After further simulations, it was decided to place struts around the cask as additional protection against inadvertent contact.

Bean lifted the hand-tool carrier off his subassembly and placed it nearby, then he flipped the subassembly onto its base and announced, "We're going to go ahead and pull down the fuel cask now, then I'll take the element out of it."

"Oh, I know what I have to do, Al," Conrad muttered. "And I'm standing here not doing it." He was to detach the scientific instruments from the subassembly before Bean inserted the fuel capsule into the RTG and that in turn began to heat up.

A lot of thought had gone into the ALSEP. It had to be lightweight and compact, both for carriage to the Moon and by an astronaut on the lunar surface. It also had to be readily set up by men wearing suits which afforded limited peripheral vision and did not permit bending down. And the deployment should take no more than an hour or so. Although it was the most intricate single task assigned to the astronauts on the lunar surface, it was designed to be as straightforward as possible. Because a suited man standing upright could reach no nearer the ground than about 2 feet, the tools would require long handles. The bolts affixing the experiments to the subassemblies were to be released using a universal handling tool. It was L-shaped, like a slender crank, with a 26-inch-long shank and a 6-inch handle at the end sticking out at right angles. This was the result of 2 years of experimentation and rationalisation. In fact, it served two roles – unfastening the experiments from the subassemblies, and then carrying them to their final deployment positions. In planning a tool for unfastening the experiments, Bendix considered off the shelf commercial equipment and rejected it. On setting out to develop a new system the company envisaged using an enlarged screwdriver, but after some testing this was rejected by NASA because it would be difficult for a suited astronaut to produce the wrist action required to operate such a tool. Instead, a nut-and-bolt system was developed. On the one hand, the bolts had to engage the experiments with 500

[16] The power level would decline, but the system was capable of supplying power for longer than the baseline operating life of 1 year because the half-life of plutonium-238 is about 88 years.

pounds of pressure to withstand the vibration of a Saturn V launch, yet they had to be able to be loosened by an astronaut with limited wrist action. The bolts were given 12 sides to make it easier for the tool to engage them; i.e. the astronaut was not required to stand in a specific orientation relative to the bolt. The initial design required a 270-degree turn to release the bolt, but trials showed this to be awkward – to achieve such an angle the astronaut had to make a partial turn, disengage the tool and then re-engage it at a more convenient angle to resume. The issue was that the man should not have to shuffle a tool, lest he drop it, because retrieving it from the ground would be difficult. Given that the suit limited the wrist action to 90 degrees for a single grip, Bendix designed the bolts to release after a turn of 72 degrees. The final design was named a Boyd bolt.[17] Meanwhile, the lifting and carrying tool had been evolving. It had been designed to function like an old-style stove-lid handle, albeit with greater technical sophistication. The shank was to poke through a hole in the top of the experiment and then metal balls on the shank would slip neatly under an internal lip to provide grip. The balls were locked into place or released by the man pulling a trigger on the handle of the tool using his index finger, with the motion of half an inch being within the constraints imposed by the suit gloves. On the basis of experience gained in testing, the shank was increased in length several times until it, too, was 26 inches. With the two tools essentially finished, NASA directed that they be combined into a universal handling tool. There was one such tool for each man, and they were delivered on the subassembly which carried the RTG.

The subpallet that Conrad was to unbolt carried the suprathermal ion detector and cold-cathode ion gauge experiments, which were physically wired together, and some items for the central station. In order to remove a Boyd bolt, he was to stand with his feet apart for stability and leverage in the weak gravity, push the head of the tool into the guide sleeve of a bolt, engage the bolt itself and turn the tool to release the bolt.[18] "Houston!" he announced, "you can log me for my first Boyd bolt on the Moon." Once the subpallet was released, he used his universal handling tool to lift it clear and put it down nearby, leaving the tool engaged ready for carriage.

At this point Bean said, "There's one thing that's pretty obvious as we're setting out the components of the ALSEP here. I just hope that these thermal coatings don't have to stay as white as they are right now, because with all this dust there's just no possibility of not getting them a little bit dirty." The packages were painted white so as to reflect most of the incident sunlight. A coating of dust would increase the heat absorption, possibly sufficiently to damage the equipment inside.

"Al, we copy your comments," Gibson replied.

With the RTG subassembly on the ground nearby, Bean was ready to transfer the fuel capsule. By pulling on a lanyard he hinged the cask down through 120 degrees, in order to gain access to its end. The two long-handled tools required to perform the transfer had been delivered on the subassembly. One of them was to remove the

[17] In February 1968 the inventor, Thomas R. Boyd, was awarded a US patent 'Self-locking captive screw assembly'.

[18] The only complication was that if a bolt was in shadow, it had to be located by feel.

Al Bean uses a lanyard to hinge down the RTG fuel cask.

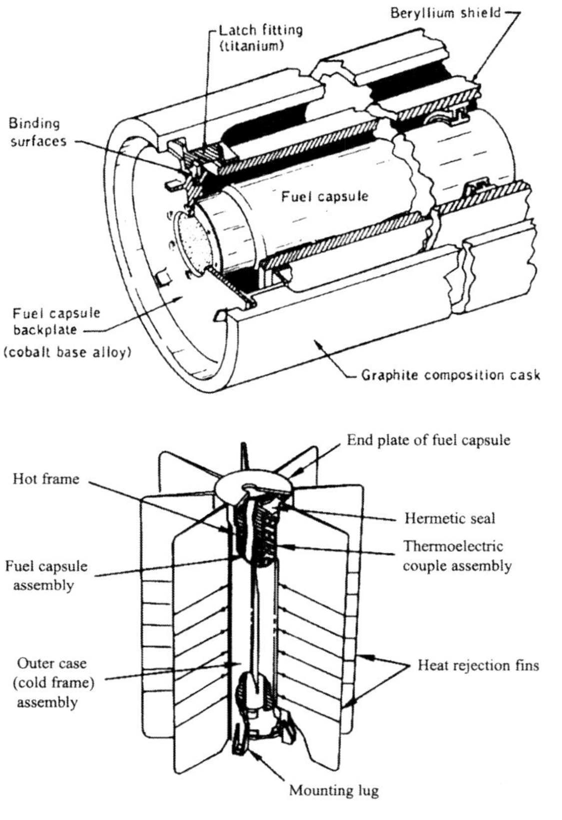

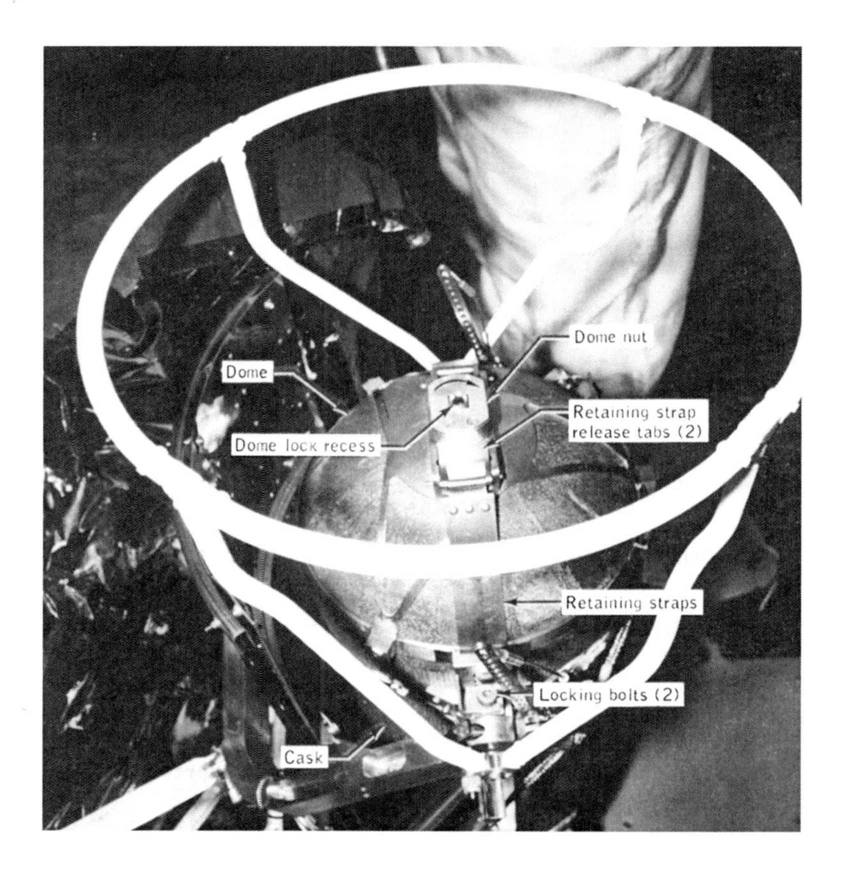

Details of the radioisotope thermoelectric generator.

Al Bean inserts the dome release tool of the RTG fuel cask.

Al Bean attempts to open the RTG fuel cask. Note that the hand-tool carrier and the subpallet have been removed from the subassembly pallet.

dome-like lid on the end of the cask, the other was to extract and transfer the capsule itself. The job was broken down into a series of steps: (1) engage the dome release tool with the dome-locking mechanism; (2) press it inward and turn the mechanism 90 degrees counter-clockwise; (3) pull on the tool and rotate the dome 60 degrees counter-clockwise; (4) detach the dome and discard both it and the tool; (5) insert the three-fingered fuel transfer tool into the capsule's head; (6) engage the tool by rotating clockwise; (7) extract the capsule from the cask; (8) lower the capsule into the thermoelectric unit; (9) release the tool by rotating counter-clockwise; and (10) discard the tool.

Bean unstowed the fuel transfer tool from the subassembly and left it temporarily on the nearby hand-tool carrier, then unstowed the dome release tool and stepped up to the cask. "Okay, I'm unlocking the cask dome right now. It unlocked perfectly. I'm shaking it, trying to get it off."

"There you go," Conrad encouraged.

"It came off beautifully," Bean reported. "I'll put the tool and the dome aside."

"Very nice," Conrad agreed.

"I'll fetch the cask removal tool." He retrieved this from the hand-tool carrier. "There you go," he continued, as he began the second stage of the process. "Sliding right in there. Okay, tightening up the lock. Hold it!" The capsule was supported by a pair of metal rings embedded in the graphite shell of cask. It was supposed to slip out easily, but would not budge. "You've got to be kidding!"

"Make sure it's screwed all the way down," Conrad advised.

The tool was engaged, but Bean decided to start afresh. "Let me undo it a minute and try it a different way." He withdrew the tool, turned it, and reinserted it with the fingers engaging the holes in a different arrangement.

"Houston," Conrad reported. "Al put the tool on, screwed it all the way down and the fuel element wouldn't come out of the cask."

After half a minute, Bean expressed surprise. "I tell you what worries me, Pete. It is a very delicate lock mechanism." He was concerned that he might damage either the tool or the capsule if he applied too much force. "Maybe I'll not push the pins in quite so far and wiggle it a little." After gaining a sense of the situation and feeling the radiative heat on his gloves, he ventured that the ring material had thickened and was gripping the capsule. He tried again. "Come out of there! Rascal." After another half-minute, he told Mission Control, "You know, everything operates just exactly like it does in the training mock-up, but it won't come out of the cask." If he could not overcome the friction by pulling using the tool, there was always another way. "I think what we can do is go get that hammer and bang on the side of it."

"I've got a better idea," said Conrad. "I want to try and put the back end in under that lip there, and pry her out." The hammer had an impact surface on one side of its head and a narrow blade on the other, and he wanted to slip the blade into the cask alongside the capsule and wiggle it in an effort to break the grip of the rings. "Let me go get the hammer. Where did you put it?"

"On the MESA."

While Conrad went to fetch the hammer, Bean removed the tool because he could feel that it was warming up.

Gibson relayed a suggestion, "Al, try to make sure you've got the pins all the way in, then try pushing down on it a little before you pull it out."

"Okay," Bean replied.

When Conrad returned, Bean reinserted the tool.

"You're not getting those pins all the way in," Conrad warned.

"They're not in now because I'm lining them up. Just a damn minute. Now they are all the way in." Reluctant to have Conrad attempt to pry the capsule loose, since it would involve working the hammer close to the tool, Bean reiterated his proposal to strike the side of the cask.

With Bean pulling gently on the capsule, Conrad hit the cask with the flat part of the hammer and the capsule moved about one-eighth of an inch.

"Hey, that's doing it!" Bean exclaimed. "Give it a few more." The capsule moved a similar amount with each blow. "Harder!" Bean urged. "Keep going. It's coming. It's coming out!" Conrad hammered harder. The capsule continued to withdraw in small increments. "Come on, Conrad!" Bean laughed. After about half an inch it began to loosen. "That hammer's a universal tool," Bean informed Mission Control. When the friction released, he was able to smoothly extract the capsule. "There, you got it!"

"Got it," Conrad echoed. In fact, he suspected that his hammering had cracked the graphite cask.

"Well done, troops," Gibson congratulated.

Gripping the capsule with the tool, Bean rotated it vertical and lowered it into the thermoelectric unit. "We got it, babe! It fits in the RTG real well!"

As Yankee Clipper flew around the corner, Gordon, listening to the MSFN relay, heard his colleagues laughing and Bean advise, "Don't come to the Moon without a hammer!"

After using the second universal handing tool to tip the RTG subassembly back onto its side, Bean manhandled it onto the carrying bar.

When Bean stood against the bar, ready to momentarily bend his knees with his arms down and grasp it for a 'static lift', he realised that his chest-mounted camera prevented him from seeing the bar, so Conrad provided verbal cues to assist. With its scientific apparatus offloaded, the RTG subassembly was much lighter than the other one, so Bean shuffled his hands on the bar to balance it up. He immediately realised that his training had been flawed. On the Moon, the ALSEP weighed about 40 pounds. As he could not have lifted it in terrestrial gravity, the training apparatus had been made to match its lunar weight, and hence had not simulated the full inertia of the ALSEP on the Moon.

When the ascent stage of Apollo 11 lifted off, its exhaust blasted insulation off the descent stage and scattered it far and wide. To protect the ALSEP from such debris, and from the dust that would be stirred up, it was to deployed at least 300 feet from Intrepid. As the individual instruments were to be laid out in a given pattern relative to the central station, the selected site had to be fairly level and free of inconvenient craters and boulders. In planning, they had envisaged locating it west of the landing site, which in this case would put it north of Head crater.

"We're making our move, Houston," Bean announced. Carrying the ALSEP was

With the RTG fuel capsule transferred to the thermoelectric unit, Al Bean joins the carrying bar to the two subassemblies.

the most arduous task yet assigned to a lunar astronaut. Bendix had estimated that to carry it 300 feet would take 10 minutes. "I can tell this is going to be a workload, so I'll take it easy."

"Take your time, Al," Conrad agreed.

"That's what I'm doing," Bean replied.

Conrad got the sampling tongs from the MESA and then set off, using a universal handling tool to transport the subpallet which he had earlier removed from the RTG subassembly.

Bean initially held the bar low with his arms almost straight but when it banged against his knees he held it out in front of his body, which was stressful. In addition, it was difficult to grip using pressurised gloves, whose neutral position was with the fingers open. After about 30 seconds, he halted. "I'm going to set it down and rest." The Flight Surgeon was monitoring his heart, which was pounding. Overheating due to the exertion, he increased the rate at which water flowed through his liquid-cooled garment. The sublimation unit shed the heat to the vacuum of space. But because the amount of water was finite, the PLSS could not be operated at this rate for very long without shortening the extravehicular activity.

Conrad, far ahead, announced, "I'm going to go right up to Head crater, I guess."

"A little bit more to the right," Bean suggested.

But Conrad had a plan: "I want to go 10 degrees off our takeoff angle, and I think I'm headed out about that way now." After rising vertically for 275 feet, the ascent stage was to pitch over sharply and fly an azimuth inclined 15 degrees north of west, aiming for the same orbit as Yankee Clipper. Locating the ALSEP off this azimuth would minimise contamination. "You just stay back there and take your time. I'll go out here and scout the area."

"You're getting pretty far out," Bean noted.

"Pete and Al," Gibson called with some helpful advice, "your LM shadow should now be about 110 feet."

"Okay," Conrad acknowledged. "I'm dying to find out what this mound is over here anyhow, Al." Then he explained, "We've got a very peculiar mound sticking up out of the ground, Houston. I want to go look at it. As a matter of fact, I think I'll go take a picture of it."

"Could you give me your position and distance with respect to the LM?" Gibson asked.

"Wait a minute," Conrad replied.

"Go ahead, Pete. Do what you're doing," Bean said. Then he explained to Gibson, "Pete's, I'd guess, about 300 feet at 12 o'clock, in the bottom of a shallow crater that you're bound to see on your map. It's sort of a doublet."

"I'm headed to the right-hand edge of the Head crater," Conrad added, by which he meant the northern rim.

"Roger," Gibson acknowledged.

In search of a fairly level patch on the undulatory landscape, Conrad veered north of the mound, away from Head crater.

"Hey, Al. Here's a neat spot to put it out up here."

"Is it flat?"

"You'd better believe it."

The site would later be determined to be 430 feet from Intrepid, and well off the launch azimuth. They were further from their vehicle than Neil Armstrong had been when he ran off-camera at the end of his moonwalk to inspect a crater. On Apollo 11 the maximum permitted walking radius from the lunar module had been 250 feet. In the case of Apollo 12, the radius would be defined by the oxygen in the OPS carried on top of the backpack.

While Bean trudged out with the ALSEP, Conrad ran back south to take a look at the mound. "I've got to photograph this thing," he announced. "I can't imagine what it is." In accordance with their photo-documentation training, he took a down-Sun picture and then moved around to take a pair from the side for stereoscopic analysis. "It's really fantastic!" In a landscape of craters and rocks, a conical mound was an oddity. "How's our timeline going, Houston?"

"At 1 plus 48 into the EVA, it looks as though you're right on there, if you've just about completed your traverse."

"We have, Houston," Conrad replied. He set off back to the site he had selected. "Over here, Al. See where I'm headed. This great big flat area."

"Hey, there's another one of those mounds over there," Bean noted. There was a smaller mound just north of the selected site.

"Hey, you're right! What do you suppose they are?"

"Houston," Bean called. "They're just sort of mounds. Don't take this the wrong way: it looks like a small volcano, only it's about 4 feet high; at the top it is about 5 feet across; it then slopes down and the diameter of the circle where it becomes level with the terrain is 15 or 20 feet. So it looks sort of like a small volcano. There's a couple of them out here. They look like they were made out of mud or something."

The scientists perked up at the possibility of volcanic activity. "Is there any hole or central vent?" Gibson asked.

"I don't know," Conrad replied. "We'll go over after we get the ALSEP out."

Still struggling on, Bean noted that there were many more rocks lying around than there were in the immediate vicinity of the lunar module.

"We could play geologist for two days and never get any further than we are right now, seeing all different kinds of things," Conrad explained to Gibson. "It's really neat. Better than any of the geology field trips." As he took a panoramic sequence to document the site, he snapped Bean making his final approach with Intrepid in the background.

"Man alive!" Bean exclaimed.

"Tired?" Conrad asked.

"No, I'm not so tired," Bean replied. "That handle, you know, when you carry this thing around in one g the packages tend to hang down. But up here at one-sixth, the RTG in particular tends to rotate the whole pallet." The RTG subassembly tended to rotate as he walked, which rotated that section of the bar, and since the two sections did not have a lock he feared it might come apart and cause the subassemblies to tumble, making the instruments even dirtier and possibly causing damage.[19] "Let me make sure that we're not going to run out into some holes." After visualising where the apparatus would be laid out, he saw that a small crater was in an inconvenient place. "Pete, I'm going to move just a little bit further to the north."

"All right," Conrad replied.

Bean pointed to various locations, "The magnetometer can sit over there, and the seismometer will sit in a good flat place. Houston, the trouble with the seismometer is we don't have any good solid bedrock or anything to set it on – all we've got is this dirt. I don't see any area around that has any rock."

"Roger, Al," Gibson replied.

"I'm afraid we're just going to have to take what we can get on this seismometer," Bean reiterated.

Meanwhile, Conrad had placed the subpallet on the ground and disconnected the universal handing tool from it. After releasing the bolts which held the suprathermal ion detector and cold-cathode ion gauge experiments, he tried to engage the tool to lift them clear. "You know," he mused, "there must be some thermal expansion or something, because I'm having a heck of a time getting this in the SIDE. It just flat won't go in there."

"Well, just pick it up with your hands," Bean suggested.

"I can't bend down that far," Conrad replied. In terrestrial training the weight of the backpack had made the suit bend sufficiently for them to reach down and grasp the instruments, but on the Moon this was not so. As he observed in the post-mission debriefing, "The ALSEP training package did its job, and I think we learned a great deal from it." It was simply that the training did not accurately represent conditions on the Moon. On finally lifting the experiments, Conrad turned to where he thought the central station should be situated. "Where'd you go, Al?"

"I'm right over here, babe."

"Oh, you're miles away!"

"Yeah, I moved over here." Bean had again changed his mind, placing the central station beyond the reach of the cable for Conrad's experiments.

"Oh, son of a gun!"

"I had to do it, Pete," Bean insisted.

Observing Bean, Conrad exclaimed, "Man, are you dirty!"

"It's from carrying the ALSEP," Bean mused. His suit was dusty from the knees down. Then he laughed at the packages. In preparing the instruments, the scientists and engineers had utilised 'clean room' techniques. The cohesive properties of lunar dust in a vacuum, augmented by its electrostatic properties, tended to make it adhere to anything it touched. "It's ridiculous," Bean laughed. "I remember how they took care of this white paint. You had to have gloves to touch it." As he noted in the post-mission debriefing, "It will just have to be designed to accept dirt and dust. If not, it will have to be packaged in some way that it can be deployed completely, then the last act is to pull some sort of pin and flip off the covering, exposing the nice clean experiment."

[19] It was decided that in future the two sections of the carrying bar should lock together.

As Pete Conrad took a panorama at a point close to where the ALSEP would be deployed, he obtained two views of Al Bean carrying the package from Intrepid. Unfortunately, the top view suffers intense lens flare. Note the offset manner in which Bean is holding the carrying bar.

Bean started by removing the RTG subassembly from the carrying bar and setting it down about 8 feet west of where he intended to position the central station.

After plugging his cable into the central station, which was standing on its edge, Conrad deployed the tripod of the suprathermal ion detector. The lid popped open. It was to have been opened by ground command once they had cleared the area. "Son of a gun. Why did that happen?"

"Never saw it do that before," Bean said.

While Conrad held the instrument, Bean closed the lid to protect the collectors of its two detectors. As Conrad stood the instrument in the dirt nearby, he laughed, "I did it! I set it down and it didn't fall over. I can't believe it."[20]

Bean's priority was to plug the RTG into the central station, and he asked Conrad to assist. "Use your tongs to hold this up a minute. It's a little hot and I don't want to touch it." The tongs were to enable a standing astronaut to lift small samples. As the tool itself was 24 inches in length, it did not need a separate handle. Conrad 'wore' it on his left hip, attached to a spring-operated device called a yo-yo which operated in a similar way to a rewinding tape measure. As he used the tongs the cable unreeled as much as required, and when he was finished he simply released the tongs and the cable reeled back in and returned the tool to his hip. As Bean recalled in the post-mission debriefing, "When I removed the bracket that carried the power cable to run from the RTG to the central station, it felt warm to the touch. I didn't want to keep my fingers there too long, so we handled it with a tool as opposed to just my gloved hands – as I had in practice." The urgency to link the RTG to the central station was to impose a dummy load on the power supply, this being a resistor in the circuitry of the central station. Once the experiments were connected, a switch would be thrown to short-out this resistor.

"We've connected the RTG to the central station, Houston," Bean reported. "And we're ready to go to work deploying the experiments."

Conrad removed the carrying bar from the other subassembly and rested it against the subpallet that he had carried out. "I've got the antenna mast," he informed Bean. "Let me tamp the dirt down underneath the central station."

Satisfied with his efforts, Conrad tipped the subassembly over onto its base, then aligned it relative to the Sun. The experiments affixed to the sunshield which formed the top of the central station would have to be unbolted before the station could be erected. While Conrad dealt with the solar wind spectrometer, Bean would tackle the seismometer.

In contrast to the solar wind collector that had been erected at the landing site and would be returned to Earth to measure the elemental and isotopic abundances of the ions that it trapped, the solar wind spectrometer (S-035) would remain in place and measure variations in the flux of the solar wind over a much longer period of time. It was a box 13 inches high, 9 inches wide and 11 inches long with a faceted dome on top. It comprised a sensor assembly, electronic assembly, thermal control assembly, and leg assembly. The solar wind was the major external force acting on the Moon's

[20] Later on, Bean would position it 60 feet south-southwest of the central station.

surface. Charged particles in the solar wind impinging on the lunar surface would stream through 'windows' in the dome of the sensor assembly, which housed seven identical Faraday cups to trap ionised particles. One cup was oriented vertically and the other cups were on inclined faces at 60-degree intervals. Each cup measured the current produced by the charged-particle flux entering its aperture. Since the cups were identical, if the flux was equal in each direction then the same current would be generated by each cup. In other cases, analysis of the currents in the individual cups would show the variations of particle flow with direction. By progressively changing the voltage on a cup and measuring the current it would be possible to obtain energy spectra for the incident protons and electrons.[21] The electronic assembly contained the circuitry for modulating the plasma flux entering the cups and for converting the data into the digital format appropriate for the central station. The thermal control assembly utilised three radiators on one vertical face of the electronic assembly. The leg assembly comprised two tubular A-frames with telescoping legs. The instrument was to be deployed about 13 feet from the central station and, in order to minimise interference, well away from the RTG. The objective was to measure the energies, densities, incidence angles and temporal variations of the solar wind that reached the lunar surface – in effect, to measure the plasma environment at the surface, and how this varied with time – to determine whether the Moon was a good conductor of electricity.[22] The experiment was developed at JPL by Conway Snyder, Douglas Clay and Marcia Neugebauer. The results were to be combined with those from the Explorer 35 satellite, which had been inserted into an elliptical lunar orbit in July 1967 to monitor the magnetic fields and plasma flows of the interplanetary medium in the vicinity of the Moon, and was still functioning. The data would also assist in interpreting the results of the lunar surface magnetometer and with analysis of the samples returned to Earth by the astronauts.

As Conrad reported in the post-mission debriefing, "The deployment of the solar wind spectrometer went exactly as advertised. I checked the four legs down, took it out the proper distance, aligned it and turned her loose." The instrument had a solar cell beneath a very narrow slit, and the manner in which sunlight passed through this slit confirmed that he had levelled the instrument well within the 5-degree tolerance specified by the experimenters.

Meanwhile, Bean had started on the seismometer (S-031). The seismometers of Apollo 11 and Apollo 12 were similar in purpose, but different in configuration. In particular, whereas the former was mounted on a pallet, the new one was to rest on a small stool. When the science support team heard there was no exposed bedrock on which to deploy this instrument, and that it would have to be placed on open ground, they relayed a recommendation via Gibson for Bean to tamp down the dirt

[21] The minimum detectable flux was 10^6 particles per square centimetre per second. It could measure protons in the energy range 45 to 9,600 electron volts, and electrons in the energy range 6.2 to 1,376 electron volts.

[22] When the phase of the Moon is 'new', the hemisphere that faces Earth is in the Moon's 'wake' in the solar wind, and when the phase is 'full' the near-side is shielded from the solar wind by virtue of being in the Earth's magnetotail.

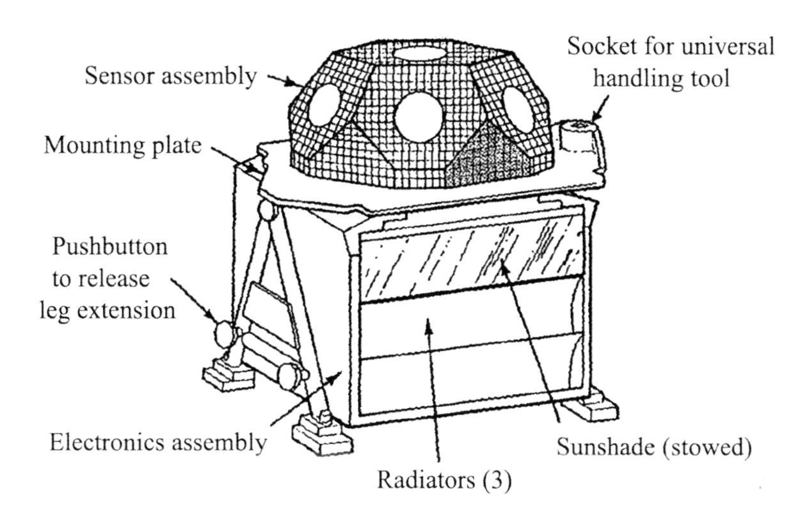

Details of the solar wind spectrometer, and in its deployed state on the Moon.

with his boots to make a firm base for the stool. Bean replied that he had already decided to do this in order to prevent a 'thermal short'. This was because the stool was merely a ring on legs. Once emplaced, it would accept the rounded base of the cylinder that formed the instrument itself. If it had been possible to stand the stool on a flat rock, this would have held the instrument clear of the ground. But if the stool were to sink into the dirt on open ground this might cause the instrument to make contact with the surface and thereby provide a path for the conduction of heat. "What I've done," he reported, "is I have sort of dug a little crater so that, essentially, the hole in the centre of the stool has full clearance between it and the ground." Nevertheless, his conclusion was that tamping was not an effective means of preparing the surface for emplacement because the degree to which the dirt compressed was small. Although the stool isolated the instrument thermally and electrically from the ground, it would readily transmit surface motions having frequencies of up to 26.5 hertz, as required for seismic studies. The three-legged stool was delivered on the RTG subassembly, and it had to be set at least 15 feet from the RTG, lest the radiated heat impair the instrument's thermal stability, so Bean had placed it on the far side of the central station.

Once Bean had the stool in position, he unbolted the seismometer from the central station and used his universal handling tool to transfer it to the stool. In the process, the cable linking it to the central station automatically unreeled. The instrument was an upright cylinder 15 inches tall and 11 inches in diameter. It consisted of a sensor assembly and an electronics assembly. The sensor assembly had three long-period seismometers with orthogonally oriented capacitance-type sensors. There was also a single short-period magnetic-type sensor located on the base of the sensor assembly. The electronics assembly was actually in the central station, and contained circuitry for attenuating, amplifying and filtering seismic signals, for processing the data, and for the internal power supplies. Bean was to align the cylinder on the stool to within 5 degrees of vertical. When Buzz Aldrin deployed his seismometer he was to level it by reference to a small ball bearing in a cup inscribed with concentric rings, and found that the weak lunar gravity slowed the rate at which the motion of the ball was damped out. Leaving the ball rolling around, he went off to do something else. Upon returning later, he saw that the ball was centred. This time an air-bubble had been added. The instrument was to be levelled by using the universal handling tool to nudge four metal tabs at the base to adjust it on the stool. As Bean reported in the post-mission debriefing, "The addition of the bubble was a good one. I noticed that it was really easy to level the experiment using that. While I was doing it, I kept an eye on the little ball bearing and it was rolling all over the place." But this manual levelling was just the first step: control motors would later refine the alignment of the sensor assembly to within 3 seconds of arc. In addition, the instrument had to be manually oriented in azimuth so that the 'zero' of a numbered sundial was in the direction of the Sun. Then Mission Control was to be told where the shadow of the mound of the levelling gauge at the centre of the cylinder intersected the dial, so that the scientists – Gary Latham and Maurice Ewing of the Lamont Doherty Geological Observatory, Frank Press of the Massachusetts Institute of Technology and

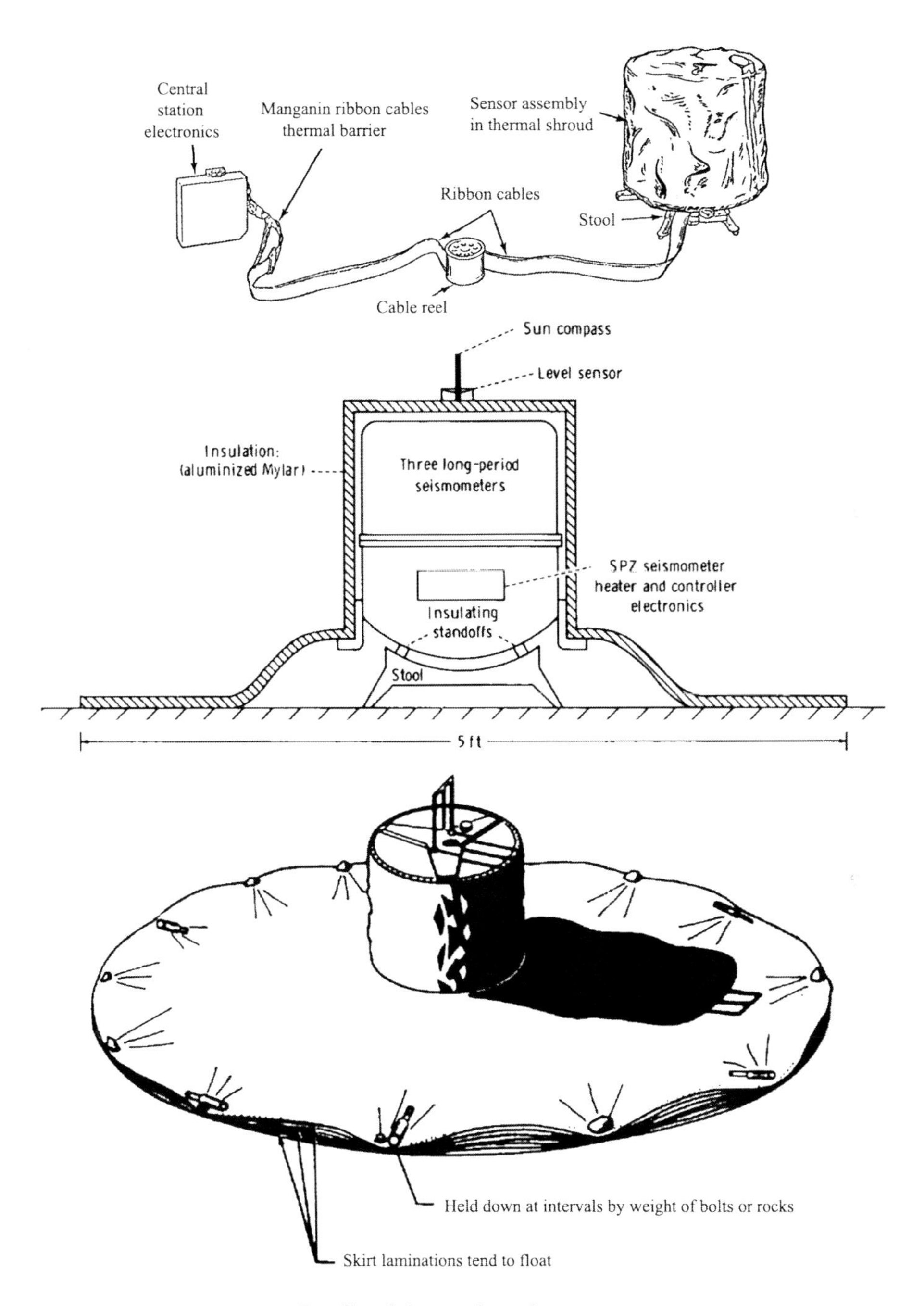

Details of the passive seismometer.

George Sutton of the University of Hawaii – could precisely calculate its orientation.[23]

With the seismometer requiring additional work, Conrad came over to assist. On seeing him stepping backwards in close proximity to the central station, which was beginning to sprout cables, Bean called urgently, "Wait a minute! Be careful. Don't move sideways or backwards. You just don't know what's there. You've always got to move forwards." The ALSEP cables were a major concern even before reaching the Moon. They ran from the RTG to the central station, and between that and the individual experiments. The layout was a compromise between the requirements of the various components of the ALSEP. The human-factors people who planned the deployment had developed a sequence for laying them out that should not require a man to walk across a cable once it had been laid. Apart from the power supply from the RTG, the cabling had to carry 32 individual wires, so it had been decided not to use twisted-wire but flat-ribbon cables which should more readily run flat across the ground.

Another way in which the new seismometer differed from that of Apollo 11 was the addition of a thermal shroud to assist in stabilising the temperature of the sensor assembly and prevent the surrounding ground from expanding and contracting with the varying thermal regime, causing the orientation of the instrument to alter.[24] The shroud consisted of ten layers of aluminised mylar which were separated by layers of silk cord wound on a perforated aluminum support. When deployed, the shroud was 5 feet in diameter. In the first design it was pleated around the cylinder, but in trials the friction of unfolding the pleats using the universal handing tool exceeded that of the cylinder on the stool, and thus disturbed the orientation of the instrument. To reduce the friction of the folded pleats, the shroud was folded neatly around the cylinder in such a way that prior to deployment it resembled a budding tulip, with the folds held in place by a wrapping which Bean had to remove. He was to use the tool to 'flip' (but not pull) the folds one by one onto the ground until the cylinder on its stool was at the centre of a pentagon of mylar forming a rough circle. The order in which the pleats were to be unfolded was predefined: the first pleat was labelled with a '1' in bright orange and an arrow showed the tab to which the tool was to be applied. The second pleat was numbered, but because by this time the direction of progress had been established the others were unnumbered. Spreading the shroud on the surface proved difficult. As he said in the post-mission debriefing, "It wasn't that it had a memory of being pleated. When I placed it near the ground, the many layers seemed to separate. It seemed to have some kind of static charge to it that would not allow it to touch the ground. It took quite a little pushing to get it to lie down on the ground. The only way I could make it lie flat was to put a little dirt on it, but that wasn't a very good idea because it was difficult to put little clods of dirt on it. I later

[23] Ewing and Press were co-investigators on the Ranger seismometer experiment that sadly never reached the Moon.

[24] On the other hand, the mylar itself generated a spate of 'noise' as it adjusted to the rapidly changing thermal regimes at sunrise and sunset.

got some Boyd bolts and made the little alignment tubes sit on it. That worked real well."[25]

Leaving Bean to finish off deploying the seismometer, Conrad unbolted the lunar surface magnetometer from the central station and placed it aside ready for Bean to deploy, then he turned his attention to erecting the central station. The top deck had been held tight against the bottom by Boyd bolts to form the structural base for the subassembly. As he released the last of the bolts, a spring-loaded mechanism raised the deck up 3 feet on four collapsible legs, in the process unfolding a thermal curtain to protect the internal hardware.

The central station was the focal point for control of the experiments and for the collection, processing and transmission of scientific and engineering data to Earth. Its data system comprised an antenna, a diplexer, transmitter, command receiver and decoder, timer, data processing system, and power distribution unit. Its antenna was a copper conductor bonded to a fibreglass epoxy tube for mechanical support, and the modified axial helix could transmit and receive a right-hand circularly polarised S-band signal. A two-axis gimbal enabled the antenna to be adjusted in azimuth and elevation for initial aiming at Earth, which would remain more or less motionless in the sky. The gimbal was mounted on a staff that had doubled up as the carrying bar. The diplexer had a circulator switch and a filter. The circulator switch coupled one or other of the two redundant transmitters to the antenna, and the filter provided the attenuation to isolate the transmitted signal from the received signal and enable both signals to use the one antenna free of interference. The carrier was phase modulated by the bit stream from the data processing system. The command receiver accepted uplink commands, and the decoder provided the digital timing and command data and issued the commands to the experiments. The timer, which consisted of a clock powered by a long-life mercury cell battery, provided predetermined switch closures to initiate specific functions within the experiments and data system when the uplink commands were not available. The data processing system had two redundant data processing channels, each of which generated experiment timing and control signals, collected and formatted experiment data, and provided data for phase modulation of the transmission carrier. The power distribution unit had the circuitry for the power-off sequencer, monitored temperature and voltage, and regulated power for the experiments and the central station itself.

After retrieving the antenna aiming mechanism from the subpallet which he had carried out, Conrad announced delightedly, "Al, those are my last two Boyd bolts." He mounted the mechanism on the staff, which he had already attached to the side of the central station, added the antenna tube, and started the fiddly task of aiming it at Earth.

"I'll take out the magnetometer," Bean said. In carrying the instrument to where he intended to deploy it, he walked straight across a well-defined crater and made an

[25] For the next ALSEP, the shroud lamination was spot-sewn together at fixed intervals around the periphery, a small weight sewn to each of the six attach-pullout points, and a blanket of teflon added to decrease solar degradation.

observation. "I'm down in a little crater now, Houston. And, sure enough, right in the bottom of the crater there is a lot softer dust than up on the rim. Not much, but it's noticeable. I don't think the sides are slippery at all. I don't think it's going to bother us going over to get our Surveyor." As he explained to Eric Jones in a 1991 interview, "Some craters were harder at the bottom. Some had glass at the bottom. Some had rocks at the bottom. I would agree that the thickest dust was generally on the rims of craters, but in this case I'm sure it wasn't. All craters are different. Some of them are newer and have raised rims. Those are the ones which are soft up there. This might have been an old one, its rim eroded by small meteorites, with some of it ending up in the middle, making a soft fill." His point was, "We were worried about the possibility that it would be slippery on the slopes of Surveyor crater – that the dust would be hanging there and you'd get on it and it would slide you down like snow."

Bean placed the magnetometer just beyond the crater, 50 feet south-southeast of the central station, as far as possible from the RTG. As he removed white styrofoam packing blocks, he yielded to temptation and tossed one to observe it pursue a high arc in the weak gravity. He called on Conrad to watch as he threw several more. But Conrad, aware of the relentlessly ticking clock, called a halt, "Stop playing and get to work." And then he pointedly remarked, "Maybe they'll extend us until 4.5 hours. I feel like I could stay out here all day." The extravehicular activity was nominally assigned 3.5 hours, but it had been agreed with the planners prior to the mission that if it went well in terms of the rate at which they consumed their PLSS resources, the moonwalk would be extended to enable them to collect some samples in the vicinity of the lunar module.

The deployment of the magnetometer was completed in about 3 minutes. As Bean noted in the post-mission debriefing, "It was easy to deploy. It was easy to align and level, requiring quite a bit less time than the seismometer because it was sort of self-contained." Once the boxy body was in place, he unfolded the three fibreglass arms, each of which had a flux-gate sensor on its tip to measure the magnetic field vector for a Cartesian axis. The sensors were 40 inches above the ground and wrapped in insulation except for the flat upper surfaces which served as heat radiators. A levelling leg was deployed from the base of each arm. Bean aligned the instrument to within 3 degrees east-west using a shadow graph on the box, and to within 3 degrees of vertical using an air bubble on one of the arms. The arm that was to be pointed due east was indicated by an orange 'E' and an arrow on the top of the box. He then removed these symbols to prevent their disturbing the thermal balance. In addition to the electronics, the box housed an electro-mechanical system designed to operate a gimbal within each arm which would reorient the sensors for calibration purposes.

"The magnetometer is deployed," Bean reported. "It's level, pointed exactly east, and the little black dot is right in the middle." However, he warned, "There's a lot of dust on top of the electronics box."

The magnetometer (S-034) was developed by C.P. Sonett and Palmer Dyal of the Ames Research Center and Jerry Modisette of the Manned Spacecraft Center. It was to measure the equatorial magnetic field at the surface, with long-term data yielding insight into the composition of the interior of the Moon. The electrical properties of

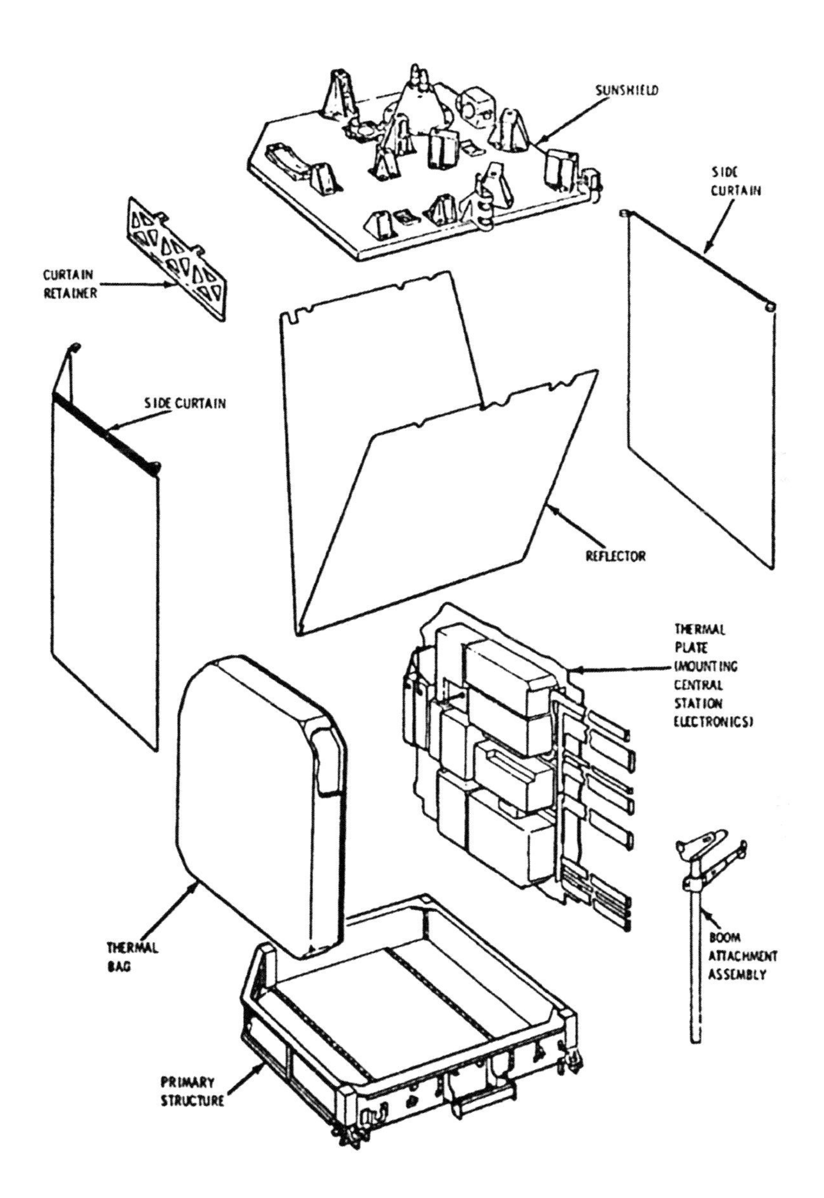

Detail of the ALSEP central station.

Pete Conrad has laid his tongs against the subpallet he carried from Intrepid. He has the carrying bar in his left hand, and is using his universal handling tool to remove the antenna gimbal from the subpallet. Note that the magnetometer in the foreground is still folded up awaiting deployment.

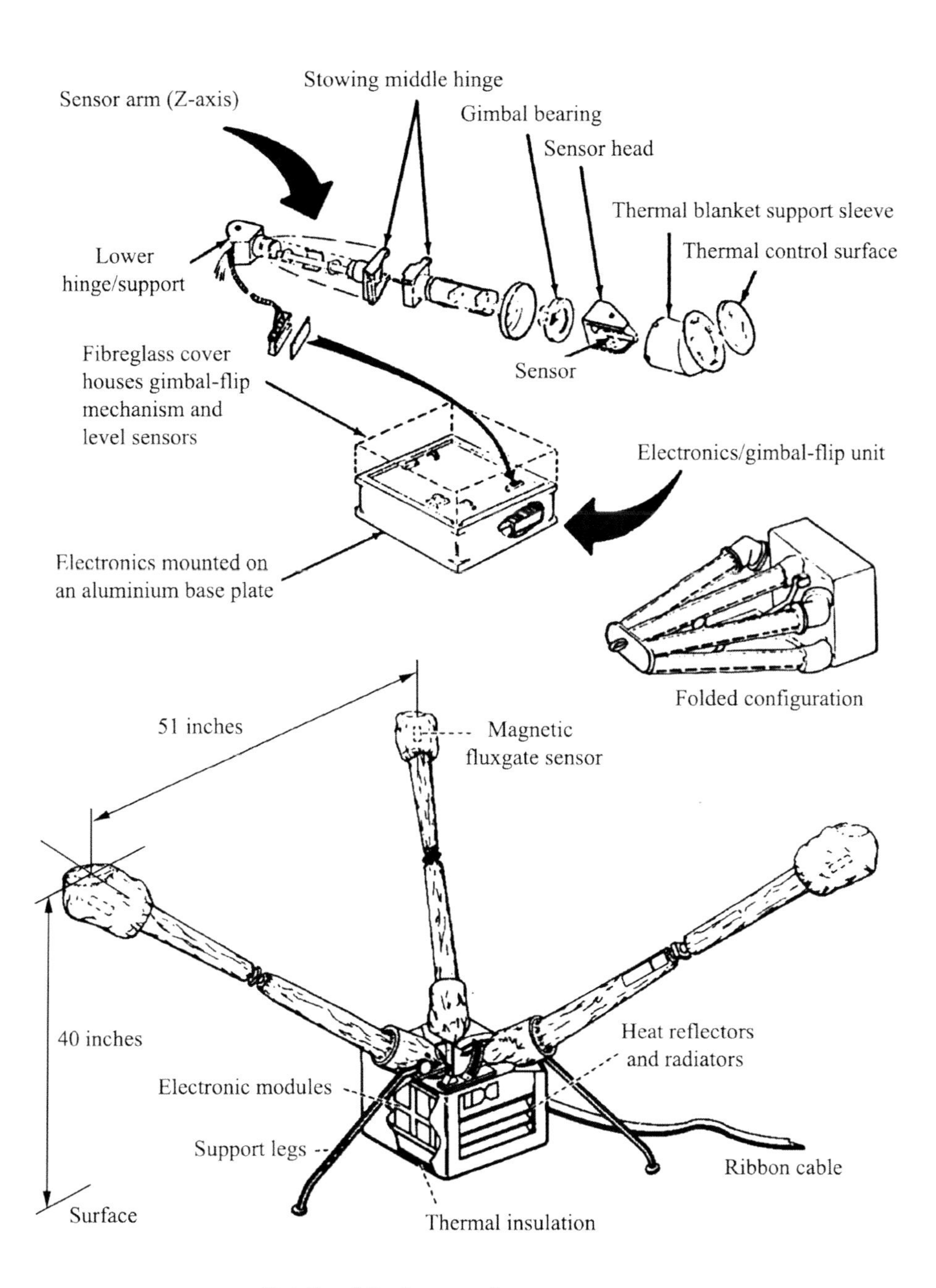

Details of the lunar surface magnetometer.

Al Bean standing by the deployed magnetometer just beyond a small crater. Note the saddlebag on his left hip. The foot pads of Intrepid are hidden by a very slight intervening rise in the terrain.

Pete Conrad is installing the antenna on the central station. The magnetometer is in the foreground. The RTG is behind the integrated suprathermal ion detector and cold-cathode ion gauge, which are awaiting deployment. The subpallet is on the left. Note the 'small mound' in the background.

The central station with its antenna pointing at Earth, the seismometer and the magnetometer. In the distance are Intrepid, the S-band antenna, the draped flag and (in the lens flare) the television camera and the solar wind collector sheet.

The ALSEP central station has its antenna pointing at Earth. The seismometer is to the left, the magnetometer beyond and the solar wind spectrometer to the right.

the material which comprised the Moon would determine what happened when the magnetic field carried by the solar wind came into contact with the surface. If the Moon were a perfect insulator the magnetic field would pass through undisturbed, but if the material acted as a conductor the incident magnetic field would make eddy currents inside the Moon and generate an induced field that would alter the magnetic field measured at the surface. The surface measurements were to be compared to the magnetic field in the undisturbed solar wind measured by Explorer 35. The results would yield insight into the electrical properties of the Moon, the temperature of its interior and whether it had an iron core. Although measurements from space implied that the Moon did not possess a global magnetic field, some of the rocks returned by Apollo 11 were magnetic and scientists wanted to know whether this was remanent magnetism representing a dipole field which was possessed by the Moon early in its history and had since decayed, possibly because the initially molten iron core had solidified.

While passing behind the Moon, Gordon had grabbed a bite to eat. Upon his reappearance on revolution 18 Gibson updated him, "Dick, the EVA is going pretty well; they're 2.5 hours into it and they've got the ALSEP a good way deployed. Apparently both of them look as though they just crawled out of a coal bin."

"Very good," Gordon acknowledged.

As Conrad finished with the antenna, Bean retrieved the suprathermal ion detector and cold-cathode ion gauge experiments and went to deploy them 60 feet southwest of the central station and also, because the package held a magnet, 80 feet from the magnetometer. The first task was to lay down a spider's web-style wire mesh shield that was stowed in the side of the 8-inch-tall rectangular box and was spring-loaded. As he wryly observed in the post-mission debriefing, "These spring-loaded devices are a real pain up there. You should have one that doesn't have any spring load to it. You should be able to open it up and drop it on the ground and it just lies flat from its normal weight, instead of having some spring-loaded device. It just adds time and work trying to get these little devices to work properly." The shield was designed to isolate the instrument from any magnetic or electric fields that might emanate from the lunar surface. The emplacement was complicated by the springiness of the shield and the short distance between the three legs of the box. As he said in the debriefing, "The legs are too close together for the height and weight of the experiment. When you try to put it on the ground, it just wants to tip over. The little place where you insert the universal tool is on the end which has a single leg, so if you put any offset of force on that attachment, that tends to tip it over." The slit atop the box had to be aligned with the plane of the ecliptic in which the Sun would cross the lunar sky. As the cover repeatedly popped open, once Bean was satisfied with the emplacement of the instrument he left the cover open, lest it fail to release when commanded to do so by Earth.

The suprathermal ion detector (S-036) was designed by John Freeman and Curt Michel of Rice University in Houston. Michel, a nuclear physicist, was a member of the scientist-astronaut group recruited in 1965, but left in September 1969 in order to resume a full-time academic career. The objective was to measure the characteristics of the ions believed to comprise a low-lying ionosphere just above the surface, and

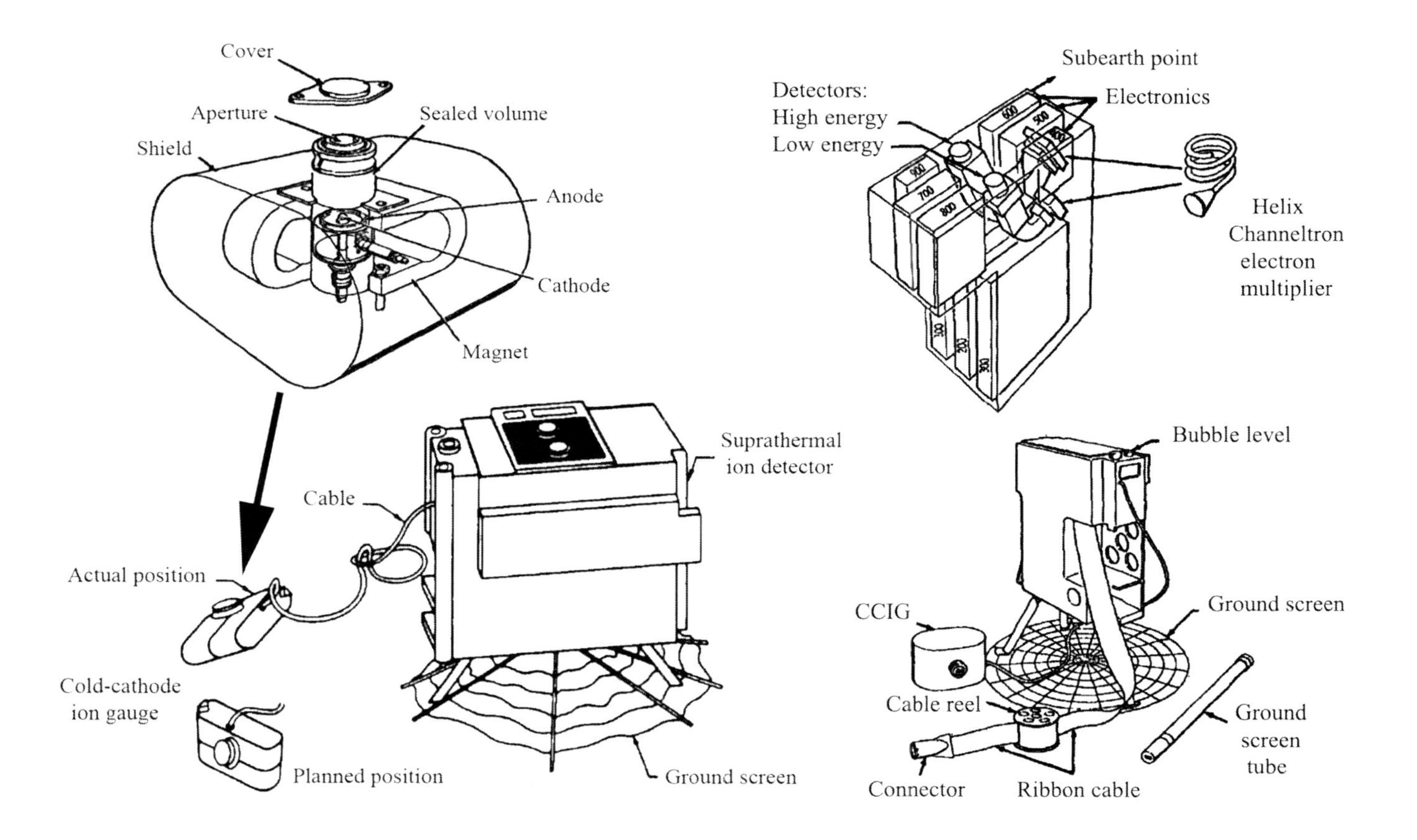

Details of the suprathermal ion detector and cold-cathode ion gauge, including the planned and actual positions of the ion gauge.

Al Bean by the suprathermal ion detector on its tripod and small cold-cathode ion gauge on the ground alongside. Note the saddlebag on his left hip.

for this reason it was also known as the lunar ionosphere detector. It contained two curved-plate analysers with which to measure the flux, number, density, velocity and energy per unit of charge of positive ions that entered the detector. The data was to provide a preliminary value for the electric field at the lunar surface. The amount of material detected was expected to be very small, but this was the first opportunity to make in-situ measurements of the density and composition of such a phenomenon. Most of it was expected to originate from the solar wind, but reports over the years by terrestrial observers of 'transient events' on the Moon suggested that gases were leaking through the crust and being ionised by ultraviolet sunlight, so it was hoped to gain insight into the chemistry, radioactivity and volcanic activity of the Moon. In addition, the impact of meteorites would vaporise both the impactor and some of the lunar surface material. However, in the weak lunar gravity most gases would readily escape to space and, upon being ionised, would be 'picked up' by the magnetic field of the solar wind and swept away. On-going measurements would also sample the tail of the Earth's magnetosphere when the Moon was in the appropriate sector of its orbit.

Next, Bean lifted the cold-cathode ion gauge from its stowage compartment in the suprathermal ion detector. This was the size of a large matchbox, but with rounded ends. The two instruments were linked by a short cable and the central station saw them as a single experiment. He set it down directly on the lunar surface about 4 feet away, oriented in such a manner that the orifice on one side did not face any other experiment, Intrepid or the lunar surface. Unfortunately, the stiffness of the cable, which was a round one rather than a ribbon, prevented the instrument from standing upright, instead forcing it to an aperture-down attitude. As Bean reported in the post-mission debriefing, "The gauge was so light and the forces in the cable so strong that the cable would just pick up the gauge and move it to an undesirable position. It ended up with Pete and I working together – he held the ion detector while I tried all sorts of deployment angles of the gauge."

As they worked, Conrad voiced his frustration, "I knew that ding-a-ling cable was going to make us spend hours trying to do that."

"Doesn't that make you mad," Bean agreed.

"Especially when you know it's going to happen," Conrad muttered. Holding the precariously mounted suprathermal ion detector upright he added, "If this falls over, I'm really going to be mad!"

After about ten attempts, Bean managed to settle the instrument tilted on its back with the aperture about 60 degrees above horizontal, which was deemed satisfactory. They had spent about 5 minutes on what he reckoned would have been a 30-second job if it had been designed properly.

The cold-cathode ion gauge (S-058) was made by F.S. Johnson of the Southwest Center for Advanced Studies of the University of Texas at Dallas and D.E. Evans of the Manned Spacecraft Center. The gauge temperature was measured directly. The neutral particles were ionised and collected by the cathode, which was one of a pair of sensor electrodes, and thereby produced an electric current at the input circuitry of the electronics that was proportional to the particle density. Any one of seven dynamic ranges could be selected to permit detection of neutral atom densities in the range 10^{-6}

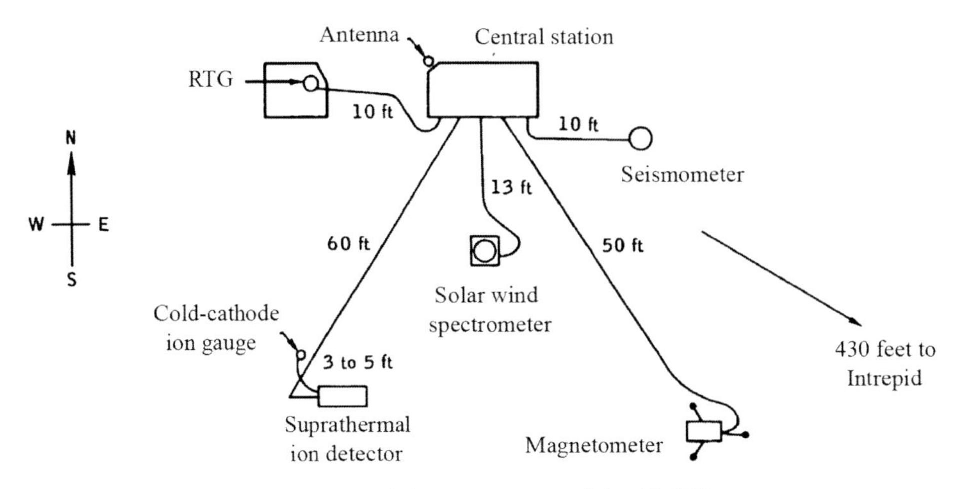

The layout of the components of the ALSEP.

to 10^{-12} torr. The sensitivity could be selected either by automatic internal adjustment or by command from Earth. In contrast to the suprathermal ion detector, which measured the composition of the ions, the cold-cathode ion gauge was only to measure the particle density and temperature of the ambient atmosphere – i.e. the total pressure of neutral particles – and so was also known as the lunar atmosphere detector. The density was expected to be very low but to vary in a cyclic manner, increasing during the day and declining at night. It was also hoped to measure the rate of dissipation of contaminants left in the landing area by the lunar module and the astronauts.

As they finished, Conrad surveyed the area, laughed and said to Bean, "You're a big litter bug. You know that?"

"I know it."

"Pete and Al," Gibson called. "A comment on picture-taking. If you would, try to take close-ups which will show the dirt we might have on thermally sensitive areas. Also, when you're done, take one or two extra pictures showing the ALSEP with the mounds that you described previously in the background. That will give us a good geometric reference." They duly obliged him by documenting each instrument in its deployed state and then several locator shots.

Back at the central station, Conrad pushed a button to redirect the electrical power from the dummy load to the suite of instruments. The final act was to throw a switch to activate the communications system. Known as the Astro Switch, this was located at the rear of the base, so he flipped it using his universal handling tool.[26] "I'll be a happy man if you tell me you're getting a signal," he prompted.

"Stand by," Gibson replied. After consulting the Experiments Officer, he said, "It looks like you did the job. We're getting data back."

"You just don't know how happy I am," Conrad replied.

[26] The switch was thrown at 9:21 a.m. EST on 19 November.

The 'small mound' north of the ALSEP deployment site.

The 'large mound' south of the ALSEP deployment site.

The central station would now transmit data to enable the scientists to verify the functionality of their experiments. Shortly thereafter, in the science support room, a pen recorder displaying the seismometer data began to show the footfalls of the two astronauts walking around.

The problems encountered in deploying the ALSEP served to show the folly of an early proposal to develop a self-deploying instrument package: consider the need to deal with the rough ground, the need to position the individual items in a particular pattern across a wide area, and the need to level and align them relative to the Sun. Furthermore, although a number of unmanned spacecraft had landed on the Moon in recent years, none carried such a sophisticated suite of instruments as this first lunar surface geophysical station.

Nevertheless, to Conrad and Bean the deployment of the ALSEP was a chore that had to be done before they could make a start on investigating their landing site.

"All right, I'm going over to this mound," Conrad said as he set off to inspect and photograph the small mound to the north. On returning to Bean, he saw "a funny rock" in a small crater. Small fragments of glass had been found in the Apollo 11 samples, but here was a rock with a "glass spatter" on it. A ring on the saddlebag that each astronaut wore on his hip held a number of smaller teflon bags in which individual samples could be sealed. Conrad lifted the rock using his tongs and gave it to Bean, who bagged and stashed it. Then, while Bean took some final shots of the ALSEP, Conrad ran south, saying, "I'll meet you over at that big mound."

The nominal duration of the first extravehicular activity had been set at 3.5 hours. If they had been on schedule, they would have left the ALSEP site half an hour ago. Although they had been on their PLSS resources for almost 3 hours by the time they were actually ready to head home, Conrad asked, "Houston, how long are you going to let us stay out?"

After consulting, Gibson replied, "You've been extended by 30 minutes, so you'll be out for a total of 4 hours."

"So how much do we have?" Conrad asked.

"You've got about 1 hour left," Gibson clarified. He then said that in a minute he would have some instructions for how to spend the extension.

"We're standing over at the Head crater," Conrad said helpfully.

"Why don't we start picking up some rocks while we wait?" Bean prompted, and pointed out some candidates.

"Yeah," Conrad agreed.

The plan called for concluding the moonwalk with 'selected sampling' back in the vicinity of the lunar module. In terms of field procedures, this was one step up from 'grab sampling' since it involved some selectivity. To further increase the degree of rigour, they took a picture to record the context of the rocks. As Conrad prepared to lift a rock using his tongs, Gibson interrupted, "There are two things we would like you to do on the way back. One is to document and take samples of those mounds. And, secondly, if you can, get over to the 1,000-foot crater which is northwest of the ALSEP and get documentation and samples there."

"Thousand-foot crater?" Conrad asked in puzzlement. "You don't mean the Head crater, do you?"

"Negative," Gibson replied. "As you're at Head crater, we'll give you a vector. Stand by."

They went over to the large mound to take close-up pictures of the position where they would take the requested sample.

"The crater we're speaking of," Gibson resumed, "is about 300 feet northwest of Head crater."

In fact, this was one of a string of similarly sized craters in a north-south crescent west of the Snowman and on gaining his first view ahead during the descent Conrad had used this arc as a visual cue. If Intrepid had landed at Site 3, they would have visited the southeastern rim of this crater after deploying the ALSEP and the route of their geological traverse would have visited a feature on the northeastern rim named Shelf. The crater was not named on the planning documents, which is why Gibson had to specify it by range and bearing from Head, but it would become known as Middle Crescent after the mission. And although referred to as a '1,000-foot crater', it was actually more like 1,500 feet in diameter. When Conrad and Bean reached the rim they would be 1,100 feet from Intrepid.

As Gibson tried to read to Gordon the data for a manoeuvre that Yankee Clipper was soon to make, Gordon cut him short, "It's impossible with those guys yakking." Rather than cease relaying the S-band downlink from Intrepid that enabled Gordon to eavesdrop on the surface team, Gibson called, "Pete and Al, we'll be talking with Yankee Clipper for about the next 5 minutes."

"Very good," Conrad replied. However, thinking that what Gibson had meant was that he would not be talking to them for a while, they continued to chat away as they sampled the large mound by using the tongs to work loose a fragment.

Bean mused, "I get the feeling that when Head crater was made, it just threw out a big blob of dirt and this is where it landed."

"Yeah," Conrad agreed.

By probing the mound, they showed it to be composed of slightly hardened clods of fine-grained material that crumbled easily, indicating that, as Bean had surmised, it was impact ejecta. As such, it did not justify the interest which his description of it as looking like a small volcano had provoked in the scientists.

Gibson tried again, this time making his request explicit, "Could we have silence for about 5 minutes."

"Yup," Conrad replied.

After updating Gordon, Gibson announced, "Pete and Al, we're back with you."

Having completed sampling the mound maintaining radio silence, they had set off running northwest. "We're almost at the 1,000-foot crater," Conrad reported. "About another 200 feet to go."

Even before they reached their target, Gibson warned, "We show you are 3 hours and 7 minutes into the EVA. We would like you back at the LM to start the closeout activities in 10 minutes – that's at 3 plus 17."

"Holy Christmas," Conrad exclaimed, using one of his toned down curses. "We're going to have to smoke there, Houston." By this he meant that all they would be able to do in the few minutes available at the crater would be to shoot a panorama across its interior and grab a sample before running all the way home.

"That's affirmative," Gibson replied.

"We're not going to get very many rocks by going this far," Conrad pointed out, meaning that the time being spent traversing could have been devoted to sampling if they had remained at Head crater. "But if that's what you want, then that's what you want."

They paused at what Conrad described as a "spanking fresh impact crater" from which they hastily lifted two samples. On setting off running again, he urged, "Let's get right to the edge of this crater and photograph it."

As the rimless shallow pit finally opened before them, Conrad exclaimed, "Look at that. It's spectacular isn't it? Wow, a monster!"

"Hey, there's bedrock down here a little ways," Bean said.

"Where?"

"It's right down the hill," Bean said, pointing at large blocks of rock about 150 feet inside the crater.

"Hey, you're right."

The scientists had urged them to be alert for exposures of what was believed to be a lava flow beneath the fragmental debris layer. During the 'window observations' they had said there did not seem to be any bedrock exposed on the open plain. If, as suspected, the debris layer was 10 to 15 feet thick, only impacts which made craters about 65 feet in diameter would even reach the bedrock. It was reasonable that for a crater of the size now being inspected there would be fragments of bedrock littering its interior.

"Pan it first?" Bean suggested.

"Yeah," Conrad agreed. He took a partial panorama that spanned the crater, then moved several feet sideways and shot another one to facilitate stereoscopic analysis. Unfortunately, from their position the opposite wall was washed out in the zero-phase illumination.

Noting that the crater looked "rather old", Bean described it for the benefit of the scientists. "There are some big boulders inside the rim, but there are none on the rim like there are on a large crater further to the west. But we don't see any outcroppings of rocks either that, you know, we could look at and say, 'well, from the top of the rim down to about 20 feet or something, there is the underlying rock'. But there are some boulders spread around. We're going to try to collect some samples."

"Al," Gibson countered, "we suggest that you hustle. We show you're 3 hours and 11 minutes, and we'd like you back at the LM in 6 more minutes."

Conrad's attempt to collect a sample from the rim of the crater was frustrated. "I can't get it using the tongs." The tines were of such angles, length and number to grasp rocks up to 2.5 inches in size, but this one was too large. His recommendation in the post-mission debriefing was, "There's no doubt about it, you need a bigger set of tongs. By the same token, you'll need bigger sample bags."

Innovating, Conrad pushed the rock over towards a boulder which Bean then used as a support in dipping down to lift the sample by hand.

"Go ahead and push another one over here," Bean suggested. "Get a couple of big ones."

A panorama spanning the width of the '1,000-foot crater' (which would later become known as Middle Crescent) from a vantage point on its southeastern rim.

Gibson curtailed further efforts, "You're 3 plus 13, and we'd like you back at the LM in 4 minutes."

"Okay," Conrad acknowledged. "We're on our way."

Intrepid looked very small, off in the distance. As they ran, Bean remarked on the fact that in the normal walking lope each stride was 3 or 4 feet, but it was twice that at a comfortable running pace in which they landed with the foot flat and pushed off with the toes. "You could go indefinitely at this pace." Although their heart rates were 150 beats per minute, this sustained run provided valuable data for the Flight Surgeon in setting his intervention limits for the long traverse that they were to undertake on the second moonwalk. "This is fun!" Bean reported.

Although they headed straight home, they avoided the ALSEP so as not to disturb it with the dust that they were kicking up.

When Conrad expressed frustration that they did not have much to put in the first day's sample return container, Bean prompted, "Ease over this way a little."

"Which way?"

"Over toward your left."

"What you want to do?"

"Why don't we grab a couple of rocks," Bean replied.

"All right. Here's one right here," Conrad said.

"Let me get a photograph of it."

"Hurry!" Conrad urged.

Bean snapped a pre-sampling picture, then Conrad lifted the rock using his tongs. When they set off running again, Gibson said that the seismometer was "picking up heavy footprints".

"Halt, halt, halt! Look at that!" Conrad observed. "I never saw one like that before. It's green." They grabbed it without taking a picture, bagged and stashed it, and then set off running again. On trying for another sample, Bean fell. "What I hate to see is an LMP laying on the lunar surface!" Conrad said as he offered Bean his hand to haul him up. On reaching Intrepid less than 2 minutes beyond the time specified by Gibson they were breathing heavily but were not tired.

"Okay, what have I got to do here?" Conrad asked himself as he flipped the pages of his cuff checklist. "I'll tell you something *you* can do," he said to Bean. "Take the pans again. I think I took them at 15 feet focus by mistake." He was referring to the 360-degree panoramas that he had taken from three points around Intrepid earlier in the moonwalk. "And I'll get the rock box out."

At the MESA, Conrad removed his Hasselblad from its bracket and put it into the equipment transfer bag, then lifted the first sample return container from its stowage slot and placed it on the work table. Also referred to as a rock box, the container was 19 inches long, 10.5 inches wide and 7.5 inches deep. It was machined from a single aluminium forging rather than using joined faces, and its hinged lid was to create a hermetic seal. On the outbound voyage it held items for use by the astronauts on the surface. One of the items delivered in this box was a bag containing very clean aluminium mesh. This was to remain in the box to enable the scientists to determine whether the container was contaminated by organic material from venting Intrepid's cabin or from the astronauts' portable life-support systems.

At the end of the first moonwalk Al Bean reshot the panoramas taken earlier by Pete Conrad. This portion of the 8 o'clock pan includes Conrad at the MESA.

"All the pans are done, Pete," Bean reported.

"Come and get the core tube," Conrad ordered.

As Bean collected the apparatus, Conrad said, "Let me get your rock bag before you go away." He unhooked Bean's saddlebag and stowed it on the MESA.

"I'll go for the core tube over near the TV," Bean announced.

"Okay, Al, good idea," Gibson praised, knowing that the engineers wanted him to vary the configuration of the television camera.

At a conference in Falmouth, Massachusetts, in the summer of 1965 scientists had debated what astronauts should do on the Moon. Hoover Mackin, a geologist at the University of Texas, had argued strongly that they should be provided with hollow tubes to obtain samples that would explore the 'third dimension' of the lunar surface in terms of layering. These tubes were 1 inch in diameter and 16 inches long. In the expectation that at shallow depths the lunar material would be loosely consolidated, and hence likely to dribble out as the tube was being extracted, the tube was given a converging bevel to compact the material upon entry. However, when Buzz Aldrin tried to obtain a core sample he discovered that the material rapidly compacted with depth and the tube would penetrate no more than several inches.[27] It had since been redesigned with a straight entrance. The bottom had a non-serrated cutting edge, and the top was designed to accept the tool extension handle, which could be hammered. The only difference to training was that it was easier to whack the tube using the flat side of the hammer's head rather than the impact face, which proved too small to be accurately aimed whilst wearing a pressurised suit.

"It's a little hard to drive in," Bean reported. "You have to auger it a bit and then pound it. But now it's in to its full length. I'll take a picture and that will be it."

Gibson was delighted, "It sounds like you have got the lunar core tube technique worked out."

"I've got the record for core tube depth, right now," Bean agreed.

The tube was easily withdrawn and, as Bean reported for the scientists interested in soil mechanics, the hole remained open. However, some fine-grained material did dribble out of the bottom of the tube.

"Houston," Bean called. "I'm coming right by the TV camera. Do you want me to do anything to it?"

"Al," Gibson replied. "First, we would like you to put the automatic light control switch to Inside and then open the aperture in steps, with 10 seconds at each step."

After advising that the camera was viewing almost directly cross-Sun, towards the lunar module, Bean performed the procedure.

While waiting, Conrad retrieved the second rock box from the MESA, put it on a foot pad, and covered it with some thermal insulation left over from the deployment of the S-band antenna.

On finishing the specified task with the camera, Bean said, "Standing by for some more instructions, Houston."

[27] It is not clear why the core tubes were designed the way they originally were, as when Surveyor 3 landed in April 1967 it was the first lander of that series to have a scoop for scraping trenches and it found that the fragmental debris layer rapidly consolidated with depth!

Upon returning to Intrepid at the end of the first moonwalk, Al Bean obtained a core sample. Here the tube is almost completely embedded in the ground, and the extension handle is visible.

"Would you pick the camera up and invert it," Gibson suggested as a last resort. "Maybe give it a shake or two and see if we can get any change."

"It's upside down and I'm shaking it now," Bean said, but it had no effect. When he suggested the fault might involve the plug that was inserted into the socket of the extension cable, Gibson asked him to wiggle the wires, but still there was no visible response.

Gibson admitted defeat. "I think we've run out of ideas here for the present time. Take the camera and put it in the shade of the LM, point it at a dark spot, the darkest spot you can find, and open the lens way up to f/2.2."

At Intrepid, Bean handed the hammer and the core tube to Conrad, who removed the extension handle and replaced the cutting edge with an end cap, then stashed the tube in the sample return container. He tossed the ring-shaped cutter into the box for later examination by the scientists.

"Pete and Al," Gibson called. "You are 3 plus 35 into the EVA and have plenty of consumables, so we suggest you go at a 'relaxed hustle' to get back in."

"It's almost all done," Conrad assured. Lifting a saddlebag in each hand, he asked Bean to use the scoop to fill them up with fine-grained material so they would pack the sample return container, but Bean was busy attending to the television camera. "I kind of feel like the guy in the shopping centre waiting for his wife," Conrad pointed out. "I'm standing here holding two bags, buddy."

"I'm coming!" Bean laughed.

After dumping the filled bags into the box, Conrad turned his attention to the lid, which had a knife-edge to penetrate a soft indium-metal strip on the bodywork. Prior to the flight, a teflon spacer had been inserted to keep the knife-edge and the indium strip from coming into direct contact. He removed and discarded this before sealing the container. With the lid in place, he engaged the locking latch.

While Bean removed his Hasselblad and put it in the equipment transfer bag with its partner and then attached the bag to the conveyor, Conrad verified the alignment of the S-band antenna.

"We ought to dust each other off," Bean suggested.

"Man, we are filthy!" Conrad agreed. He tried brushing his gloves across his partner's suit. "It does dust off a little bit."

Once his torso had been brushed, Bean jumped up onto the bottom of the ladder and stamped his boots to shake the dirt out of the deep treads of the overshoes, then Conrad moved in and cleaned his legs. Unfortunately, Bean had not brushed off his commander's suit.

After Bean ascended the ladder, Conrad provided verbal cues for ingress, but the tricky task was to get up off his knees. He crawled in until his head was hard against the ascent engine cover, then, keeping his rear as low as possible, pushed up with his arms until his head was above the cover and he could lever himself up onto his feet.

Bean suggested switching communications from the steerable dish on the roof to the S-band antenna deployed on the surface while Conrad was outside and available to attend to any problem. Gibson agreed that this would be wise. The new radio link was satisfactory. Although there was no television for the antenna to transmit, the increased gain would enable Mission Control to monitor the state of Intrepid during the sleep period using telemetry in the low-power mode, and it was also expected to improve the quality of communications during the traverse that they would perform on the second moonwalk.

After moving things around to make room in the cabin, Bean kicked a bag of trash out through the hatch. Conrad was meant to retrieve this and shove it out of the way beneath the vehicle, but the bag did not clear the end of the porch. Rather than have Conrad waste time going up the ladder specifically to fetch the bag, they proceeded with the conveyor. After the bag of cameras, Bean effortlessly hauled up the sample return container. He placed both items temporarily on the ascent engine cover where they would not impair Conrad's ingress.

At this point, the communications relay was switched from Honeysuckle Creek to the station near Madrid in Spain. Normally, Mission Control would advise the crew of an impending handover but on this occasion did not. Although, Gibson continued to hear the astronauts, they ceased to hear him – without knowing why.

"I wonder what happened to Houston," Conrad mused after a minute and a half of silence on the uplink during which he had expected a 'Go' to head up the ladder. To determine whether the deployable antenna was the problem, Bean switched back to the roof antenna and the link was immediately re-established, at which point Gibson explained that the loss of communications had been a temporary problem associated with the handover. Conrad chastised him for failing to provide advance notice. Bean reinstated the deployable antenna.

With nothing more to do, Conrad headed up the ladder. On reaching the porch, he pushed the trash bag to the ground. "Anything else in there you want to get rid of?"

"Not a thing," Bean replied.

To enable the hatch to form a hermetic seal, they had either to draw in the entire length of the conveyor or send it outside. Given the filthy state of the line, Conrad asked Bean to pass the 'inside' end out and he tied it to one of the handrails of the porch. On seeing the tear in the hatch cover which he had made while egressing, he exclaimed, "I can't believe I did that." He considered applying tape, but decided to ignore it. Once he was upright, Conrad stood clear to enable Bean to swing the hatch shut.

"Can you give me a push-me-down," Bean asked. Conrad shoved down on Bean's shoulder as Bean bent his knees in an effort to operate the handle of the hatch, but it was just beyond his reach. "Back up and let me tilt further forward, which is just as good."

"I can't go back any further," Conrad said, feeling his backpack jammed against the panel behind. "Why don't you let me close it."

"See if you can reach it," Bean agreed. "Let me get back here in the corner."

"Hatch closed," Conrad announced, having succeeded on his first attempt.[28]

"Good show," Bean congratulated.

[28] From hatch opening to hatch closure, the duration of the first period of extravehicular activity was 3 hours 56 minutes. However, there were various definitions of the duration of an EVA. In particular, between letting the cabin pressure vent down from 3.5 psi to the restoration of this level upon ingress was 4 hours 1 minute.

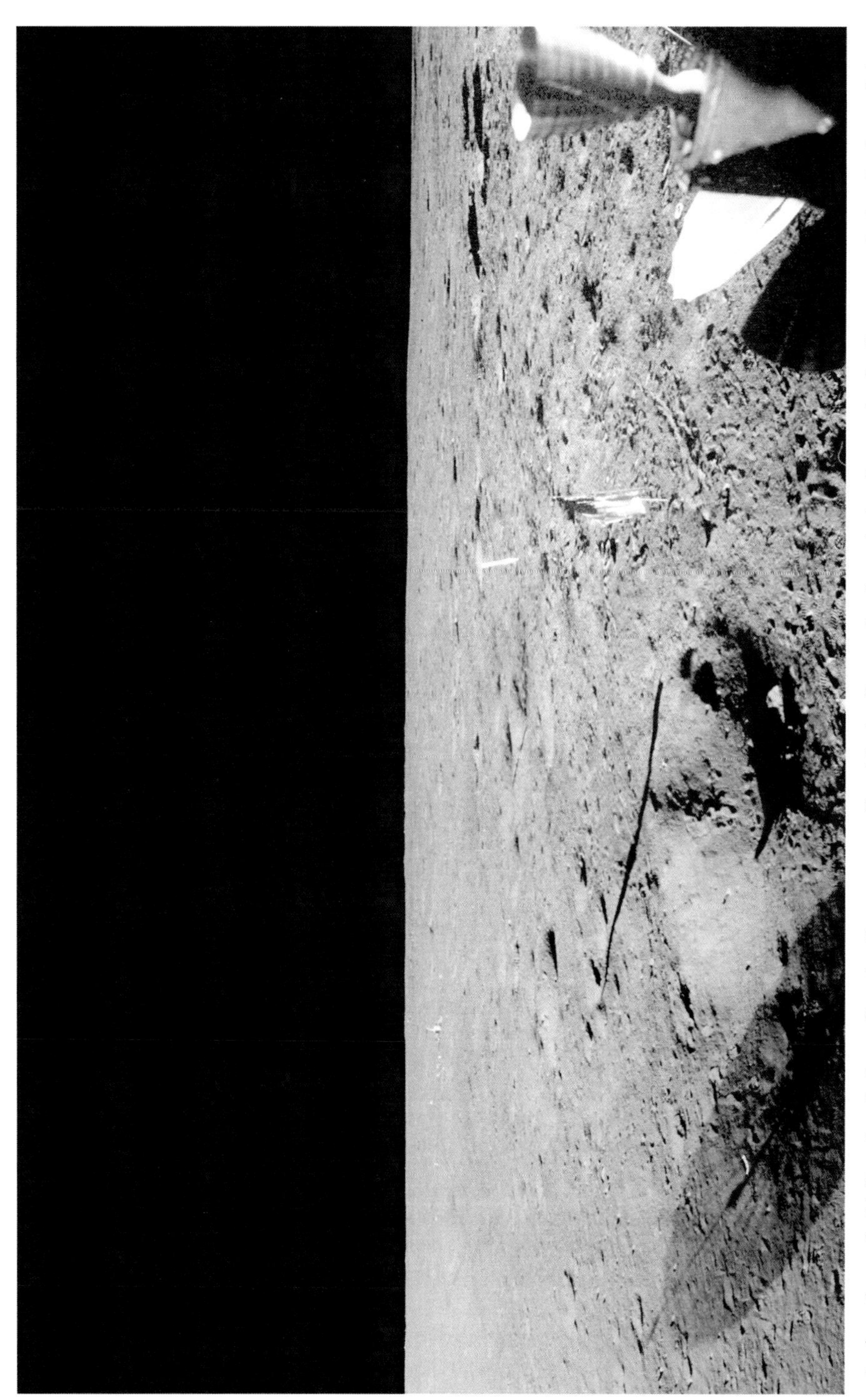

A view through Al Bean's window after the first moonwalk with the flag and solar wind collector in the foreground, the shadow of the S-band antenna, and the ALSEP some 430 feet away.

SLEEPING OVER

After Bean had verified the setting of the dump valves on both the front hatch and the roof hatch, the cabin was rapidly repressurised up to 4.6 psi. They then switched off the oxygen flow from their backpacks, opened purge valves to deflate their suits, and opened valves to enable Intrepid's environmental control system to feed oxygen to the umbilicals which they then plugged into their suits. After doffing their gloves and helmets they switched the supply of water for their liquid-cooled garments from their backpacks to the cabin system. And to conclude the process, they plugged back into the spacecraft's communication system.

The next task was to replenish the oxygen tank of each PLSS. This was a sensible precaution against the development of a problem with the cabin supply. Intrepid's environmental system had outlets to replenish a PLSS with oxygen and water. The oxygen tanks were readily refilled, because it was simply a matter of connecting the hose and pumping up the tanks until the gauge on the backpack indicated the desired value. After shedding their backpacks they replaced the batteries and carbon dioxide scrubbers with the ones that Conrad had retrieved from the descent stage and sent up using the conveyor. Then they stowed the rock box in a slot behind the ascent engine cover, expended the remaining Hasselblad colour film on window photography and installed magazines of black-and-white film for the second moonwalk.

Meanwhile, Yankee Clipper appeared around the limb on revolution 19 with only a few minutes remaining to an SPS burn. "Hello, Houston," Gordon called, "Clipper here."

"You were sort of the forgotten man for a little while," Gibson admitted. "But all eyes are on you now."

This was the first firing of the SPS engine to be made by an astronaut flying solo. With a full crew aboard, each man monitored specific systems. While on the far-side Gordon had performed the checklist to the 6-minute mark. As he explained in the debriefing, "Realising that it had been a really long day and I was tired and therefore more prone to make mistakes, at acquisition of signal I chose to go to Vox and read the checklist as I performed it, so that the ground could monitor exactly what I was doing and would be abreast of the status of the spacecraft at all times."

The spacecraft's orbit was inclined at about 15 degrees to the lunar equator with its most southerly point east of the meridian, and because the Moon's axial rotation was synchronised with the period of its orbit Yankee Clipper's track drifted further west with each revolution. As a plane change would erode the ascent stage's limited delta-V, it had been decided that Gordon should make the change in preparation for the rendezvous.[29] The manoeuvre was made at an altitude of 62.2 nautical miles at selenographical coordinates 14.0°S, 77.7°E. After using two thrusters for ullage, the computer fired the SPS for 18.23 seconds to produce a delta-V of 350.0 feet per second that shunted the orbital plane 'backward' along the equator to realign it with

[29] Apollo 11 flew in an orbit less inclined to the lunar equator and therefore had not been required to make a plane change.

the landing site. Specialists in Mission Control would study the radio tracking from the remainder of the near-side pass to assess the manoeuvre.

"Say, Dick," Gibson called. "That was a fantastic job that you did, picking up the Surveyor and the LM. That plus the plane change on the burn. You've been doing a good job."

"Thank-you, sir."

Meanwhile, Intrepid's crew were working to complete the replenishment of their backpacks. As the rate of coolant consumption was not monitored by telemetry, the only way to determine how much was actually used was to pour the remaining water into a bag and weigh it. The first step was to calibrate the spring-scale by weighing one of the remote control units worn on the chest during extravehicular activity. But zeroing the scale proved difficult.

"Houston," a frustrated Conrad called, "you'll never believe what we have been doing for the last 35 minutes."

"Go ahead," Gibson prompted.

"I am going to take this 35-cent scale that they sent out here to weigh these bags and break it over somebody's head!"

"I take it you're having a malfunction with the bag and the scale," Gibson replied.

"Just the scale. The nut came off the top of the adjustment," Conrad clarified. The zero-adjustment nut had come off during the calibration for lunar gravity. He asked Gibson to check the spacecraft documentation and tell him where the pliers had been stowed, so that they could try to fix the scale.

Gibson did as asked, then said, "Our suggestion is, if you don't think that you're going to be able to do it in a short period of time, you should look for some way you can accurately measure its volume and we'll calibrate that when you come back. The other option is to just plain guess at the volume."

"Let us think about it a minute," Bean replied. "We're still working on the scale." Half a minute later he was back, "Eureka; we did it! Got the nut back on."

"Well done, Intrepid," Gibson praised.

As Gordon would inform his colleagues later, he "laughed his ass off" listening to Conrad's grumbling about the scale.

"The scale ought to have been set to zero," Conrad explained in the post-mission debriefing. "If anybody had thought about it, including myself, the spring tension in the scale itself was never zero in one-sixth g. As I unscrewed it to zero, I unscrewed it all the way and the screw, the spring and everything disappeared into the bottom of the scale. We had some difficulty putting that baby back together again. It would be wise to put in a reasonable scale that can be zeroed in one-sixth g. Let somebody think about it a little bit."

With the scale repaired, they weighed the remote control unit. "One RCU weighs 3.8 kilograms on this nickel-and-dime scale," Conrad announced.

"3.8 kilograms?" Gibson asked incredulously.

"I'm sorry. Make it nought point three eight."

"You had the metric fellows down here wondering," Gibson said.

"Me, too," Conrad laughed.

"This bag is really filling up, Houston," Bean observed. "Pete's water bag weighs 0.26

kilograms." This, of course, was its weight in lunar gravity. A few minutes later he added, "And 0.17 kilograms for the lunar module pilot's water." The difference mostly reflected the times spent with the flow rate increased for additional cooling. In particular, Bean had to overcome the exertion of carrying the ALSEP for several hundred feet.

In fact, the workloads experienced during the first moonwalk were an average of 975 BTUs per hour for Conrad and 1,000 BTUs per hour for Bean, significantly less than the 1,166 and 1,142 respectively predicted prior to the flight.[30] In terms of beats per minute, Conrad's average heat rate was 105, with a high of 150 and a low of just under 80; Bean averaged 121, with a high of 151 and a low of 82. Conrad's PLSS had 42 per cent of its oxygen remaining, 44 per cent of its coolant water and 34 per cent of its battery power. Bean's was similar. They could have remained outside for another hour, but NASA had a conservative safety margin. This was understandable. In effect, Conrad and Bean were calibrating the PLSS to obtain the data required for planning future missions.

"I'm a little bit puzzled about that TV camera," Bean mused. "Do you think it had some sort of mechanical failure or did we point it at the Sun too much?"

"We have got the mechanical and the vidicon-burn people taking sides," Gibson replied. "We're not sure right now."

"We'll bring it back for you," Conrad offered.

"That's a possibility," Gibson agreed.

As they replenished the PLSS tanks with water, Conrad asked, "Do you have any questions that you want to ask us about the EVA?"

"We'd like to get your comments first," Gibson pointed out.

"The EVA went pretty well as planned," Conrad began. "I think that once we got to a task the way we'd practiced it, we got it done. It was kind of the unforeseen, as usual, which almost got us behind. I will say one thing. It very definitely took about 10 minutes or so to adapt to what was going on, but as soon as I did, I really got the hot-foot and I think that Al felt the same way."

Gibson replied, "I think you did a tremendous job. You were able to go along, as you said, on the nominal things and take care of the off-nominal. There were quite a few points where we mightn't have met the objective had you not played 'heads-up' ball."

"Yeah, that's the advantage of having a hammer aboard," Bean observed.

"My heart was in my throat when he couldn't pull the element out of that cask," Conrad pointed out, referring to fuelling the RTG.

"Al, you should have been a surgeon," Gibson joked.

"That was me that was beating with the hammer, not Al," Conrad corrected. "As far as the geology goes, we really didn't have a chance to look very hard; but I think it's very obvious that there are a variety of different kinds of rocks." Interpreting the harsh shadows as indicating that Surveyor 3 was standing on a sleep slope, he said,

[30] The imperial unit of 1 British Thermal Unit, which is approximately the amount of energy required to heat 1 pound of water by 1 degree Fahrenheit, equates to just over 1 kilojoule in the metric realm.

"I would also like to say that I think we're in a most favorable position to get to the Surveyor. I don't think we want to walk down the wall on the side that the Surveyor is on. What we should do is walk down in the crater right from the LM across the bottom and walk up to Surveyor. It looks far too steep to approach from the other side."

When Conrad recommended that the route for the geological traverse combine the best options for the different landing sites, Gibson replied, "We're leaning right now towards the traverse for Site 4, although we wouldn't take it necessarily in the same order it's spelled out there. If you want, you can get out your notes onboard for that and I'll give you a tentative order in which you would hit those points." Site 4 was southwest of Head and the return leg of its traverse route passed very close to where Intrepid stood.

On checking his map, Conrad replied, "That ought to work out pretty darn clever, actually."

Options were discussed, but the geologists would work out the details overnight. "We'll get back to you in the pre-EVA briefing," Gibson promised, "and talk about the location of the sampling, the core tubes and the trench site."

"Okay," Conrad agreed.

"We have several questions for you related to the EVA," Gibson continued. "We would like to move through these pretty quickly, so you can get to bed." After some discussion of why the preparations for egressing had taken longer than expected, he asked, "Al, do you now think that it is feasible to join two core tubes together and perhaps get at least one and a half lengths in? Something on that order?"

"It was getting harder as I drove it in, just like it does back on Earth," Bean said. "But I think if you wanted to stand there and pound maybe three times as long as you would have to drive in one, you could do it." The only issue was the 'follower' which was pushed up inside the tube by material entering the aperture, compressing it. The upward motion of the follower was limited by a small pin near the top of the tube. To obtain a double-length core sample this pin would have to be removed.

Next, Gibson sought confirmation that they had sampled the large mound north of Head crater, and Conrad said they had. Gibson asked if they had sampled the surface around the mound in addition to the mound itself. "That's right," Conrad said. "We can get a documented sample tomorrow, if you want."

"We'll talk to you about that in the briefing before the EVA, Pete," Gibson said. "A question about the number and sizes of rocks: what was the ratio of fines to rocks that you finally ended up with?"

Conrad described the packing, "I put two of the large scoop's worth of fill in one bag that has three rather large rocks in it; I think it's three. The other bag of rocks fills half of the rock box, and I guess there were 10 or 12 rocks in there. The box is completely full. I couldn't get anything more in and still get the core tube in there, I'll tell you that."

"Could you give us an estimate of the number of rocks that you have on board?"

"I really didn't get a count," Conrad pointed out. "I guess it would be about 15 to 20 rocks."

"We're really looking for the weight of rocks."

"That rock box is heavy, I'll tell you that," Conrad replied. "I think it is right up to maximum." The sample return container weighed about 15 pounds empty, and when weighed on arrival at the Lunar Receiving Laboratory it was 44 pounds.

"That is good enough," Gibson said.

To draw what had been a long and memorable day to a finish, Conrad and Bean grabbed a bite to eat.

Although they were running behind the timeline as a result of having spent longer than intended on the window observations, delays in the preparations to depressurise the cabin, the extension of the first moonwalk and the subsequent in-cabin activities, Gibson announced that the wake-up call for the second day would occur as per the flight plan.

After Yankee Clipper appeared on revolution 20 and Gibson read up the latest batch of data, Gordon said, "I guess that's it for the night, huh?"

"That's it, Dick," Gibson confirmed. "We'll talk to you in the morning."

As Gordon noted in the post-mission debriefing, "I got extremely tired at the end of that first day of solo activities." After the SPS plane change he had to perform all the housekeeping and pre-sleep tasks by himself, which took much longer than when there was a full crew. As a result, he was late in starting his scheduled 9.5-hour sleep period.

On manning the communications console, Paul Weitz called Intrepid, "We have a couple of questions to get rid of some Irish pennants here, and then we can turn you loose to go to sleep." He was using naval slang for a line onboard ship which isn't stowed properly. "First off, what are your intentions regarding your suit-hose configuration for sleep?"

"Al says that he is hot and so is going to leave his connected to blow air," Conrad replied. "I'll probably sleep with my hoses off."

"Secondly, how about a crew status report for our friendly Flight Surgeon?"

"The crew is in super shape," Conrad replied enthusiastically. "No medication."

"I had one of those decongestant pills just prior to the EVA," Bean corrected.

"Understand, Al," Weitz acknowledged.

After providing a reminder of the need to exchange the lithium hydroxide canister that scrubbed carbon dioxide from the cabin, Weitz said, "Let us know when you're getting ready to turn in, Pete, and we won't bother you any more."

"We're just getting ready to rig the hammocks now," Conrad pointed out. Several minutes later he signed off with, "Nighty-night."

Having been awake for fully 24 hours, during which they had landed on the Moon and made a moonwalk, they were utterly exhausted. By the time they retired for the night they were approximately 2 hours behind the flight plan.

After their moonwalk, Neil Armstrong and Buzz Aldrin had rested in the cramped cabin for a few hours prior to initiating preparations to leave. As a safety precaution against airborne dust, they donned their helmets and gloves and plugged their suits into the cabin's environmental control system. Aldrin settled on the floor across the cabin with his legs bent because it was not wide enough to stretch flat. Armstrong reclined on the circular cover of the ascent engine, leaning against the aft wall and with his feet suspended above Aldrin in a sling improvised by hanging one of the waist tethers from a convenient fixture. With the windows shaded for

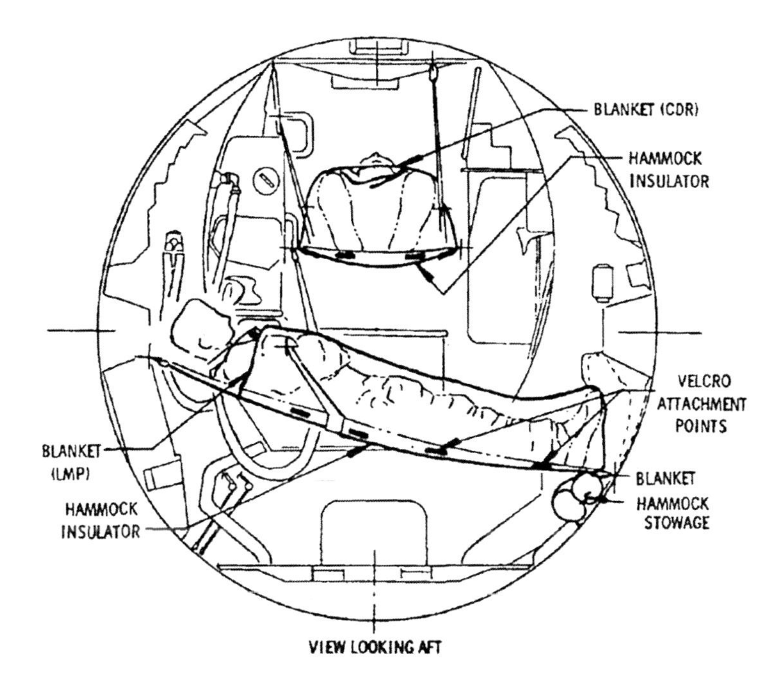

The arrangement of the overnight hammocks in the cabin.

darkness, the temperature in the cabin dropped and, even with the circulation for the liquid-cooled garments at its minimum rate and their oxygen supply at its warmest, they were too cold to sleep comfortably.

As yet, NASA was unwilling for astronauts to doff their suits in the lunar module, in part owing to concern about the very fine and highly abrasive dust impairing the working parts of the suits, in particular the zippers, hose connectors, and helmet and glove rings; but also due to the time required to put them on again in an emergency. Nevertheless, this time the astronauts were allowed to shed their helmets and gloves for greater comfort.

Apollo 12 introduced hammocks of beta cloth. Once one had been slung at a high level, oriented fore and aft, and Conrad had scrambled into it with his head towards the rear, Bean installed his at a shallow slope sideways across the front of the cabin with his head at waist level on his side. The only issue that Bean faced was that he had to temporarily unstrap the beta cloth covering at the low end and pull that back before he could tighten the hammock satisfactorily. The hammocks were effective in the reduced gravity and Conrad soon fell asleep. Bean, however, had trouble. As he explained to Eric Jones in a 1991 interview, "I didn't sleep well because I was just nervous and excited. I think it was just being hyper and being worried about it going well." The cabin remained comfortable, with the temperature and pressure fixed, but both men were awakened from time to time by what they presumed to be irregularity in the pitch of the sound of the water-glycol pump mounted on a bulkhead in the aft cabin floor area. However, the performance data for the pump showed that it was not varying. It was subsequently concluded that the fluid lines and supporting structure downstream of the pump must have experienced thermal changes that altered their vibrational harmonics.

With Glynn Lunney as Flight Director for the graveyard shift, the geologists in the science support room refined the details of the geological traverse. The selectors for the first Apollo lunar landing had avoided craters but Neil Armstrong had set down near a small crater. Although he ran over to photograph its interior he did not have the time to collect any of the rocks that littered it. For Apollo 12, craters were one of the scientific attractions and the intention was to sample as many of them as possible. And the rule for planning the traverse was that if Intrepid set down within 2,000 feet of Surveyor 3, every effort would be made to include it on the traverse. Thanks to Emil Schiesser's 4,200-foot 'range correction', this was on the cards. The planners benefitted from Surveyor 3's study of the crater in which it had landed, but unfortunately it had not been able to view the ground beyond the rim. Nevertheless, the resolution of the overhead pictures provided by the Lunar Orbiter missions was sufficient for a traverse to be planned in terms of specific sampling objectives. The limited time available to astronauts on the Moon obliged them to collect samples at a hectic pace, completely at odds with how a terrestrial field geologist would operate. 'Documented sampling' had been developed to obtain the best sampling rigour on the Moon. Neil Armstrong was to have obtained at least one such sample, but due to limitations of time was able only to collect a representative suite of rocks in an area which was included in a photographic panorama of the landing site. Documenting a rock with stereoscopic views prior to sampling would enable scientists to reconstruct how it was situated on the surface. Each sample was to be stored in a separate teflon bag with an identification number, which the astronauts would state, and Conrad and Bean had trained until this procedure was second nature. The field geology traverse would therefore rebut critics who decried the Apollo program as merely an exercise in 'flags and footprints'.

Meanwhile, the ALSEP geophysicists were eagerly activating and checking their experiments.

The electrical power from the RTG was 56.7 watts when the central station was activated at 9:21 a.m. EST on 19 November, but it steadily increased to 73.69 watts as the generator warmed up. The lunar dust detector (M-515) was an engineering experiment to measure the rate at which dust accumulated on the ALSEP, in order to model the degradation of its thermal surfaces. In the Apollo 11 version this was on the seismometer, but for the ALSEP it was on the central station. The measurement apparatus consisted of three calibrated solar cells, one facing east, one west and the third vertically upward, in order to monitor the passage of the Sun across the lunar sky.[31] The presence of dust on a cell would reduce its electrical output.

The seismometer was activated immediately and recorded the astronauts moving about nearby, but it suffered early instrumentation problems. In particular, the

[31] In fact, the 'top' and 'east' cells showed a slight disagreement that probably implied a tilt of the central station.

short-period vertical sensor operated at a reduced gain and failed to respond to calibration pulses. A comparison of signals observed by both the long- and short-period sensors suggested that the inertial mass of the short-period seismometer was rubbing against its frame. Signals violent enough to produce inertial forces on the suspended mass which exceeded the restraining frictional forces resulted in a normal response. And whereas the thermal control system was meant to maintain a temperature of 52°C to within 1 degree, it actually ranged between 56°C during the day and 30°C at night. This did not impair the quality of the seismic data, but it reduced the probability of obtaining useful very-long-period data. Nevertheless, in addition to the impact of meteors and seismicity within the Moon, the instrument did manage to detect the surface tilt of tidal deformations resulting, in part, from periodic variations in the strength and direction of the forces acting on the Moon as a result of the eccentricity of its orbit around Earth, and the associated libration effects.

The magnetometer was activated at 9:39 a.m. EST on 19 November and data was received immediately. Ground commands were sent to establish the proper range, field offset, and operational mode for the instrument. Temperatures measured at five different locations in the instrument were approximately 20°C higher than expected owing to dust on the thermal control surfaces. It had three operating modes: (1) Site survey mode. This was to identify and locate any permanent magnetic influences so that these could be taken into account in interpreting the data. Although no global magnetic field had yet been detected, it was possible there would be localised fields in the rocks. (2) Calibration mode. This was to be performed every 12 hours in order to determine the absolute accuracy of the sensors and enable any drift away from the laboratory calibration to be taken into account in interpreting the data. (3) Scientific mode. This would make continuous measurements of the strength and direction of the magnetic field. The three magnetic sensors provided signal outputs proportional to the incidence of the magnetic field components parallel to their respective axes. Each sensor recorded the intensity three times per second, faster than any field was expected to vary. The instrument had a dynamic range of several hundred gammas and a fixed sensitivity of 0.2 gamma.[32] The initial measurements indicated the Moon was passing through the 'bow shock' of the Earth's magnetosphere. It went on to cross the transition region during 20, 21 and 22 November into the magnetotail, the part of the magnetosphere that is distended by the pressure of the solar wind. The site survey was made while in the tail. A direct correlation was identified between signals recorded by the magnetometer and those of the short-period sensor of the seismometer, in particular in the transition region when there were rapid variations in the magnetic field strength. An intriguing result was the discovery of a field of approximately 30 gammas that was attributed to a discrete but unidentifiable source nearby.

[32] A gamma is the unit of intensity of a magnetic field. At the Earth's equator, the planetary magnetic field is 35,000 gammas. In the Earth's vicinity, the magnetic field originating from the Sun had been measured as 5 to 10 gammas.

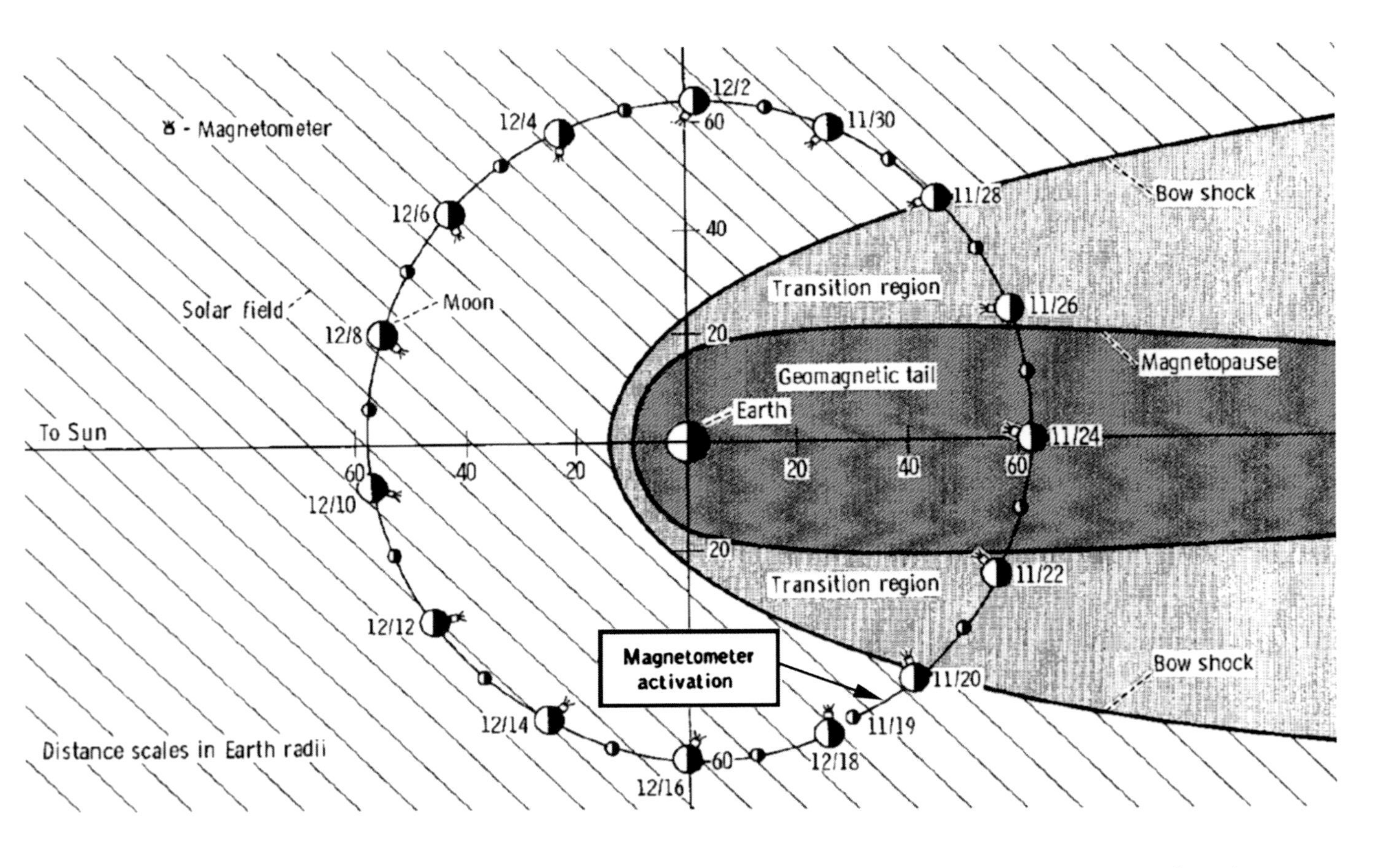

The activation of the ALSEP magnetometer in the context of the progress of the Moon around its orbit of Earth.

The solar wind spectrometer was activated at 1:40 p.m. EST on 19 November. All background plasma and calibration data appeared normal. The covers of the Faraday cups were opened at 10:25 a.m. EST on 20 November, an hour after Intrepid lifted off. Plasma ions characteristic of the transition region of the Earth's magnetosphere were observed. This was consistent with indications from the magnetometer that the Moon penetrated the bow shock at about the time that the ALSEP was deployed. As expected, the plasma was highly variable in the transition region, and essentially no solar plasma was present in the magnetotail. On re-emerging from the magnetotail, the Moon again crossed the transition region and, 9 days after observations began, crossed the bow shock into the interplanetary solar wind. With the subsequent onset of lunar night, the plasma activity decreased to below the measurement threshold of the instrument.

The suprathermal ion detector was activated at 2:18 p.m. EST on 19 November (a few minutes after Conrad and Bean signed off for the night). It functioned normally until 14.5 hours after activation, when both the 3.5-kilovolt and 4.5-kilovolt power supplies and the voltage sequencer for the low-energy curved-plate analyser flipped off. The 3.5-kilovolt supply and the sequencer were able to be restored, but not the 4.5-kilovolt supply. Operations continued until about 29 hours after activation, when the instrument changed its data accumulation mode and both the high-energy and low-energy sequencer voltages fell to zero. The instrument was commanded into its normal mode and the sequencers restarted. At this time, the total background counts were close to 200 per accumulation interval and rising, indicating that outgassing in the interior of the electronics package was causing arcing in the 3.5-kilovolt power circuit, so the power supply was switched off to allow the outgassing to diminish to a negligible level. The 3.5-kilovolt supply was restarted near sunset, and the experiment functioned normally thereafter.

The cold-cathode ion gauge was activated with the suprathermal ion detector, to which it was connected and from which it drew power. Its aperture was meant to be horizontal and aimed away from Intrepid to minimise the influence of outgassing by the descent stage. Although its aperture was actually left angled upward, the azimuth was favourable. The instrument indicated a full-scale response on activation because of outgassing by non-metallic materials within the instrument, but after about half an hour of operation the reading started to diverge from the full-scale value and after 7 hours had fallen to about 3×10^{-9} torr. At the time of cabin depressurisation prior to the second moonwalk it rose to at least 7×10^{-8} torr, the exact value being difficult to determine owing to calibration issues whilst acclimatising. Whenever an astronaut approached the experiment during that moonwalk, its response went off-scale (i.e. to 10^{-6} torr) due to the gases released by his PLSS. Unfortunately, the instrument was crippled by the failure of the 4.5-kilovolt power supply shared with the suprathermal ion detector. The primary objective of measuring the lunar atmosphere was therefore unable to be achieved.[33]

[33] As a unit in meteorology, 1 torr is approximately equal to the fluid pressure of 1 mm of mercury. Hence atmospheric pressure at sea level on Earth is 760 torr.

After sleeping soundly for over 4 hours, Conrad woke up with an aching shoulder. Each Apollo crewman was issued with three space suits – one for training and two flight-worthy suits, one prime and the other backup. Because he had worn long-john underwear instead of the bulkier liquid-cooled garment for a fitting session, one leg of his suit was too short. The discomfort was minimal with the suit pressurised, but it awakened him from an otherwise satisfying sleep. As he described the discomfort to Eric Jones in a 1991 interview, "it was like my shoulder was in a vise". Realising that Bean was awake too, Conrad got off his hammock and sat on the ascent engine cover. Bean took down the hammocks to gain room and then methodically 'let out' the cords laced around Conrad's right calf to extend the leg of the suit a little, a task which took about an hour. As Conrad admitted in the post-mission debriefing, "That was my own fault. I should have insisted on fitting it while wearing a liquid-cooled garment."

With only an hour of the sleep period left and little chance of falling asleep again, Conrad decided to pursue his 'get ahead and stay ahead' strategy. "Hello, Houston," he called. "How are you this morning?"

Weitz was still on duty,[34] and asked, "How did you sleep?"

"Short, but sweet," Conrad replied. "We're hustling right now, and we're going to eat breakfast, have a little talk with you, and then get about our business." But Weitz had an important request about the cold-cathode ion gauge. When it initially yielded a full-scale reading, the experimenters worried that the instrument had been flipped over by the cable that linked it to the suprathermal ion detector, aiming its aperture at the ground. He asked Conrad to take a look using their monocular to see if it was as they had left it. The line of sight was blocked by the more boxy instrument, but Conrad offered, "We'll run over there after we get out and look."

"Okay. Thank-you, Pete," Weitz acknowledged.

Conrad sought clarification, "I'd like to know if there are any restrictions on when we go over the sill."

"Stand by, Pete," Weitz replied.

"I'd like to go as soon as I can get ready without hurrying," Conrad emphasised. "I've kind of got the suspicion, looking over the prep card, a good bit of this stuff is done and it's pretty much a case of hooking up the PLSS and going."

"Affirmative, Pete," Weitz came back a little later. "Whenever you're ready, at your own pace, you can go over the sill. Of course, we do want to talk to you about the briefing on the traverse before you go out."

"I'll give you a call in a few minutes," Conrad replied. "We've got some sprucing up to do, and then while we eat breakfast you can give us the hot word on geology."

"Also give us some word on our families, if you would," Bean requested.

Weitz promised to find out.

Half an hour into the day, Ed Gibson came on the line, "Pete and Al, we're ready to go with the traverse plan."

"Ho, ho, ho. Good morning," Conrad greeted.

[34] It was 8:22 p.m. on 19 November.

Once everyone was looking at the map of the traverse route planned for a landing at Site 4, Gibson explained the revisions requested by the field geologists. "First of all, the two prime sites are Bench and Sharp craters. We could pretty much follow the traverse that we discussed earlier. What I'd like to do is give you the additional information that you don't have on your sheet and also discuss how we will fit in a revisit to the ALSEP." As the cold-cathode ion gauge was drawing high voltage, if it was decided to move it they were to wait for commands to be relayed via the central station to make the instrument safe to handle. "Your first point along the traverse is Head crater." On the original route they would have sampled the northeastern rim of this crater and then skirted its southern rim, passing north of Bench crater. "What we'd like to do, in view of the fact that you are going out towards the ALSEP, is to move that first sample site over to the northwest rim of Head crater. You'll carry out what we already have outlined for Head. That's the two partial pans across the crater and documenting the slumps and ledges. In addition to that, seeing as we have the seismometer so closely located, we'd like to see if we can get a known signal; so, if possible, could you roll a large rock into the crater and take a stereo-pair of the rock prior to rolling and a stereo-pair of the track that it made. That's sampling point one. Do you copy?"

"Yes, sir!" Conrad laughed. "We'll rock and roll!"

Bean chipped in, "We've had a lot of training for that sort of thing on the geology trips we had."

"We've got some happy-looking geologists," Gibson said. "Uel Clanton is betting that somewhere along the traverse, you will find some 'stuff'." When Buzz Aldrin described a fragment of lunar material during his moonwalk as being "like biotite", this drew criticism from some scientists because biotite is a mineral that forms in the presence of water, and the Moon was arid. They failed to appreciate that Aldrin was not making a mineralogical identification, he meant only that it was visually similar to a sample he had been shown in training. In response, Conrad and Bean eschewed geological terminology, and in training Conrad had referred to everything merely as "stuff". As a result of a bet, each time Conrad described anything as "stuff" Clanton would lose $1 to his colleagues. On the other hand, Jack Schmitt had appended two pages to Conrad's cuff checklist with handy geology phrases, just in case he decided to dazzle the science team.[35]

"Oh, I think there's stuff all over the place," Conrad laughed.

When Gibson said that after this first sampling point they should inspect the cold-cathode ion gauge, Conrad countered, "I think we should do it the other way around. Let's go to the ALSEP and then to Head crater."

"Fine," Gibson conceded. "Either way you want it."

"Okay, let me tell you another thing," Conrad continued. "I want to tell you, I do have Bench crater in view from the window; Sharp crater, I do not. So it looks to me like it would be relatively easy to go from the ALSEP to the coordinates you gave me

[35] One example was for describing bedrock exposed in the bottom of a crater: "The bedrock surface resembles Moshia breccia, with a consistent organisation to its internal fabric."

Pete Conrad got the second day off to a fine start with a panorama to the southwest through his window.

on Head crater. I'm looking right now, and I can see several rocks that we might be able to roll down into that crater – and followed by one astronaut, probably! But anyhow, we will give that a whirl. Is the next point you want us to go Sharp?"

"No, Pete," Gibson replied. "The next one is Bench crater." The new plan was to skirt the western rim of Head and rejoin the route north of Bench. "And then move on to Sharp."

"I'd kind of like to do it the other way around," Conrad suggested. "That way, I'm going around in a circle." If they had landed at Site 4, around 200 feet southwest of Sharp crater, the first sampling point would have been Sharp, followed by a point on the southwestern rim of Bench and skirting that crater's southern rim on an easterly heading. As originally planned, they would have passed north of Bench on the way home. Conrad was suggesting that they go from Head straight to Sharp, then sample the southwestern rim of Bench whilst heading east, as otherwise they would end up walking over the same ground to and from Sharp, which seemed silly. But Mission Control had already considered this, and Conrad had cut Gibson off before the latter could supply a refinement, which he now did.

"What we want to do," Gibson explained, "is to move your point on Bench crater over to the northwestern edge of that, as opposed to the southwestern edge."

"I'm with you," Conrad acknowledged, realising that he had interrupted Gibson's exposition. "Great minds think alike."

Gibson continued. "There are things we'd like you to do, in addition to what's on your plan. Take stereo-pairs of the features of interest in Bench crater, especially of the bench structure." The thickness of the fragmental debris layer that blanketed the lunar bedrock was believed to vary. With overhead imagery of sufficient resolution it was possible to estimate this by correlating the diameters of craters with the size of the blocks on their rims. In general, larger impacts would penetrate deeper and excavate wider craters, and impacts which reached the bedrock would make craters with blockier rims. The thickness of the regolith could therefore be inferred from the minimum size of crater to possess a blocky rim. In the vicinity of the Snowman it seemed to be 10 to 15 feet thick. Bench crater was so-named owing to an irregularity in its interior known to geologists as a bench, and the suspicion was that this marked the depth of the bedrock. However, the fact that there were no benches in the larger craters nearby cast doubt on this interpretation. It was hoped the astronauts would be able to settle the issue. "Determine whether the bench is bedrock, or breccia near the base of the regolith," Gibson said. The alternative interpretation of the bench was a consolidated mass of fragmented rock that had been buried by the regolith and was poking out of the wall of the crater. "If it is bedrock, sample the ejecta representative of the bench; or sample the bench itself, if possible. And lastly, look northwest and southwest from the rim of the crater to see if Copernican ray material is obviously different from other units."

"Okay," Conrad agreed.

"Moving on to Sharp crater," Gibson continued. "First, we'd like a full trench site sample in the crest of Sharp, and we want to make sure you also add to that the gas analysis sample." Their field geology tool kit included a scoop for scraping a trench and the three remaining core tubes. To supplement the trench they were to obtain a core. And since Sharp would probably mark the furthest point of the traverse from Intrepid, it presented the best opportunity to gain a sample of lunar material free of micro-organisms from Earth to test for indigenous life, so they were to put a sample of regolith into a canister that they would seal with a cap. There was also a smaller sealed canister to preserve any gases that might be present in a sample of regolith. A full panorama was to be taken on the rim of Sharp. If they had not already sampled material which looked as if it might be the ray from Copernicus and they could see such material a little further west, they were to continue in that direction and collect a sample.

"I can tell you now, it is going to be pretty darn hard to do that," Conrad warned. The albedo differences suggestive of contacts between different materials that were evident when viewed from overhead were not so visible at ground level. "It's really weird. I'm sure you can see the stuff from far out, but down here it might as well all be the same until you get right up on top of an individual rock."

"I understand," Gibson said. "But if you could keep your eyes open."

"We will," Conrad assured.

"The fourth point is Halo crater," Gibson announced. This was the third sampling point on the Site 4 route, after Sharp and Bench. Halo was so-named because it was surrounded by a splotch of bright material. It was suspected of being fairly young in terms of the lunar timescale, and a sample might yield the least-weathered materials of the traverse. The bright halo was not expected to be very thick, and the plan was to obtain a core sample through it into the older material beneath. To play safe, the geologists had decided to ask for a double-length core "Try to join two core tubes together and core through the thin ejecta of Halo." To achieve this, they would have to select one tube to serve as the penetrator, which would have the cutting bit on it, and after adding the upper tube they would require to extract the pin from the top of the bottom tube to enable its 'follower', driven by the penetrating material, to cross over the junction and progress into the upper tube. "We would like you to avoid the rockiest parts of the crater," Gibson pointed out, for otherwise there would be a good chance of the exercise being frustrated by the tube hitting an obstacle. "If the tubes cannot be joined, just take one on the rim and then one about 100 feet west of that location." If the sample on the rim failed to reach the deeper material, the one taken further out almost certainly would. "If possible, give us a panorama at that location, too."

The instructions continued: "Here is a comment which is really applicable to the entire traverse: document patterned ground and fillets on different slopes and blocks, especially any asymmetric fillets you may run into. We'd find it most interesting to get this type of information on the youngest material, so that's why we're calling for it at Halo crater. The best way to photograph patterned ground is into the Sun, near-field, and that way the pattern should show up in an optimum way."

"Okay," Conrad replied.

"And the last one," Gibson said, meaning the final site on the traverse, "is you go down into Surveyor crater." After cutting parts off Surveyor 3 they were to visit a small crater on the northeastern rim of the main crater. This had been photographed by the unmanned lander, and the fact that it was littered with rocks had prompted its moniker of Block crater. Such an impact could not have penetrated to the bedrock, but Surveyor crater must have. During the process of crater formation the material excavated from the deepest point is deposited on the rim. Although most of this was now buried by the accumulating regolith, the later impact had exposed some of it. Block would therefore enable the astronauts to sample lumps of bedrock which, although no longer in-situ, were clearly from the deepest point of the excavation. "At Block we would like samples of major rock types and a partial panorama across Surveyor crater."

"We may have a little trouble getting to Block crater," Conrad warned. "I'm not sure whether it's an optical illusion, but the wall that the Surveyor [spacecraft] is on looks one whale of a lot steeper than 14 degrees." He was referring to the impression gained from the shadow in the eastern part of the interior of the crater. He was open to the possibility that the steepness of the slope might be illusory, but, "it gets pretty rugged over on that side, especially in the Block area. But we'll give her a go."

On leaving Block on the Site 4 traverse they would have visited the northeastern rim of Head crater, but since that was no longer appropriate they would head straight home. Involving a walking distance of 6,000 feet, the Apollo program's first geological traverse was very ambitious.

Conrad was well aware that if they fell behind schedule they would put at risk the visit to the Surveyor lander, pictures of which would be the highlight of the mission for public relations. There were two constraints on the duration of the extravehicular activity. One was water coolant in the PLSS, but the data from the first moonwalk would not only allow the full planned duration but also make Mission Control look favourably on an extension, possibly as much as an hour. The second constraint was immutable: they had to be back inside Intrepid with sufficient time to lift off at the planned time. But beginning the moonwalk early would weaken the need to return to Intrepid by a specific time being used as an excuse for denying an extension.

As the astronauts began to suit-up for extravehicular activity, Gerry Griffin took over as Flight Director. They worked through the checklist rapidly, and without any mistakes. Just before they reached the point of checking the PLSS communications, Gibson reported, "Pete, Jane sends her congratulations for a job well done. And Al, Sue has followed it all, and is thrilled that you've really made it on the money. The children are fine – tired but happy – and they are going to continue following all the way through the second EVA."

"Thank-you, Ed," Bean replied.

It had been decided to attempt a transmission to determine whether the television camera had somehow recovered. When Gibson reported that the situation had not improved, Conrad said, "I'm sure sorry the TV didn't work. It's a beautiful sight to see Intrepid and Surveyor sitting here on this crater."

"We'll be waiting for those Hasselblad pictures," Gibson pointed out.

The decision was to return the television camera for the manufacturer to examine. The astronauts had shears with which to cut parts off Surveyor 3, and were asked to use these to cut the television cable in such a way as to retain the plug and socket. The camera was to be sent up to the cabin in the equipment transfer bag along with the Hasselblads at the end of the moonwalk and then wrapped in a towel and stowed on the bracket on the ascent engine cover where a replacement lithium hydroxide canister for the cabin environmental control system had been carried.

As Conrad and Bean worked the final steps of suiting up, Gordon, having just emerged from behind the Moon on revolution 25, called Houston, "Hello Houston, Yankee Clipper."

"Good morning," Gibson replied, using a separate channel so as not to distract the surface crew.

"Reveille has been held on the Yankee Clipper, we have swept down fore and aft, the batteries are charged and the crew is ready for work. Yankee Clipper is reporting for duty, sir."

"Roger, Dick," Gibson said. "Your friends are just about ready to egress."

"Good," Gordon replied.

A minute later Conrad asked Mission Control, "Are we Go for EVA?"

"You're Go for EVA, Pete," replied Bill Pogue. Selected as an astronaut in April 1966, Pogue was not actually on the support crew for this mission but was handling Intrepid while Gibson was on the other channel.

"Okay," Conrad acknowledged. They were 1 hour 40 minutes ahead of the flight plan.

SECOND MOONWALK

As previously, Bean opened the valve on Intrepid's forward hatch, waited until the pressure declined to 3.5 psi, and then closed the valve to verify the integrity of their suits. The dust from the first moonwalk had resulted in a pronounced increase in the operating force to attach the oxygen hose connectors and glove rings, but neither suit leaked and so he resumed the depressurisation. By now, the cold-cathode ion gauge had dipped off its full-scale reading, and it detected a slight increase in the ambient pressure from this venting.

Gibson called Gordon, "Pete and Al are ready to egress. If you like, we'll give you a MSFN relay?"

"Yes, I'd like to listen to them," Gordon replied.

This time Bean was able to reach the handle and open the hatch, and immediately shoved out the bag of trash which held food waste, urine bags and the expired PLSS batteries and lithium hydroxide canisters.

"Did it make it?" Bean asked.

"No," Conrad replied. "It's on the front porch."

Conrad got down on his hands and knees and reversed out, this time not scraping against the hatch. He nudged the trash bag off the porch and retrieved the conveyor, which he had left tied to one of the hand rails, and passed this in through the hatch to Bean.

"I'm heading down the ladder," Conrad announced.

"All right," said Bean, as he attached the conveyor to the fixture on the overhead.

Jumping down onto the foot pad, Conrad mused, "Whoops, long step." Then as he stepped off the pad he announced, "Houston, Mark. I'm on the lunar surface."

Conrad went to the MESA to make a start on loading the hand-tool carrier. This had been obtained from one of the ALSEP subassemblies the previous day. It was a truss structure of formed sheet metal. It had two sides with a hinge and three 6-inch legs. It had attachments for the hammer, scoop, core tubes, the two sample canisters, an extension handle and a dispenser for cup-like teflon bags for individual samples. A beta cloth bag with a triangular aperture clipped onto the top edges of the frame, to occupy its interior. A gnomon was provided for documented sampling. This was a tripod that supported a weighted staff on a two-ring gimbal. The staff rose 12 inches above the gimbal and was painted with a greyscale in bands 1 cm wide (they were using black-and-white film). The staff was to indicate vertical to within 20 minutes of arc. Magnetic damping was used to reduce oscillations. The gnomon was to be placed alongside a sample to provide angular orientation relative to the local vertical and metric control of near-field (10 feet) stereoscopic photography. Information on the range of objects, and on the pitch, roll, and azimuth of the camera's optical axis were thereby included in each picture. Photogrammetric techniques were to be used to produce three-dimensional models. For transport, the gnomon was affixed to the hinge of the hand-tool carrier. It was also possible to mount a Hasselblad on the carrier.

Meanwhile, Bean hooked the equipment transfer bag onto the conveyor. When it was ready, Conrad retrieved the line and kept it clear of the porch while Bean played it out to lower the bag. As Conrad shuffled about, he complained, "Boy, I'll tell you two things I'm going to learn to dislike: the TV cable and the S-band antenna cable. They are constantly underfoot." Once he had the bag he stowed the line, taking care to ensure that it would not interfere with Bean's egress. The bag contained the two Hasselblads and a replacement colour magazine. Leaving the bag on the MESA, he resumed loading the carrier. Next, he placed the second rock box on the table of the MESA ready to accept the samples upon their return.

After verifying the circuit breakers and the operation of the 16-mm movie camera in his window, Bean egressed. Stepping onto the surface 10 minutes behind Conrad, he reported, "LMP is off the foot pad."

"Okay, LMP," Conrad said. "Let's get a Surveyor parts bag here." On retrieving a large bag, Bean attached it to the rear of Conrad's backpack.

"Hey, look at that Surveyor, Al," Conrad exclaimed.

"Houston," Bean called, "that Surveyor looks a lot better today."

"Yeah, now that the Sun is on it," Conrad explained. The elevation of the Sun had increased sufficiently to banish the shadow in the eastern part of the crater, revealing the slope to be less intimidating than it had previously seemed. By now, each man's shadow was only 18 feet long.

On retrieving the Hasselblads, they affixed them to their chest brackets. The bolt cutters were used to cut the television cable. The television camera was placed in the equipment transfer bag and the cutters were stowed in the Surveyor parts bag. Then they completed loading the hand-tool carrier with items delivered in the second rock box. This time only Bean wore a saddlebag, and it contained the colour magazine from the transfer bag and a 30-foot tether which had been supplied so that one man could stand on firm ground while his secured partner tackled a tricky slope.

Half an hour into the moonwalk, Conrad left Bean at Intrepid and ran off to report on the state of the cold-cathode ion gauge. "Can the guy with the seismometer hear me running?"

"We're watching you down here on the seismic data," Gibson replied. "It looks as though you're really thundering right by it."

Upon approaching the cold-cathode ion gauge, Conrad slowed down and walked carefully so as not to kick dust onto the instrument, then halted about 5 feet away. It was as they left it. "The cover is off, and it is pointed up at about a 60-degree angle on a down-Sun azimuth." The gases released by his PLSS drove the sensor reading off-scale high.

Gibson relayed the verdict of the experimenters, "Pete, there is no need to change the configuration; let's press on."

"Where's my handy-dandy LMP?" Conrad asked.

"He's contrast charting," Bean replied.

"Oh, okay," Conrad acknowledged. "I will meet you at Head crater." He left the ALSEP, heading southwest towards their first sampling point.

Bean was taking pictures to calibrate their black-and-white film. While working at the MESA early on the first day Conrad had stowed three calibration charts on the table and accidentally knocked off the colour chart, which landed in the dust edge-on and remained upright. As he explained in the post-mission debriefing, "I lifted it up and tried to brush it off, but it was impossible. I just made a complete shambles of it." Using the tongs, Bean carefully laid one of the greyscale charts on the ground in sunlight near a small crater and photographed it from up-Sun, then he put the second chart on the shadowed inner slope of the crater and photographed it from down-Sun at an exposure calculated to 'see into' the shadow.

"Man, have I got the grapefruit rock of all grapefruit rocks!" Conrad exclaimed. To enable the scientists in the mission support room to visualise the rocks described by astronauts, there was an informal scale from pebbles, to grapefruits, to cobbles, to boulders, with the latter also being called blocks. "That's got to come home in the spacecraft, but it'll never fit in the rock box." He lifted it without first documenting its context. "Houston," he resumed, "I'll tell you what I'm going to do. I'm going to the right place at Head crater, and will roll a boulder for you while I'm waiting for Al."

"Sounds good, Pete," replied Jack Schmitt, standing in while Gibson switched to an independent channel and interrupted Gordon's breakfast in order to provide some data.

On reaching the northwestern rim of Head, Conrad saw that despite the increased elevation of the Sun the eastern wall was still in shadow. "That crater is, by golly, a rather steep crater; a lot steeper than it looks from out the LM." There was a litter of rocks. "Let me ask you a question, Houston. How big a rock?"

"I presume whatever's a convenient size for you," Schmitt replied.

There were both rounded and angular rocks present on the rim, and observing that the large ones were partially buried and hence likely to be difficult to start rolling, he laughed and said, "How about a grapefruit-size rock? That's what I'm holding in my hand."

"Any size is fine," Schmitt agreed.

"Okay," Conrad replied. To prevent the seismic signal from rolling the rock being confused with signals generated by their movements, he and Bean would have to be immobile, "Al, are you standing still?"

"I'll stand still," Bean replied. "Go ahead."

"Houston, on my mark I'm gonna roll it." He discarded the sample which he had collected. "Mark. It's starting down." He counted the bounces at roughly 1-second intervals, "Hit, hit, hit, hit. Now it's just rolling: roll, roll, roll, still rolling, roll, roll, roll."

"Pete," Schmitt called, "We've got some jiggles that I can see here."

"Still rolling. Still rolling. Very slowly still rolling. And Mark! It just stopped." It had taken 30 seconds for the rock to skip and roll down the slope in slow motion in the weak gravity.

Finished with the calibration photography, Bean left the charts on the ground and went to inspect the solar wind collector.

Meanwhile, Schmitt urged, "Pete, if it is convenient and you can find another rock there and give it a heave, Experiments sure would like to see another one."

"Okay," Conrad agreed. He had been examining the rocks on the rim of the crater. Most fragments were angular and dark grey. Some appeared to be coarsely grained, and even when coated with dust their crystals were clearly visible. He used his boot to flip some over to expose their undersides. "I was looking at a rock that has small crystals in it. One of them is shining very, very bright green; like ginger-ale-bottle green." He was well aware that such a hue was suggestive of the mineral olivine but was reluctant to be specific. "Al, are you on your way?"

"That's affirm," Bean replied. Having returned to Intrepid, he was checking their tool kit. "Houston. As I was working on the hand-tool carrier, the central bag came loose. It took me about 2 minutes to put it back together again. It came off the metal sides. It looks like those clips that hold it on are going to be completely inadequate, and I expect we're going to have trouble with it all day." The bag was affixed to the frame by three aluminium spring clips and its weight was shared by these clips and three hangers. The grip was weak in order to enable the loaded bag to be easily lifted off. The two spring clips on the left side remained attached, but the single spring clip on the right side continued to slip off until the bag had been loaded with a significant weight of rocks.

"Al, where are you?" Conrad persisted.

"I'm just leaving the LM," Bean replied, as he lifted the hand-tool carrier and set off for Head crater. "Boy, this HTC is nice and light compared to carrying it around on Earth. I think we might be able to just slip it right down inside Surveyor crater with us. Piece of cake." However, as he soon realised, the wedge-shaped frame kept bumping against his leg. As a result, he was unable to run as fast as on the return leg of the visit to the large crater at the end of the first moonwalk. And, of course, as the bag was loaded up, the carrier would become ever heavier and require more effort to lug around.

Surveying the rocks as he ran, Bean offered a geological commentary, "I can see everything from fine-grain basalt to a few fairly coarse-grain ones. I see some sort of

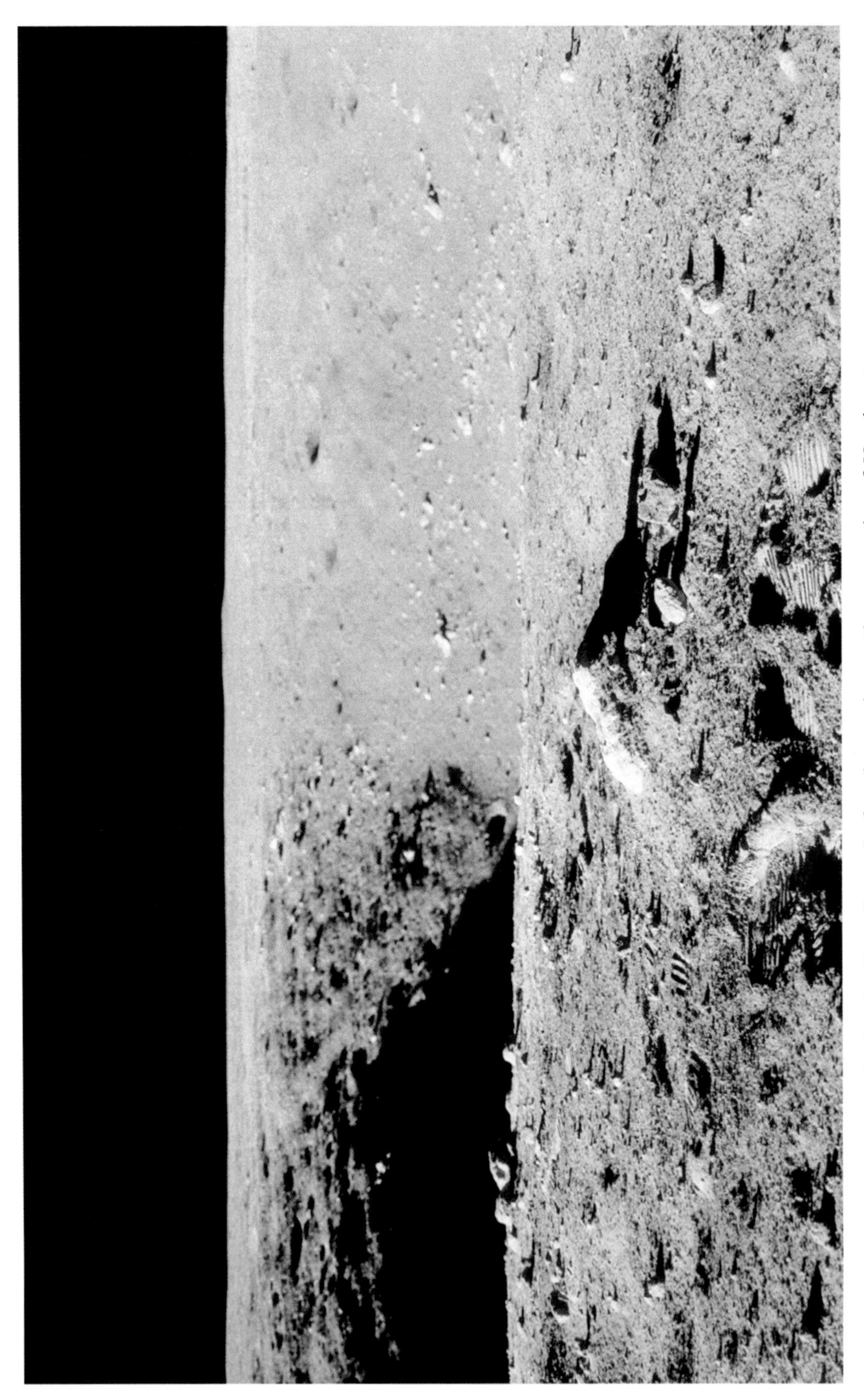

A view by Pete Conrad of rocks on the northwestern rim of Head crater.

light-reddish-grey coloured rock that I would call... I don't know really what I would call it. It looks almost like a granite. Of course, it probably isn't, but it has that sort of texture – the individual constituents aren't such large crystals, but it has that same appearance." An acidic rock like granite would be remarkable on the Moon, but the field geologists had trained him to describe a rock in terms of its colour and texture, and so that is what he was doing, comparing it to something he had been shown as a point of reference.

"Roger, Al," acknowledged Gibson.

Along the way, Bean halted at a crater that was about 3 feet in diameter in order to sample the glass-coated rock fragments in its interior. What had attracted his attention was that the glassy material appeared different to that which they had seen in other craters. After taking a stereo-pair for documentation, he used the tongs to lift the glassy material, which he placed in one of the small bags. For comparison he added other fragments that were not glassy and which tended to fall apart as he lifted them.[36]

Meanwhile, Conrad, whose camera had a polarising filter, was taking pictures of rocks which he had flipped over using his foot.

On approaching Head, Bean explained, "I've been concentrating, Houston, to see if there were any changes in either texture, slope, colour or anything I could think of that would suggest I was walking on a different surface than when I started. And I haven't seen a thing yet. It all looks the same. It all looks like it's covered with this black rock." On joining Conrad, Bean assisted in documenting the rocks that Conrad had flipped over. With the polarisation photography finished, they removed the filter and tossed it away, with Bean expressing surprise at the length of the ballistic arc in the weak gravity.

Picking up on Schmitt's request, Conrad announced, "On my mark, I'm going to send a slightly smaller rock into the crater. Are you ready?"

"We're watching," Gibson replied, meaning that the scientists were watching the seismometer.

"Mark!" Conrad called, booting a rock. "I didn't quite kick it hard enough. Wait a minute. I'll do it again." However, "It's a funny thing, Houston, even though the slopes are steep and everything, in one-sixth g these rocks just don't want to go anywhere."

"Roger, Pete," Gibson acknowledged. "We haven't been able to pick it up on the PSE."

"That was too small a rock," Conrad decided.

Now that they were able to work as a team, they set to work. "We probably ought to do a little trench and compare some of the soil profiles," Bean said.

Having already appraised the site, Conrad suggested, "I've got an area right over here that looks good." What had attracted his attention were "little white spatter-type craters" which looked very fresh.

"That's a good idea," Bean agreed.

[36] Sample 12030 in bag 1D.

"This is the northwest corner of Head crater that they wanted us to work," Conrad said, indicating the route marked on his map.

When Gibson, reviewing the checklist, reminded them that they were to take two partial panoramic sequences spanning Head crater, Conrad replied, "All right." And Bean assured, "We'll get them."

But first they wanted to collect some material from one of the small fresh-looking craters. Handing the tongs to Conrad, Bean fetched the gnomon from the hand-tool carrier. Having selected a rock, they set about the ritual of documentation, with Bean viewing down-Sun and Conrad standing to one side in order to take a stereo-pair. As they worked, Bean saw his partner's boots expose light-toned material just beneath the dark surface and pointed this out. "That's interesting; that's the first time we've seen that." As Conrad photographed the material he had disturbed, Bean explained to Mission Control that they had discovered "a much lighter colored soil". All five Surveyor landers had found the subsurface to be darker than the uppermost several millimetres of the regolith. This was the opposite – in precisely the location where a ray from Copernicus appeared to cross the site. The news stirred excitement in the science support room. Indeed, Aaron Waters, a member of the geology team, leapt to his feet and shouted, "That's it!"

After taking a sample,[37] they decided to find out how deep the light material was. While Conrad attached the extension handle to the scoop, Bean offered Houston a broader observation about Head crater, "One of the interesting things about this side of the crater is that the boulders aren't uniformly distributed around it. For some reason they're predominantly over here on the western side."

"Al," Conrad interrupted, "quit baloneying and help me."

"Okay," Bean replied.

As Conrad scraped a trench, Bean looked back along their tracks near the rim of Head and noted "the same thing's occurred in several places where we've walked".

When Gibson asked whether the difference in albedo might perhaps be merely an effect of the Sun angle, Bean replied, "No, not at all. This is definitely a change to a light grey as you go down, and the deeper Pete goes, and it is about 4 inches now, it still remains this light grey."

Pausing to let Bean take a picture of the trench, Conrad expressed his agreement, "It's different here; we've kicked up all kinds of stuff around the spacecraft and it's all the same colour."

"I tell you what we should do here, Pete," Bean offered. "Dig as deep as you can, then give me a sample right out of the bottom."

As Conrad resumed digging, Gibson sought clarification, "What is your location relative to the centre of Head crater. Specifically, are you on the west side where we have the triple crater?"

"We're right where you told us to go," Conrad replied.

"Pete's down now about 6 inches," Bean reported for Mission Control's benefit.

[37] Sample 12031 in bag 3D.

He held out a sample bag, and Conrad tipped a scoop of material into it.[38] Observing that some of the darker material was mixed in with the light material in the sample, Bean pointed out, "There is a little of the top soil mixed in, because the sides of the trench collapsed. The angle of repose is about 85 degrees, but the minute you touch the side, it falls in. Even although it remains almost vertical, it isn't cohesive. I guess it must be the low gravity." To conclude, Conrad dug a small rock out of the bottom of the trench.[39]

In '*To a Rocky Moon*', Don Wilhelms says isotopic dating of the trench material gave an age of 810 million years, which is now generally presumed to be the age of Copernicus, so this 'ticked' one item on the list of objectives for the traverse.

In the post-mission debriefing, Bean pointed out, "The entire lunar surface was covered with this mantle of broken-up material, fine dust of varying depth, with the result that everything looked pretty much the same." Consequently, "If we're going to do any good geology, it's going to take a lot of trenching to get down below the surface." However, owing to the length of the extension handle and the inflexibility of the suit at the waist, the maximum depth of trench they could achieve using the scoop was about 8 inches. Accordingly, "I'd like to recommend that we get a better trenching tool. Maybe all we need to do is lengthen the extension handle about six inches."

When Conrad stepped back and inadvertently flipped over a rock, Bean noted that the newly exposed face was white and suggested they investigate further.

"Houston," Conrad called. "How long do you want us to spend at Head crater?"

"It looks like we could just spend all our time here if we wanted to," Bean mused.

"That's what's bothering me," Conrad noted. As he admitted in the post-mission debriefing, "I knew this was going to happen. [...] There was no point to which they sent us that we couldn't have spent at least an hour very easily." Although there was a natural temptation to continue sampling what appeared to be the ray material, he appreciated that the geologists wanted samples from as many sites as possible. And he was also aware that if they were to slip significantly behind the timeline, it would jeopardise the visit to Surveyor 3.

"Pete," Gibson replied, having consulted the science support room, "we show that you're 58 minutes into the EVA. We'd like to get you over to Bench crater."

Inspecting his map, Conrad announced, "By the way, this is the smartest idea we came up with, Houston. This map just works great out here." In addition to having the names of craters specified, the photographic maps were coloured in with how the geologists had interpreted the terrain in terms of ejecta blankets and other interesting characteristics. Unfortunately, although craters were the most striking features when viewing from overhead, at ground level it was difficult to use holes in the ground as navigational references. From the elevated vantage point of Intrepid's cabin, a crater as large as Head was merely an elliptical oval in the middle distance. A man on the surface had an even more oblique perspective. The advantage of the map was that it

[38] Sample 12033 in bag 5D.
[39] Sample 12034 in bag 6D.

The first formal sampling exercise near Head crater using the gnomon. Pete Conrad is about to use his tongs to lift a rock. Note the light-toned material exposed by the boot prints.

The trench excavated near Head crater to investigate the light-toned subsurface material.

provided an azimuth for an objective. As they would discover, this was satisfactory for large craters but less so for smaller ones.

"I'm trying to find the triple craters they're referring to," Conrad said, inspecting his map.

"That triple crater is just south of your present position," Gibson assisted, but then suggested that they just go straight to Bench crater.

Before moving on, however, they documented and lifted the rock that had a white underside.[40] Taking a final look around, Bean noted, "Even the rocks that are almost completely covered with the soil, when I look at them I can see glints of crystals or something."

"Yeah, every one of them," Conrad agreed. As they checked that they had all their tools, he announced, "All right, we're going to head for Bench crater."

"We didn't get our pans here," Bean pointed out.

"I'll take them when we're at the triple craters, right over here," Conrad said. Ten seconds later, he announced, "Here they are. Ho, ho, ho!" The three craters were in a line oriented radially to the west-northwest of Head, with the largest, the one nearest Head, being about 35 feet in diameter. "Houston, do you want a pan of Head crater from the triple? Is that what you want? Or do you want the triple craters?" In fact, as Gibson had reminded them when establishing that they were at the first designated site, they were to have taken two partial panoramas spanning Head crater.

"Pete," Gibson reiterated, "we suggest you just move on to Bench."

"It's really a shame," Conrad lamented. "We could work out here for 8 or 9 hours. The work is no strain at all!"

"I took three quick pictures of the triple craters, Houston," Bean announced.

"Bench, is it," Conrad conceded.

"We need to get a pretty large-sized rock here, before we leave this area," Bean prompted, since, as he put it, they had to be "bedrock from somewhere".

"I'll tell you what, I'll stop right here and take a pan," Conrad decided. He took a full 360-degree sequence, but because he was standing west of Head crater the view was of the shadowed eastern wall. The best coverage was obtained earlier, from the planned place, but the crater was merely in the background in the pictures he shot of rocks from various angles using the polarising filter and the fact that he was some distance from the rim meant the floor of the crater was not visible.

"There's an interesting rock," Bean persisted.

"All right, Al," Conrad yielded.

Bean indicated a rock that had "kind of an oblique edge on it".

"This one?" Conrad enquired. "The big one?"

"The big one," Bean confirmed. "I'll just grab it and put it in the bag."

"Ho, ho, ho! Wait until I get the pictures," Conrad said.

"I think it's a different-looking rock, Houston, in the way it is shaped. It's partly rounded and has some oblique angles on it. Maybe under all that dirt is something a little bit different." Once Conrad had photographed the rock, Bean retrieved it. But

[40] Sample 12055 (unbagged).

A view by Al Bean along the line of a triplet of smaller craters west-northwest of Head crater.

to his amazement, "The thing that was giving it that unusual shape was the dirt that was adhering to it. But we'll take it back with us. Good rock." Furthermore, "This is probably typical of the rocks around this crater." He put it into the bag in the centre of the hand-tool carrier.[41]

As they ran south, heading for Bench, Conrad spotted a boulder with a lot of fine-grained material accumulated around its base. "Look at the fillet around that rock!"

The main source of the material for a fillet was ejecta from nearby impacts that splashed against the side of a rock and fell down to accumulate at its base, but a rock could also shed material as a result of direct micrometeoroid strikes and/or thermal stress during the diurnal cycle. As the scientists had asked the astronauts to be alert for well developed fillets, this 'ticked' another objective.

"That's a beauty," Bean agreed.

"We'd better stop and get that," Conrad decided.

As they took their photographs, Bean explained that the rock was "about 3 feet in diameter, about 2 feet thick and well-rounded" and had "obviously been struck a lot by meteoroids" because it had "a lot of surface pits".

Commenting on the fillet, Conrad said, "The rock has got dirt built up on all sides of it, all directions."

"It looks about equal too," Bean added.

"That's right," Conrad agreed.

Bean resumed his description, "If you look real closely at the rock, there are some pits that are maybe even up to three-eighths of an inch in diameter, but most of them are smaller. It doesn't look like a basalt, although the grains are too small for me to identify any individual one. Some pits have glass in, which isn't too surprising, but many of them don't." In fact, for basalt, most of the cavities were more likely to be vesicles than micrometeoroid strikes. These marked where bubbles of gas became trapped in the molten lava as it cooled. If the lava cooled rapidly, the minerals in the gas would have rapidly lined the spherical wall of a bubble with a coating of glass, but a slower rate of cooling would have allowed crystals to form, making the cavity irregular and forming a type of vesicle known as a vug. "That's about all we can say about that rock, Houston, and that's typical of the ones in this area." Although they documented the fillet, they did not sample it because the material looked similar to that previously sampled.

Leaving Bean at the fillet rock, Conrad set off south. "I'm coming up on Bench crater." As the pit opened up before him, he said, "This looks like a very interesting crater. It's different. And I see some really different rocks. Big ones. Hey, that looks like bedrock. Gee, what a crater! Oh, boy! Hey, Al, come on over here!"

"I'm coming," Bean replied. "Boy, there's some big fragments around here." The area north of the crater was littered.

Bench was the first crater they visited that had a substantial amount of rock in its interior. Viewing from the northwestern rim, Conrad speculated, "It looks to me like all that stuff is melted in the bottom of it, but I can't swear to that. I'll get you some

[41] Sample 12052 (unbagged).

The gnomon is alongside the filleted boulder north of Bench crater. The hand-tool carrier is in the background. Pete Conrad has his scoop (on the extension handle) standing in the ground.

pictures." He shot a partial panorama spanning the crater, then moved several feet to take another one to facilitate stereoscopic analysis. "I'm just sorry you guys aren't all here," he told Mission Control. "What a fantastic sight. Al, look in the bottom of that crater."

"Hey, look at that!" Bean exclaimed upon reaching the rim.

Presuming that shattered rock fragments would be angular, Conrad mused, "But don't they look melted on the top? Don't they look like they were molten? They're not completely jagged."

"No, they're not," Bean agreed. "It's hard to tell. I noticed when I was looking up real close at that rock back there, that it had been hit by meteorites so much, I guess, it had gained a rounded appearance something like those in the hole." Just because a rock was rounded did not necessarily mean that it had been melted: an angular rock could be rounded off by the erosional processes which operate on the lunar surface. "We ought to grab one of these pieces of rock."

"Let's get with it," Conrad prompted, wishing to make up time. "They'll baloney about it all day long in the LRL. The name of the game is to get the business done." Their task was to collect samples as rapidly as possible. Bean should not spend time providing descriptions. There would be ample opportunity after the mission for the scientists to examine everything.

Conrad selected a rock, "How about that baby? It looks a little different." Bean tried to fit the rock into one of the individual sample bags but it was too large and so he stashed it in the bag in the centre of the hand-tool carrier, confident that it would be able to be identified by the documentation.[42]

"Why don't you pick up two or three little ones from that same area," Bean said. "Get some here that we've already taken pictures of."

"Yeah," Conrad agreed.

As they sampled some of the smaller rocks that seemed typical of the area,[43] Bean said, "You notice that underneath this soil on the rim, it too is light grey."

Conrad, looking around for their next target, did not reply. Instead, he said, "Let's go over to that corner and try to break off a piece of that big rock, huh? It looks like bedrock to me."

"Can do," Bean agreed.

"Holy Christmas!" Conrad exclaimed on reaching the big rock. "Look at this, Al."

"You're kicking up the same sort of light grey," Bean reiterated as he followed in Conrad's trail.

"Huh?" Conrad said, momentarily confused. "No, no. Look at this stuff!"

"Hey, that's interesting," Bean agreed as he reached the rock.

"What do you suppose that is?" Conrad asked.

"Here's something interesting, Houston," Bean reported. "It kind of looks like a surface coating. What we've got is what looks like a semi-buried rock. And there's a

[42] Sample 12053 (unbagged).

[43] Sample 12032 in bag 4D.

A view of Bench crater. The intimidating interior shadow suggested a slope so steep as to dissuade the astronauts from using their tether to enable one man to venture into the crater to sample the rocks on its floor.

The first sample site at Bench crater.

small piece of it over there to the left. See it, Pete? We'll be able to catch it and put it in the bag."

"Yeah," Conrad said.

Bean resumed for Houston, "It's a buried rock, not unlike the others around here, except that it appears to have some sort of coating on it that's very iridescent. It has lots of crystals shining in it."

"I'll tell you what happened," Conrad speculated. "It's been laying in the ground and it's been hit by another fragment."

"Think so?"

"Yeah," Conrad decided. "Look at the glass beads, too."

"They're all over the place."

Working a fragment loose using the blade of the scoop saved fetching the hammer to break off a chip.[44]

"It's got kind of an interesting coating on it," Bean reiterated. "It's different from what we've seen."

As they sampled the adjacent regolith for comparative purposes,[45] Conrad asked, "How long have we been going, Houston?"

"You're 1 hour and 14 minutes into the EVA. We'd like you to move on from this crater in about 13 minutes. If you could, go on down and take a look at the bedrock on the bench." The planners had envisaged one man using the tether to venture into the crater in order to sample the material of the bench, but the average slope of the wall was 25 to 30 degrees.

Conrad was reluctant, "I'd hate to try and get down into the bottom of this fellow. It's awfully steep."

"Okay," Gibson acceded. "Hold off on that request."

If more time had been available for experimenting with the tether, one man could have gone into the crater, but as Conrad said in the post-mission debriefing, "I'm not sure how well we could have gotten back out of it."

But the scientists should not despair because, Conrad explained, "We're going to get you some of the bedrock. It looks like it's up in the lip here." He ventured a short distance over the rim to examine a large rock on the slope. "Boy, this is interesting. I want to see if I can sample it without falling in the crater. Well, look at this, Al. This is different. Look at the glass all over those rocks. We've got to get some of this."

After obtaining some fragments of glass-spattered rock,[46] Bean said, "Okay, Pete, what's your next pleasure?"

"What do you think, Houston?" Conrad prompted.

"You're looking in good shape," Gibson replied. "You can press on over to Sharp crater."

"Okay," Conrad agreed. After checking his map he looked for the crater. "I can't seem to locate it."

[44] Sample 12035 in bag 7D.

[45] Samples 12036 and 12037 in bag 8D.

[46] Sample 12038 in bag 9D, and samples 12039 and 12040 in bag 10D.

"It's 400 feet southwest from your present position," Gibson advised.

"Al, it's got to be over that hill right there," Conrad reasoned.

"Okay," Bean said. "Let's try it."

"Now we want to get the core tube, that gas sample and a bunch of good things at Sharp, right, Houston?" Conrad checked, as he raced ahead on the specified azimuth with Bean trailing behind lugging the hand-tool carrier.

"That's affirmative, Pete," Schmitt confirmed.

"We've got to find it first," Conrad muttered.

"Have you got it pinpointed, Pete?" Bean asked.

"I can't find it, but we're going in about the right direction," Conrad replied. "The trouble is, I'm looking down the zero-phase." The cleavage planes in the tiny fragments of rock in the regolith were backscattering sunlight, washing out the terrain. "That's got to be it right there. I see it."

"There's some big fragments out here," Bean mused. They were still in the ejecta blanket surrounding Bench crater.

"No, that's not it," said Conrad, deciding that the crater he was heading to was not Sharp.

"Let's stop here and look at the map more closely?" Bean suggested.

Halting, Conrad turned to give Bean the map and exclaimed, "Man, does that LM look small back there!" The slight rise over which they had passed was masking the lower part of the craft. When Gibson suggested that he take a picture of Intrepid in the distance, Conrad readily acceded, "We're so darn far out, I might as well make it a full pan."

"A full pan at Sharp," Gibson corrected, because they wanted that crater included in the coverage. "You're 1 hour and 23 minutes into the EVA, and we're looking for you to leave Sharp in 28 minutes, so you've got lots of time."

"We got to *find* Sharp crater first," Conrad emphasised.

"It's pretty small," Bean pointed out, after inspecting the tiny splotch on the map.

"Okay, I've got it," Conrad announced. "It's right here in front of me!"

"Yeah, that's it," Bean agreed.

As Conrad took a 360-degree panorama from a point short of the crater, Bean told Mission Control, "This has a nice white rim on it." Sharp was so-named because its rim was well defined in the overhead imagery, and it had been selected for sampling because it looked fresh. "In fact, the rim looks pretty much like the areas we kicked over on the previous craters."

Conrad was having second thoughts, "I'm not sure this is Sharp crater."

"Let's use it anyway, because it's the only one out here," Bean argued.

"We're estimating a diameter for Sharp crater of about 40 feet," Gibson pointed out helpfully.

Assessing the size of the crater, Conrad changed his mind again, "Al, this may be it!"

"Forty feet, huh" Bean mused. "This is it."

"It's got to be it," Conrad insisted. It was in the right area and it was the right size.

"It's got a nice raised rim on it," Bean noted. "What would you say, 2 feet?"

"Ooh, yeah! Oh-hoo-hoo-hoo!" Conrad exclaimed as he approached the rim. "It's awfully soft in here; watch it."

"Okay," Bean replied.

"Holy Christmas! Look at the bottom of that," Conrad said upon peering into the pit from a vantage point on its eastern rim. "Hey, Houston, it looks like blast effect coming out of it. It looks like it has got blast effects radial all around. This has got to be fairly fresh."

"Boy, the rim is soft here," Bean agreed, as he reached the crater. "I guess I'm to do a what? A double core tube here?"

"Al," Gibson interjected, "we'd like you to get the trench sample. Hold off on the double core until you get over to Halo crater."

"Okay," Bean acknowledged. "Aren't we supposed to look west for that ray from Copernicus?"

The planners had given the astronauts the option of continuing west to sample ray material if they could see a distinct contact, but the view westward was impaired by the zero-phase glare. "Houston," Conrad warned, "there's no way to tell a difference contact-wise."

Bean advised Gibson that although there was ejecta radiating out from Sharp, this material was indistinguishable from the local material, which was the western fringe of the ejecta from Bench.

When Gibson suggested that they get the environmental and gas samples from the trench, Bean replied, "We'll do the whole smash here for you."

"You want it right in the crater rim?" Conrad asked.

"That's affirmative," Gibson replied. "That would be perhaps the easiest and best place to do it. And you can also get the core tube down in the bottom of the trench."

While Bean went to fetch items from the hand-tool carrier, Conrad took a partial panorama cross the crater then moved to obtain a second for stereoscopic analysis.

After they had documented an undisturbed site on the rim, Conrad used the scoop to excavate a trench in the soft material, making rapid progress. "Now what do we want to do?"

"Fill the big container with dirt," Bean said, having already fetched the special environmental sample container from the hand-tool carrier. This was a cylindrical can some 8.5 inches in length and 2.5 inches across its exterior diameter. Owing to the way in which the shoulder joints of the suit incorporated internal cables, it was difficult to manipulate the scoop to station it above the narrow container, so Bean, who was holding the container, grasped the scoop and tapped it against the aperture to dribble in some of the material.

"Houston, this dirt came from about 8 inches down," Conrad reported.

The lid was connected to the canister by a wire so that it could not become lost. In training on Earth the lid had screwed on readily but it now tended to stick, possibly owing to thermal expansion – Bean had reported that upon grasping items which had been in sunlight for a while he could feel the heat through his gloves. The rock box from Apollo 11 had been opened in the Lunar Receiving Laboratory in vacuum and the 15 psi pressure differential had made using the glove-chambers tiring work. For Apollo 12, the box with the samples from the second moonwalk was to be

Pete Conrad's panorama from Sharp crater included this view looking back towards Intrepid, the lower portion of which is masked by a slight rise in the intervening ground.

The interior of Sharp crater. Unfortunately, because Pete Conrad photographed it from directly up-Sun the far wall is lost in the zero-phase glare.

At Sharp crater, Al Bean holds the special environmental sample container, into which Pete Conrad has just poured some loose material. The lid is hanging by a cord from the bottom.

opened in an atmosphere of sterile nitrogen at ambient pressure. Storing this sample in a sealed canister would ensure that the material remained in its pristine state, ready for study in a laboratory vacuum chamber.

"Now, we need a core tube in the bottom of that trench. Is that right, Houston?" Conrad checked.

"Affirmative," Gibson replied.

"This is core tube number 2," Conrad reported.

While Bean detached the extension handle from the scoop and affixed it to the top of the tube, Conrad photographed the trench. Then Bean placed the coring bit of the tube on the floor of the trench and started to hammer the tube in. "It's driving in real easy, Houston."

"Stop. That's it," Conrad decided.

Bean put the hammer back on the hand-tool carrier and then documented the tube in the floor of the trench. "Boy, this dirt has gotten on my camera and I can't see the settings anymore. I'm going to have to do something about that." As Bean withdrew the tube from the ground, Conrad stood by with the end cap.

"I hope that soil stays in there," Bean said.

"It's full," Conrad announced.

Once Bean had unscrewed the coring bit, Conrad handed him the cap and he put it on to seal the tube.

"Good sample," Conrad congratulated.

"Yeah, it is," Bean agreed.

As Conrad swapped the extension handle back to the scoop, he called, "Houston. What else do you want here?"

"The gas analysis sample," Gibson replied.

"We need some little rock fragments for that, Pete," Bean reminded.

"That's *surface* rock fragments," Gibson clarified.

"Coming up," Conrad promised.

Bean pointed out some candidates, "See those bright shiny ones there?"

"Yup."

"Let me get a shot of them," Bean said, and took a picture. "We need some more, Pete," he urged after Conrad had supplied several tiny rocks. "There's not enough in there to do anything with." The gas sample return container was 3.8 inches in length and 1.5 inches across its outside diameter. The bottom had a thin skin that was to be punctured in a laboratory vacuum chamber so that scientists could analyse any gases released. Bean wanted to fill the can.

"Come on," Conrad laughed, "I'm getting tired of picking up those little things!"

"Get that big one!" Bean urged.

"This one?" Conrad asked.

"Yeah."

"I don't think that will fit," Conrad warned.

"Let's try it," Bean urged.

"Come on, Al, we're wasting time," Conrad complained.

As Bean screwed the lid on the canister, Gibson prompted. "As soon as you've finished up there, you can head on east to Halo crater."

A context shot showing the trench excavated in the soft material of the raised rim of Sharp crater.

A core sample was obtained from the floor of the trench at Sharp crater. This view shows the extension handle, the tube itself is entirely in the ground.

"We're heading for Halo crater now," Conrad reported. He set off with the tongs on his waist and carrying the scoop and gnomon. Bean followed with the hand-tool carrier. Even at Sharp, their maximum range from the ALSEP, the seismometer was able to detect their footfalls.

Gibson advised them to skirt the southern rim of Bench crater "and continue east until the LM is at your 9 o'clock, then a couple of more steps ought to take you right to Halo crater".

"Sounds like a pretty good vector," Conrad agreed. "That says that we're running right into the Sun. Does that agree with you?"

"That's affirmative," Gibson said.

When travelling down-Sun in search of Sharp crater they had been blinded by the zero-phase glare from the surface. Now heading in the opposite direction, they faced the glare of the Sun low on the horizon ahead.

"You know what I feel like, Al?"

"What?"

Conrad laughed and asked, "Did you ever see those pictures of giraffes running in slow motion?"

Bean, trailing behind Conrad, agreed, "That's about right."

"That's exactly what I feel like."

"Say," Gibson cut in, "would you two giraffes comment on your boot penetration as you move across there, as compared to back at Sharp crater?"

"Oh, it's much firmer here," Conrad reported. "We don't sink in anywhere near as much."

"Yeah," Bean agreed. And, being able to observe Conrad's prints, he added, "Your toes sink in a bit, Pete, as you push off. You land flat-footed, so your heels don't sink in. But as you push off with your toes they sink in down about 3 inches. Your heels are only sunk in perhaps an eighth of an inch."

"Thank-you, Al," Gibson replied.

Bean continued, "Every time he lands, he sends little particles spraying out ahead of him, beside him and everywhere else to distances of 2 to 3 feet."

"Pete, Halo crater is just about the same size as Sharp crater, and should resemble it," Gibson advised.

"I think I have it in sight, but I'm not sure. There's a couple of them up ahead," Conrad replied. By this point Bean had edged slightly ahead and Conrad, seeing his colleague suddenly draw to a halt, announced to Mission Control, "I'll tell you what I'm going to do, Houston. I'm going to take an EMU break." Then he asked Bean innocuously, "How are you doing, Al?"

"Okay," Bean replied somewhat distractedly.

They had run 700 feet in 3 minutes 20 seconds, were short of breath and had heart rates of about 160 beats per minute. Conrad's statement that they were going to take an 'EMU break' was NASA-ese for a halt in place to let their heart and respiration rates relax. However, as Bean explained after the mission his suit had 'coughed', by which he meant he had sensed a sudden pressure change. It was this that caused him to halt in order to check his cuff gauge. Seeing this, Conrad had made the excuse of taking a break and asked Bean, admirably ambiguously, how he was; to which Bean replied that he was okay. However, neither man immediately reported the pressures, oxygen quantities and status flags as required during an EMU break. Afterwards, the suit engineers concluded that as Bean was running, his body must have made contact with the oxygen outflow port in the suit, momentarily closing it and causing an over-pressure which he felt in his ears.

Gibson refined his earlier statement, "Pete, the dimension on Halo crater is about 20 feet, so that would make it half of what you saw at Sharp."

"Okay," Conrad replied. Looking at a crater ahead, he asked, "You suppose this is it, Al?"

"It doesn't have any halo around it," Bean pointed out, but this would be difficult to discern in the up-Sun glare.

Although they drank water prior to suiting up in order to stave off dehydration on the moonwalk, breathing pure oxygen at a high respiration rate made their mouths and throats dry. "One thing I'd go for," Conrad said, "is a good drink of ice water."

"Good thinking," Bean agreed. On walking up to Conrad, he said, "Let me look in the map."

Gibson offered further advice, "Halo crater should be right on the rim of Surveyor crater, and you ought to see Surveyor off directly to the northeast."

"I know where we are," Conrad insisted.

At Gibson's prompting, they belatedly provided the information to complete the EMU break. On spotting a glassy spherule about one-quarter of an inch in diameter on the ground, Conrad decided to scoop it up.[47]

With the experience of lugging the hand-tool carrier around, Bean estimated that they had collected only about 5 pounds of samples so far. "I would hate for us to get back to the LM and then have to fill up around there again." But Conrad was confident that they were going to reach Surveyor crater, where they would find more than enough material to fill the rock box.

"I'll tell you what," Conrad said. "Look over here at me and smile." He snapped a tourist picture of Bean.

Bean did likewise, "I'll get you right there by a crater."

"Where's the LM?" Conrad asked.

"Right in the background," Bean replied, meaning that it was in the frame behind Conrad.

"All right," Conrad decided. "Let's ease off at a slower pace. I know where we're going now. I think this is Halo crater right up here in front of us."

As they set off again, Bean offered advice to his counterpart on Apollo 13, "Hey, Ed, you might tell Fred Haise he ought to quit working on running and start working on holding things in his hands for long periods. Your legs don't get a bit tired, but your hands get tired carrying these tools, particularly the hand-tool carrier." In order to prevent the samples from bouncing out of the large bag within the frame he was holding the carrier against his chest, which was stressful because it required his arms to constantly overcome the suit's reluctance to adopt an arms-bent configuration.

[47] Sample 12041 in bag 11D.

"I'm sure he is listening," Gibson replied.

"And tell Jim Lovell to practice digging!" Conrad told Gibson. In the debriefing, he explained, "I didn't notice that my hands got tired as much as I noticed that they got sore. When you work for 4 hours and use your hands, you have a tendency to press the end of your fingertips into the end of the gloves. Although my hands never got stiff or tired, they were quite sore when we started the second EVA. As soon as I got working again, I forgot it. It wasn't until we got back in Yankee Clipper that we noticed our hands were sore again. But this was because we did almost 8 straight hours of EVA work, which we'd never really done before. And I think in one g you don't have the tendency to thrust your hands as far down to the bottom of the gloves as you do in one-sixth g. You really ought to hang onto something up there. It's not as apparent when you're working up there that you are pressing you fingers as far out in the gloves."

Noting that they were crossing an area with "a completely different texture", Bean urged, "Let's take some pictures here. We've run across a sort of a textural contact. We're suddenly on an area that's not so smooth. It's got dimples and wrinkles in it." This 'ticked' another item on the list of objectives for the traverse.

"Let's take a couple of good dirt samples of this stuff," Conrad decided.

Bean continued for Mission Control's benefit, "It is interesting. You know, I think this looks like that material that we saw on the first day in front of the LM." Conrad placed the gnomon at the selected spot, and Bean observed, "It looks almost like this material is more cohesive and forms clumps, instead of being so nice and smooth."

"I wanted to get my footprints in it too, so they can see that," Conrad pointed out. Once Bean had taken the documentation, Conrad said, "If you'll get some sample bags, I'll scoop this stuff." As he lifted some of the material, he noted, "Boy, it sure is fine. It's kind of like over at Sharp crater."

"Except it looks almost finer," Bean opined. "If you saw this on Earth, you would think it was a real soft dirt that had been rained on recently; not hard rain, but just a sprinkle." They filled a bag with two scoops.[48]

Wondering whether a nearby crater was Halo, Conrad asked if it was shallow with several dimples on its southern side, and Gibson replied that he would check it out.

"We can collect a rock while we wait, Pete," Bean prompted.

"I think this is Halo crater right here," Conrad said, "so let's go over and get some rocks and everything." If his identification was correct, then they had already obtained a soil sample. "But this isn't 20 feet in diameter," he mused. "Is Halo right on the rim of the Surveyor crater, Houston?"

"Affirmative," Gibson replied. "And from your comments on the three dimples, we show that you're there." But Mission Control was wrong. They were actually at a small crater about 200 feet northwest of Halo, with both roughly the same distance away from the rim of Surveyor crater. It was a striking example of the difficulty of trying to use a map to locate small holes in the ground whilst not knowing precisely where you are.

[48] Sample 12042 in bag 12D.

Al Bean in the vicinity of Halo crater, southwest of Surveyor crater. Note his cuff checklist.

Pete Conrad in the vicinity of Halo crater, southwest of Surveyor crater. He is holding the extension handle, without a tool affixed. His cuff checklist is open at a page with a Playboy image.

"Okay, what do you want?" Conrad asked.

"We'd like to get the pan and the double core tube," Gibson replied.

"I can't believe we're at the right place," Bean warned.

"I'm not sure either," Conrad admitted.

Surveyor crater was obvious, and its position indicated that they were in roughly the desired place but, as Bean warned, "There's hardly a crater worth looking at where we are."

"Okay, Pete," Gibson called, aware of the clock. "It's your call there. You're the local experts. If you see a better location for that double core tube, go ahead."

"We're trying to find the right crater, Houston," Conrad reported. As the scientists had specified a particular crater, he considered it his duty to try to find it.

The only way to correlate the map to the actual terrain was to look for patterns of craters and, having done so, Bean announced, "Hey, Pete, I think it's that area right over there. Halo is this first one right here, the little one, and then all those others are next over according to the chart."

"Which one's Halo?" Conrad asked.

"You see where I'm pointing?" Bean replied. "As I see it, it's that one right over there."

"Okay. Let's go," Conrad said. "And you want what, Houston? A partial pan?"

"A full pan, Pete," Gibson corrected. Then, referring to the experimental double-length core sample, he went on, "And, Al, could you give us some sort of an estimate of how hard it is to get the core tube in; that is, what the force history is."

"Sure will," Bean replied.

When Conrad noticed glass in the bottom of a small crater and suggested that they pause to sample it, Bean countered that they already had a lot of glassy material.

"I think that's Halo right there," Bean said as they drew up to a crater.

"It's too big," Conrad decided, then suggested another crater, "Let's take this one right here."

"All right," Bean agreed.

The lesson learned from the search for Halo was that trying to figure out which of a group of similarly sized craters was the specified target was wasting time which could be better spent sampling any one of them!

To keep tabs, Gibson asked for a frame count on each Hasselblad. On discovering that his counter was lower than he expected, Conrad complained, "You know what's happened? This thing hasn't been taking every picture!"

Gibson naturally responded, "We'd like you to get the pans taken on the LMP's camera. You can either have Al do the pans or switch cameras. Your choice."

But the core sample came first. This involved removing the coring bit from the upper section, joining the tubes and pulling the pin from the top of the lower section to enable the internal 'follower' to cross the junction, and then affixing the extension handle to the top.

As Yankee Clipper had just appeared around the limb on revolution 26, Gibson updated Gordon, "Dick, they are pressing right on with the EVA and are about two-thirds of the way through the traverse, working their way over towards Surveyor."

"Okay," Gordon replied. Once the MSFN relay was established, this would enable him to listen in.

"Where are you going to drive it?" Conrad asked, referring to the core tube.

"Where would you recommend?" Bean invited.

Conrad looked around and decided, "Let's go over to this crater right here."

"I hope this is a good soft place," Bean said. He drove the tube in by hand until it encountered "something solid" about 7 inches down, then overcame the obstacle by wielding the hammer. Ever observant, he reported, "Hey, Houston. When you hit on the side of this hammer, it knocks little chips of metal off the side of the hammer. I don't think that's too good." This was proving to be the first heavy work performed by a lunar surface hammer.

"Is it damaging the hammer or the core tube?" asked Schmitt, who was standing in while Gibson dealt with Yankee Clipper.

"I'm afraid some of the fragments will damage the suit. It's not damaging itself," Bean replied. For a change, he switched the hammer from one hand to the other, and announced, "Hey, I'm better left handed than right. There goes another fragment. Do you see it, Pete?"

"Yeah, I'm watching," Conrad replied. In fact, he was taking pictures of Bean at work.

Bean continued, "Even when you hit it with the front end, some of them pop off. They're flying all over the place."

"He's really driving that baby," Conrad confirmed. "He has reached the bottom of the handgrip portion of the upper tube."

After a minute and a half of pounding, and with the double-length tube fully in the ground, Bean inspected the tool. "Look at that! It looks like it has got a coating over the hammer, Pete, and I'm knocking the coating. Instead of being a steel or aluminium hammer, it's some sort of coated arrangement."

"That's affirmative, Al," Schmitt confirmed. "There is a coating on that hammer, and that's probably what you're knocking off."

"Why don't I trade you cameras?" Bean suggested.

"All right," Conrad agreed.

On removing his camera, Bean discovered it to be loose on its mount and it came apart as he examined it. There was a screw hole in the base of the camera and a screw atop the handle underneath. The screw passed through a hole in the support bracket and a fork in the assembly that enabled a gloved hand to trigger the shutter. It was tightened using a thumbwheel which compressed a spring washer. As Bean attempted to adjust the wheel, the mechanism came loose and the trigger fell to the ground. Inspecting Conrad's camera, Bean warned him, "Your nut's loose too, Pete!"

This explained Conrad's surprise on reading the frame counter. The intermittency was the result of excessive trigger play caused by the loose assembly. "You're going to have to help me get this camera off," Conrad said.

"There's no need to get it off now," Bean replied. Once the mechanism had been tightened up, Conrad would be able to continue using his camera.

"No, I want to give it to you," Conrad insisted.

Al Bean taking a core sample in the vicinity of Halo crater. The lower tube is entirely in the ground and he is hammering the extension handle to drive in the upper tube. Note the saddlebag on his left hip.

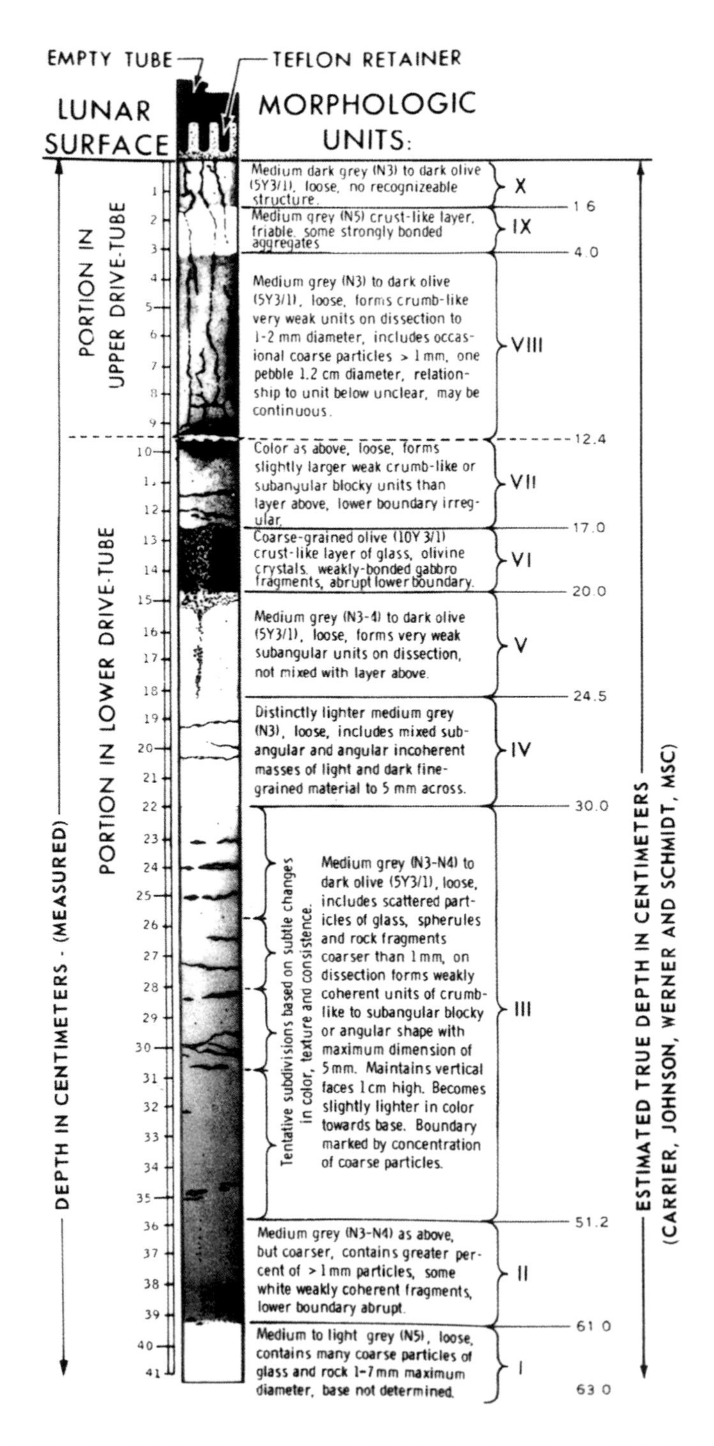

An analysis of the double-length core sample obtained in the vicinity of Halo crater.

"And you're going to use this one?" Bean asked, doubtful that Conrad would be able to operate the detached camera's trigger manually.

"Al," Conrad explained, "you've got to take the Surveyor pictures, so why don't I give you my camera?"

"Okay, good enough," Bean agreed. He passed his camera to Conrad, removed his colleague's camera, including the bracket from the chest unit, then tightened up the mechanism. Once Bean had discarded his own bracket he was able to don Conrad's camera. The unusable camera was stowed in the centre of the hand-tool carrier. This incident was costly in terms of time, as 3.5 minutes had elapsed since Bean first suggested exchanging cameras.

"All right, I'll go get your core tube," said Conrad, glad finally to be able to move on. After examining the top of the extension handle, he pointed out, "Hey, you sure beat on it."

"That's what it took to get it in the ground," Bean replied, stating the obvious.

Conrad, his voice straining with the effort of tugging the T-shaped handle, spoke the opposite of the truth, "It's coming up real easy."

Seeing Conrad's effort, Bean gave a hearty laugh. "It looked for a minute like you were going *down* real easy." With the tube refusing to budge, Conrad's boots were digging themselves into the soft surface.

With the tube finally in his hands and ready to be broken down, Conrad realised, "We made a tactical error."

"In what fashion?"

"I think we dropped an end that we shouldn't have dropped," Conrad replied. To join the tubes together they had discarded the top of what became the lower section, and it was now required to seal the tube.

"We'll have to pick it up," Bean said.

"Where is it?"

"Right over there." Bean pointed to the boot prints where they had assembled the tube. "We'll find it."

Conrad went over and looked, "It's someplace buried in the dirt. I see it! Ha, ha, right here." He used his tongs to lift the cap.

On unscrewing the two tubes, Bean noted that the material at the top of the lower section looked the same as on the surface. Then, with Bean firmly holding the upper section, Conrad detached the extension handle, which he returned to the scoop while Bean finished capping the core tubes. On realising that the core was safely stowed in the hand-tool carrier, Gibson prompted, "Have you gotten the panorama?"

"No," Conrad replied. "I'm going to get Al to do that. He's using my camera. His camera's had it." As Bean shot the panorama, Conrad asked, "Okay, Houston. What else would you like here?"

"Pete," Gibson replied. "You're 2 hours and 7 minutes into the EVA. We show you leaving Halo in around 8 minutes. And that's for a full 4 hours, because we've extended you by 30 minutes." This extension guaranteed that they would be able to reach Surveyor 3 for the 'photo opportunity' eagerly awaited by the public relations people. "Before you move on, we'd like to regroup and figure out a plan of attack on the Surveyor. One thing: we'd like to make sure that you remain away from directly

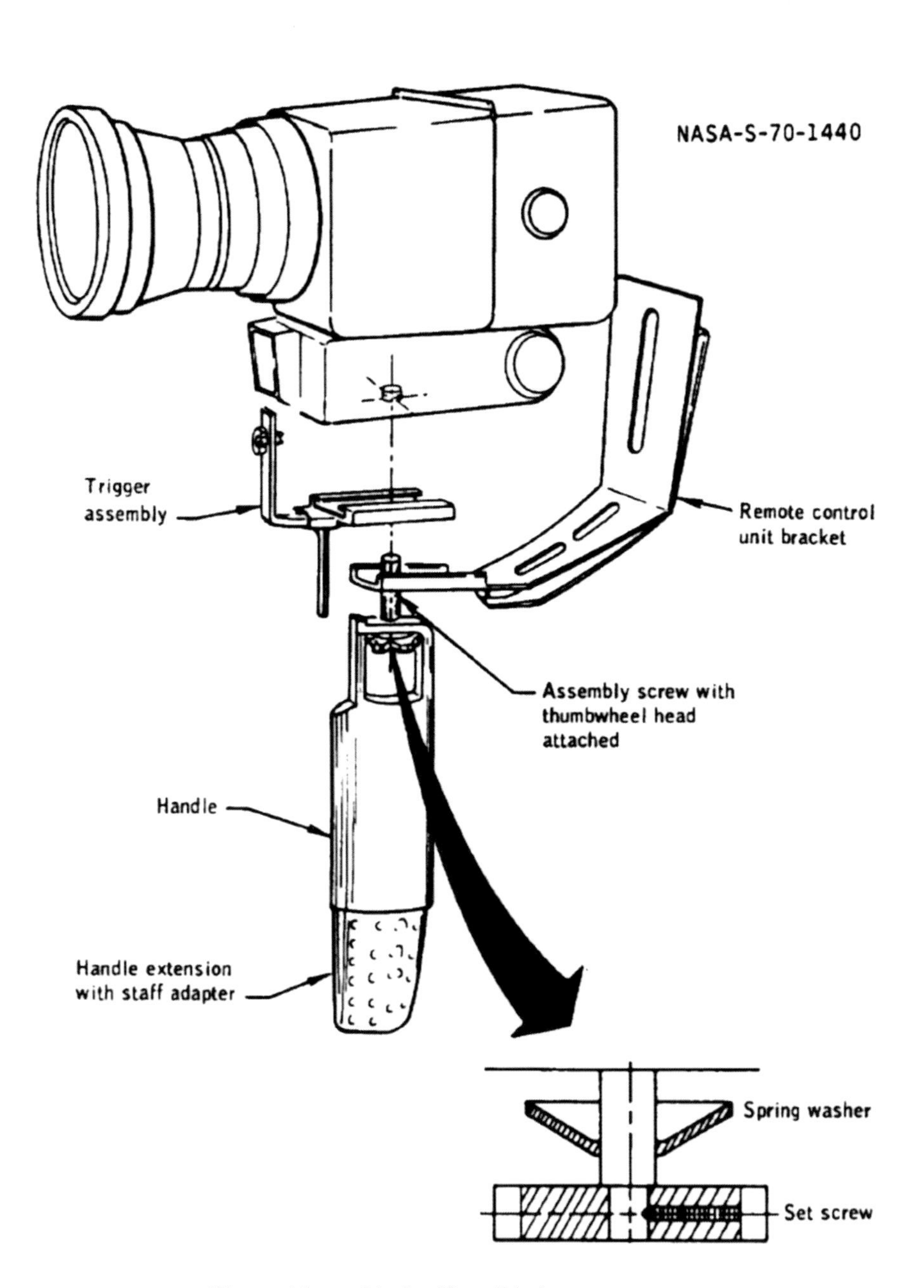

The problem with the Hasselblad camera mount.

below the Surveyor as you move up to it." There was concern that if the inert lander were to be precariously established on the interior slope of the eponymous crater, it might slide when disturbed and possibly trap an astronaut.

"We were talking about it last night," Conrad replied. "We're going to approach it from the side."

"Roger," Gibson acknowledged.

On completing the panorama, Bean said, "We probably ought to get one of these rocks here and just throw it in the bag."

"Yeah," Conrad agreed. "Let's sample a couple of these laying right over here." After scooping a small rock and some soil,[49] he decided, "Al, let's move up on the rim of Surveyor crater and start getting some rocks." On reaching the southern rim he drew Bean's attention to some rocks which appeared "a little bit different".

"Let's grab some," Bean urged, still concerned that their haul of samples was not very substantial.

"They look like granites, don't they?"

"They do," Bean agreed, having earlier compared the texture of a rock to that of granite. Seeing a nearby rock with a splotch of glass on it, he called out, "Here's a beauty! Let's get this one."

"Okay," Conrad agreed.

After lifting the rock, Bean turned it over in his hands to examine it and almost dropped it.[50] It was too large for an individual sample bag, so he dumped it straight into the hand-tool carrier.

"Now I want some of these granites over here," Conrad said, then to play safe he added, "Or what looks like granite."

Bean held the strap of the Surveyor parts bag to enable Conrad to lean over and grasp another rock by hand, and then hoisted him upright again.[51] "Look at the sheer-face on that rock," Conrad observed.

"It's got some pretty interesting fracture marks," Bean agreed. "It also has what look like abrasion marks on it. Maybe that's just hard-packed dirt."

"What I recommend we do now is change film packs," Conrad said, because the Surveyor lander was to be comprehensively documented.

"That's a good idea," Bean agreed.

"Use up that film first," Conrad said. "Get the Surveyor."

"It's a bad place to shoot," Bean noted, because the robotic lander was obliquely up-Sun, "but I'll try it." After several frames the film was exhausted, so he detached the magazine and put it in the hand-tool carrier, then replaced it with the black-and-white magazine from the other camera. According to the mission plan, at this point he ought to have replaced his black-and-while magazine with the colour one that he was carrying in his saddlebag specifically for documenting Surveyor 3. But this was not actually stated on their cuff checklists and, distracted as they were by the camera problem, they had forgotten.

[49] Bag 13D; for some reason it never made it back to Earth.
[50] Sample 12054 (unbagged).
[51] Sample 12051 (unbagged).

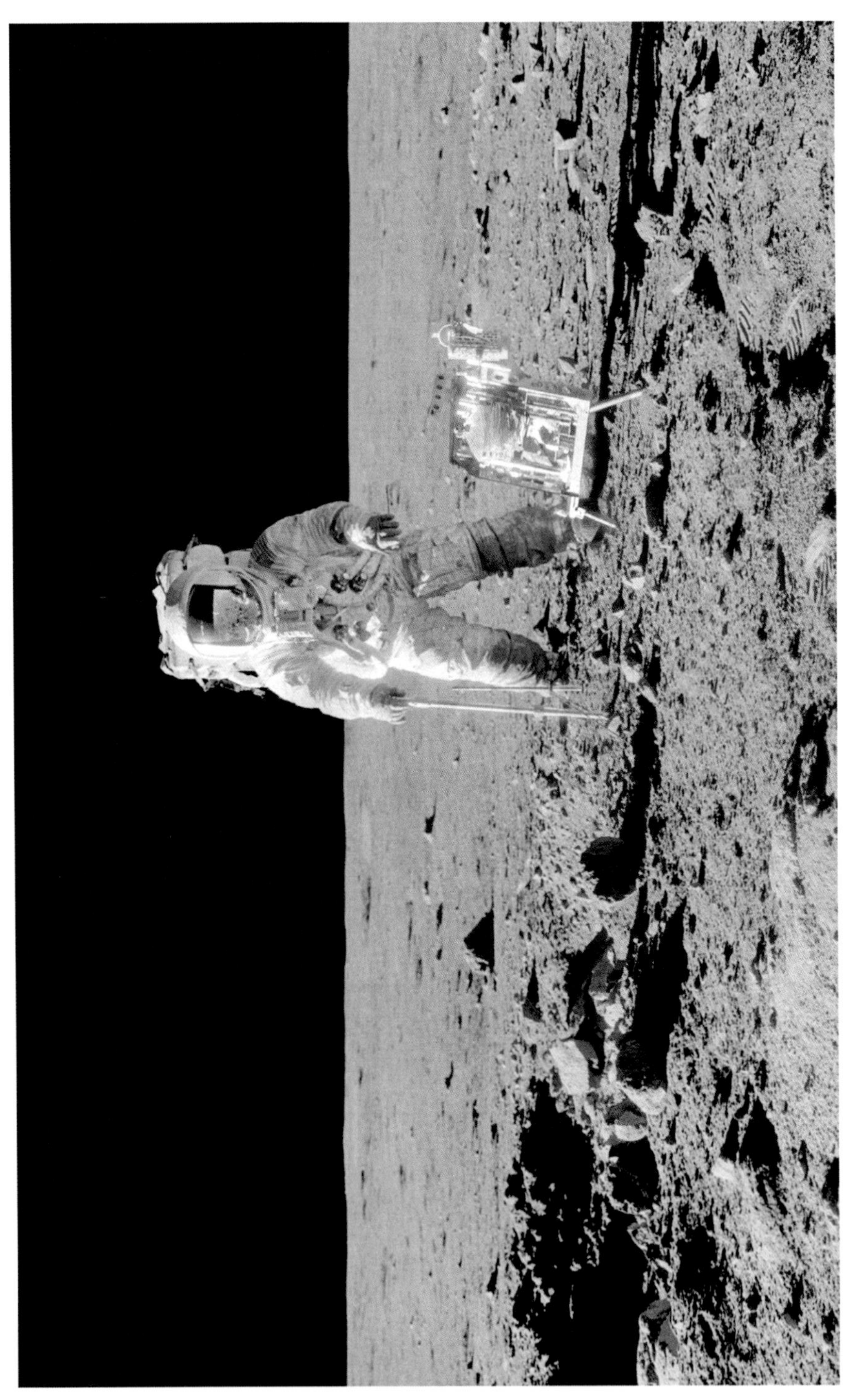

Pete Conrad sampling a small crater on the southern rim of Surveyor crater.

Surveyor 3 was about 150 feet inside a crater that was about 600 feet in diameter, with its centre 50 feet below the level of the surrounding plain. There was a smooth inflection in the profile from the concave floor to the convex upper wall about half way to the rim in both radial and vertical directions, with the lander on the inflection where the slope was about 12 degrees. On the first day, the shadow had given the impression that this slope must be considerably steeper than this, but once the rising Sun had banished the shadow it seemed less intimidating.

The traverse had brought them to the southern rim of the crater, and Conrad suggested that the best way to approach the lander would be to walk along the rim to the southeastern side and then start into the crater, still moving mostly parallel to the rim but slowly descending. This would allow them to remain 'above' their objective, as Mission Control wished. Bean readily agreed. The geologists had suggested that because the crater was fairly old and degraded, thick dust would be slowly migrating down its wall and accumulating on the floor. This led Conrad and Bean to expect to find the footing inside the crater to be like walking on thick sand. A radial approach had been recommended in which Conrad would remain on the rim and play out the tether while Bean assessed conditions on the interior slope, but judging the situation benign they decided not to use the tether. Furthermore, there was a communications consideration. If Intrepid had landed far from the crater, as they descended into it the rim might have blocked the line of sight for the VHF link. If the link broke, Conrad was to walk upslope until it was restored and then stand in place to serve as a relay while Bean went all the way down to Surveyor 3, whereupon Conrad would read the list of photographic assignments and Bean would take the pictures – using the colour film he was supposed to have placed on his camera. But with Intrepid standing right on the opposite rim there was no problem with the line of sight.

"Let's wander over here," Conrad said, starting east and leaving Bean to fetch the hand-tool carrier. To Mission Control he explained, "We're going to the area where we can stop and case the joint."

Pausing to take a sample,[52] Bean opined, "You know, we keep collecting a lot of the same type of rock because there just doesn't seem to be any other kinds around. I haven't seen any microbreccia the whole day, and I've looked around for it; all I've seen is basalt." A breccia is a mechanically assembled rock made of fragments of shattered rock bound together by a fine-grained matrix. With so many impact craters nearby, some such rocks could have been expected but they were noticeable by their absence.

"We're going to jog on here for a little bit, Houston," Conrad reported, "and get a little bit closer to the Surveyor and look her over." The lander was gleaming in the sunlight.

Once they were due south of their objective, Bean said, "Hey, coming in from the south looks like a good way, Pete."

"I'll tell you what, Al, I am just going to lope right around here," Conrad replied, and with that he started down the slope.

[52] Samples 12043 and 12044 in bag 14D.

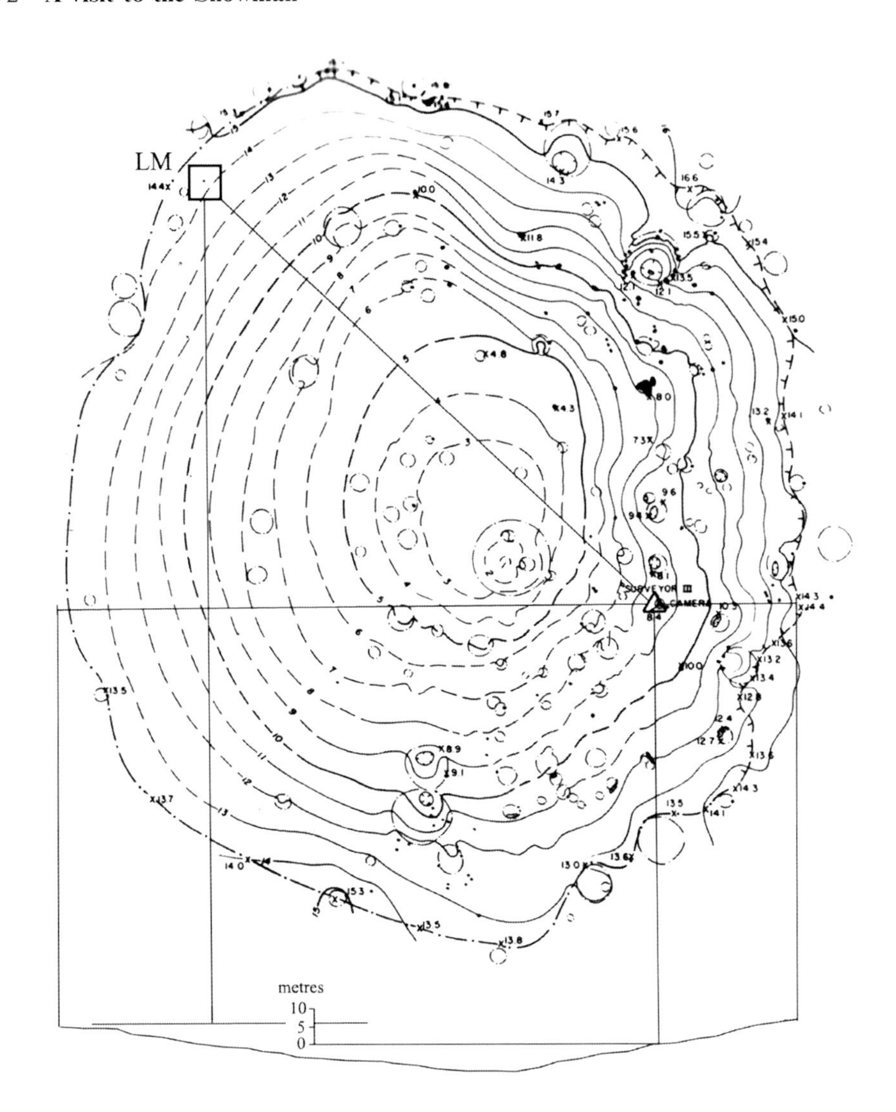

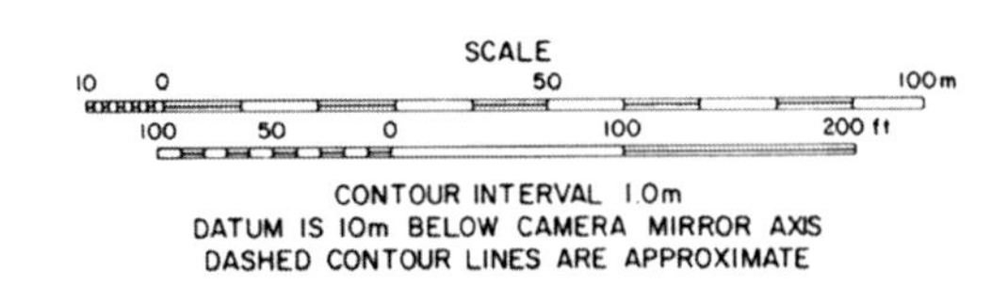

THE MAP IS BASED ON SPACECRAFT LOCATION BY RESECTION
ORIGINALLY MADE BY E A WHITAKER

The contour map of Surveyor crater, annotated to show the horizontal and vertical relationships of Intrepid and the Surveyor 3 lander.

A vista of Intrepid on the far rim of the 600-foot-diameter crater, Surveyor 3 on the interior slope and Block crater beyond.

A view of Surveyor 3 as Pete Conrad and Al Bean made their approach. They are inside the 600-foot-diameter, the rim of which forms the horizon.

As Surveyor 3 came to rest, the foot pad on the downslope side (left) dug itself into the loose material. The television camera is the cylindrical object mounted on struts. Note the pantograph-style robotic arm with the scoop.

A view of Surveyor 3 with Intrepid on the far rim of the 600-foot-diameter crater.

Realising they were in the crater, Gibson called, "Could you give us a comment on how far you're sinking in?"

"I'm not sinking in very far at all," Conrad replied. "It's fairly firm stuff." At almost the level of the lander, they veered right to make their final approach on the contour. "Don't worry about it, Houston, because it's no strain." To calm concern in Mission Control about exiting the crater, he added, "I'd be able to go right back up the way I came down with no strain at all."

"I don't think there'll be any sweat about it," agreed Bean, who was lugging the hand-tool carrier. As he said in the post-mission debriefing, "It wasn't very difficult to operate on that slope. It wasn't particularly slippery. One of the things I wondered about beforehand was once we got down the slope we wouldn't have a good sense of vertical and we'd tend to lose our balance. But that wasn't the case at all. It was just like a 12-degree slope on Earth." The ability to work on a slope was good news for future missions.

On nearing their objective, Conrad, mindful of the belief by Mission Control that the slope would be tiring, announced, "Al, I'll tell you what we'll do. We'll park all of our gear, have ourselves a little rest and go over your photo plan, then have at it."

"Let's go right over here," Bean replied.

"Why don't you get a photograph of it right now," Conrad prompted.

"It's a good place," Bean agreed.

Gibson asked whether there was any evidence that the dust displaced by Intrepid's engine plume had coated the inert lander. Recalling the speed at which the dust was blown radially outward Conrad ventured that it had flown a long way, "I'll tell you, the way that dust was going, it probably went right over top of it."

"That's right, dust from up there would never fall into this crater," Bean agreed. The crater was 50 feet deep, and the Surveyor lander was half way down in terms of vertical distance.

Noticing the imprints of Surveyor 3's foot pads, Conrad said, "You can see which way it came in."

The normal touchdown procedure for a Surveyor was for it to turn off its engine when the radar altimeter indicated the pads were 12 feet off the ground. In this case, as it descended through 30 feet the radar lost its 'lock' and was unable to issue this command. The vehicle touched down at a vertical rate of about 6 feet per second. It was level at this time, but the ground sloped down to the west and caused leg no. 2 to make contact first. In response to the tilt induced by the other two legs touching down, the flight control system, which was in attitude-hold mode and did not realise it was on the ground, increased the thrust of engines no. 1 and 3 in an effort to regain a level attitude. In the weak lunar gravity, this additional thrust caused the vehicle to lift off. After peaking at about 38 feet, it made second contact some 50 feet west of the initial point, this time at a vertical rate of 4 feet per second. As before, the slope caused leg no. 2 to touch down first and in attempting to hold its attitude the vehicle lifted off again. After it peaked at a height of 11 feet and was falling, the engines were cut off by a command from Earth. At that time it was at a height of only 3 feet. As a portion of the thrust had been aimed laterally at each liftoff, this had built up a horizontal component. Consequently, when the vehicle struck the surface its vertical

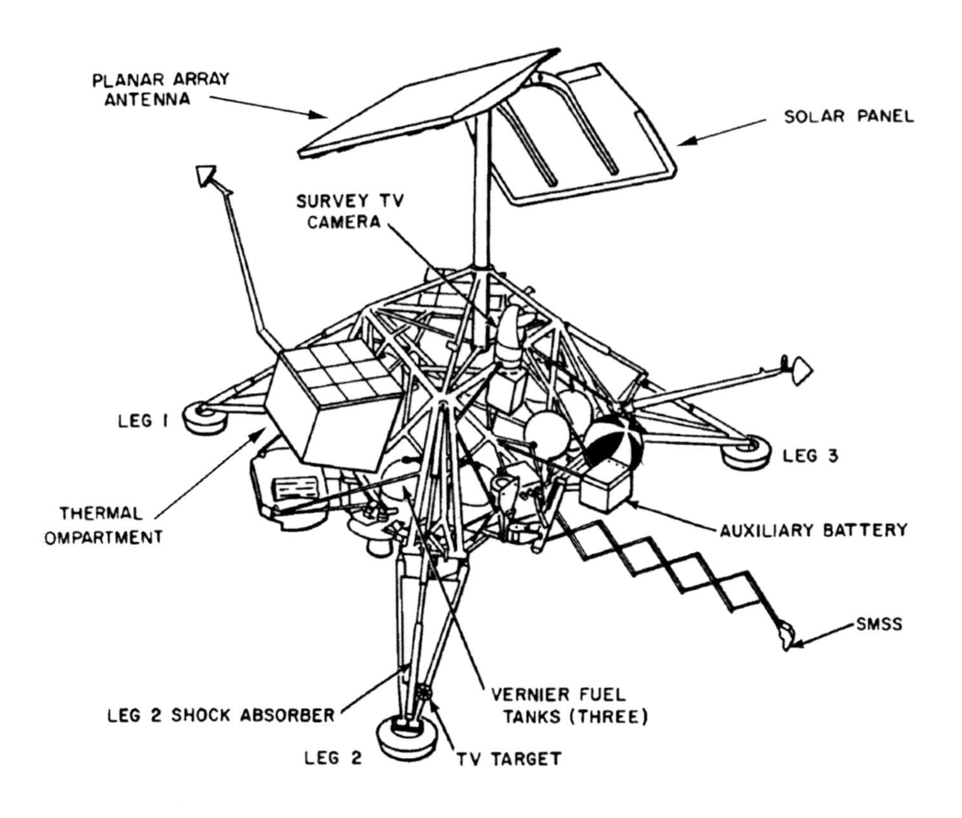

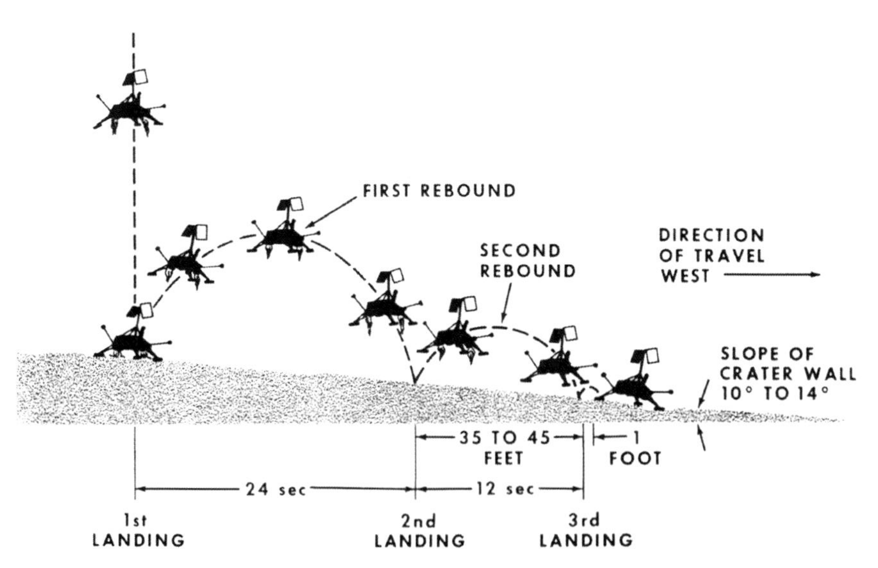

Details of Surveyor 3, the first in the series to have a robotic arm, and how it bounced upon landing.

rate was only 1.5 feet per second but its horizontal rate was 3 feet per second. The elasticity in its legs caused it to rebound several inches and hop another 18 inches further downslope before it settled some 36 feet west of its second point of contact, coming to rest with loose material piled against its foot pads.

"It's a beautiful sight," Bean said, taking pictures from upslope to document the lander. "You know, this one is brown." The training mock-up was white. Wondering if it might have degraded in the harsh environment, he asked, "What colour was this one when it started out, Houston? White?"

"Stand by on that," Gibson replied.

"It looks a light tan now," Bean noted.

Gibson replied, "The equipment bays were white on the side, and the scoop itself was a light blue."

"What was the general colour of the structure?" Bean persisted. "All the struts and the like?"

"That's all white," Gibson said, referring to the thermal paint.

"We will have to look at it more closely," Bean said. They were maintaining their distance in documenting the site, but would be able to inspect the state of the vehicle more carefully when they moved in to use the cutters.

"The Sun's cooked that paint brown," Conrad speculated.

The lander had settled with its robotic arm upslope. As Bean manoeuvred, he took care not to kick dirt into the area where the arm had scraped its trenches, so that the scientists could compare his pictures with the television pictures transmitted by the lander itself.[53]

What Mission Control did not know was that Conrad and Bean had a secret plan. As Andrew Chaikin related in '*A Man on the Moon*', Conrad "had one of the support crew [...] buy an automatic timer for the Hasselblad, a little spring-loaded gadget. [The] idea was [to] mount the camera on the tool carrier and then pose, side by side, next to the Surveyor. It would take only a minute for them to fire off a few shots – saluting, waving, shaking hands, whatever – and Conrad was sure that when they got home one of those pictures would end up on the cover of *Life* magazine." In fact, in '*Flight – My Life in Mission Control*', Chris Kraft explained that Conrad had "help from *Life* magazine and Deke Slayton in sneaking a camera timer to the Moon". As Chaikin continued, "Conrad managed to smuggle the timer in the pocket of his space suit. He had remembered to bring it into the LM with him [prior to undocking from Yankee Clipper], and just before they headed out on the traverse that morning he dropped it into the tool carrier – which was now full of rocks and tenacious lunar dust. Bean rummaged in the bag for a moment, looking for the glint of chrome, but saw only grime. The only solution was to take all of the rocks out of the tool carrier. They could not talk about it; the whole world would know what they were up to. So they made hand signals. While Conrad held the tool carrier, Bean rummaged among the samples."

[53] Post-mission analysis found that in the 31 months since Surveyor 3's arrival, no meteoritic craters larger than 1.5 mm had been created in the areas disturbed by the scoop. Nor were there any other signs of weathering in that time.

"Forget it," Bean said in frustration on drawing a blank in the search for the timer.

Accepting defeat, Conrad said, "I'll tell you what. Why don't you mosey down there and start taking some photographs?" He meant that they should resume work, which at this point was for him to run down his checklist calling out specific parts of the lander that Bean was to photograph in detail.

When Bean again remarked on the tan colouring of the lander, Gibson wondered, tongue in cheek, "Do you think there's a chance you're at the wrong Surveyor?"

"No, sir!" Conrad insisted.

Bean momentarily raised his gold-coated visor to verify that the white paint really was tan, which it was.

As there was obviously no risk of the craft sliding, they decided to ignore the rule that they remain upslope of it. Knowing that the camera on the lander had been unable to view the rear foot pad, Conrad suggested that as a bonus for the scientists they should document how it had dug itself into the dirt, which Bean duly did. This completed the set of three.

Having circled the lander, Bean moved in to photograph the television camera in detail and reported, "That TV mirror is brown!"

Conrad suggested, "It's brown because it's *looking at* brown, isn't it?"

"It's no longer a mirror," Bean insisted.

"You stay right there and I'll come in and wipe it," Conrad said. He ran a gloved finger across a section of the mirror, removing a fine layer of dust and exposing the reflective glass. "It's just got a fine dust on it."

Noting that they had plenty of film, Conrad suggested that they take tourist shots of each other by the lander. He moved in close and had his picture taken, then they exchanged the camera and reversed roles.

With the camera back on his chest, Bean reached into the bag that was strapped to Conrad's PLSS and extracted the cutters and a large vacuum canister in which some parts of the lander were to be stowed for return to Earth.

As Conrad told Eric Jones in 1991, "We worked with an exact mock-up (exact, at least, in dimension) maybe three or four times in training. We would practice with it and even cut the appropriate tubing; then they would replace the tubing and we'd do it over again." During this training, the astronauts also learned what *not* to cut on the lander, because although the high-pressure helium tank had been vented there would still be hypergolic propellants onboard that were individually toxic and would ignite if allowed to come into contact.

The 17-pound television camera had transmitted over 6,000 pictures of features in the crater in which the lander resided. On examining the electrical wiring which he was to sever Conrad mused, "I think that cable's arranged a little bit differently than the one we trained on." The fact that the cable was routed behind some tubing would complicate the retrieval of the camera. As Conrad used the cutters to obtain a length of cable, Bean stood by to insert the samples into to the can. "Ah! Here's something else they didn't tell us about either," Conrad said upon finding more wires than in training. He cut a pair of painted tubes that were supporting the camera, which was easier than in training, and a short section of cable whose insulation disintegrated as

A 'tourist shot' of Pete Conrad standing alongside Surveyor 3, with his right hand on the television camera.

A 'tourist shot' of Al Bean standing alongside Surveyor 3.

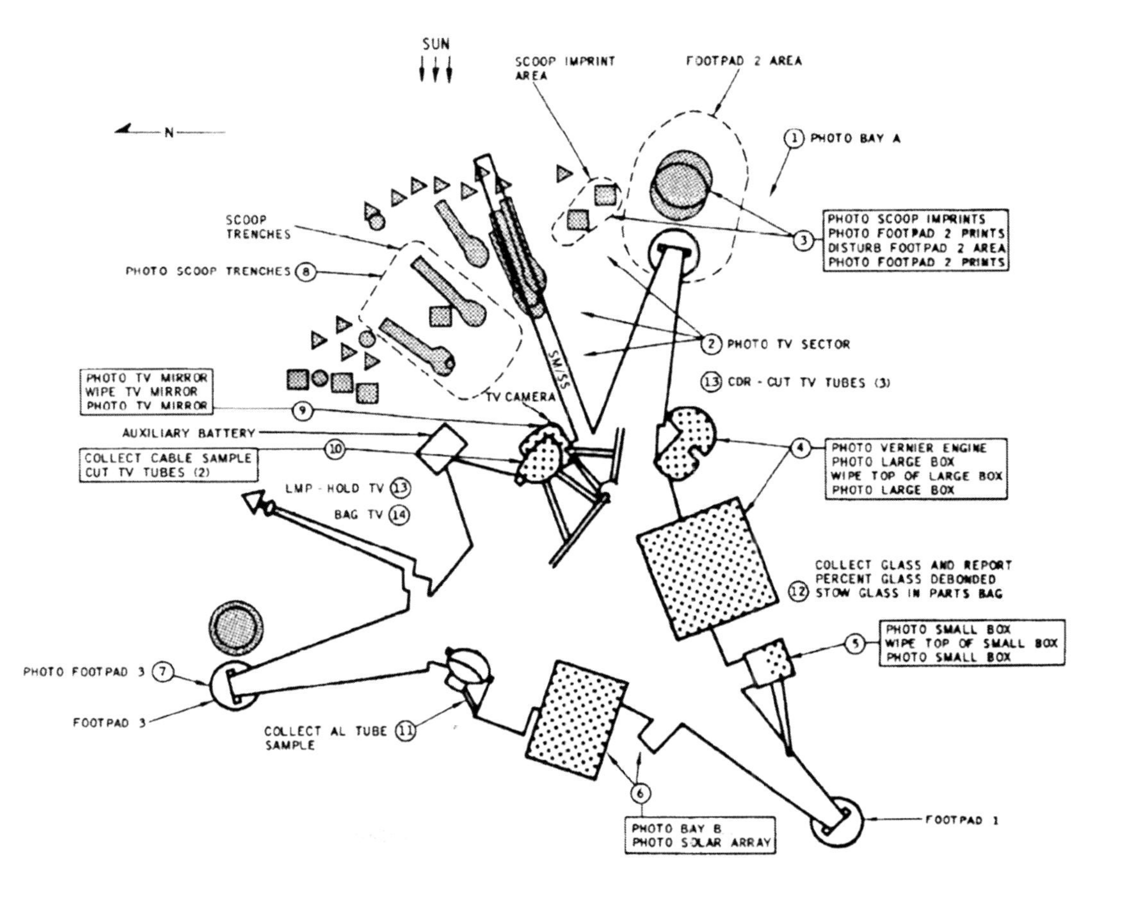

The formal objectives for the visit to Surveyor 3.

he applied the cutters. "Okay, Houston, we've got the cable and the tubing for the sterile sample," Bean reported, and returned the canister to the bag.

The next task was to get an unpainted tube, but this was much tougher than that of the mockup. As Conrad said in the post-mission debriefing, having decided that the wall of the tube must have been thicker than expected, "I worked on it as hard as I could, but hardly made a dent on it."

"We're going to have to pick another tube," Bean said, and pointed at one of the supports for an electronics compartment.

"That's even thicker!" Conrad complained. "Forget the tube."

In the expectation that such a prolonged exposure to the harsh environment would have cracked the glass mirrors of the radiators on the electronics compartments, they were to have gingerly collected a sample of the glass. Although the segments were a little warped, the glass was intact. Bean suggested that Conrad tap the cutters against the glass to break it. Conrad obliged, but without effect. Bean laughed and reported, "That's pretty good glass, Houston." Conrad successfully curled the corner of one of the segments and fractured the glass, but the pieces remained bonded to the segment of metal sheeting. As he explained in the debriefing, "All we got were fine slivers of glass, so we stopped messing around with it." But then, spying an unpainted tube, he gave it a try, and easily snipped it off.

"Okay, let's get that camera," Conrad said. Moving into position, he snipped one of the three remaining support tubes.

"Attaboy. Good cut," Bean praised.

"Two more tubes on that TV camera and that baby's ours," Conrad promised. He snipped another one.

"Let me get a grip on it," Bean said.

With Bean holding the camera, Conrad snipped the final supporting tube and then announced, "It's ours!"

"We got her!" Bean laughed.

"Beautiful!" Conrad agreed.

After wrapping the remaining length of cable around the cylindrical body of the camera, Bean stuffed it into the bag on Conrad's back. The engineers who had built the camera were looking forward to finding out how well its various mechanical and electrical components had survived the environment.[54]

"Let's get the scoop while we're at it!" Conrad suggested.

"Okay, grab the scoop," Bean agreed.

The scoop was not a formal item on the checklist, but they had time available and in training had been shown how best to snip the scoop off the robotic arm. Bean held the scoop as Conrad wielded the cutters. Although the steel tape was thin enough to flex in the shears and could not be cut, a twisting action caused it to debond. "One scoop!" announced Conrad delightedly.

"Outstanding! It's even got dirt in it!" Bean reported.

[54] The television camera retrieved from Surveyor 3 is now at the National Air and Space Museum in Washington D.C.

The surplus cutters were discarded.

"Well done troops," Gibson congratulated. "You're 3 hours in, and something like 10 minutes behind the nominal traverse for a 4-hour EVA."

At this point Bean exclaimed, "Hey! Look there, Pete!"

"What?"

"We thought this thing had changed colour, but I think it's just dust," Bean said. For Mission Control, he explained, "We rubbed that battery, and it's good and shiny again." He took a picture which recorded the mark. "I think maybe this thing is just collecting all this red dust."

On discovering that the Surveyor was discoloured, they had reasoned that the dust displaced by Intrepid's engine plume was unlikely to have fallen into the crater. The fact that there was dust on all sides of the inert lander also argued against its having been contaminated by their own arrival. However, whilst there was dust on all sides of the craft, it was not uniform around each specific item. Generally, it was thickest on the areas that were most easily viewed when walking around the perimeter. For example, the side of a tube or strut that faced the interior of the craft was relatively clean in comparison to a side facing out. The key to understanding how this coating of dust was acquired is the anomalous nature of its touchdown. When the television pictures proved to be impaired by a veiling glare, it was realised that either engine efflux or dust stirred up by the engines during the multiple 'hot' landings had coated part of the camera's mirror. Later, intermittent stickiness of the mirror in its azimuth and elevation motions indicated dust had penetrated into the mechanism. It can therefore reasonably be concluded that the lander coated itself in dust!

Gibson reminded them to collect some of the rocks which had been in view to the television camera. Conrad used his scoop to lift several,[55] and then declared, "That's about enough rocks."

"They're all big rocks, Houston," Bean explained, "at least 6 inches in diameter." He had dumped them straight into the carrier. "I think they are some of the ones you wanted. It is kind of hard to tell without having a photograph on hand or something, and standing there and studying it for a lot longer than I think we care to do it."

"Roger," Gibson acknowledged.

As they set off for Block, up on the rim of the main crater, Conrad raced ahead as usual. Bean was weary of the hand-tool carrier, whose load of samples made it bang against his legs. His heart rate increased to 140 beats per minute during the climb. As he observed to Eric Jones in a 1991 interview, "I felt like I was really running out of gas." In contrast, the bag on Conrad's back with 20 pounds of Surveyor parts was no burden. Their recommendation was for crews to sling tools and large sample bags on their backpacks.

Breathing heavily, Bean pointed out, "I thought it'd be tough down in the crater, with losing your balance, but it doesn't seem to be; it's just harder walking, that's all." In the post-mission debriefing he said of leaving the crater, "It took a lot more

[55] Samples 12056, 12062, 12063 and 12064, all unbagged and undocumented.

A view of the Block crater provided in 1967 by Surveyor 3's television camera. (Courtesy Philip J. Stooke)

A similar view as the previous image, but with Surveyor 3 in the foreground.

An excellent panorama of Surveyor crater taken from the blocky crater in its northeastern rim. Surveyor 3 is on the slope (left) and Intrepid on the rim.

work, because you could not bounce from side to side and spring off your feet in the way that you could on level ground. It just wore you out a little bit more."

Relaying a request from the Flight Surgeon, Gibson called, "Pete and Al, could we have EMU check?"

"We're right at the top of the rim," Bean reported, meaning that that they were in a good place to take a break.

"I'll tell you what we're going to do, Houston," Conrad said. "We're going to get an EMU check here. Then we're going to pick one sample out of this blocky crater and give you a partial pan, because it's a pretty fantastically interesting crater with a lot of bedrock. Big chunky rocks blown up out of it."

"Very angular. Very sharp," Bean added, referring to the large rocks that littered the 30-foot-diameter crater. On inspecting one more closely, he said, "We ought to pound one of those things with the hammer." But first, because he was wearing the camera, he shot a pair of panoramas for stereoscopic analysis. "This is probably the most spectacular crater we've come to, I think." The rocks placed much of the crater into shadow. "These blocks are a lot more sharp-cornered than anything we've seen anywhere else. I guess this must be the most recent one we've been around."

Their field geology tutors had urged the astronauts to try to read the landscape in terms of a sequence of events. Until now that had been impracticable, but the reason for such a small crater being so blocky was obvious – it had re-excavated rocks from the rim of the much larger crater.

"Let's get a sample of rock and get out of here," Conrad urged.

"I think it's going to be the same," Bean mused, meaning that all the rocks in the Snowman appeared to be a single type of material. Nevertheless, he suggested that after taking the prior-to-sampling documentation they should grab "a couple of the big pieces". Dispensing with the gnomon, Conrad used his tongs to lift several small rocks.[56]

"There's some of that light-colored undersoil," Bean observed. "Do you want me to get another sample bag?"

"Nope," Conrad replied. "I want to start moving out."

"I'll just pick up this one big rock here, Pete, and stick it in the bag," Bean said, hastily tossing another rock into the hand-tool carrier.[57]

Conrad set off running around the northern rim of Surveyor crater. Trailing him, Bean predicted, "I'll just bet you that everything we got here is really black basalt."

After less than 2 minutes, Conrad reported to Mission Control that he was back at Intrepid. Two hours 45 minutes had elapsed between his setting off for Head crater and his return to Intrepid. As he said in the post-mission debriefing, "We covered a lot of distance. We were told that we went more than a mile." The most remote point was the eastern rim of Sharp crater, 1,450 feet from the lunar module. Their average heart rates were 108 beats per minute for Conrad and 122 beats per minute for Bean. Their workloads were 875 BTUs per hour (as against a predicted

[56] Samples 12045, 12046 and 12047 in bag 15D.

[57] This might be sample 12065, unbagged and undocumented.

As Al Bean passed around the northern rim of Surveyor crater, he paused to take this picture showing Pete Conrad already alongside Intrepid. The bright object on the right is the solar wind collection sheet reflecting sunlight.

1,210) for Conrad and 1,000 BTUs per hour (as against 1,134) for Bean. Hence, despite spending a lot of time running, they expended 10 per cent less energy than predicted on the basis of training, which was excellent data for planning the next mission. Interestingly, their metabolic rates were higher while deploying the ALSEP than during the geological traverse.

Bean dumped the hand-tool carrier behind Intrepid and, pausing to rest, began to describe how the engine plume had disturbed the surface. However, Conrad, having prepared the sample return container, said, "Al, get those rocks over here. Come on. We can't baloney all day." When Bean joined his commander at the MESA, Conrad said, "I want you to put the tool carrier right here." This done, Conrad, very much in charge, told Bean to place the Hasselblad in the equipment transfer bag and retrieve the second film magazine. "I'm going to start packing up the gear." Once Bean had done as instructed and also removed his saddlebag, Conrad said, "I'll tell you what, go get me the solar wind." This was to go into the box. But a moment later Conrad said, "Wait, take the Surveyor TV camera off." Once Bean had lifted the bag off his partner's backpack, he put it on the foot pad. On noticing that Conrad's boots were becoming entangled in the discarded television cable, Bean made an effort to push it under the vehicle, but the cable snagged on the lunar equipment conveyor and he had to spend some time separating the two.

Leaving Conrad to resume packing rocks, Bean went to retrieve the solar wind collector. Unfortunately, as he explained in the post-mission debriefing, "That didn't work like I was hoping it would. When I reeled the collector out (at deployment) it came out very nicely, and the foil itself seemed pretty flexible. At the beginning of the second day, after we woke up, it looked like the foil had taken a set around the pole. [Based on its behavior when I tried to roll it up, I'd say] it definitely had taken a set of some sort. When I tried to roll it up following the second EVA, it rolled up about 6 inches and then didn't want to roll any more. It wanted to crinkle and tear, although I was very careful with it and tried to recycle it several times like a roller shade. Finally, it did tear about a 6-inch longitudinal rip, and I realised then that it just wasn't flexible enough and didn't want to roll up. So I let it go, and let it sort of window-shade all the way around and then tried to roll it up by hand and not get my fingers on any of the foil. I'm sure I wasn't able to do this entirely. I expect there is some dust from my gloves on the foil, but I did the best I could. I understand that it's possible to take that dust off and it doesn't bother the experiment. As a result of this, the rolled up experiment was larger than the bag it was supposed to fit into, so I had to squeeze down on the foil. It made it look sort of full, but I don't think it degraded the experiment at all." The collector had been exposed to the solar wind for 18 hours 42 minutes, as against 77 minutes for Apollo 11.[58] Between them, Bean and Conrad managed to stuff the crumpled sheet into a teflon bag for return to Earth.

"Give me a hand getting this rock box closed," Conrad said.

[58] The solar wind collector was deployed at 7:35 am EST on 19 November and exposed for a total of 18.7 hours. The ALSEP data showed that during this interval the Moon was passing through the bow shock of the Earth's magnetosphere.

"That's a nice full box," Bean pointed out. It contained 21 individual samples, a single-length core, a double-length core, two vacuum-sealed samples and the solar wind collector.

At this point, Conrad drew Bean's attention to an item he had retrieved from the bottom of the bag on the hand-tool carrier: the delayed-action timer.

"Just what we need," Bean observed wryly.

In disgust, Conrad threw the timer away.

Only after the mission did it occur to them that they really ought to have installed the surviving camera on the hand-tool carrier and used the timer to snap a picture of themselves alongside Intrepid. As Bean lamented to Eric Jones in a 1991 interview, "We could've shaken hands in front of the LM. It would have been a great picture. Maybe better than the Surveyor."

In loading the loose samples into the box, the seal of the lid became contaminated with a considerable amount of dust. Although they used a brush with steel bristles to try to clean it, the fine particles prevented the box from maintaining a vacuum on the trip back to Earth.[59]

Several weeks prior to the mission, it had been pointed out to Jim McDivitt, the Apollo Spacecraft Manager in Houston, that it was likely that more rocks would be collected on the geological traverse than could be accommodated by a sample return container. It had therefore been decided the largest rocks would be left in the bag of the hand-tool carrier and this would be taken aboard and lashed to the cabin floor for the rendezvous with Yankee Clipper.

Conrad hung the bag on the spring-scale that he had installed on the MESA and announced, "I've got about 1 inch worth of rocks in the bag."

Converting this displacement on the scale to a terrestrial weight, Gibson replied, "That's 10 to 15 pounds. No problem, pack it up."

The Apollo Lunar Surface Closeup Camera was designed to take high-resolution stereoscopic photographs of the lunar surface to provide fine-scale information on the soil and rock textures that could *not* survive sampling. Its inventor, Tommy Gold of Cornell University, was particularly interested in the pore spaces of the powdery surface material. The camera had a long handle to enable an astronaut to operate the shutter standing up. It was fixed-focus, with a stand-off hood to position itself at the proper range and an electronic flash for illumination. It was preloaded with a roll of 35-mm colour film with a capacity of 100 stereo pairs. It photographed a 3 × 3-inch area in 24-mm-square pairs and could record detail as fine as 40 microns. When Neil Armstrong used it on Apollo 11 he reported its handle to be awkwardly angled, but persisted and obtained 17 stereoscopic pairs. The handle had been modified, and for this mission the photography had been assigned to Bean.

When Gibson recommended that they start the ingress procedures, Conrad said, "I would like to offer my congratulations to all the people who are involved with this EVA."

[59] When weighed upon arrival at the Lunar Receiving Laboratory the sample return container from the second moonwalk weighed 52 pounds.

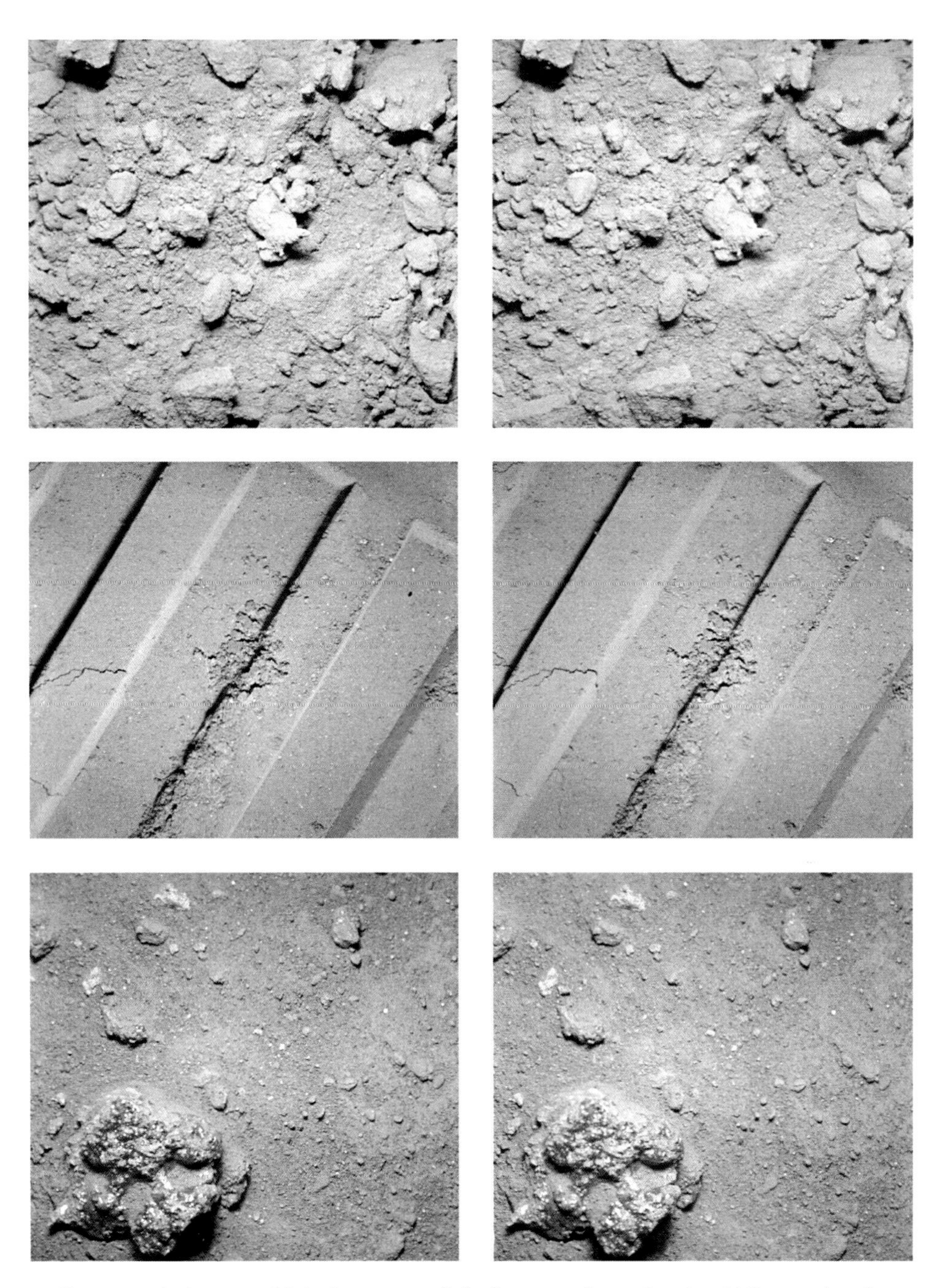

Stereoscopic images of 3-inch squares of the lunar surface taken by Al Bean using the close-up camera devised by Tommy Gold in order to investigate the fine structure of the undisturbed regolith, although in one case Bean used it to photograph the imprint of his boot.

"I think you two folks did an excellent job," Gibson replied.

"I'll keep taking these pictures until you give me a call, Pete," Bean suggested.

"Why don't you start working your way over here," Conrad replied. "We have got an awful lot of gear, and we'd better start getting it up."

"Will do," Bean acknowledged. He took several more pictures, retrieved the film magazine and ditched the camera. In the post-mission debriefing, he said of this, "As a result of having to hustle, I don't think I really put in a good enough effort on this camera. I did as much as the time allowed, but that wouldn't allow me to go out onto the three different types of soils that we'd seen and take pictures. I couldn't go down in some of the craters that had some good glass in them, or any of the glass-topped rocks." He stowed the film in the equipment transfer bag. One of the 15 stereo pairs that he obtained showed how the fine-grained surface material could bear the imprint of a boot.

"Okay, I'll get in," Bean announced, and headed up the ladder into Intrepid. After using the conveyor to haul the miscellaneous items aboard, he tossed the line out through the hatch.

"Have I forgotten anything?" Conrad asked, and proceeded to recap: "The closeup stereo film was in the ETB, two black-and-white magazines and one Hasselblad. I threw the other one away because it was broken. And the television camera went up. So I believe we've got everything."

"Also the camera which had the third film pack on?" Gibson prompted.

"The third film pack never got used," Conrad replied.

"It did too, Pete," Bean corrected.[60]

"Oh, I'm sorry. Okay, we have got three film packs and one camera up there right now. How's that?" Conrad said.

"Did you send the film up?" Bean asked.

"Yeah," Conrad replied.

"Okay, then we have all three film packs back up," Bean informed Gibson.

In fact, they had erred. In preparing for the geological traverse, Bean had put the colour magazine into his saddlebag and added the 30-foot tether. Upon their return to Intrepid, he dumped the saddlebag on the MESA. When Conrad was stowing the samples he ignored the saddlebag because he knew they had not placed any rocks in it. As a result, this magazine was left on the Moon.

Conrad retrieved the map from the hand-tool carrier as a souvenir, but it slipped from his fingers as he ascended the ladder.

By the time the hatch was closed and the cabin repressurised, the extravehicular activity had lasted 3 hours 49 minutes. Although they had been granted a 30-minute extension to the full 4 hours, the fact that they had started the moonwalk early meant they were nearly an hour ahead of the flight plan upon ingressing.

[60] Bean was under the impression that he had used this third colour magazine to photograph Yankee Clipper after undocking and the subsequent Earthrise, but these pictures were actually at the start of one of the magazines employed during the first moonwalk.

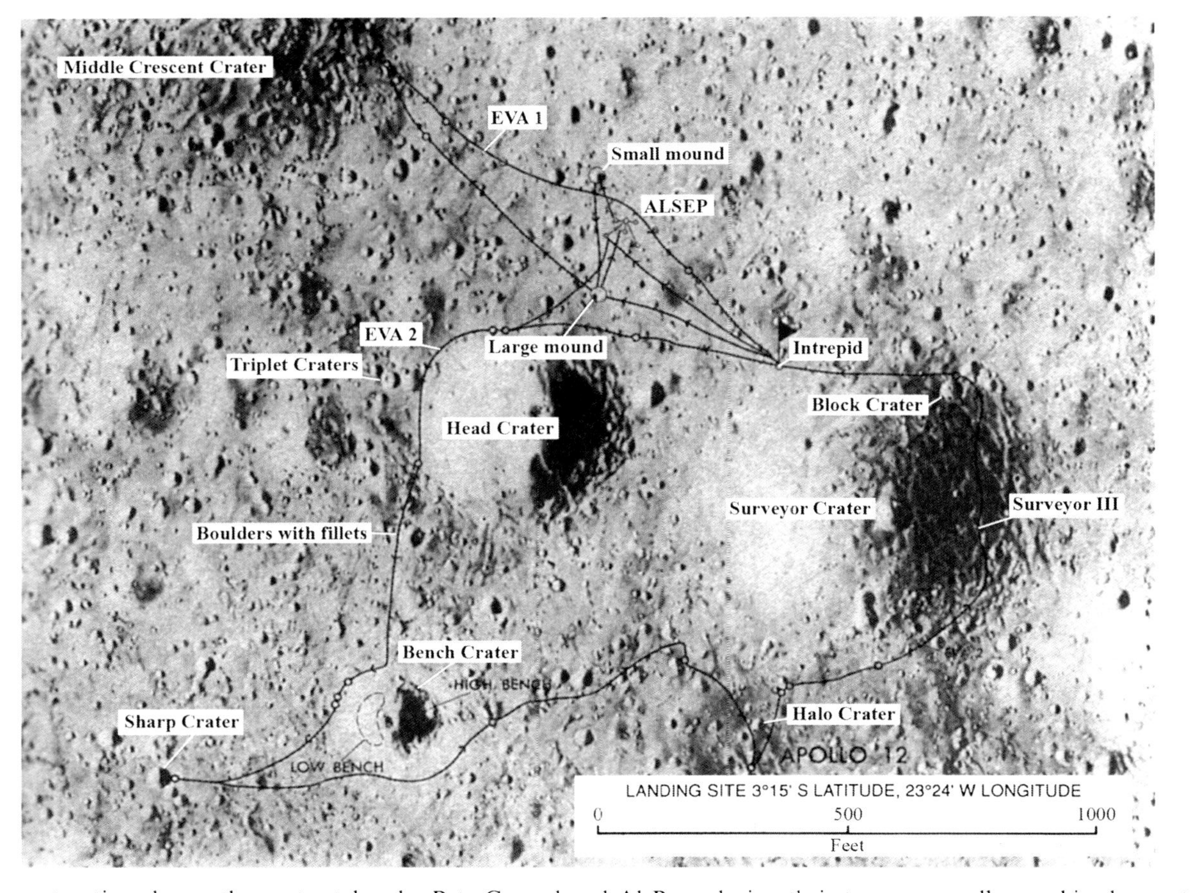

As this reconstruction shows, the routes taken by Pete Conrad and Al Bean during their two moonwalks combined aspects of the activities planned for landings at Site 3 and Site 4.

A view of the Snowman by Lunar Reconnaissance Orbiter on 29 November 2009 showing Surveyor 3, the descent stage of Intrepid and the tracks of Pete Conrad and Al Bean.

IDLE TIME

Upon acquiring Yankee Clipper as it flew around the limb on revolution 27, Gibson called to inform Gordon, "Your two cohorts are back in. They just finished another 4-hour EVA, accomplished all the objectives and did an excellent job."

"Okay, thank-you very much," Gordon replied, busy with the lunar multispectral photography experiment (S-158), the goal of which was to photograph the surface from a vertical perspective at four wavelengths in the green, blue, red and infrared parts of the spectrum. Four 80-mm Hasselblad cameras, each with black-and-white film and an appropriate filter, were mounted on a ring that was affixed to the hatch window and a timer operated the shutters in synchrony at 20-second intervals. The photography was to be obtained in continuous swathes during the daylight passes of revolutions 27 and 28.[61]

As Gordon said in the post-mission debriefing, "The updates were well planned. It was all conducted on GET, so there was never any doubt as to what exposures, what times the camera should be operated, and so forth. Once I'd manoeuvred to the right attitude so that the hatch window was pointing at the nadir, the orbit-rate torquing allowed me complete freedom and did not require any attention to the spacecraft as far as flying was concerned. I virtually forgot about the spacecraft. It never got more than a half a degree off the proper attitude. It was an excellent control system to allow someone in there by himself to devote his entire attention to other things. I don't think this experiment could have been done without it." As regards the camera apparatus, this "was easy to handle, easy to install, easy to remove from the window when a change of f-stop was required, and easy to put back in place".

The first pass for this experiment was accomplished with the same camera setting, but in two parts, the first from a local Sun elevation of 10 degrees to 60 degrees and the second from 60 degrees down to 10 degrees. The second pass covered the track near the subsolar point using a different setting. The photography was completed in accordance with the plan, and Gordon doubted that the state of the window would have degraded the pictures. Continuous vertical strip coverage was obtained along the ground track from 118°E to 14°W, spanning both highlands and maria. A total of 141 frames were obtained in each of the red, green and blue, and 105 in the infrared. Afterwards, the equipment was used to obtain two photographs each of several targets of opportunity, including Descartes, Fra Mauro and the northern wall of Theophilus. However, an inadvertent misalignment of the spacecraft meant that in these cases the designated points were 10 to 15 degrees off the optical axis.

The experiment was devised by Alexander F.H. Goetz of Bellcomm Incorporated, T.B. McCord of the Massachusetts Institute of Technology, Edward Yost of Long Island University and F.C. Billingsley of the Jet Propulsion Laboratory, and the goal was to investigate subtle variations in the colour of the lunar surface which could not be recorded by standard colour film. As the variations seemed to be associated with

[61] A similar multispectral photography package was flown on Apollo 9 in Earth orbit as experiment S-065.

compositional differences, the idea was to enable the 'ground truth' of the samples retrieved from the surface at several points to be extrapolated to areas which either would not or could not be visited by Apollo missions. However, as Don Wilhelms wryly pointed out in '*To a Rocky Moon*', the results did not distinguish the contacts between the maria and the highlands!

As Gordon passed over the centre of the Moon's disk, Mission Control asked him to use a Hasselblad with an 80-mm lens to obtain pictures of the crater Alphonsus, some 110 nautical miles south of his track. By using the left-hand window he would not need to interrupt the attitude which the spacecraft was maintaining for the S-158 photography. While Apollo missions were orbiting the Moon, amateur astronomers maintained a watch on sites with a history of 'transient activity', and Alphonsus was high on their list. Its floor was generally flat but had irregular rilles and a number of small 'dark-halo' craters which some scientists thought might be of volcanic origin. Following reports by Dinsmore Alter in America in 1956 of a slight "veiling" of the floor of Alphonsus, Nikolai Kozyrev kept watch on the crater, and on 3 November 1958 obtained using the 48-inch reflector of the Crimean Observatory a spectrogram of a "glow" which obscured the central peak and which he interpreted as a release of gas. In March 1965 NASA sent Ranger 9 plunging into Alphonsus to seek evidence of volcanism, with inconclusive results. The request to Gordon was in response to a report of a glow, but he saw nothing remarkable. "There's a dark area in between the central peak and the west wall, but I can't tell what it is. It's a little bit darker than its surroundings. The crater is a ways away though, so I really can't see much down there."

The immediate task for Conrad and Bean was to jettison everything that was now surplus. They kept their helmets on, but doffed their gloves for improved dexterity. After switching their umbilicals to the cabin systems, they shed their chest units and backpacks, then checked the OPS, which they were to retain for an external transfer to the command module in the unlikely event of a docking failure.[62] After unloading and stowing the items from the equipment transfer bag, this was put aside with their filthy overshoes for jettisoning. Rather than return with unexposed film, they shot a bonus panorama through the windows, stashed the magazine and added the camera to the pile of trash. Before ditching their PLSS packs they poured out the remaining coolant water and weighed it.[63] The spring-scale and the bags of water were added to the trash.

"As soon as we get through our suit integrity check, we'll depressurise the cabin for the jettison," Conrad informed Mission Control, just over an hour after the hatch had been closed following ingress. "The equipment's jettisoned," he reported several minutes later.

"We copied two impacts on the seismometer during jettison", Gibson pointed out, undoubtedly referring to the PLSS units.

[62] If either of the OPS had indicated a problem, they would have retained the corresponding PLSS for the spacewalk contingency.

[63] On the scale (in lunar gravity), Conrad's PLSS had 0.32 kg of coolant water remaining and Bean's had 0.26 kg.

"As a matter of fact, to help you calibrate that thing, I got a clean kick on both of them. In other words, when they left the hatch they fell freely to the ground without touching the ladder or anything on the way." One by one, they had stood each PLSS unit in the open hatch, tipped it slightly and then kicked it out.

Gibson sought clarification, "So they pretty much went straight out on a ballistic arc and hit the ground. They didn't arc up at all?"

"That's right," Conrad confirmed.

Knowing that the seismometer was some 430 feet from Intrepid, the mass of the PLSS, and the geometry of the trajectory, the scientists would be able to estimate the force with which it struck the surface.

Conrad and Bean then set about stowing all the loose items, to prepare the vehicle for flight.[64]

Meanwhile, Yankee Clipper had passed around the far-side. When Gordon reappeared on revolution 28, Gibson told him, "Pete and Al are just finishing up the post-EVA, and it looks as though they're pretty far ahead. They'll have a little bit of time to sit back and relax."

"Those guys deserve a rest," Gordon replied. "They did a great job." A moment later, he added, "It sounds like they didn't have any problems down there at all."

"They ran into a few things," Gibson admitted. "But, as they did yesterday, they overcame them."

"I expect most of the problems are human," Gordon ventured.

"Maybe so, Dick," Gibson agreed, "but not many of them theirs."

At this point Conrad reported, "We have everything stowed and are ready to start the launch countdown at the proper time. If you'll give us about 15 or 20 minutes to chow down here, we'll come back and have a little chitty chat about the EVA."

"Sounds like a good plan," Gibson replied.

Clearly delighted with their success, Conrad reported, "Man, oh man, it is filthy in here. We must have 20 pounds of dust, dirt and all kinds of junk."

"That'll be interesting in zero g," Gibson warned sagely.

"Right at the moment Al and I look just like a couple of bituminous coal miners!" Conrad laughed. "But we're happy."

"So are a lot of people down here," Gibson noted.

"How's the SIDE doing?" Conrad asked. "Did that cold jabber-do get running or not?"

"Stand by on that, Pete," Gibson replied.

"Also, how are the package temperatures doing?" Conrad asked, concerned about the fact that it was impossible to keep the ALSEP as clean as the scientists expected.

Having consulted with the Experiments Officer, Gibson said, "The cold-cathode ion gauge came up and the temperatures are looking nominal."

"All the equipment is running, huh?"

[64] The television camera and scoop from Surveyor 3 and the contingency sample and the rocks in the bag were not sealed in vacuum for the trip home, and so were exposed to the cabin environment.

"That's affirmative," Gibson replied. "And I'll tell you, watching the plots down here, that seismometer is sure doing the job."[65]

"Great!"

After leaving Intrepid in peace while the crew ate lunch, Gibson asked a number of questions, including the risk posed by the cables of the television camera and the S-band antenna, the damage done to the hammer while driving in the double-length core, the tendency of the teflon bags to split, and the manner in which the handle of one of the Hasselblads came apart. Then he asked about the third magazine, "How much of it was used on the inside of Intrepid, and whereabouts on the traverse was it exchanged onto one of the cameras?"

"Well, I've got some bad news for you and some good news," Conrad replied. "In the first place, the third magazine was a colour one and all it had on it were some shots of Earthrise and some things like that, taken coming around on descent.[66] And, unfortunately, Al and I got our signals crossed and it's outside on the lunar surface right now. What we did, was take the black-and-white magazine off of Al's camera when it failed and put it on my camera and used it up so that we have two complete black-and-white magazines of the second EVA and two complete colour magazines of the first EVA."

"You did get the Surveyor, though?" Gibson asked, because on the plan they were to have documented the unmanned lander in colour.

"Oh, yeah," Conrad replied. "We have all the Surveyor pictures, but they are all black-and-white."

Gibson checked some points relating to jettisoning the trash and the asked, "How does the inside of the cabin look about now?"

"It's very neat and orderly except for the fact that it's very dirty," Conrad replied.

"Kind of an orderly coal mine," Gibson mused.

"That's about the size of it," Conrad agreed.

In the post-mission debriefing, Bean said, "When we got back to the LM we tried to dust each other off. Usually, it was just Pete trying to dust me off. I would get up on the ladder and he would try to dust me off with his hands, but we didn't have a lot of luck. We should have some sort of whisk broom on the MESA." Indeed, while Conrad was trying to clean Bean's legs the displaced dust merely made his own suit even dirtier!

A little while later, Gibson asked, "Say, did either one of you kneel down in order to get anything off the surface? Or did you use the newly-developed Bean technique of holding onto the Surveyor parts bag and lowering the commander to the surface?"

"We used all kinds of things like that," Conrad replied. "We'd stick the shovel in the ground and do a one-arm pushup, then just lean down and pick up a rock off the

[65] In addition to the astronauts' activities during the moonwalks and jettisoning the trash afterwards, the seismometer was able to detect them moving around inside Intrepid.

[66] As became evident after the mission, the pictures they believed to be on the lost magazine were at the start of a magazine used on the first moonwalk. The magazine that they inadvertently left on the Moon was almost certainly blank.

ground with the other hand. It's really a ridiculous way to do it. If you had a suit that would bend, you'd have the whole program wired. I fell over once out there, and Al picked me back up again. It's no big deal."[67]

Although the waist of the suit was inflexible, Bean had tried to kneel down. As he explained, "I knelt down and picked stuff up. It is particularly easy if you have that hand-tool carrier with you. But we really do need to come up with some sort of strap that'll allow you to lean over and grab a rock that is too large for the tongs." In the post-mission debriefing he said, "The suit was completely adequate for the mission objectives, but the efficiency of the overall lunar surface work could be enhanced by 20 or 30 per cent if it were possible to bend over and retrieve samples." The verdict from the first lunar geological field trip was that the tongs were about 5 inches too short, and the desirable rock samples were larger than the tongs could grasp and than the individual sample bags would accommodate.

As a result of being ahead of schedule at ingress, they found themselves cooped up in the lunar module with essentially nothing to do. As Conrad mused in the post-mission debriefing, "I know everybody, including ourselves, agreed to a criterion of one EVA extension, but it broke my heart when Mission Control hustled us back in after 4 hours and then we found out that the PLSS had a capability of 6 hours. We sat on our rear ends and did nothing for 2 hours! We weren't tired. It's a real shame we didn't get a second extension. We hustled past the blocky crater and back to the LM. I was hustling Al because I felt that he'd committed us to get in at 4 hours, and nobody changed their tune. So we were [hounded] up the ladder, and that's really a shame. I think we've got to be open-minded in future. If the guys are in good shape, then let's not hustle when we don't have to. As a matter of fact, we could have gone another (2-hour) revolution down there on the lunar surface before liftoff and it wouldn't have perturbed anything." Following on, Bean said, "I agree 100 per cent. That was a real shame. If there was anything in the whole flight that we should have done differently, we should've gotten in maybe an hour later. Like Pete said, we got in and we sat around for a couple of hours waiting for the time to start working on the ascent checklist." But this was still 'early days' for lunar excursions, and NASA was being cautious. Apollo 12 had provided a valuable 'data point' on the ability of a PLSS to sustain a highly active astronaut, and future explorers would benefit from this experience with longer excursions.

[67] It is unclear when Conrad fell.

4

The voyage home

RENDEZVOUS

With 2 hours 50 minutes remaining to the scheduled time of liftoff, Pete Conrad and Al Bean started the preparatory checklist. In Mission Control, Pete Frank took over as Flight Director and Gerry Carr as Capcom for the ascent and rendezvous.

As Conrad reported in the post-mission debriefing, "Al and I went through this by ourselves, absolutely per checklist. But I called the ground a couple of times to find out if they were still around."

The procedure began with powering up the prime and backup guidance systems, and aligning the inertial platform by performing a P57 in the same manner as during the rehearsal shortly after touchdown, using the lunar gravity vector and a sighting on one star, in this case Procyon.

When Yankee Clipper appeared on revolution 29, Carr made contact and Gordon said, "Well, hello there, stranger. How are you?"

"Morning, Dick," Carr replied. "We are fine. How are you?"

"Well, pretty good," Gordon said.

"Got the house clean?" Carr asked.

"As a matter of fact, I just finished," Gordon replied. "Keeping this thing clean is quite a chore."

"You've got a couple of coal miners coming up to see you," Carr warned.

"I'll be glad to see them," Gordon assured.

At a break in the checklist, Conrad enquired, "Has Yankee gone overhead yet?"

"Not yet, Pete," Carr replied.

"Give me an overhead time, so I can watch him go by." By looking up through his rendezvous window Conrad would be able to observe the overflight.

"Clipper should be overhead at 140:04:10," Carr advised.

"I'm going to check the plane change," Conrad explained. He was referring to the manoeuvre which was supposed to have driven Yankee Clipper's ground track back over the landing site.

Gordon had a landmark tracking exercise scheduled for his pass. As he explained in the post-mission debriefing, "One thing that I wanted to do, just to see if it could

D.M. Harland, *Apollo 12 – On the Ocean of Storms*, Springer Praxis Books,
DOI 10.1007/978-1-4419-7607-9_4, © Springer Science+Business Media, LLC 2011

be done, was to try and track the LM from liftoff to insertion. The pass before liftoff, we were to do simultaneous radar tracking, but it was decided we wouldn't do that. I did try to sight on the LM during that pass. It was left up to me whether I wanted to track landmark 193 or the LM. I made the decision to track the LM, since I knew that I could find it. If we had really needed that P22 information, it would have been a bad decision on my part. It turned out that I really didn't track the right landmark, and I'm certain I would have if it had been 193. The reason I didn't track the LM on this particular pass, I guess, was I got overconfident and forgot the procedures I had established before: namely find the target first through the telescope and make sure, with the wider field of view, that the sextant was on the right target. Well, I didn't do that this time. I had confidence in the state vector and the Auto optics and P22. I relied on that to initially point the optics at the LM. Instead of going to the telescope to confirm that it was on the target, I went straight to the sextant and, unfortunately, neither the LM nor the Snowman were in view. I played with the sextant in Manual for a while, but couldn't even find the Snowman. So, I went back to Auto optics and looked in the telescope, but by that time it was too late. [...] It was a bad decision. I should have tracked landmark 193, and I recommend that anyone requiring a P22 on the revolution before LM liftoff not to take a chance on finding that LM. Go back to a known landmark and do it the right way. I was far too confident at this time, and I shouldn't have done it."

After Yankee Clipper had passed over, Conrad prompted, "Are you ready for my RCS hot-fire?"

"Fire away," Carr invited.

"Here you go, Houston, with roll, pitch and yaw," Conrad called. Mission Control monitored the telemetry to verify that the system was functioning properly, and was alarmed to see the transmission cut short.

After communications had been re-established, Conrad explained, "We just blew over our S-band erectable!"

As Bean said in the post-mission debriefing, "When Pete fired the thrusters on the right side, it knocked over the S-band erectable antenna, so we switched over to the spacecraft's S-band, which didn't even lose lock. We got some good movies, I think. I took 16-mm movies out the window as he fired some of the thrusters. Hopefully, the geologists will be able to get some feel for the effect of those engines on the dust and maybe extrapolate to the descent engine."

When Conrad asked if Mission Control wanted him to repeat any of the thruster firings, Carr asked him to go back to the beginning. After doing so Conrad reported, "Pitch-up didn't sound right." He fired it again. "How'd the hot-fire look?"

"It went kind of fast. Give us a chance to take a look at our tapes," Carr replied. A moment later he reported, "The passive seismometer just verified that you did do the hot-fire."

Meanwhile, having crossed the terminator into the Moon's shadow, Gordon took star sightings to realign his inertial platform. Shortly thereafter he flew around the corner. If all went to plan, by his next pass over the landing site Intrepid would be in the final minutes of its countdown.

The ascent propulsion system was built by Bell Aerosystems using an injector

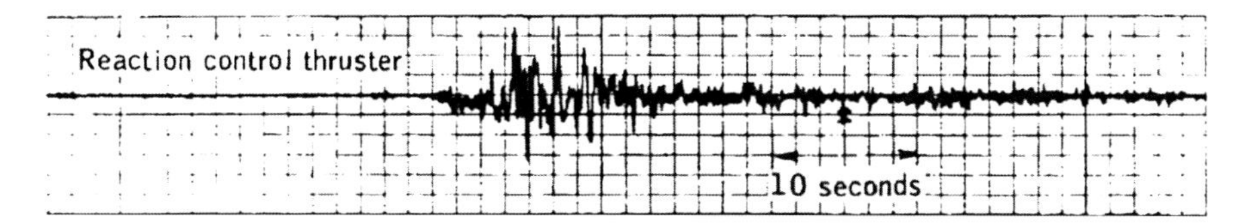

The seismometer detected the hot-fire test of Intrepid's thrusters.

plate supplied by the Rocketdyne division of North American Rockwell. In all, the engine stood 4 feet 6 inches tall. Its combustion chamber was actually located inside the cabin with a cylindrical cover that rose up from the floor, and its short nozzle sat over the central part of the descent stage. Its only moving parts were the ball valves to feed propellants to the injector. Each primary valve had a backup with a bypass line. It was required only that the valves open. There was no ignition system, since the hypergolic propellants would ignite on coming into contact in the chamber. If the computer command did not reach the valves, the circuit could be bypassed.

Liftoff was to occur with Yankee Clipper 80 nautical miles west of the landing site. It there was a slight delay, Intrepid would have to adopt a lower orbit in order to catch up with Yankee Clipper more rapidly. If liftoff were so delayed as to render this impracticable, then Intrepid was to climb to a high altitude and go passive, while Gordon lowered his spacecraft and flew an 'extra' orbit to get into position to mount a chase. There were many variations on this theme, depending on the circumstances, and Gordon had rehearsed them all in simulations.

In '*A Man on the Moon*', Andrew Chaikin relates that with an hour remaining to liftoff Conrad said, "Beano, are you worried about the engine?" Bean replied with a quiet "Yep" and Conrad said, "Well there's no sense worrying about it, Al, because if it doesn't work we're just going to become the first permanent monument of the space program." He wasn't sure whether his humour had improved Bean's outlook. Yankee Clipper could remain in lunar orbit for two more days before Gordon would have to head home. However, the fact that Intrepid's remaining power and oxygen would last no longer than 24 hours meant that by then his colleagues stranded on the surface would be dead.

Meanwhile on Earth the back row of the Mission Operations Control Room was starting to fill up with senior managers, including George Low, Rocco Petrone, Jim McDivitt and Chris Kraft. Clustered around Carr at the communications console were Deke Slayton, Al Worden, Jim Lovell and Tom Stafford.

When Yankee Clipper appeared on revolution 30, Carr asked Gordon about the scheduled landmark tracking on Intrepid. Smarting from his failure on the previous revolution, Gordon said, "Can we skip it? The reason I say that, Gerry, is that since I can't see them at these high Sun angles it's pretty academic to do this one."

Carr agreed, then, moving on, said, "Here's the communications plan. About the time when you get VHF communications established with Intrepid we're going to end the MSFN relay. If for some reason you lose the VHF and you want to hear the liftoff, let us know and we can reconfigure in about 20 seconds, but we'd prefer to leave the relay out as long as you've got VHF."

"I think that'll be fine," Gordon agreed.

In Intrepid, Conrad and Bean were repeating the P57 in order to measure the drift rates of their inertial platform, which were well within the acceptable limits.

In the post-mission debriefing Conrad recalled that as they donned their helmets and gloves, "it was all we could do to get our wrist locks to work; they were clearly beginning to clog with dust."

After Pete Frank had polled his flight controllers for a Go/No-Go for liftoff, Carr relayed the result, "Intrepid, you're Go to cast off on this revolution."

"Roger-roger," Conrad acknowledged. This marked the T-30 minute point, but in accordance with his 'get ahead and stay ahead' mantra they had been holding at this point in their checklist for some time.

At the same time as the propellant tanks of the ascent stage were pressurised using helium stored at ambient temperature, valves were opened to vent the residual supercritical helium in the descent stage.

Intrepid had been living off the four batteries in its descent stage, but in serving as a launching pad this would be discarded and so the two batteries in the ascent stage were now brought on line.

With the clock counting down the final few minutes, Conrad called up P12 on the computer, which was the guidance program for the ascent. Bean started the 16-mm movie camera that was viewing through his window in order to document the ascent to orbit.

With 2 minutes remaining, Gordon reported to Mission Control, "I have the LM." He had changed his mind about attempting the landmark tracking exercise, had his computer aim the optics at the landing site, and then used the telescope to refine the alignment before switching over to the sextant. He was hoping to be able to observe the liftoff.

"Intrepid," Carr called, "Clipper's watching you."

"Very good," Conrad acknowledged. Thirty seconds later he announced, "On my mark, Yankee Clipper, it will be 1 minute." As he called out the "Mark!" the Master Alarm sounded, but seeing there was no light to indicate a specific fault he simply pushed the button to cancel the alarm, which was later attributed to a momentary out-of-tolerance condition.

Pushing on, Conrad told Bean, "You watch the ALSEP and I'll fly the bird." By estimating the azimuth of their ground track relative to the vector to where they had deployed the package, Bean would be able to make a crude check on the accuracy of the guidance system. The observation would also back up the movie camera in determining whether dust and debris interfered with the scientific instruments.

In the final seconds the Abort Stage button was pushed to mechanically sever the linkage between the two stages, the ascent engine was armed, the computer flashed Verb 99 and Conrad pushed the Proceed button to authorise P12 to ignite the engine as planned. At the scheduled time, he backed up the computer by manually pushing the Engine Start button.

"Liftoff," Conrad reported, as the engine ignited to conclude a lunar surface stay of 31 hours 31 minutes 11.6 seconds.[1] "And away we go."

"Boy, did it fire!" Bean added, impressed by the rate of climb. The engine had a fixed-thrust of only 3,500 pounds, but the ascent stage was light and there was about 0.5 g of acceleration. The exhaust plume shredded the foil covering of the descent stage, sending fragments radially out on shallow trajectories. In addition to concern that the thermal characteristics of the scientific instruments could be impaired if they were to become coated by dust, there was some concern that the boxy suprathermal ion detector might be knocked over by the blast effect.

"It didn't get the ALSEP," Bean said.

In fact, although the ALSEP had been deployed well away from Intrepid in the hope that the dust stirred up by their liftoff would not coat the instruments, the dust detector on the central station registered contamination at the moment of liftoff. The seismometer detected the departure, the suprathermal ion detector noted the engine exhaust, and the magnetometer observed field fluctuations for about 12 minutes.

The ascent involved a vertical climb for terrain clearance, then a 45-degree pitch manoeuvre onto the desired azimuth for the orbital insertion phase. This transition occurred 10 seconds into the burn, when the vertical rate reached 50 feet per second at an altitude of about 275 feet.

"Pitch over is looking good," Conrad reported.

"Nice and quiet, isn't it?" Bean said. In the post-mission debriefing he would say, "As you lift off, there is a large bang as you separate the ascent and descent stages, and then you just move rather rapidly as the ascent engine burns. It makes no noise."

At the 30-second mark they were climbing through 1,900 feet with a vertical rate of 985 feet per second and a horizontal rate of 177 feet per second. "That's pretty good," Bean noted, comparing the actual values against those on the cue card.

"We're on our way," Conrad agreed.

Since parts of the vehicle's structure could block the steerable high-gain antenna's view of Earth in a windows-down attitude, Conrad, as per the flight plan, rotated through 20 degrees on the thrust axis in order not to lose communications.

"It kind of wobbles around," Bean noted, several seconds later.

"Looking good at 1 minute," Gerry Carr advised.

"Right down the pike," Conrad replied, meaning they were on the desired profile. "What a nice ride!"

The engine was not gimballed, so the thrusters had to overcome the oscillations caused by the sloshing of the propellants, and there was a burst of thruster activity as Intrepid climbed through 9,000 feet. As the profile became ever flatter in order to increase the horizontal velocity, the vertical rate eased off. On Earth, the atmosphere made such an early transition to the orbital insertion phase impossible, but the Moon was airless.

"This thing is right on the pitch profile," Conrad confirmed at the 2-minute mark. By now the horizontal rate was 1,061 feet per second.

[1] It was 9:26 a.m. EST on 20 November.

After confirming that the pressures of the propellant tanks were satisfactory, Bean compared the primary and backup guidance systems and reported, "PGNS and AGS agree perfectly."

"Okay," Conrad acknowledged. After giving the 2-minute 30-second status report at 19,700 feet with a horizontal rate of 1,373 feet per second, he said, "Houston, you better clear me for flight level 600." A flight level expresses altitude in thousands of feet without the least significant two digits, so Conrad was referring to 60,000 feet, which was the nominal altitude for orbital insertion.

"Roger," Carr replied. "Squawk 21." In aviation, this specified the code signal for a transponder.

Conrad laughed, "Okay, squawking 21."

At the 3-minute mark they were at 25,000 feet with a horizontal rate of 1,752 feet per second.

Making his first call since the ascent began, Gordon asked Mission Control to request Intrepid to activate their VHF ship-to-ship communications system. When Carr relayed this, Conrad said they were already transmitting on VHF. Gordon could hear Carr talking to Intrepid, but not the response. Because he was supposed to be able to hear Intrepid on VHF, Houston was not repeating Intrepid's voice downlink to him by S-band. If Intrepid were to find itself in serious trouble, then this communications issue would make a rescue difficult.

Bean noticed that the 16-mm movie camera mounted in his window to document the ascent was no longer operating, so he restarted it. "I hope it got the ALSEP," he said, because the scientists wished to observe the extent to which the liftoff affected their instruments.

"You're looking good at 4 minutes," Carr advised.

Peering down from 37,000 feet, Conrad admired a rille crossing the landscape. On checking the movie camera, Bean saw that it had stopped again. "Forget it," Conrad said.

At the 5-minute mark they were climbing through 47,000 feet with a horizontal rate of 2,403 feet per second. "The harbour master has cleared you into the main channel," Carr said, referring to achieving orbital insertion. If the engine were to cut off early, they would be able to make up the velocity shortfall by a lengthy firing of all four downward-facing thrusters.

On the nominal flight plan the ascent engine was required only to achieve orbital insertion. The rendezvous manoeuvres would be made by the thrusters. Although the two systems burned the same propellants, they were isolated. But as soon as it was clear that they would attain the desired orbit, Bean was to open valves to enable the reaction control system to draw propellants from the ascent propulsion system. As the final minutes passed, he set the computer to continuously display their velocity relative to that for insertion. As the vehicle burned off mass, with the engine thrust fixed, its acceleration increased rapidly. "It's sure picking up fast," he noted. When the computer indicated a differential of 200 feet per second he operated the valves.

"System A didn't open," Conrad pointed out.

Bean recycled the switch. Telemetry indicated that the fault was not the valves but the 'talkback' indicator on the control panel.

Distracted by the issue of the valves, Conrad missed his cue and failed to prepare the engine for automatic shutdown when the velocity was 100 feet per second short of orbital insertion. When the computer commanded the engine to shut down having achieved the desired velocity, this had no effect. To add to the excitement the Master Alarm sounded, again without an indicator light. On realising his mistake, Conrad shut the engine down manually.

"Everything looks okay," Bean noted.

The target orbit was 9 × 45 nautical miles, with insertion 166 nautical miles from the liftoff point. However, because Conrad missed his cue the 434-second burn was 1.2 seconds longer than intended and produced a 32.5-foot-per-second excess which raised the apolune to 51.9 nautical miles. He promptly used the thrusters to trim the residuals and enter a more desirable orbit of 8.8 × 46.3 nautical miles from which to initiate the planned rendezvous profile.

"Does that look satisfactory to you?" Conrad asked.

"Looks good, Pete," Carr replied.

"I took it all out," Conrad confirmed, referring to the over-speed, and explained to Carr how he had been distracted by events. In the post-mission debriefing, he was apologetic, "I trapped myself by getting interested in Al's problem." Conrad had not appreciated the pace of events in the final minute of the burn, "It was not apparent to me in the simulator that the vehicle was accelerating as rapidly as it was." But it was not just a training issue. The procedure ought to be changed. "I feel that when you're within 400 feet per second [of orbital insertion], you're well within RCS capability, and inadvertent shutdown [of the ascent engine] is not going to do anything to you. You could relight the engine for that matter – I have a suspicion that the ground can check to see that all those relays are closed, and give you a Go; the ground can verify that you have normal arming through the normal system, which is something we never thought of." Bean recommended operating the valves at the 300-foot-per-second differential in order to provide a little more time in which to overcome a problem before dealing with the engine.

In fact, the burn was approximately 9.8 seconds shorter than predicted owing to a better than expected engine performance, an error in predicting the centre of mass, and a lower stage weight. The calculated nozzle erosion was 3 per cent greater than predicted but, as with the descent engine, this discrepancy reflected the difficulty of numerical modelling and the data served as a reality check.

As hoped, while Intrepid was on the surface the gravitational perturbations of the mascons had drawn Yankee Clipper into an almost circular orbit at an altitude of 60 nautical miles, ideal for the rendezvous. Having "screwed up" his landmark tracking of the lunar module on the previous revolution owing to overconfidence, Gordon did it correctly when flying overhead shortly prior to liftoff. As he explained in the post-mission debriefing, "I found the Snowman in the telescope and put the cross-hair on the Surveyor crater – actually on the LM – so that when I went to the sextant there it was." Once he had the target centred in the sextant, he switched the optics to manual control, placed the digital autopilot in Free and attempted to use the thrusters at their minimum impulse level to watch liftoff, but this proved difficult and he lost sight of it.

The ascent stage had a beacon on its front face to enable Gordon to track it while in darkness. This lamp was to flash for 20 milliseconds once per second, and was designed to be visible from 400 nautical miles by sextant and 130 nautical miles by naked eye. After the post-burn checklist was completed, the lamp was activated in readiness for the rendezvous.

"Gee, it's a dirty spacecraft in here," Bean observed.

The cohesive properties of lunar dust in a vacuum, augmented by its electrostatic properties, made it adhere to the suits, sample return containers and other items that had been brought in after the moonwalks. But the adherence of the dust diminished in the presence of the cabin atmosphere and a lot of it had settled on the floor. Now that they were weightless, some of the dust began to float and disperse. As there was enough material to make breathing potentially hazardous, they opted to remain fully suited.

When Gordon told Mission Control that he was still not hearing Intrepid on VHF, Carr relayed this to Conrad, who confirmed they were using the required antenna. It was not evident why Intrepid could hear Gordon on VHF but he could not hear them. As a temporary measure Mission Control provided Gordon with an S-band relay, but the ship-to-ship would be essential after they passed around the limb onto the far-side.

Throughout the ascent to orbit the AGS had remained in excellent agreement with the PGNS. In contrast to Apollo 11, where the AGS had operated passively during the rendezvous, available for use if the PGNS failed, this time the system was to be updated by manually inserting a succession of range and closing rates obtained from the radar. The goal was to simulate a worst-case scenario to assess both the ability of the AGS to compute the manoeuvres and the workload it imposed upon the operator. As Bean put it in the post-mission debriefing, "The plan was for us to align the AGS independently on the surface, keep it independent all the way through rendezvous, and then make a comparison and see how it did."

On crossing the terminator into darkness, Gordon realigned his inertial platform. Mission Control gave him a state vector for the lunar module so that he could adopt the attitude required for sextant tracking of the flashing beacon. Meanwhile, Conrad and Bean used their two-man procedure to take star sightings to update the platform of the PGNS, and when Bean checked the AGS he was pleased to find that it had not drifted significantly.

With no indication of a problem with Intrepid's VHF, Carr asked Gordon to turn down the volume on the S-band relay and check whether he could hear anything on the ship-to-ship; he could.

A moment later, having sighted the lunar module in his sextant, Gordon called by VHF, "Intrepid, Clipper. I have you in my big eye."

"Okay, that's good," Conrad replied.

Upon orbital insertion, Intrepid was trailing its target by 250 nautical miles. The rendezvous radar worked in conjunction with a transponder on Yankee Clipper and locked on immediately after activation, which was at a range of 235 nautical miles. When Conrad later advised Gordon that the radar measured the range at 200 nautical miles with a closing rate of 351 feet per second, Gordon replied that this agreed with his own VHF ranging system.

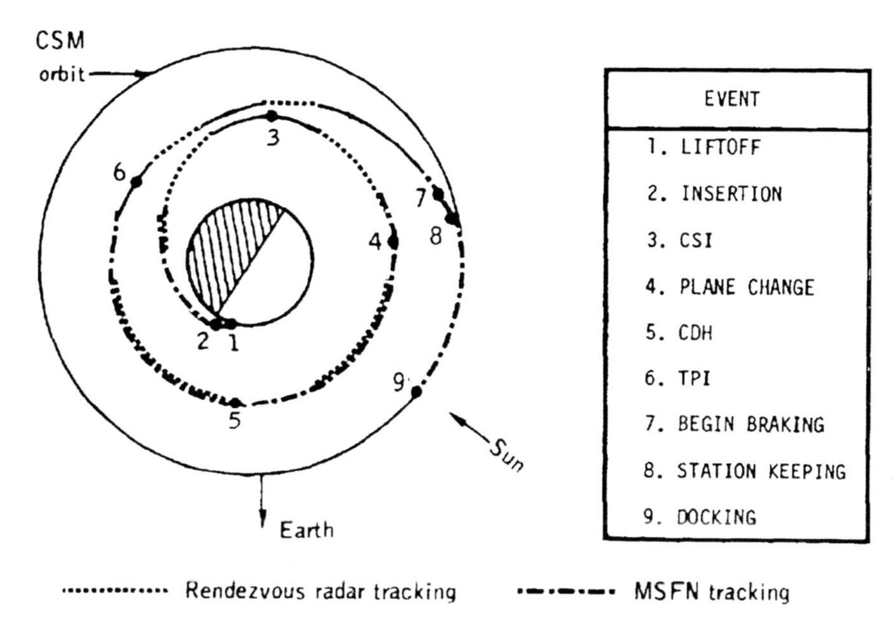

The sequence of manoeuvres to be performed by Intrepid during the rendezvous.

The rendezvous sequence called for coelliptic sequence initiation (CSI), constant differential height (CDH) and terminal phase initiation (TPI) manoeuvres timed to yield station-keeping shortly after sunrise on Intrepid's second revolution. The CSI burn was to be performed near the apolune of the elliptical insertion orbit, when the lunar module was 15 nautical miles below its target. The CDH burn half a revolution later would shape the orbit to precisely match that of the target. If desired, the CSI manoeuvre could initiate corrections for any out-of-plane dispersions resulting from the azimuth at lunar liftoff, and a plane change was available at the nodal crossing of the two orbits some 30 minutes later to ensure that the vehicles were coplanar at the time of the CDH burn. Being in a lower orbit, Intrepid would slowly catch up with its target in terms of central angle, and at the appropriate point would execute TPI in order to climb and intercept it. In effect, it was to be a rerun of the Apollo 11 rendezvous.

In the final seconds prior to loss of signal, Carr provided Mission Control's radio-tracking solution for the CSI burn as about 46.5 feet per second. Each vehicle was to track the other and compute its own solution, and Conrad would execute the one that seemed most appropriate.

In the post-mission debriefing, Gordon spoke of his experience, "My first solution for CSI was bad. I got 38.8 feet per second using 9 VHF marks and 14 optics marks, which I thought was plenty to get a fairly decent solution, but it didn't converge. I continued marking through that whole time period and ended up with 14 VHF marks and 21 optics marks for my final CSI solution, which converged and compared very favorably with the ground and the LM. I got 45.9 feet per second. The LM was 45.3 and the ground was 46.5, so we were all right there in the same ballpark."

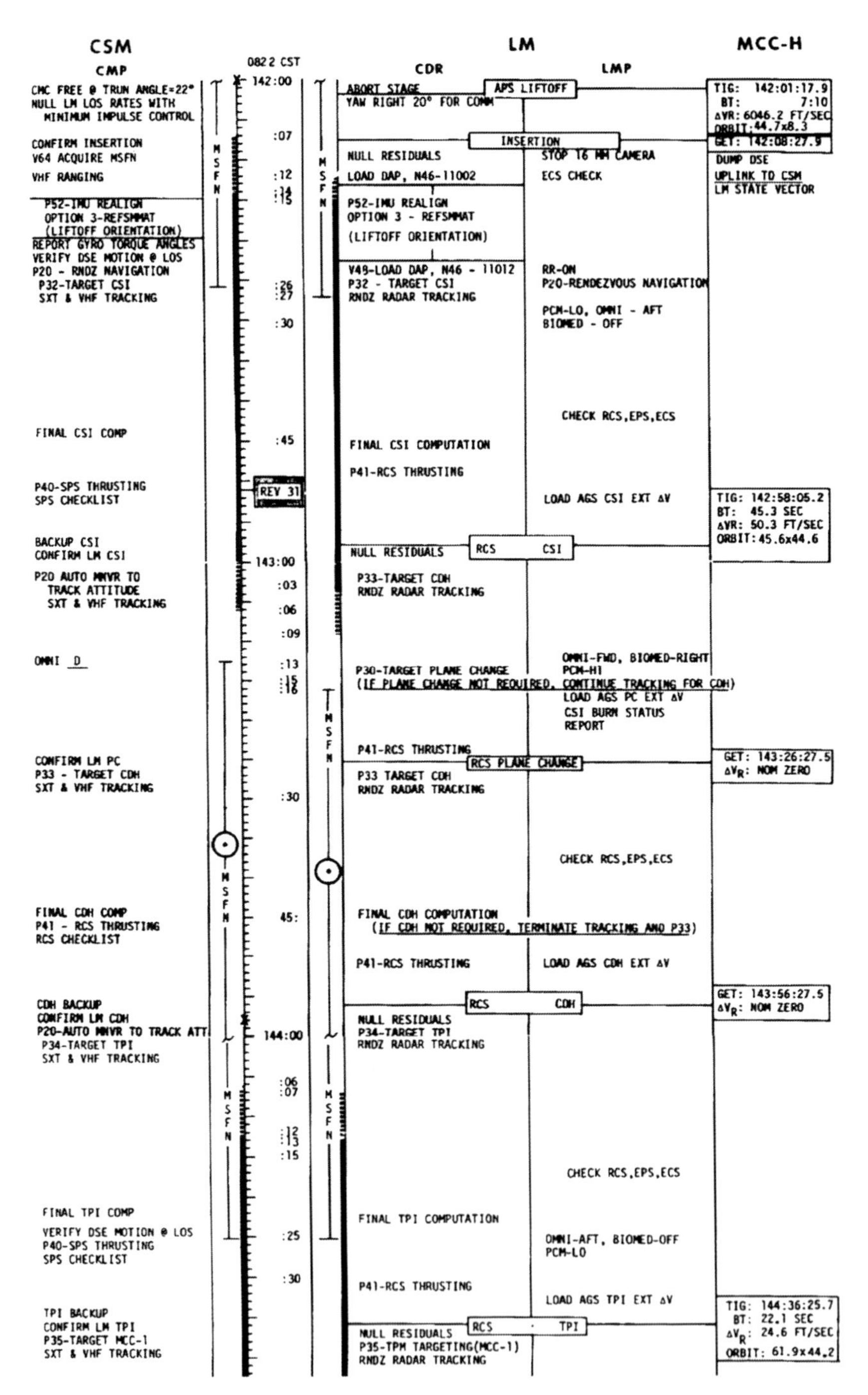

The flight plan detailing crew activities between ascent stage liftoff and the initiation of the terminal phase of the rendezvous.

On the far-side, taking their own tracking data, Conrad noted, "Something smells terrible in here."

"It's probably all this stuff flying around," Bean suggested, referring to the dust.

"But the suit loop's buttoned up," Conrad said. "It smells like ammonia."

"I don't smell it, and I've got a real good sniffer," Bean said. However, he had a head cold.

As Bean followed a routine of updating the range and range-rate, he said, "I hope I get a good AGS solution here." Several minutes later he noted, "Two more marks and I'll have a solution for you." However, in his haste he erred. "I did that wrong! I bombed that damn AGS, right there."

"What did you do?" Conrad asked.

"I entered a range instead of a range-rate."

"Well, put it right back in again," Conrad suggested. "It won't bomb it. Hurry. Put a range-rate in."

"It won't work," Bean replied. The false input would prevent the solution from converging. If they had been running on the AGS, he would now have had to resort to using paper charts to compute the CSI manoeuvre. Because it was essential they have a backup to the PGNS whilst conducting the experiment with the AGS, he had been writing the AGS data into the timeline book as a precaution in case he made a mistake such as this or the AGS computation using his data diverged. By plotting on a previously prepared chart the data already taken and the data he would continue to collect, he was able to obtain a CSI solution that "differed from the PGNS by only 0.1 foot per second, which was pretty good".

Looking ahead, Conrad saw a flashing light. "Have you got a flashing light on, or anything?" he asked Gordon.

"No," Gordon replied. "I have got the emergency EVA lights and the rendezvous lights on."

Conrad was amazed, because the two vehicles were still 150 nautical miles apart. A little later he would realise that what he had presumed to be Yankee Clipper was actually a nearby speck of debris from the ascent stage that was reflecting his own beacon.

With the CSI burn computed, Conrad opted to ignore the negligible out-of-plane component. In the final minutes leading up to the manoeuvre, VHF communications became intermittent. In a moment of good reception, Conrad asked, "Dick, have you been lonely without us?"

"Well just a little bit," Gordon admitted. "I've been cleaning up for you."

"We're going to dirty it up, pal," Conrad warned. "We will do our best, but we are filthy." After a moment's thought, he added, "Most of it should stay in the LM."

"I hope to heck it does," Bean agreed.

With the burn imminent, Conrad asked Gordon, "How do you read?"

Not having heard, Gordon asked, "Hey, are you guys burning?"

"Not yet. No. Another 15 seconds," Conrad replied, but Gordon did not hear.

Not appreciating that communications were out, Conrad provided a countdown to ignition and then called, "We're burning, we're burning. We're burning now, Dick. How do you hear? Burning good."

Not having heard, Gordon enquired, "How are you doing?"

"Just fine," Conrad replied. "Just fine. Burn is good, Dick. Burn is good. How do you hear?"

Gordon, still not having heard anything since before the burn, repeated, "How are you doing?"

"Just fine," Conrad assured, but still Gordon did not hear him.

The manoeuvre was performed at selenographical coordinates centred on 4.9°N, 163.6°E. The 41.1-second thruster firing produced a delta-V of 45.2 feet per second which raised the low perilune of the insertion orbit.

As Gordon noted in the post-mission debriefing, "I never knew for sure that Pete was burning CSI. I naturally assumed that he was, but I never had any confirmation and I kept asking. I'm sure I bothered him by asking him whether he was making it, but I really felt that I ought to know. He could hear me, but I couldn't hear him. The only thing I could do, or course, was to assume that he had made the burn."

When conditions improved just after the burn, Conrad informed Gordon that they had indeed made the manoeuvre, and observed that there was "something screwy" with the communications.

"You're cutting in and out pretty bad," Gordon replied.

"God damn it," muttered Conrad.

Bean offered to recheck the VHF configuration, but they both knew that it was as specified in the checklist. "Okay. I'll just leave it like it was."

A minute later, Conrad tried again, "Dick, how do you read?"

"Loud and clear."

"I'll be goddamned if I know what's wrong with the stupid son of a bitch," mused Conrad. What puzzled him was that the VHF had worked flawlessly during the lunar decent. Conrad speculated that the thrusters were interfering with the signal from the forward antenna, the one they were using, but Bean pointed out that communications had deteriorated shortly prior to the burn. After another unsuccessful attempt to talk to Gordon, Conrad decided, "Al, let's screw around with something. How about the VHF squelches or something?"

"The squelch only applies to us hearing him," Bean pointed out.

"Well, he's got a squelch over there, too," Conrad argued.

"That's right, he should screw with his," Bean agreed. "It doesn't have anything to do with our transmitter." But with communications so poor it was not possible to pass this suggestion along.

Later analysis would determine that command module reception of lunar module voice was impaired during ascent and the early part of the rendezvous as a result of the squelch-sensitivity setting in the command module's VHF system. At times, the received signal strength was sufficient to maintain the squelch circuit open, but at other times, such as CSI, it dropped in and out. Designed to ignore an empty carrier wave, the squelch was ignoring Intrepid's signal.

On acquiring Earth on revolution 31, Gordon relayed the news that the CSI burn had been made. Two minutes later, Intrepid appeared around the limb.

"Here comes Earthrise, Conrad said. "Too bad we don't have a camera."

"We do," Bean replied. "Where is the Earthrise?"

"Right out your window!"

When Bean passed him the 16-mm movie camera, Conrad said, "It'd be better if you took it out your window."

"It's all frosted, Pete," Bean noted.

At that point, Carr interrupted their photography by requesting details of the burn so that Mission Control could incorporate it into their tracking. Meanwhile, the computers on both vehicles were computing the CDH manoeuvre, and Bean, having recovered the AGS by reinitialising it to the PGNS, had resumed his evaluation of its utility.

As Conrad peered ahead, he saw a constant speck of light. "Ah! I see him now," he told Bean. The vehicles were 122 nautical miles apart and closing at 95 feet per second. "I finally have a visual on Yankee Clipper, I think," he reported to Mission Control.

"Roger," Carr acknowledged.

With the rendezvous going like a smooth simulation Conrad said, "I've done this so many times, I'm getting bored."

"I'm not!" Bean laughed as he tended to the AGS.

With 26 minutes remaining to the CDH manoeuvre, Gordon, taking advantage of an improvement in communications, asked Conrad, "Did you enjoy your hot meal last night?"

"Hot meal? We didn't eat any hot meal," Conrad replied, then realised that he was being teased because Yankee Clipper had a water heater whereas Intrepid did not.

Idle awaiting CDH, Conrad tried to relax. After laughing at the filthy state of their cabin, he told Gordon, "Old Beano is over here updating his AGS, and he's all over it." Keying in a succession of range and closing rates without making a mistake was a demanding task in and of itself, but he had also to keep his backup paper plot up to date. Then Conrad said, "Wait until we tell you where we landed."

"I saw," Gordon replied.

"You don't know *exactly* where we were," Conrad laughed.

"Pretty close to that rim?" Gordon pointed out.

"Oh, yes," Bean chipped in.

"Like 20 feet!" Conrad cackled.

Then communications deteriorated.

Speaking to Bean, Conrad mused, "Shall we tell Houston that we're going to burn CDH on the AGS?"

"No," Bean retorted. "It'll scare the hell out of them."

But Conrad wanted to test the idea, "Houston, Al's working so hard keeping his AGS updated here, why don't we let him burn CDH?"

"Roger, Pete," Carr replied.

"Say again, Gerry?" Conrad asked in astonishment.

"Stand by on that," Carr laughed. "We're consulting our oracle, right now."

"I'm only kidding," Conrad insisted.

"Come on, babe," Bean said, urging the AGS to produce a good solution to prove that it was a viable backup system.

In an effort to remove some of the floating dust, the environmental control system was activated to circulate cabin air through the lithium hydroxide canister which was normally used to prevent a build up of carbon dioxide. The crew remained isolated on their suit loops. However, the filthy suits were continuously shedding more dust and all that was achieved was a steady state of airborne particulates.

When Carr read up a sequence of instructions for the computer, Conrad, tired and yawning a lot, first noted them down incorrectly and then, once he had corrected the list, made a mistake in keying them in. Bean offered to help by reading the list while Conrad punched the keys.

With Gordon once again garbled on VHF, Conrad advised him, "Don't sweat it if you don't hear from me during this burn." And when Gordon tried to offer his CDH solution Conrad cut him off, saying, "Wait, wait, wait, Dick. Call Houston and have them relay it to me. I can't understand you."

The thruster firing for CDH lasted 12.5 seconds and the delta-V of 13.8 feet per second produced an orbit of 40.4 $\times$ 44.4 nautical miles shaped to match the orbit of Yankee Clipper. The fact that the differential height was 17.5 rather than the ideal 15 nautical miles was a minor issue. As Conrad said in the post-mission debriefing, "I think the reason we didn't have a nominal delta-H rendezvous was my screw-up on the ascent shutdown. But the PGNS is quite capable of handling that, and it was no problem."

Ten minutes after CDH, the vehicles were 75 nautical miles apart with Intrepid closing at 141 feet per second. The next task was the TPI manoeuvre which would start the climb to the target's altitude. As Intrepid caught up with Yankee Clipper in terms of central angle, the elevation of the target above the horizon would increase, and the burn would be performed when this angle indicated the desired geometry.

As they flew towards the terminator Gordon found it difficult to take sightings of Intrepid because sunlight was blinding his sextant, so he waited until both vehicles were in the Moon's shadow, at which time, to his surprise, he lost sight of his target. "Hey, would you guys check your tracking light?" When Conrad said that the switch settings showed that the flashing beacon was on, Gordon replied, "Well, I sure don't see you."

"Hey, Houston, it looks like our tracking light's burned out," Conrad announced. "On the first night-side pass we had little bits and pieces floating along with us and we could tell that the tracking light was flashing on them. And we still have, I've presumed to think, bits and pieces floating along and nothing's flashing on them, so I'm pretty sure it burned out." By monitoring the telemetry while the circuit breaker for the lamp was cycled, Mission Control verified that the high-voltage circuitry was drawing current.

"Well you may have current but you don't have any light," Gordon insisted. In calculating his solution for TPI, he would have to rely on VHF ranging to augment the optics marks obtained prior to sunset. Meanwhile, Intrepid's computer had been processing the range and closing rates from the rendezvous radar to compute its own solution.

The TPI burn was made shortly after passing around the far-side and occurred at an altitude of 44.5 nautical miles over selenographical coordinates 14.6°N, 129.0°W.

The thrusters were fired for 25.75 seconds and the resulting delta-V of 28.5 feet per second included a 1.5-foot-per-second out-of-plane component.

For some reason, VHF communications had improved. "Have you got that probe extended?" Conrad asked Gordon.

"Yes, sir."

"Let's hope it works," Conrad said. When Gordon started to play a music tape over the radio, Conrad joked, "Al and I are dancing."

The terminal phase had two minor burns scheduled to refine the approach. These were calculated by the computer. After the first, Conrad observed, "Man, this thing is right on the nominal line of sight."

Several minutes earlier, watching Bean slaving over the AGS even although the rendezvous was almost complete, Conrad had suggested that he "just quit working and sit back and enjoy the flight". In the post-mission debriefing Bean pointed out, "After CSI we realigned the AGS to the PGNS. I made all the AGS marks after that just as we planned to do, and I got solutions that all compared very favorably. This shows that the AGS would get solutions – which, of course, we suspected it would. But the whole point is that you don't want to use the AGS as the normal rendezvous mode. It requires that every 2 or 3 minutes you make a lot of entries in the AGS. It requires that you point the spacecraft exactly at the command module, which takes time and effort. The LMP is working continually and isn't able to sit back and think through what is going on in the rest of the spacecraft." "I got to thinking about it later, and that was the first time I'd really looked out to see what was going on. The rest of the time, I'd just been working my fanny off trying to get all those marks into the AGS, and that's not the way you want to fly a spacecraft." They had operated the AGS as its designers had envisaged, just to prove that it was possible. But it was far too much work. The manual charts had also worked, but what was really needed was an automatic backup system that could operate closed-loop without so many manual data inputs. However, no such device was on the drawing boards.

As Bean relaxed, Conrad made him an unprecedented offer, "Take a minute and fly this vehicle."

"You're on," Bean replied enthusiastically. The thrusters were rated to manoeuvre the vehicle with the descent stage attached and a full load of propellant, so with only the ascent stage left and almost depleted it had the feel of a fighter jet. In '*A Man on the Moon*', Andrew Chaikin writes: "For a few minutes Bean had his hand at the crisp, responsive ascent stage. It was a moment that he would always remember as pure Pete Conrad, that in a small craft somewhere over the far side of the Moon, he had taken the time to share with Bean a flying experience that even most astronauts would never know." Despite being referred to as a lunar module pilot, the astronaut on the right-hand side was really a flight engineer. Owing to Conrad's gesture, Bean became the first and only lunar module pilot ever to truly fly his spacecraft. When finished, he nulled the velocities that his manoeuvring had introduced, in order not to disturb their trajectory.

"I've got you at 14 miles, Pete," Gordon reported. Without the flashing lamp, he could not see Intrepid in darkness; his only aid was VHF ranging.

"10 miles, Dick," Conrad reported 3 minutes later. Then onboard to Bean he said, "I'm cold. Are you cold?"

"Yes, damn cold," Bean admitted, and switched off the system that was feeding cold water to their liquid-cooled garments.

When Yankee Clipper flew into daylight, dazzling Conrad, he observed, "He gets brighter and brighter and brighter!"

With the range down to 5 nautical miles, Conrad performed the second correction, which he described as a "little tweak".

Gordon first spotted Intrepid at a range of about 2 miles and as he observed using an optical sight in the left-hand rendezvous window he satisfied himself that it was making a straight-in approach. His VHF ranging system indicated that it was closing at 38 feet per second.

During the far-side pass, Pete Frank handed the Flight Director's console over to Glynn Lunney several hours early, but his team remained in place to complete their shift. When the vehicles emerged from behind the Moon on revolution 32, Gordon had the television camera in the right-hand rendezvous to show the final approach and docking. The transmission was received by the 85-foot antenna near Madrid in Spain and routed to Mission Control via a geostationary satellite.

"What do you have on the tube?" Gordon asked.

"Nothing yet, Dick," Carr reported.

Gordon rolled Yankee Clipper to improve the high-gain downlink sufficiently for the television transmission.

On the basis of his experience in approaching the Agena target vehicle during his Gemini 11 mission, Conrad expected Yankee Clipper to become dazzlingly brilliant and so he momentarily took off his helmet in order to put on sunglasses.

At 6,000 feet Conrad reduced the rate of approach to 31 feet per second, and upon reaching 3,000 feet he slowed to 19 feet per second.

"You're looking pretty good," Gordon said, watching Intrepid on the monitor of the television camera.

"You look pretty darn good yourself!" Conrad replied.

At 2,000 feet Conrad fired his thrusters for 38 seconds for the main braking burn. He was to make lateral adjustments to hold Yankee Clipper 'fixed' against the stars, but it was not necessary to make such line-of-sight corrections until 1,000 feet and even then they were minor. In the post-mission debriefing he said, "From 1,000 feet, I took 1 foot per second off per 100 feet of range. We slid right on in there, and that was that. It was Mickey Mouse."

"Stand by to receive the skipper's gig," Carr prompted, upon hearing that Intrepid was stationary 15 feet away from Yankee Clipper.

"Aye-aye, sir," Gordon replied.

As Yankee Clipper manoeuvred into the attitude for docking, Conrad drew Bean's attention to a brown stain on the cover of the umbilical which ran around the edge of the command module's heat shield and into the service module. "Do you suppose that is where it got hit by lightning?" When Bean said this seemed likely, Conrad suggested that they take a picture for the engineers, and Carr heartedly agreed. They translated Intrepid to one side to obtain a better line of sight and shot several frames using the 16-mm movie camera. On spying a curved strip of material about 3 feet in length swinging from the rear of Yankee Clipper, they added some pictures of what was almost certainly tape left over from the pyrotechnics which had separated the service module from the launch vehicle adapter.

In preparing for docking, Gordon asked Conrad to move Intrepid, "Can you drop down a little, Pete."

"Can I do what?" Conrad asked.

"Drop down."

"Down?"

"Yes. Towards the Moon."

"Oh, yes; okay."

"No," Gordon complained as Intrepid moved in the opposite direction. "Towards the Moon."

"He's upside down," Bean observed.

"Oh, I'm sorry: when you say down, that's up to me, pal," Conrad said.

"I said towards the Moon," Gordon pointed out.

"I don't know where the Moon is," Conrad admitted.

With the confusion resolved, Conrad moved Intrepid to the requested position and announced, "Dick, I'm going to pitch over 90 degrees now."

"Okay, let's go," Gordon agreed.

After rotating to face his docking system towards Yankee Clipper, Conrad yawed left through 120 degrees in order to enable Gordon to view the docking target on the roof.

Peering up through his overhead window to refine the alignment, Conrad noticed that Gordon had not placed a docking target in the left-hand rendezvous window as an aid to this final manoeuvre. "Richard," he called. "You didn't set up the target for me."

"You're having your picture taken instead," Gordon replied, having installed the 16-mm movie camera in that window.

With the vehicles aligned and Intrepid in Attitude-Hold with Narrow Deadband, Yankee Clipper eased forward.

"Capture!" Gordon called, when the three latches on the tip of the probe achieved a soft docking by engaging the apex of the lunar module's drogue.

"Okay, go Free," Conrad ordered.

"Free," Gordon confirmed.

With both vehicles floating passively on the extended probe Conrad pronounced himself satisfied, "Just as stable as a rock."

"Wait a minute," Gordon replied. "I want to pitch down just a little bit." Although there were no oscillations, the vehicles were slightly misaligned and he wanted them precisely lined up. Satisfied, he asked, "Are you ready to retract?"

"I'm ready," Conrad confirmed.

"Here you go." Gordon threw a switch to hydraulically retract the probe and draw in the lunar module. After the twelve main latches had achieving a hard docking, he called, "And you're home free, boy."

The Mission Operations Control Room shortly prior to docking. Glynn Lunney has taken over as Flight Director. Capcom Gerry Carr has Deke Slayton to his right and Tom Stafford to his left, with Dave Scott and Jim Irwin also present. Note the television from Yankee Clipper reflected in the VIP gallery window.

"Super job, Richard," replied Conrad.[2]

Docking occurred at selenographical coordinates 14.5°S, 47.0°E, at an altitude of 58.1 nautical miles in an orbit of 58.3 × 62.3 nautical miles. The vehicles had been separated for 37 hours 42 minutes 17.9 seconds. If a docking had not been feasible, Conrad and Bean would have made an external transfer using their OPS (which is why these were retained when the backpacks were jettisoned) while dragging along their cargo. There was a handrail on Intrepid to facilitate a transfer from the forward hatch to a position from which they could reach across to a rail near the command module's side hatch, and there was an external lamp in case the transfer was made in darkness. As Gordon had decided not to remain suited throughout his solo mission, such a transfer could not have been made until he had suited up and stowed the centre couch to create a clear isle into the lower equipment bay to accommodate two suited newcomers and their goods, but he estimated these preliminaries would have required no more than 20 minutes.

"That's the end of the TV show," Gordon called down to Earth. "We've got some work to do."

"We can see Pete through the window now," Carr pointed out.

"I'll leave it on for a little while then," Gordon replied.

"Hello there!" said Conrad, looking up and waving at the camera, showing that he was wearing sunglasses inside his bubble helmet.

Several minutes later Gordon called, "I expect the folks are a little bored with that scene now. I'll turn the television off."

"We concur," Carr replied.

"How was the show?" Gordon asked.

"Very, very good, Dick," Carr assured.

As Gordon set about pressurising the tunnel and Conrad and Bean attended to their post-docking checklist, Conrad idly enquired, "What time is it back there, anyhow?"

"It's just about high noon," Carr replied.

"Oh, I've completely lost all sense of night and day," Conrad admitted. And then he relayed a follow up query: "Al wants to know what *day* it is."

"20 November," Carr said.

"Roger," Conrad acknowledged.

"And it's Thursday," Carr added for good measure.

DISCARDING INTREPID

When Carr called shortly thereafter with the data for jettisoning Intrepid, Gordon cut him off, "I'm busy up here in the tunnel."

"Gerry, don't let him go over the hill without getting it," Conrad warned.

[2] This was in marked contrast to Apollo 11, when "all hell broke loose" as Mike Collins retracted the probe and the two vehicles fought for control.

The flight plan called for Conrad and Bean to take off their helmets and gloves, but as Conrad stated in the post-mission debriefing, "The LM had so much dust and debris floating around in it that when I took my helmet off I almost blinded myself. I immediately got my eyes full of junk and had to put my helmet back on. I told Al to leave his on. We left the helmets on and took off our gloves."

Once Gordon had disconnected the probe from its mount Conrad opened his roof hatch and hauled in the entire docking assembly, which was to be discarded with the ascent stage. Meanwhile, Carr read to Bean the data that would enable the computer to perform a manoeuvre after Intrepid had been cast loose. On Apollo 10 the ascent stage had fired its main engine to boost itself into heliocentric orbit, in order not to leave behind 'orbital junk' that might threaten a later mission. The ascent stage of Apollo 11 had been left in orbit, but the mascons would have rapidly caused this to decay and the craft would have crashed in an uncontrolled manner. For Apollo 12, it had been decided to have the ascent stage perform a deorbit burn and crash near the ALSEP in order to stimulate the seismometer with a 'known' seismic event.

Finding a spare moment, Gordon copied down his own data, which included how to orient Yankee Clipper for tracking Intrepid using the sextant. When Carr offered the limb-crossing times for the forthcoming far-side pass, Gordon cut him off again, "Hey, Gerry, can we skip this. We're kind of busy right now."

"Sure can, Dick."

"Just holler at us when we reappear around the corner," Gordon suggested. "We'll be looking for you."

"Okay, that takes care of the paperwork for this pass," Carr pointed out.

Ten minutes later, Gordon warned, "Pete, I'm going to manoeuvre to the jettison attitude. Don't let it worry you."

"I won't," Conrad replied.

As Apollo 12 passed behind the Moon, Conrad and Bean were bagging items in an effort to limit the transfer of dust to the command module. In addition to the two sample return containers they had the lunar surface television camera that they were returning for failure analysis, the Surveyor souvenirs, the films from the Hasselblad, closeup and 16-mm movie cameras, the magnetic tape from the data recorder, their personal preference kits and the rocks in the large bag. "Don't bother attempting to dust them," Bean said, having discovered this to be counterproductive.

When Conrad retrieved the contingency sample, he complained, "Look at all the shit that comes out of this!"

Ideally there was a stowage nook assigned to each item passed through the tunnel, but upon receiving the contingency sample Gordon asked, "Where the hell does this go?"

"I don't know," was Conrad's reply.

In the post-mission debriefing Gordon said of these intravehicular transfers, "To me, this was a period of 'hustle, hustle'. I believe we had a bit more gear to transfer back and forth than did the Apollo 11 crew."

In addition to the items returned from the Moon, Bean bagged up and transferred their documentation, unused food, medicines, tissues and clean towels. Meanwhile, Gordon was passing into the lunar module the items that he wished to discard. When

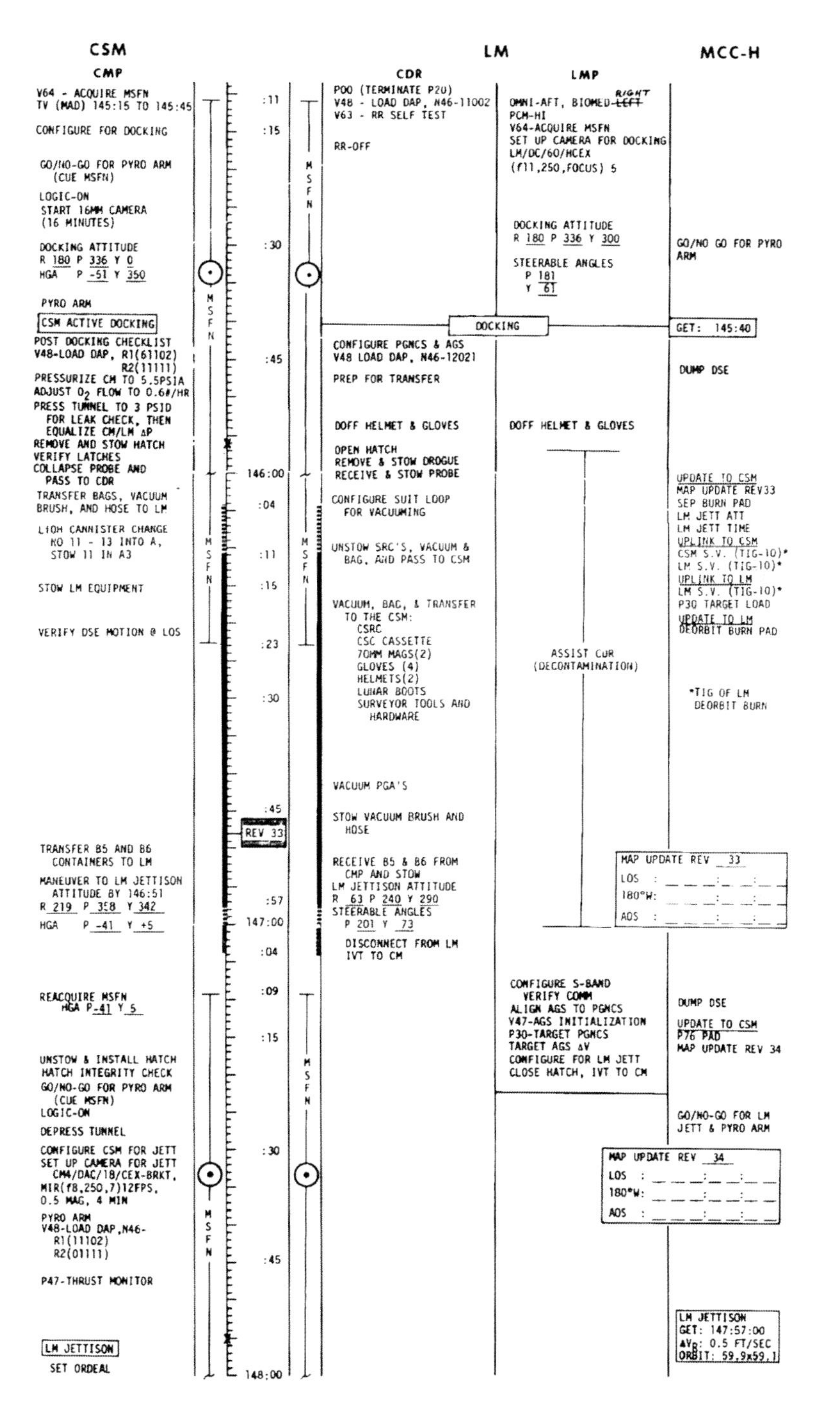

CSM — CMP

V64 - ACQUIRE MSFN
TV (MAD) 145:15 TO 145:45
CONFIGURE FOR DOCKING
GO/NO-GO FOR PYRO ARM (CUE MSFN)
LOGIC-ON
START 16MM CAMERA (16 MINUTES)
DOCKING ATTITUDE
R 180 P 336 Y 0
HGA P -51 Y 350
PYRO ARM
CSM ACTIVE DOCKING
POST DOCKING CHECKLIST
V48-LOAD DAP, R1(61102) R2(11111)
PRESSURIZE CM TO 5.5PSIA
ADJUST O_2 FLOW TO 0.6#/HR
PRESS TUNNEL TO 3 PSID FOR LEAK CHECK, THEN EQUALIZE CM/LM ΔP
REMOVE AND STOW HATCH
VERIFY LATCHES
COLLAPSE PROBE AND PASS TO CDR
TRANSFER BAGS, VACUUM BRUSH, AND HOSE TO LM
LiOH CANNISTER CHANGE NO 11 - 13 INTO A, STOW 11 IN A3
STOW LM EQUIPMENT
VERIFY DSE MOTION @ LOS
TRANSFER B5 AND B6 CONTAINERS TO LM
MANEUVER TO LM JETTISON ATTITUDE BY 146:51
R 219 P 358 Y 342
HGA P -41 Y +5
REACQUIRE MSFN
HGA P -41 Y 5
UNSTOW & INSTALL HATCH
HATCH INTEGRITY CHECK
GO/NO-GO FOR PYRO ARM (CUE MSFN)
LOGIC-ON
DEPRESS TUNNEL
CONFIGURE CSM FOR JETT
SET UP CAMERA FOR JETT CM4/DAC/18/CEX-BRKT, MIR(f8,250,7)12FPS, 0.5 MAG, 4 MIN
PYRO ARM
V48-LOAD DAP,N46- R1(11102) R2(01111)
P47-THRUST MONITOR
LM JETTISON
SET ORDEAL

Time: :11, :15, :30, :45, 146:00, :04, :11, :15, :23, :30, :45, REV 33, :57, 147:00, :04, :09, :15, :30, :45, 148:00

LM — CDR

P00 (TERMINATE P20)
V48 - LOAD DAP, N46-11002
V63 - RR SELF TEST
RR-OFF
DOCKING
CONFIGURE PGNCS & AGS
V48 LOAD DAP, N46-12021
PREP FOR TRANSFER
DOFF HELMET & GLOVES
OPEN HATCH
REMOVE & STOW DROGUE
RECEIVE & STOW PROBE
CONFIGURE SUIT LOOP FOR VACUUMING
UNSTOW SRC'S, VACUUM & BAG, AND PASS TO CSM
VACUUM, BAG, & TRANSFER TO THE CSM: CSRC, CSC CASSETTE, 70MM MAGS(2), GLOVES (4), HELMETS(2), LUNAR BOOTS, SURVEYOR TOOLS AND HARDWARE
VACUUM PGA'S
STOW VACUUM BRUSH AND HOSE
RECEIVE B5 & B6 FROM CMP AND STOW
LM JETTISON ATTITUDE
R 63 P 240 Y 290
STEERABLE ANGLES
P 201 Y 73
DISCONNECT FROM LM
IVT TO CM

LM — LMP

OMNI-AFT, BIOMED-~~LEFT~~ RIGHT
PCM-HI
V64-ACQUIRE MSFN
SET UP CAMERA FOR DOCKING LM/DC/60/HCEX (f11,250,FOCUS) 5
DOCKING ATTITUDE
R 180 P 336 Y 300
STEERABLE ANGLES
P 181
Y 61
DOFF HELMET & GLOVES
ASSIST CDR (DECONTAMINATION)
CONFIGURE S-BAND
VERIFY COMM
ALIGN AGS TO PGNCS
V47-AGS INITIALIZATION
P30-TARGET PGNCS
TARGET AGS ΔV
CONFIGURE FOR LM JETT
CLOSE HATCH, IVT TO CM

MCC-H

GO/NO GO FOR PYRO ARM
GET: 145:40
DUMP DSE
UPDATE TO CSM
MAP UPDATE REV33
SEP BURN PAD
LM JETT ATT
LM JETT TIME
UPLINK TO CSM
CSM S.V. (TIG-10)*
LM S.V. (TIG-10)*
UPLINK TO LM
LM S.V. (TIG-10)*
P30 TARGET LOAD
UPDATE TO LM
DEORBIT BURN PAD
*TIG OF LM DEORBIT BURN
MAP UPDATE REV 33 — LOS : 180°W: AOS :
DUMP DSE
UPDATE TO CSM
P76 PAD
MAP UPDATE REV 34
GO/NO-GO FOR LM JETT & PYRO ARM
MAP UPDATE REV 34 — LOS : 180°W: AOS :
LM JETTISON
GET: 147:57:00
ΔV_R: 0.5 FT/SEC
ORBIT: 59.9x59.1

The flight plan detailing crew activities between docking and jettisoning the ascent stage.

the transfers were complete, Conrad and Bean used a small vacuum cleaner on their suits, but as Conrad said in the post-mission debriefing this was "a complete farce".

Peering through the tunnel at his colleagues in their filthy suits, Gordon came to a decision. The plan called for them to doff their suits once in the command module, but it was evident that the dust would overwhelm the environmental system. "Why don't you take those suits off over there," he suggested.

Conrad agreed. To keep their eyes from burning and their noses from inhaling the floating particles, they disconnected their bubble helmets from their suits and, as they stripped off their suits, kept the helmets on their heads. The pressure garments, which hopefully would not be required again, were bagged and stowed beneath the side couches of the command module. After Bean suggested that they discard their liquid-cooled garments along with the ascent stage, they stripped to the diapers that they had worn in case of bowel movements whilst confined to their suits in the lunar module.

Several minutes after Conrad returned to Yankee Clipper, it made its appearance on revolution 33 and Carr alerted them. Shortly thereafter, Bean sealed Intrepid and joined his colleagues. After reinstalling the command module's apex hatch Gordon initiated the slow process of venting the tunnel. Lunney gave the Go to jettison the ascent stage on time, some 45 minutes into the near-side pass. In too much of a rush to unpack replacement long-johns, Conrad and Bean finished the stowage of the transferred items prior to making an effort to clean themselves. When Conrad finally took off his diaper, he discovered a rash on his buttocks that itched for several days.

As Gordon reported in the post-mission debriefing, "We just barely got the hatch back in and the tunnel vented in time for separation. I realise we could've separated without having the tunnel completely vented, but I wanted that." As Conrad pointed out, "The tightest part of that timeline is the fact that it takes about 20 to 25 minutes to vent the tunnel down."

Prior to leaving Intrepid, Bean had set its computer to hold inertial attitude after it was cast loose – this attitude being almost local vertical with Yankee Clipper below, so that the ascent stage would be in the attitude required for the deorbit manoeuvre that it would perform some three-quarters of a revolution later. The act of jettisoning was performed by a pyrotechnic charge which severed the tunnel, discarding surplus hardware such as the docking collars. Five minutes later, Yankee Clipper fired its thrusters for 5.4 seconds to withdraw radially downward at 1.5 feet per second into an orbit of 57.5 × 62.0 nautical miles. This separation burn occurred at an altitude of 59.9 nautical miles at selenographical coordinates 1.4°N, 43.3°W, and was designed to preclude a collision.

As Apollo 12 approached the limb, Gerry Carr handed over the Capcom's console to Don Lind. When Lind called to say that the Flight Surgeon had some advice about Conrad's biosensors, Gordon replied, "Why don't you stand by until they get some coffee?" But then communications were impaired by the recurring problem with the high-gain antenna and the spacecraft passed around the far-side before Lind had an opportunity to supply this. Whilst out of contact, the astronauts had supper and made a start on the pre-sleep checklist. On reappearing on revolution 34 there was a delay in acquisition due to the fault with the high-gain antenna.

When Lind finally established contact, the lunar module's deorbit burn was about

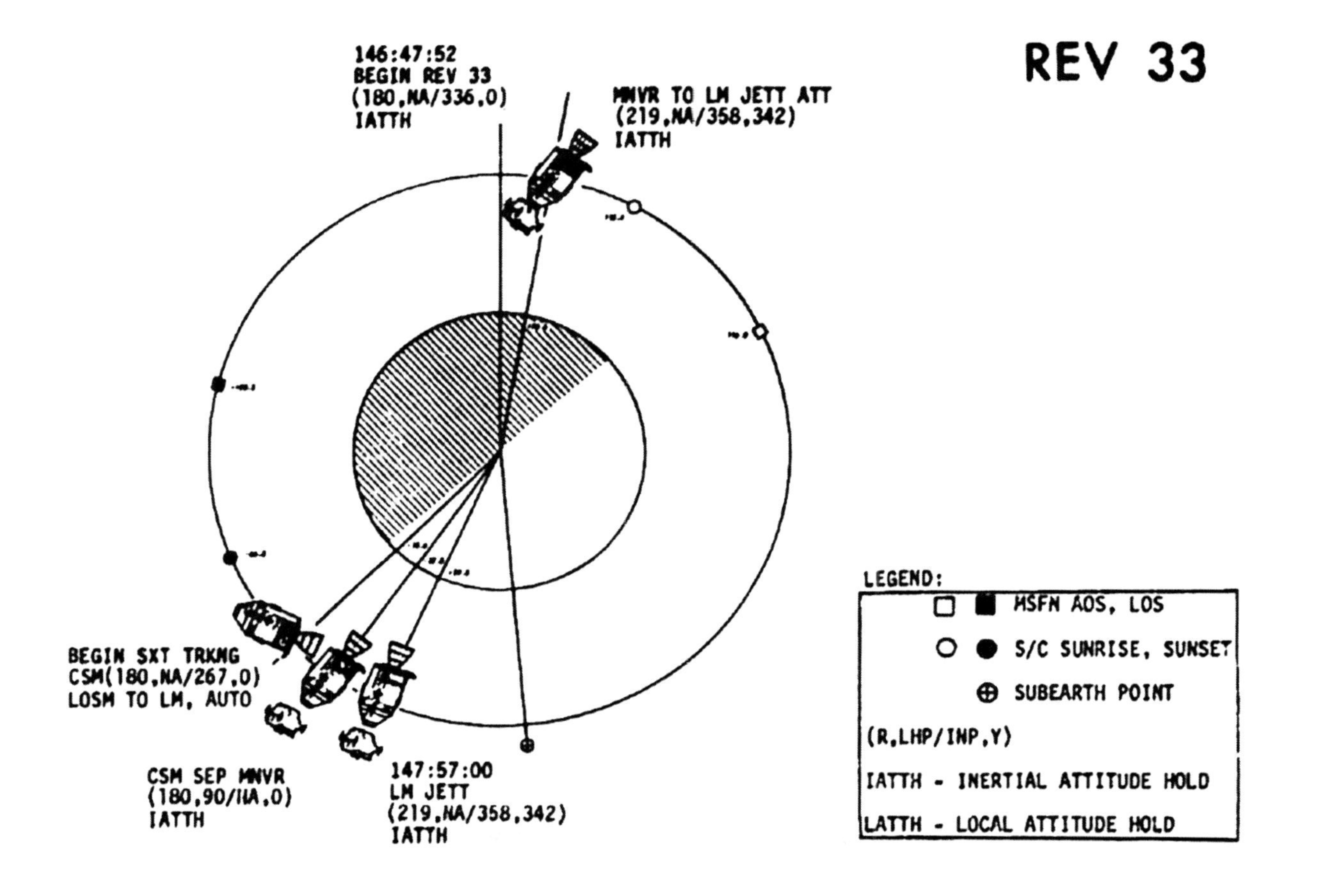

A page from the flight plan showing the jettisoning of the ascent stage, which would perform its deorbit manoeuvre about three-quarters of a revolution later.

12 minutes away. Gordon said he had his spacecraft oriented to view the other craft, could see it in the telescope, and had the 16-mm movie camera on the sextant ready to try to document the impact. The manoeuvre was to be initiated on radio command from Earth, and include an out-of-plane component which would drive the trajectory north towards the track flown by Intrepid during its landing. The burn was executed at an altitude of 57.6 nautical miles, at selenographic coordinates centred on 14.4°S, 60.8°E, and the 83.5-second firing of the thrusters produced a delta-V of 196.3 feet per second that yielded an orbit which had a notional perilune 66.3 nautical miles *beneath* the surface.

"The LM is on its way down," Lind announced.

"Roger," Gordon acknowledged. "Is it still burning?" he asked 45 seconds later.

"That's affirmative," Lind replied.

"Let me know when the burn is finished," Gordon requested. But communications degraded and when he finally got through he was informed by Lind, "It's been over about a minute and a half." Gordon had lost sight of Intrepid at some point during its burn. "Let me know when that thing impacts so I can shut off this camera, because I can't see it through the telescope."

While they were waiting, Lind picked up on the item inherited from Carr, "Does Pete want to hear recommendations from the Flight Surgeon on his skin irritation?"

"No," Gordon replied. "He doesn't need any information on that."

The plan was for the 5,254-pound ascent stage to smash into the Moon at a speed of about 5,500 feet per second and at the shallow angle of 4 degrees at a point some 5 nautical miles south of the Snowman, this being the minimum attainable distance from the ALSEP. However, a delay in transmitting the cutoff command resulted in an over-burn of 5 feet per second, producing a steeper descent which impacted about 36 nautical miles short. It would have had the energy equivalent of detonating a ton

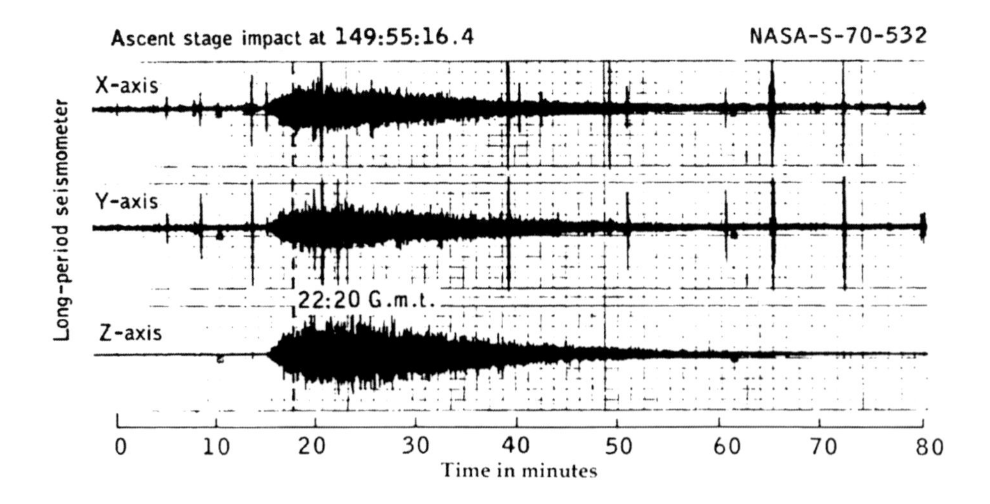

How the impact of the ascent stage appeared to the seismometer.

of TNT on the surface. The prediction was that it would excavate an elongated crater of about 20 feet by 40 feet with a maximum depth of 2 feet. Crashing an object of a known mass and velocity at a given distance from the seismometer would serve to calibrate the instrument. As the Public Affairs commentator explained, "We've had indications from the previous seismic experiment left on the Moon as to what sort of signals we get from the instrument. We don't, however, have a good handle on how a given energy is transmitted through the lunar surface. It is hoped [that this impact] will help us in interpreting previous and subsequent seismic signals from the passive seismometer." Mission Control was displaying data from the seismometer 'live' on a wall screen.

Lind recited a countdown based on MSFN radio tracking, and the telemetry from the lunar module ceased precisely on cue. "LM impact," he announced.[3]

"Okay, thank-you," Conrad acknowledged. As a result of the trajectory error, they had not known where to look for the impact.

Signals on all three long-period sensors of the seismometer lasted an hour. As the Mission Report said, "Distinct phases within the wave train are not apparent [...] and no phase coherence between components is evident. The spectral distribution of the signal ranges from approximately 0.5 hertz to the high-frequency limit of 2 hertz for the long-period seismometer. The seismic wave velocity, corresponding to the first arrival, ranges between 3.0 and 3.8 kilometres per second." The fact that this signal was similar in character to several noted by the short-lived instrument left by Apollo 11 suggested that they were caused by impacts. The signal was so unlike anything seen on Earth that seismologists were baffled. Nevertheless, there was 'instant science' speculation at a press conference. One suggestion was that the 'ringing' was due to debris thrown up by the impact raining onto the surface downrange, in the direction of the seismic sensors. Another idea was that the impact had triggered rock slides inside existing craters. After a detailed study, geophysicists realised that they had gained an important clue to the makeup of the Moon. The protracted signal could be caused by the shock passing through a rock structure that was fragmented to a considerable depth, and it was speculated that the shock waves from Intrepid's impact "may have penetrated the Moon to a depth of about 6 miles".

This was clearly a fruitful line of research, but higher energies would be required to pursue it further. The velocity and shallow angle of approach of Intrepid's impact were equivalent to a 'secondary' impact caused by the fall of ejecta excavated from the Moon by the 'primary' impact of a celestial body. That is, the ascent stage struck at less than lunar escape velocity, whereas a primary impact will typically occur at ten times that speed. It was therefore decided that in future the S-IVB stage would be made to strike the Moon instead of a slingshot into solar orbit. This would deliver a greater shock than dropping an ascent stage from lunar orbit, not only because the spent stage would weigh 30,000 pounds and strike at 8,500 feet per second, but also because the approach would be almost vertical. When Apollo 13 did this, the impact site was some 73 nautical miles from the Apollo 12 seismometer and the energy was

[3] Impact occurred at 5:17 p.m. EST on 20 November.

equivalent to detonating 11 tons of TNT. The signal lasted 3 hours 20 minutes, and was so intense that the gain control on the seismometer had to be reduced in order to keep the recording on the scale!

During the remainder of the near-side pass, Conrad, Gordon and Bean worked on their end-of-day tasks. Yankee Clipper's disappearance around the limb marked the start of the sleep period. Meanwhile, Cliff Charlesworth took over as Flight Director for the graveyard shift. The day of the lunar landing had been a long one for Conrad and Bean in particular, fully 24 hours. After only about 3 hours of fitful sleep on the Moon they had been awake for 21 hours. Nevertheless, once safely back in Yankee Clipper they slept for a mere 4 hours, which left them in poor shape for the 18-hour day of activities which would precede the crucial transearth injection manoeuvre to head for home.

ORBITAL PHOTOGRAPHY

The assignment for most of the remaining time in lunar orbit was photography of potential landing sites for future Apollo missions. As there would be no more Lunar Orbiter satellites, this 'bootstrap' method of site selection meant the sites that could be considered for future landings were strongly correlated with the ground tracks of their predecessors. Specifically, the task for Apollo 12 was to obtain: (a) stereoscopic strip photography by a Hasselblad fitted with an 80-mm lens of the ground track from terminator to terminator on two passes, with concurrent 16-mm movie footage through the sextant on the first such pass; (b) subsequent landmark tracking of four landmarks that bracketed three sites in the vicinities of the highland craters Lalande, Fra Mauro and Descartes, obtained on two successive revolutions; (c) high-resolution vertical photography by a Hasselblad with a 500-mm lens and additional high-resolution westward-looking oblique photography.

In making the first call of the day, Ed Gibson got straight down to business with the data for the plane change manoeuvre due in just over an hour and a half which was intended to optimise the ground track for the photographic assignments. He told Conrad that he could dispense with the bioharness whose sensors were persistently irritating his skin. And he gave the latitude and longitude of where Mission Control calculated Intrepid to have impacted. Although such a small crater would be difficult to resolve, they were to take pictures using a hand-held Hasselblad with a 250-mm lens. The best time would be immediately after the high-resolution photography of the crater Lalande on the next revolution, when the impact site would be 22 nautical miles south of their track. Cramming in as much as possible before Yankee Clipper flew around the far-side, Gibson recommended that they switch off the power to the steerable S-band antenna while behind the Moon to let it cool down, as the difficulty it had in locking on was suspected of being thermal in nature. Barely a minute later, the spacecraft disappeared around the corner.

Mission Control had uplinked a revised state vector to the spacecraft's computer in preparation for the plane change manoeuvre, and while they were in the Moon's shadow Gordon took star sightings to update the inertial platform. Meanwhile, his colleagues placed the 16-mm movie camera in the left-hand rendezvous window and a Hasselblad with a 500-mm lens in the other one for high-resolution photography. After Gordon had oriented the vehicle for the burn, he sighted on a star to check. As they flew into daylight, they began the checklist to prepare the SPS engine for firing. Less than 2 minutes after appearing around the limb on revolution 39, four thrusters were fired for 11 seconds to settle the main propellants in their tanks and then at the appointed time the computer ignited the SPS engine. The manoeuvre occurred at an altitude of 58.8 nautical miles above selenographical coordinates centred on 6.75°S, 110.0°E. It lasted 19.25 seconds and yielded a delta-V of 382 feet per second out of plane to drive the orbit 3.2 degrees along the equator so that the ground track would pass directly over the three candidate Apollo landing sites.

In the post-mission debriefing Gordon noted, "The only anomaly I noticed at all in any of the SPS burns was in this one, and I'm not sure it was an anomaly. During the burn, even although guidance looked good and it was tight, it felt to me like the spacecraft was doing a Dutch roll throughout the burn – a typical aircraft Dutch roll-type thing that oscillated in roll and yaw." Nevertheless, the residuals did not require trimming. Conrad suggested, "I guess that we had some condition where the centre of gravity was passing through our low-stability point."

With the post-burn checklist completed Conrad asked, "That hot engine of ours didn't by any chance buy us enough gas to come home a day early, did it?" As per his own plan, he was asking whether their remaining SPS propellant would permit a more energetic transearth injection manoeuvre which would advance splashdown by 24 hours.

"Stand by," Gibson replied.

"I knew it was touch and go with fuel reserves," Conrad admitted, "but seeing as the engine has been so hot I'm wondering whether we've wound up with a little bit more."

"We'll see what we can work out," Gibson assured.

"Attaboy," Conrad said.

"That SPS is a real hummer," Gordon agreed, but pointed out, "That's the first time I've seen it wallow; it is starting to wallow through the sky now."

The astronauts were now to combine breakfast with the first of their photographic tasks. They were to take a series of high-resolution pictures of the approach routes from the east to potential landing sites near the crater Lalande. This data would be used to identify surface relief that would be visible to a landing radar. They were then to cancel the surface-tracking pitch rate and roll 20 degrees left in order to snap some hand-held pictures of the area in which Intrepid had impacted.

Shortly thereafter Conrad called, "We goofed on Lalande and got you some neat 500-mm pictures of Herschel." Because the long lens had a narrow field of view, the procedure was to orient the spacecraft to gimbal angles supplied by Mission Control and then refine the aim using the optical sight mounted in the left-hand rendezvous window. But they had lined up on the wrong crater.

"I fouled this one up completely," Gordon noted in the post-mission debriefing. I sighted on a crater that looked like Lalande. I didn't use my head. I wasn't paying attention to the times that the ground gave us. These times were accurate and were the ones we should have been using. I was looking at the [lunar surface] and this crater looked just exactly like Lalande but lo and behold it wasn't, it was Herschel, and we got some high-resolution photography of its southern lip."

"Now," continued Conrad, "I've got a question for you."

"Go ahead," replied Gibson.

"Would you prefer us to take the 500-mm of Lalande on this next pass rather than the stereo-strip, or go on and get the things in order?" They were scheduled to obtain a sequence of pictures on the next daylight pass to facilitate stereoscopic analysis. Conrad was asking which was more important. He also had an offer: if they were to extend the photography into the revolution preceding transearth injection then they ought to be able to pick up whichever task was postponed.

"Okay," Gibson came back, "Lalande is the lowest of the photo priorities, so we recommend you continue with the flight plan. If you think it possible to pick that up on the last revolution before TEI then go ahead and give it a go."

"We'll give her a go," Conrad promised.

"Were you able to pick up anything on the target of opportunity?" Gibson asked, referring to Intrepid's impact point.

"No, we weren't," Conrad replied. They had been distracted by the realisation that they had missed Lalande.

With that, Apollo 12 flew around the far-side. When it emerged on revolution 40 the 16-mm movie camera had been installed on the sextant to shoot a movie at one frame per second and the Hasselblad in the right-hand rendezvous window had been fitted with an 80-mm lens for the terminator-to-terminator stereo-strip photography of the ground track, in this case extending from 122°E to 52°W. An intervalometer fired the shutter every 20 seconds to obtain overlapping frames. As Gordon would explain in the post-mission debriefing, this was easy because once the spacecraft had been oriented with its apex facing directly down and a pitch rotation set up to match the orbital rate, "nobody had to fly the spacecraft". In this attitude the Hasselblad would view vertically and the movie camera would view at an angle of 45 degrees to capture oblique views along the line of approach.

"We've got Dick working this pass, and Al and I are sightseeing," Conrad called. "This is really the first chance we've had to get a good look at the Moon. So we're enjoying this pass with the map, checking off all the craters."

"Does it look any different after being down there," Gibson asked. "Do you get a little better feel for what it's like?"

"I personally think it's more spectacular from orbit," Bean said. "When you get out on the surface it's interesting down there, naturally, but it's not too much unlike just being out in a big field of clumps on Earth." He found observing the Moon from orbit to be more like science fiction than walking on its surface, in part because on the surface he had been too busy to really appreciate being there.

"Do you think that will satisfy Dick?" Gibson asked.

"No," replied Bean. "He keeps talking about making a low pass over the landing site before we go!"

"I may get another chance; you can't ever tell," Gordon chipped in. He hoped that the program would last long enough for him to command his own mission and walk on the lunar surface.

"Hope so," Gibson agreed.

Several minutes later Bean pointed out, "Ed, I was just looking at the flight plan, and noticed the time back there." It was almost 4 a.m. in Houston. "How come you get all the good deals?"

"I guess I'm just lucky," Gibson joked. "No, I would just as soon be up during the middle of the night rather than be up during the day and watch you guys sleep!"

"That's true," Conrad said.

"Paul Weitz has become a 'sleep expert'," Gibson said, prompting laughter from the spacecraft. In terms of the circadian rhythm of the crew, Weitz was Capcom on the graveyard shift.

As they flew over the Snowman, Gibson came back on Conrad's question about a fast transearth coast. They could do it, but the delta-V margin of about 20 feet per second was too small for comfort. If the SPS performed poorly on its final burn, it would have to be fired to depletion and then the velocity shortfall made up using the thrusters. It would be better to perform the burn as planned, and have a significant amount of propellant in reserve.

"Okay. No problem," Conrad replied. "We weren't sweating it. We just knew that if we had a better engine or something, there was a chance we might have enough to do it."

"That's one less day in the LRL," Bean pointed out. The quarantine period began on the day of the rendezvous, not the day of their return to Earth, and so returning one day early would mean spending an additional day cooped up in the Lunar Receiving Laboratory.

"I think you've probably got a little better place to spend it there," Gibson agreed, and then added, "I bet you there's lots of guys sitting over in the LRL that wouldn't mind trading you a day or two." The staff who were to be quarantined with them had already spent sufficient time in isolation to identify any bugs *they* might be carrying, to prevent these from passing to the astronauts upon their arrival and being mistaken for lunar infections.

The stereo-strip photography for this pass was concluded as the spacecraft crossed the terminator into darkness. Just before they flew around the corner, Gibson relayed a request from the Flight Dynamics Officer that from now on they should perform waste water dumps only whilst behind the Moon. The far-side pass was fairly quiet. The 16-mm camera was moved back to the left-hand rendezvous window alongside the optical sight, and the 500-mm lens was reinstalled on the Hasselblad in the other rendezvous window ready to resume high-resolution photography.

One unwelcome development was that Conrad had contracted Bean's head cold. "Shit," he complained. "I'll let them figure out I've got this cold. I ain't going to tell them. I'll just tell them: CDR took one decongestant." A moment later, he wondered out loud, "How did I get the world's greatest cold *on the Moon*, for Christ's sake?"

Bean replied, "Because you had the world's greatest LMP with you on the Moon, who had a cold."

"And I kept bullshitting Houston," Conrad recalled, "saying, 'Aw, it's just stuffy up here.'"

"That's the way to play it, though," Bean pointed out.

"That's right. I'll tell them I took the decongestant but I ain't got a cold," Conrad reiterated, then sneezed. His frustration was aggravated when he went to urinate and realised he needed a towel to clean the receptacle, "Oh, shit! I keep forgetting to get my rag for this goddamned pisser."

"What the hell difference does it make to those guys, where you dump?" Gordon wondered, referring to the request not to vent water overboard on near-side passes.

"They can see it on their MSFN data," Conrad explained.

"Tell me what difference it makes?" Gordon persisted.

"When you come out on the front side they pick you up and they track you all the way across, and they can project ahead. If you dump on the back side, you'll not be where you're supposed to be."

Gordon was still puzzled, "So?"

"Right there, they would know what the error was that you put in," Conrad said. "Apparently, it freaks up the way their processors work if you disturb it during their tracking – it assumes you're not making any delta-V. If you make it on the back side they can account for it."

"That's right," Gordon realised. Mission Control had to precisely determine their orbit to calculate the transearth injection manoeuvre. He checked their contingency data and joked about breaking orbit a couple of revolutions ahead of schedule.

Conrad laughed at the prospect of doing an unannounced TEI and appearing early at acquisition of signal. "Wouldn't that fool the shit out of them! Come whistling out from behind the Moon." Thinking about the 90-hour transearth coast he complained, "Shit, I'll be dead by then with this cold, the way it's coming on! These things come on fast and go away slow." A moment later he wondered, "Hey, are we recording on tape out here?"

"Yes, sir," Bean replied.

Oops!

As they checked the timing of the photography for the next near-side pass, Conrad spurred Gordon on, "All right, Dick. You've had your one error, now for Christ's sake you had better get Fra Mauro and Descartes because I'm going to take this big map here and check your ass."

"I had my error," Gordon explained, "because I didn't read this freaking time they gave me; that's why I had my error."

Several minutes later Conrad decided that the pictures of Herschel represented a bonus for the scientists. "You know what's going to happen, don't you, Gordon. I'll make a prediction: they're going to look at the freaking high-resolution photography of Herschel and will make some great discovery."

As they flew into sunlight, Gordon looked out, "Isn't that a spectacular sight."

"It's unreal," Conrad agreed.

"Like it's a different world," Gordon ventured.

Conrad said, "If you saw it at the movies, you'd say, 'God, that's faked.'" By way of an explanation he said that because the unilluminated lunar far-side was utterly black it gave the impression of being right outside the window.

Moments later, they emerged from behind the Moon on revolution 41. While they were out of contact Pete Frank had taken over as Flight Director. When Gerry Carr established contact Bean reiterated, "Pete and I are finally getting our first look at the Moon. Just kind of skylarking. We didn't get a chance before we went down. We always had something going on and as a result we just got glimpses." However, it was time to get back to work with high-resolution photography of potential landing sites.

After 20 minutes of radio silence Conrad reported, "We've got Descartes and Fra Mauro."

"Good show, Pete," Carr congratulated.

"I guess you've got a pretty 'full' Moon down there right now, Gerry?" Conrad asked as they approached the terminator.

"That's affirmative, Pete," Carr replied. "I got a look at it coming in this morning. It is almost completely full and is beautiful."

As Yankee Clipper passed into the Moon's shadow, Gordon took star sightings to update the inertial platform.

"Bye-bye, Houston," Bean signed off just before they passed around the far-side, and then announced onboard, "I gotta go pee."

While behind the Moon they transferred the 16-mm movie camera back onto the sextant in readiness for landmark tracking.

On examining the frame counter of the Hasselblad with which he had obtained the high-resolution photography, Bean began to suspect a failure. He had monitored its progress by watching the rotation of a small wheel on the side of the magazine, but the counter showed fewer frames. "Shit, we've taken more that 50 photos. I've taken 80 on the last two targets alone."

"Take another one," Conrad suggested.

Bean did so, and observed the wheel rotate and the counter increase. "Everything moves."

As Gordon manoeuvred Yankee Clipper into the attitude for landmark tracking he pointed out, "Three more revs until we get the hell out of here."

Conrad mused, "Hey, we ought to get some pictures inside this son of a bitch. We haven't got any. We didn't take any views in the LM." It had not occurred to them to take pictures of each other inside their vehicles. A moment later, after coughing and sneezing, he complained, "Shit, I'm really catching a cold." And then, realising that he was tired with the transearth injection manoeuvre looming, he mused, "The smartest thing for me to do would be to crawl into the bilge down there and sleep for a couple of hours." But this was not feasible. Instead, he announced, "I'd like to dig into some of that chow." He retrieved a tuna salad from the pantry. Bean joined him.

"What's today, Friday?" Gordon asked.

"What do we care?" Bean laughed. "We ain't going anywhere!"

Back in daylight, Gordon used the telescope to aim the sextant so that the 16-mm movie camera could document landmark CP-1 in the highland terrain on the far-side about 15 degrees east of the Sea of Smyth. This control point was the northern crater of a doublet of small craters on the rim of a crater about 22 nautical miles wide in a cluster of relatively shallow craters. At acquisition of signal on revolution 42, Carr, knowing the crew were busy, left them alone. When he heard Gordon admiring the view of Earth, Carr established contact and read up some data. Bean wrote it down and asked, "Have you heard from our families lately?"

"I expect they're just up and having breakfast, getting the kids off to school," Carr replied. "When things slow down, what do you say I give them a call and get some words?"

"Thanks," Bean said.

"Did you get the word that Amy was visited by the tooth fairy?" Carr asked.

"No," Bean replied.

"It happened when you guys were getting ready for your descent. We didn't have time to get that up to you but Amy wanted me to be sure and tell you that."

"Thank-you," Bean acknowledged.

Meanwhile, Gordon had tracked the second of the series of four landmarks. CP-2 was the southern crater of a doublet of craters near the eastern edge of a larger crater in the eastern portion of the Sea of Fertility. The third one, DE-1, was the westerly crater in a doublet near the large crater Descartes in the rugged terrain of the central highlands. Beyond that was FM-1, a small crater on the rim of a large shallow crater in the hummocky terrain north of Fra Mauro. After using the specified shutter speed of 1/60th second for Fra Mauro, Gordon advised Carr that the area was so bright that 1/125th would have been more appropriate, and after checking with the experts Carr relayed the recommendation that Gordon should follow his instincts while tracking the landmarks on the next revolution.

As Yankee Clipper passed into the Moon's shadow Gordon took star sightings to update the inertial platform yet again. In the final minutes before loss of signal, Carr said to Bean, "Would you pass the word to Dick that the P22 marks are looking real good and they're very consistent. About the only thing that we might have to offer is that he is starting his marks just a bit too early." Each target involved shooting five frames, the first at 39 degrees from vertical during the approach, then at 22 degrees, then from overhead, and finally two more at corresponding angles beyond vertical. The angles were specified in terms of time intervals.

"How much early," Bean asked.

"About 11 seconds."

Gordon responded, "I'll just put in a comment about starting these a little early. I felt that I had to, since by that last mark the target is going out of sight because the field of view in the telescope isn't what it is in the sextant and I think we've just got to start a little bit early."

"Okay, that's fine," Carr conceded.

With that, Apollo 12 flew around the corner. The main activity behind the Moon was lunch.

Having replaced the 500-mm lens on the Hasselblad with a 250-mm lens to shoot a target of opportunity Bean accidentally opened the magazine and fogged the high-resolution pictures of Descartes and Fra Mauro. As he explained in the post-mission debriefing, "I had the camera in my hands and was rolling it about. All of a sudden, the side popped off the magazine. I tried to clamp it shut quickly but was unable to do so. I was unable to tell how much of the film was ruined by having been light-struck from the opening of the side. We taped it up. I think that earlier, when I took some of the 500-mm photography, I may have operated the unlocking mechanism instead of the film winding mechanism, which is on the other side – they both look the same." Indeed, subsequent analysis of the film showed precisely the overlapping exposures that would arise if the mechanism that opened the magazine was actuated, releasing the entire film holder portion of the magazine from its housing, because in that state the drive would fail to transport an entire frame of film for each exposure. This explained why the counter showed fewer frames than the number of pictures he knew he had taken. One way or another, the film was useless.

After discussing options, they decided to recommend breaking off from landmark tracking early in order to retake the high-resolution pictures of Descartes and Fra Mauro.

On flying into daylight, Gordon lined up on CP-1 to initiate the second series of 16-mm movies of the landmarks. Watching Earth appear on revolution 43 Conrad mused, "Man, there's not much Earth out there, I'll tell you that." Because the Moon was at almost 'full' phase only a thin crescent of Earth was illuminated. After Carr made contact Conrad said, "Listen, Gerry, I've got something real important. We were taking some target-of-opportunity photographs with the same film pack that we had the 500-millimetre shots on, and that magazine back popped off and we're not sure we didn't wipe out that film magazine. Now, what we suggest is that we dump the landmark tracking, seeing as we got good ones last time on Descartes and Fra Mauro, and [instead] get some more 500-millimetre of them on this pass on another magazine."

"Stand by," Carr replied. A few minutes later he was back, "We concur with your plan, Pete." It was also decided to apply tape to the other film magazine to prevent it from opening.

After Gordon finished tracking CP-2, Carr said that the telemetry from the optics showed he had mastered the art of landmark tracking.

"We lucked out," Gordon replied.

"Anything is simple with the best crew!" Bean said onboard.

"You should have told them that last July," Conrad pointed out. The first moonlanding mission was in July.

Next, Gordon reoriented Yankee Clipper to perform high-resolution photography of Descartes and Fra Mauro. Later analysis indicated that the results were, at best, of only moderate use for mission planning. The quality of the pictures was impaired by window and lens transmission effects and by camera instability, but the main factor was that the photography was obtained on a later revolution, since the higher Sun angle reduced the shadow definition. However, Fra Mauro had been photographed at a lower Sun angle shortly after Apollo 12 arrived in lunar orbit, and even though the resolution from the 80-mm lens was inferior to that produced by a 500-mm lens the shadows were more representative of those which would be encountered in making a landing at this site.

Once the Descartes and Fra Mauro photographs had been obtained, Carr offered a suggestion, "We've had a meeting of the minds here and the tentative plan is to drop the stereo-strip on rev 44 because it has a lower priority than the landmark tracking, and so we want landmark tracking on the next rev."

"You want Fra Mauro and Descartes landmark tracking. Is that right?" Conrad asked, seeking clarification that Mission Control did not want another pass over the easternmost pair of control points.

"That's affirmative," Carr said.

Instead, Conrad suggested that they obtain the landmark tracking of Descartes and Fra Mauro on revolution 44 and then "on the next pass, which is the pass we burn TEI on, we can stereo-strip her up to that point, get Lalande, keep on going with the stereo-strip and burn TEI around the corner."

"Stand by, Pete," Carr replied. While awaiting advice, Carr provided an update on the astronauts' families. "Pete, Jane says you are doing a great job and she's really proud. They are all waiting anxiously for all three of you to get back. Dick, Barbara said you're doing a fabulous job but she sure wishes you could get some rest, so she could. And, Al, Sue says she spent last evening with Jane and had a lot of fun. They both feel real good about everything. She's now in a watching-and-waiting mode."

Meanwhile, because Yankee Clipper had flown into the Moon's shadow, Gordon took star sightings to update the inertial platform.

Carr said that Mission Control was reluctant to delay the stereo-strip to the final near-side pass because that was assigned to preparations for the transearth injection. "The plan is to do your stereo photos up until the landmark tracking time, and then terminate the strip photography and do your tracking."

"Okay, Gerry," Gordon acknowledged.

Shortly thereafter Apollo 12 disappeared behind the Moon. The main item on the flight plan for the far-side pass was to install the Hasselblad with an 80-mm lens for stereo-strip photography. In doing so, they discussed the possibility of switching to the 500-mm lens after curtailing the stereo-strip and then, in between Descartes and Fra Mauro, manoeuvring to get the high-resolution pictures of Lalande rather than squeezing this in on the final pass but they decided not to jeopardise getting the Fra Mauro tracking.

Gordon oriented the spacecraft apex down for the stereo work and initiated a pitch rotation at the orbital rate. He started the automatic timer of the camera on crossing the terminator into daylight. This task was curtailed on reaching 37°E to perform the postponed landmark tracking of the control points for Descartes and Fra Mauro. On flying into the Moon's shadow he updated the inertial platform and shortly thereafter the spacecraft passed over the hill.

While Apollo 12 was behind the Moon on what was intended to be an idle period for the astronauts, Gerry Griffin took over as Flight Director in order to handle the transearth injection manoeuvre. At acquisition of signal on revolution 45, which was expected to be their last, Don Lind established contact.

"Hello, there!" Conrad greeted Lind. "How are you today?"

"Just fine," Lind replied. "How's things up near the Moon?"

"Oh, not too bad but I think we're about ready to leave," Conrad said.

"We'll be glad to have you back," Lind assured. Then, getting down to business, he supplied the data for the high-resolution photography of Lalande, emphasising, "We want to make sure that you understand that this is your option. We don't want to press you too much on this last pass before TEI; if you want to do it, fine."

"Don," Gordon replied, "we want to do it because I messed it up this morning. I want to get it."

Next, Lind provided the data for the scheduled transearth injection manoeuvre at the end of this revolution and for the contingency of their missing that opportunity and having to make an additional revolution. Meanwhile, Mission Control uplinked a revised state vector into the spacecraft's computer. After the pre-TEI checks were complete, Gordon aimed Yankee Clipper's apex to the lunar surface and snapped the Lalande pictures. As they crossed the terminator into darkness several minutes later he adopted the inertial attitude for the forthcoming manoeuvre and verified this by sighting on a star.

"The star check's okay," Gordon reported.

"Very good," acknowledged Lind, who had been joined by Deke Slayton, Tom Stafford and backup crewmen Dave Scott and Al Worden.

With 2 minutes remaining to the onset of what would hopefully be the final pass around the back of the Moon, Lind gave Mission Control's verdict on the telemetry, "Everything's looking good to us down here."

"Roger," acknowledged Conrad.

"We've got a nice spot in the South Pacific all reserved for you," Lind assured.

"See you on the other side," Conrad promised.

"Have fun," Lind said just before loss of signal.

The transearth injection manoeuvre was to occur 22 minutes into the far-side pass. As they waited to start the final preparations, Bean enquired if anyone wanted "to go around one more time".

Gordon replied, "I'll go if you'll go."

As mission commander, Conrad said, "Would you mind if I vetoed that!"

Reflecting on the difference between the simulator and reality, Bean said, "Flying is better than I ever imagined it."

"What is weird," Conrad mused, remembering Yankee Clipper's overflights while Intrepid was on the Moon, "is being down on the lunar surface and watching him go whistling overhead. Whoosh!"

Bean agreed that Gordon had passed overhead "like an Earth satellite".

"You weren't as bright as I thought you'd be," Conrad said. "You were bright, but very small – just like a star." After a moment, he laughed, "But there was no doubt who it was!"

LEAVING ORBIT

Upon checking the flight plan Bean said casually, "Hey, Richard, I'll tell you what, why don't you give it a GDC Align."

Gordon duly pushed the button to align the stabilisation and control system with the guidance and control system. This was a precaution against the primary system failing during the transearth injection manoeuvre and their then having to rely on the backup system to finish the engine burn. Then their mood became serious as they performed the familiar routine of preparing the SPS for ignition.

At the appointed time, the computer initiated an 11-second firing by four thrusters to settle the propellants of the main engine in their tanks.

"We have ullage," Bean confirmed.

With 5 seconds remaining to the burn, the computer flashed Verb 99 requesting final authority to light the main engine and Conrad pushed the Proceed key.

TEI was a crucial manoeuvre. If the engine failed to fire, Yankee Clipper would remain in lunar orbit. If it ignited and shut down prematurely there were a number of contingency options available and the action would depend on the timing. If the burn lasted for less than 80 seconds it would leave the vehicle in an elliptical lunar orbit. A revised transearth injection would be attempted after an additional revolution. If the planned SPS burn was interrupted between 80 and 93 seconds the vehicle would depart the Moon on a hyperbolic path and the plan was to attempt two SPS burns to establish a viable transearth trajectory. If the planned SPS burn lasted for longer than 93 seconds, a single additional burn would suffice. If the engine were to fail to make an abort manoeuvre after escaping from the Moon, the spacecraft would fly off into deep space. Ignition was on time and at full thrust, imparting an acceleration of two-thirds g.[4] The chamber pressure was "solid as a rock", as Gordon put it. There was a slight oscillating roll within limits but no Dutch roll such as during the second plane change burn. When the desired velocity had been attained, the computer cut off the engine with the burn having lasted 1 second longer than expected. The residuals were trimmed by a small thruster firing. The burn occurred at selenographical coordinates centred on 8.0°N, 174.3°W, at an altitude of 63.6 nautical miles, lasted 130.3 seconds, and increased the velocity by 3,042.4 feet per second to 8,350.4 feet per second, boosting the spacecraft out of lunar orbit after 3 days 17 hours 2 minutes.

On finishing the post-burn checklist Bean announced, "Okay. That's it, my good friends." After checking the flight plan and seeing that they were to start a television transmission 15 minutes after making contact with Earth, he suggested that they get ahead on the timeline, "Why don't we break out the TV camera?"

"All right, we'll do that," Conrad decided.

When Gordon unstrapped from his couch he noticed that his feet were wet. The water vapour which had condensed in the apex had literally rained on him during the burn. After connecting the television camera to the monitor screen and plugging in the power and signal cables, they unpacked a Hasselblad and checked which kinds of film they had available.

Meanwhile, Mission Control was eagerly awaiting acquisition of signal. If all was well, the vehicle would reappear 12 minutes earlier than if the SPS engine had failed to ignite. As the Public Affairs Officer told his audience with 1 minute remaining to the predicted acquisition, "The TV lines are up. It is conceivable that the crew could

[4] It was 3:50 p.m. EST on 21 November.

have the TV on as they come around the Moon. It isn't scheduled for that time, but we're prepared to take a TV picture should the camera be on."

Lind put in the call, "Apollo 12, Houston."

After clearing his throat, Conrad replied, "Hello, Houston. Apollo 12's en route home."

As Conrad provided the burn status report, the initial MSFN tracking confirmed the trajectory was acceptable. Flight Dynamics Officer William Boone calculated that this was an excellent start to the transearth coast because it would produce an entry interface at 244:21:14 with a velocity of 36,116 feet per second on a flight path 6.69 degrees below local horizontal.

"If you have a camera out already, we have a target of opportunity for you," Lind suggested.

"We have a camera out," Conrad confirmed.

Lind said that the scientists had requested using a Hasselblad with an 80-mm lens and black-and-white film to photograph the high lunar latitudes in order to assist in small-scale mapping. They were to shoot at 30-second intervals during the time they were making the television transmission.

"Are you ready to receive TV?" Conrad asked.

"Anytime," Lind said.

The transmission was received by the 85-foot antenna near Madrid and relayed to Mission Control by geostationary satellite.

The departure trajectory provided a view of the evening terminator on the trailing hemisphere.

"It looks like we're climbing straight up," Conrad noted. They were already at an altitude of 426 nautical miles.

"We really get the impression that you're on a fast elevator," Lind agreed.

Gordon explained that everyone was working, "Al's busy getting the [Hasselblad] black-and-whites and I'm holding the TV monitor for Pete."

"It looks great," Lind assured.

Of the dawn terminator, Bean said, "The first time we passed over it, we said to ourselves, 'Now there's a real rough part of the Moon.' And the next day, when the terminator moved 14 degrees, we found that the part that was now in a higher Sun looked fairly smooth, or at least like the rest of the Moon as you see it, and the part that was now into the terminator looked the roughest."

"That texture really comes through loud and clear on your picture," Lind agreed.

Bean made another point, "It's really useless for you all to have colour [on your screen], because it is pure black-and-white."

By 10 minutes into the transmission their altitude had increased to 1,000 nautical miles. Lind commented on this, "It's really amazing how much the size of the Moon has changed just in the few minutes you've been on the air."

Bean pointed out that the camera was showing the Sea of Smyth, which could be observed only heavily foreshortened from Earth. The Sea of Fertility came into the shot as more of the illuminated portion of the Moon became visible.

Conrad offered a comment for the scientists. With the Moon near 'full' phase, the bright rays of Copernicus were prominent. One crossed the Snowman. Earlier in the day, they had inspected the rays with the monocular. Although the rays were readily visible from an altitude of 60 nautical miles, "I think the difference in texture is so slight that down on the surface Al and I had the impression that we just couldn't see any contacts whatsoever anywhere we went."

"What about the white and grey differences you saw around the west side of Head crater?" Lind asked.

Conrad was willing to speculate, "Al and I talked about where our boots turned up the lighter material, and it seemed to us it was still the same material; it's just that it hadn't weathered on the surface." As to the rays, "It's pretty much the same general material, but it came at different times and it's had different amounts of exposure to the weather."

Referring to the long-standing debate about the colour of the lunar surface, Bean explained their own observations, "As we started at the [evening] terminator and went around the Moon it changed from grey to white, and finally to brown. Then the next day it did the same thing, except the part that used to be more white was now grey because the shadows were over there more as the terminator moved in that direction."

When Lind asked about the difference between the maria and the highlands, they suggested that he imagine scattering water onto a bed of Portland cement.

As they climbed through 1,700 nautical miles, Bean said, "Let's move over to the hatch window, Pete."

"That's a very impressive picture," Lind observed.

Conrad explained, "Dick just manoeuvred so that I can see the whole Moon."

"You know," Lind said, "The most amazing thing is that you were in orbit down there just a few minutes ago."

"More so to us than to you, I'm sure, Don," Gordon pointed out.

"I'm sure that's true," Lind agreed.

A couple of minutes later Lind relayed another query from the scientists, "Since you're the experts on lunar rock rolling, how does it work?" One suggestion for why the impact of Intrepid had caused the Moon to 'ring like a bell' was that the seismic shock triggered landslides in the walls of craters. The presumption was that rocks on the airless Moon would be precariously balanced.

"It goes very slowly," Conrad explained. "I guess the impression you have is the same way as if you throw something. It sort of moves out, not too rapidly, but just keeps going. And that's exactly what happens when you roll a rock down the side of a crater. It was hard to get them going. I was surprised. I think everybody had the idea that because you're in such light gravity things would roll down easily, but that really wasn't the case. Once you got it going, it just sort of went along in animated slow motion but it kept going for a long, long time."

"Did it bounce?" Lind asked.

"They bounce and slide, a little bit of everything, just like they do on Earth, but just stretch it out."

Reflecting on the slow-motion nature of movement in lunar gravity, Bean chipped in, "You know the funny thing is, if you try to walk on Earth in the pressurised suit with anything close to the weight that you have on your back on the Moon, you get

A view of the eastern hemisphere of the Moon taken shortly after Apollo 12 broke out of orbit in order to head home.

tired very rapidly and after 300 yards you're ready for a rest. But in the light gravity your legs never seem to get tired. I guess when you run up the side of a steep slope it could be tiring, but running on level ground you assume some kind of normal pace and you are able to go for long distances without your legs tiring." This was useful information for planning future missions, because Conrad and Bean had been much more active than their predecessors. Neil Armstrong had made only one serious run, which was about 65 yards and marked the furthest that either he or Buzz Aldrin ever ventured from their lunar module.

Conrad added, "I don't think we approached anywhere near the heart rates that we had in just our normal walk-through on Earth. I agree with Al. You could go for 8 or 9 hours out there. Another thing that I think is interesting, is everybody got worried about us falling over and going down slopes and things. I fell over once up there, but I didn't have any problem getting up [using] a one-hand push-up." In fact, both men had fallen at some point or other.

"Your heart rates were just about as expected," Lind said, passing on information from the Flight Surgeon. Then he warned, "We are about to lose the satellite that is relaying this TV to the States."

Conrad swung the camera inside to show the cabin interior for whatever time they had remaining and then signed off with, "We'll see you in about 3 days."

By the end of the 38-minute broadcast Apollo 12 was 2,500 nautical miles from the Moon. Mission Control uplinked a new REFSMMAT for the transearth coast directly into the spacecraft's computer.

"Once you guys get bedded down," Lind said, "we're not going to awaken you in the morning. So whenever you get up and want to start a new day, just give us a call. You've earned a good long night's sleep."

"No problem," Conrad acknowledged. He and Bean in particular had just endured several long and busy days with very little sleep in between. The Flight Surgeon was concerned about their health.

As they prepared their supper, Gordon informed Lind, "I think I gained weight on this trip. They've accused me of being a chowhound."

"How come you're not getting out and doing your mile a day?" Lind teased.

"He does it running from his couch to the food compartment," Conrad retorted, prompting Lind to laugh.

As the day drew to a close, Lind called, "I checked with your wives and I have a short status report on the families whenever you get a minute."

"Let's hear," Bean replied.

"Pete. I talked to Jane. She saw the TV show and enjoyed it mightily. Also, she sent a letter for you out to the carrier. It will be there with all the family news when you arrive."

"Very good," Conrad said, understanding the reference to the aircraft carrier that was the prime recovery ship.

"Dick, Barbara said that they also saw the show and they thought it was great. She says the family is in great shape."

"Okay, Don," Gordon replied.

"Al, I talked to Sue, and it seems that when the network put on the TV show they had Pete's name under your picture, and she said it's been so long since she's seen you that it even confused her for a moment! But the family has been watching the flight. They're looking forward to splashdown and everybody's fine."

"Thank-you for checking, Don," Bean said.

"You might wear a nametag or something so that she will recognise you," Lind suggested.

"There won't be any worry about it for another 20 days or so," Bean pointed out, referring to the forthcoming quarantine.

After tidying up the cabin Gordon and Bean retired and Conrad set up the passive thermal control roll, then he too settled down for a good night's sleep.

Cliff Charlesworth took over as Flight Director and Paul Weitz as Capcom. After a very quiet 8-hour shift they handed over to Pete Frank and Ed Gibson respectively. As the MSFN continued to track the spacecraft, the Real-Time Computer Complex recalculated its projection of the entry interface. The crucial parameter was the angle relative to local horizontal, which was steepening. It was not a case of the spacecraft mysteriously straying off course, just that further tracking improved the accuracy of the projected trajectory. The flight plan for the transearth coast had options for three midcourse corrections. The first one was scheduled 15 hours after injection, and its purpose was essentially to 'tidy up' after that burn. It was calculated that firing the thrusters to refine the velocity by 2.2 feet per second would delay the entry interface by about 1 minute and reduce the angle from the currently projected 7.95 degrees to the ideal 6.5 degrees. With the time for the manoeuvre looming without any sign of activity onboard, Pete Frank decided to postpone it until the astronauts were awake. In the meantime, Apollo 12 crossed the gravitational 'neutral point' into the Earth's sphere of influence.

SCIENCE QUESTIONS

When telemetry indicated the crew were awake, Gibson broke 12 hours of silence by piping up some music to start flight day 8. Upon receiving an acknowledgement, he asked, "How are you folks?"

"Fine," Conrad replied. "We really sacked out last night, I guess."

After a joke about Paul Weitz sleeping through his shift, Gibson got straight down to business. The first item was the midcourse correction, but the timing was flexible. Conrad decided that as they had overslept they would postpone breakfast until after the burn. In preparation, Gordon terminated the passive thermal control roll and took star sightings to update the inertial platform. The manoeuvre was made 180,030 nautical miles from Earth. Four thrusters burned for 4.54 seconds, and the delta-V of 2.0 feet per second adjusted the entry interface to 244:22:02 at the optimal depressed angle of 6.49 degrees.

"Ed," Gordon called, "for your information, this EMS is useless for this kind of thing." As he had discovered during the transposition and docking sequence early in the translunar coast, the entry monitor system, which was designed to measure the major changes in velocity that would occur during atmospheric entry, was incapable of accurately measuring a delta-V of only a few feet per second. The computer had monitored the midcourse burn by integrating the delta-V from the accelerometers in the inertial measurement unit, then cut off the thrusters with residuals of precisely zero. Gordon's recommendation was that in future the EMS should be used only for its intended purpose.

During the postponed breakfast Gordon performed a P23 by using the sextant to measure the angles of stars relative to the Earth's horizon as the first of six planned cislunar navigation exercises. Observations by his predecessors had established that if a spacecraft were to lose communication with Earth, it would be able to calculate the midcourse corrections required to achieve a safe atmospheric entry. The reason Gordon had so many observations on his flight plan was to conduct an experiment. The fact that the lunar landing site was so far west of the lunar meridian meant that by the time they headed for home the phase of the Moon was almost 'full'. This, in turn, meant that for a spacecraft crossing cislunar space the Earth and the Sun were close together in the sky. Gordon was to assess the extent to which sunlight entering the sextant impaired his ability to see the stars required for these horizon sightings.

In listening to Conrad and Bean discuss what they had done on the Moon, Gordon lost track of his progress and asked Mission Control, which was monitoring his use of the sextant by telemetry, "How many marks is that on the second star? I've lost count. Is it two or three?" Gibson told him to stand by. "Also," Gordon added, "ask if they want another trunnion bias before I finish this set." After a protracted silence, Gordon called, "Ed, you still with me?"

"We're still with you," Gibson replied. "They're still scratching their heads. Just a minute, Dick."

"Well, tell them to quit scratching. I'll say that was three and I'll press on and do another trunnion bias," Gordon decided. "They don't have to worry about it."

"Okay, Dick," Gibson replied, sensing Gordon's frustration.

"Now, Dicky-Dicky," Conrad said onboard. "Control yourself."

"Freak you!" Gordon replied.

"Hey, that's *our* fault, Dick," Conrad pointed out. "Don't get mad at them."

Gibson explained, "Dick, the problem down here was that we didn't have data when you were taking your first mark and so we really weren't sure where you stood in the total flow."

"It's our fault, Houston," Gordon admitted.

Conrad reminded Gordon, "We all know that they don't respond that fast down at MCC anytime." Several minutes later, he made an effort to soothe any irritation on the ground, "Say, Ed, you're getting pretty doggone good at this Capcoming."

"Oh, well, I enjoy it," Gibson replied. Then he harked back to Weitz's graveyard shifts, "Paul came in here and was waiting six and a half hours for the big moment, but then you overslept and he went out of here with a long face again!" Picking up a moment later, he added, "He'll be on for re-entry though, and is assuming you're not going to be sleeping through that."

Conrad laughed and replied, "I don't think we will be."

When some sort of a problem impaired the optics during Gordon's final sightings, Conrad joked that it must be a glitch thrown in by the simulation supervisor.

Recalling the exciting liftoff, Gibson ventured that the simulation supervisor had struck "about a minute after launch". This prompted Bean to ask about the results of the investigation. Gibson had the news sheet which had been prepared for reading to the astronauts later in the day: "A tentative analysis shows that Apollo 12 was struck twice by lightning during liftoff. Don Arabian told a press conference that the strike occurred at 36 seconds after liftoff and again at 52 seconds after the Saturn left the pad. According to Arabian, the rocket and the engine plume exhaust acted just like a wire that ran from the clouds to the ground in the first instance and then from cloud to cloud in the second." Donald Arabian joined NASA in 1960 and was currently chief of the test division of the Apollo Spacecraft Program Office. Gibson added that there were photographs of the lightning.

Conrad said they were all looking forward to seeing the film, adding, "I wonder if they are going to revise the weather rules for launch?"

"What's been said so far," Gibson replied, "is that under identical conditions with an identical spacecraft, they wouldn't do it over again."

"These three guys would!" Conrad insisted. He meant that the launch vehicle had not been affected by the lightning and the glitch to the spacecraft had not required an abort, and after the spacecraft had been restored to health the mission had been able to continue.

"You left out one item: the crew," said Bean pointedly.

"I guess we hold the world's record as the world's fastest lightning rod," Conrad mused.

"The world's tallest," Gibson agreed. Several minutes later he was back to report that the problem in the optics was a computer sequencing issue that could be easily remedied. Once Gordon had completed the navigational sightings and incorporated the data into the state vector, Gibson reported that the entry angle differed from the MSFN value by a mere 0.035 degree.

Soon after Apollo 12 entered lunar orbit the large steerable S-band antenna had begun to experience difficulty in locking onto Earth. Engineers suspected a thermal problem in the electronics. A test had been devised. Gibson explained the procedure, "What essentially we're doing is looking at the effect of primary versus secondary transponder, primary versus secondary electronics, and wide beam versus narrow beam." Since the spacecraft would not be rolling for passive thermal control during the test, they were to aim the rear of the service module sunward; an attitude which he described as being "backed up to the fire". Maintaining this attitude would place the antenna under thermal stress, yet enable the radiators on the periphery of the cylindrical module to shed heat to the vacuum of space. As the test started Gibson warned, "This could go on for a little while."

While the crew prepared lunch, Gibson read them the daily news, most of which was sports, and then an update on their families. The standard response whenever a reporter asked an astronaut's wife what she thought of him flying in space was 'I'm proud, thrilled and happy'. The previous day the three Apollo 12 wives had paraded out of the Conrad house in stunning white pant-suits, each holding a sign, stating in turn: 'Proud', 'Thrilled' and 'Happy'. "Dick, Barbara will attend Mass this morning. And, Al, Sue will attend a luncheon today at the Lakewood Yacht Club and will visit

Mission Control this afternoon at about 3 p.m. Pete, Jane will shop for Chris's birthday present. She is also to go to the Yacht Club for the luncheon. After church, they plan to picnic at Cloverfield with the Rice's. Your father-in-law is expected to arrive here sometime Sunday and will remain until after splashdown."

"Thank-you, Ed," Conrad replied on behalf of the entire crew.

When Gibson asked if they were interested in news about their ALSEP, Conrad replied enthusiastically, "Go, babe."

The failure of a high-voltage supply had degraded the suprathermal ion detector and crippled the cold-cathode ion gauge but otherwise, as Gibson reported, "The RTG output is around 73 watts. The central station is performing well, and has been sent a total of 382 commands as of a short while ago. The seismometer has reached a stable temperature of around 126 degrees [Fahrenheit], and at three different times the tracings have shown some seismic activity. The magnetometer is increasing in activity as the Moon enters a magnetic zone between the Earth's two solar shock waves. That is, the Moon is approaching the centre of the Earth's magnetic tail near lunar noon, where the field is the lowest. And at that point the 'site survey' will be accomplished. And the solar wind spectrometer is perking right along and doing real well."

"Say, Ed," Bean asked, thinking about the dust on the white thermal paint, "from all they know now about watching the temperatures, are they forecasting it will last for 2 years?"

"Folks down here are pretty optimistic," Gibson replied.[5]

During the meal, they offered Gibson some thoughts about their moonwalks. Prior to the mission they had received tuition in field geology from scientists of the US Geological Survey. Bean said, "You know we talked with Al Chidester and the guys before we went, about how the main objective of the geology wasn't to go and grab a few rocks and take some pictures but to try to understand the morphology and the stratigraphy and what-have-you of the vicinity you were in – look around, and try to use your head along those lines. Well, I'll tell you, there was less than ten times I stood in spots, including in the LM both times we were back in, and said, 'Okay now Bean, can you fill that square? Is it possible to try and determine where this came from? Which is first? Which is second? And all that.' And except for deciding which craters looked newer than others, which we knew from the earlier observations, I was not able to see any special little clues, like for example over in Hawaii and out at Meteor Crater and other places. That whole area has been acted on by meteoroids or something else so that all the features that are normally neat clues to you on Earth aren't available for observation. I didn't find any way to fill those big squares, you know. I never was able, when walking up to a crater, to determine when the normal ground stopped and the ejecta started – except for the

[5] The ALSEP was designed to operate for at least 1 year. By December 1972, five ALSEPs had been deployed with a variety of instruments. There were issues with individual instruments but the overall success rate was high. On 30 September 1977 the entire network was irrevocably commanded to shut down in a measure designed to save a paltry $200,000 per annum!

difference in slope or that it got a little bit more powdery under my feet, and that's not a very good index. We never saw anything of a different colour or a different amount of rocks or anything other than when we kicked up that very light grey as opposed to the more dark cement-grey material. There's just no contrast to look at."

Conrad had come to the same conclusion, "I think even a trained geologist would have trouble doing a whole lot of field geology that way on the Moon. I think what you're going to have to do is pick your traverses like we did, and sort of select your rocks at a regular interval as you go along, then come back and analyse the stuff to find out differences." Because everything was thick with dust, it was not possible to say much about a rock without picking it up and brushing off some of the dust. This might be feasible if poking around for something interesting, but it was incompatible with the rigour of documented sampling. As to interpreting the landscape, "I've kind of got the idea that a lot of it is the same, the only difference being the relative ages by being blasted by a meteor and getting thrown out at different times."

Bean had a suggestion, "I think you are going to want to increase the number of core tubes so you can get down into these areas you're interested in and find out what's going on under there, because it's covered with this layer and there just ain't no way to figure it out. Before the EVA, during the EVA, and afterwards, we talked about it and thought about trying to get the big picture, trying to be more than rock collectors and picture takers. And, believe me, we worked at it, and I think we were pretty doggoned good at getting that sort of thing in training, not just grabbing a few rocks. But it's difficult to do because the clues just aren't right there on the surface. It's got this big blanket of all-beat-up soil over every single thing. I think maybe you want a better trenching tool. The one we had was just that shovel and Pete could go down only about 8 inches without falling on his head. Now, if you don't want to get a lot of core tubes but you want to see what's going on, you need a longer trenching tool."

Conrad agreed, "We were really hindered in the fact that we couldn't bend over. It wasn't as apparent in training as it was up there because in training when you weigh 285 pounds with all the stuff on your back it is fairly easy to sort of scrunch down or lean quite a bit. You can't do that on the Moon. I'm short and low to the ground to begin with, so somebody that's taller than I am is going to have difficulty trenching as deep as I did using the same length tools. We've got a whole bunch of ideas and [...] in the MQF we'll put down on paper some suggestions for how to improve the tools to do a little better job."

"Yes," said Bean, "I think those tools can really be worked over. They seemed pretty good before we left, but once we got up there and started working in that one-sixth gravity you can't always do the same things: you're leaning in a different way, and things are a little different. I think we've thought about it enough and observed it enough that we can give some pretty good suggestions for improvements that will help the next guys get more rocks, and better rocks, and faster, and trench deeper, and do more core tubes, or whatever else they want to do."

Conrad continued, "We concluded that everything is too delicate to start with. The extension handle was almost wiped out by the time Al finished driving in the double core tube using the hammer."

"Thanks for your comments," Gibson replied.

In selecting the landing site for Apollo 11 the planners had avoided craters, but a trajectory error resulted in Eagle flying towards a crater littered with boulders. Neil Armstrong intervened and passed over both this crater and a smaller one in search of an open area in which to set down. Towards the end of his moonwalk he ran back to photograph the interior of that smaller crater. For Apollo 12 craters were a scientific theme. As Bean said, "We tried to do crater morphology and all that business. We could see what we thought was bedrock on the outside of the craters, and before we went EVA we said to ourselves, 'Great! We're going to look in those craters and we're going to see a deep contact between the regolith and the bedrock.' But when you look into those craters, what it looks like is just like the surface except there's a few rocks that seem to be resting on the wall and in the bottom. Now, if you went down there and dusted away all that material maybe you'd find a contact between the regolith and the bedrock, but we couldn't see it." Referring to the adventurous ideas for missions to canyons and mountains he made a prediction: there wouldn't be any exposed stratigraphy, all the slopes would be masked by loose material and dust.

Having no personal insight into moonwalking, Gordon informed Gibson that the chocolate pudding and butterscotch pudding were delicious. It was not surprising he was putting on weight.

Mission Control curtailed the high-gain antenna test after 2.75 hours and Gordon reoriented the vehicle to perform his second set of cislunar navigation observations. When he reported that he could not make out one of the stars on his list, Don Lind, who had taken over as Capcom, said, "We wanted to determine whether or not you could see that star with the light shafting."

"It's light out there," Gordon admitted. "I don't necessarily see shafting but I sure can't see the star in the field of view."

The star was 20 degrees away from the Sun. "It's in that area where it is kind of marginal, Dick," Lind said.

"I looked all over and couldn't see any star in the field of view," Gordon pointed out. "In fact, it surprised me. I thought I was going blind."

"Maybe that star didn't get turned on today," Lind joked.

Gordon selected another star from the list to continue the experiment. In doing so he reported, "For the navigation experts' information, the airglow is starting to show up, so I'm going to use the top part of the airglow for a visible horizon." At a range of 168,000 nautical miles the angular diameter of the planet's disk allowed him to discern the stratification of the upper atmosphere. On finishing the sextant work he reinstated the passive thermal control roll.

As the crew had their supper, Lind posed a list of moonwalk questions which had been drawn up by the mission science team. "Let's start with two quick ones on the Surveyor. Are you bringing back any glass from the thermal switch plate?"

"No," replied Conrad. "The glass, apparently, which we didn't know, was bonded onto metal and the metal, in turn, was bonded onto the little standoff things on top of the box. Everybody told us it would have separated from its bonding, but it hadn't. It was in great shape. It was bonded to the metal just perfect. We beat on it, but all we did was break it into little teeny tiny pieces that remained fastened tenaciously to the metal. That was that. We just couldn't get it."

"Whoever made that bonding material will appreciate that testimonial," Lind said. "Did you get any soil from the Surveyor's trenching area, other than what may be in the scoop itself?"

"The scoop has some material left in it, I believe," Conrad replied. "And that'll be it."

When the astronauts had described mounds near Head crater as looking like little volcanoes this had excited the geologists. Lind asked, "Did you ever climb one of those mounds? What more description can you give us of them? In particularly, was there any apparent orientation or elongation to the mounds? Also, anything about vent holes?"

Conrad realised that the scientists did not appreciate that the mounds were only a few feet tall. "The mounds weren't so big that you'd climb them. We just stood and looked at them," he pointed out. "There were two of them. One was bigger than the other. Both appeared to be oriented in an east-west direction. And, no, there weren't any vent holes. We sampled all around one and have brought back stuff from it – material, excuse me Uel Clanton." His apology was because the mention of 'stuff' had cost Clanton another dollar in the bet.[6] Conrad went on, "I guess you're hunting for anything volcanic in nature. The mounds appeared like big globs of something that had been pitched into that particular area, either by the craters that were formed nearby or something else further away. We looked around for any evidences of vent holes or anything coming out of it that might be scattered around – you know, rocks from itself or some ejected pyroclastics around on the ground near it. We couldn't find any of those either. I was kind of wondering at the time why you didn't ask us to take a core tube through it, but you didn't."

"Thanks," said Lind. "The next one is whether or not you noticed any preferential distribution of the glass beads and the glassy material?"

"Generally speaking, it was all over the place; in the bottom of even the smallest craters that we came across," Conrad replied. "I think we have three or four samples of glass that all looked the same. They were taken from different places and should be documented. One of them isn't, but I remember where we got it."

"I'm not sure I understand. Do you mean that the glass beads were in the bottoms of all craters, or that they were on the level surface as well as in the bottoms of the craters?"

Bean explained, "As we walked around on the level surface we found beads here and there. We'd run across big ones every once and a while. By big ones, I mean up to about three-eighths of an inch in size. But generally there were a lot of little ones around. Now the craters up to about 4 foot in diameter and 1 foot deep which didn't look as if they had been made by very fast particles, looking into those you'd usually find glass-covered rocks."

Conrad added, "If I remember, we have a rock which is some 2 inches or so in size that is spattered with glass and we brought it back for that reason."

[6] In a 1995 conversation with Eric Jones, Clanton explained that the bet had cost him a total of $14.

Bean resumed, "One time when we were walking around outside a big crater, we saw a rock about 3 inches in diameter, I guess, that was almost completely covered with glass."

"Very good," Lind replied. "Next question: Can you give us some more detail on the material that appeared to be melted in the bottom of Bench crater? Did this just cover the central peak or did it appear more extensively spread around down there?"

"It appeared a little bit lava-like in nature, but I don't mean to imply that I thought the crater was volcanic in origin," Conrad said. "It looked more to me like we were seeing the effects of some high-speed impact that melted some of the material down there. I wish we could have gone down in that crater to get you a sample but it was too steep and too rugged for us to attempt it. We did take some partial stereo pans of the whole crater and we sampled material from the top, but that didn't resemble the material in the bottom."

Bean expanded by referring to the volcanism that they had been shown in Hawaii, where lava from vents spattered and built up knobby-looking mounds. Although the material in the bottom of Bench looked similar, he emphasised, "neither of us think of it as a volcanic material".

Moving on, Lind asked, "On the northwest side of Head crater you talked about a rock that you kicked over and found the bottom to be different from the top. In what way was it was different. Remember that one?"

"I remember it," Conrad replied. "I should have clarified it then because it wasn't that big of a deal. I guess it was the first time that we kicked over a rock and found a difference in colour. Before, we'd kick one over and it'd be the same all over. This one looked a little bit lighter grey on the bottom. After I thought about it, the reason was because we were marching around in that same area where we noticed that there were two different types of soil. The dark topsoil was a thin layer of an eighth of an inch or something, and below that was light grey. It was causing the rock to appear white. So, I don't think it's a big thing."

"Roger," acknowledged Lind. "Let me give you the last two questions so you can cover them together. Are there any special or unusual features that you remember, thinking back on it now, that you didn't have time to describe? And can you recap the traverse along each leg and recall what you think was the significant feature that you saw at each of the stations where you stopped?"

After an onboard discussion, Conrad said, "We can't think of anything right now that we didn't mention to you at some time or another. Al's only comment, which he already talked about this morning, was the fact that the colour changed with the Sun angle between the first day and the second day. As far as the traverse goes, there was nothing unusual at Head crater other than it was where we first saw the difference in soil below the ground and at surface level. Our second stop was Bench crater, and we've just discussed the texture of the rocks at the bottom of that crater. I guess the next most significant thing was that somewhere between Bench and Sharp crater we obviously ran over a contact, because the ground very definitely changed to a softer, finer dust. We sank in deeper out there, not only right at Sharp crater but leading up to it. We both found it very difficult to walk slowly, ever. We always went at a lope. That just seemed the natural way to go. So, Al sort of spotted it first because he was behind me and could see that I was kicking up more dust."

"That's right," Bean agreed. "It was obvious that Pete had started running on a different kind of ground."

Conrad resumed the replay of the traverse, "We're not sure that we ever did get to Halo crater. There turned out to be about five little craters, any of which could have been Halo, but we were awful darn close. I guess the next significant thing was that in coming up to Halo we really got on a third type ground. We discussed this when we got to Surveyor crater. It seemed to be the firmest, especially down in that crater. It still had dust. We still sank in, but we sank in the least in that crater; both going down to the Surveyor from the one side and going up towards the LM through that blocky crater on the other side. The blocky crater was also an interesting feature, and I think we did discuss as we stood there that we felt that Surveyor crater was an old crater. I guess it had been impacted very early and hit bedrock, and this bedrock had been weathered down to where the crater was very smooth. Then came another one and made this small blocky crater in the side of it. This indicated to me that bedrock was not too far below the surface there. And, of course, we have samples of that."

After a brief pause, he continued, "Al just asked me to mention that the Surveyor, except for the fact that it had changed colour, looked to be in very good shape. This is true. But there's something that I noticed using the cutters. Supposedly, the tubes we used in practice were exactly the same metal and aluminium that the Surveyor was made out of. If this is the case, something very definitely happened to the metal, like it crystallised, because it was much easier to cut the Surveyor tubes – except the one tube that I simply couldn't cut; I think that must have been much stronger than they indicated, maybe a much thicker walled tube. But the wire bundles that we cut also had the appearance of being very brittle. They cut very, very easily. The coating flaked off the insulation. There was one wire bundle that had a cloth insulation on it which was not on our mock-up, and there was a wire bundle that wasn't in quite the same configuration. But these seemed to cut quite easily also. And I don't think it's because I was juiced up; there was very definite crystallisation or something there. You'll get to see that with the stuff we bring back."

"Very good," Lind said.

Bean offered a comment about the pace the traverse, "The thing that kept us from studying more details at each site was that we had to keep pressing on. So what's going to happen when we get back is, we're not going to know all the details of each site because we just weren't able to stay as long as we would have liked at any site. We could have spent that whole time in any of those craters, trenching around them and collecting rocks, and going back and forth to check a blanket to see if we could discover any difference in texture, and all that sort of thing. But the time just wasn't available."

"What Al's saying," Conrad added, offering an analogy in terms of the field trips during training, "is we did Big Bend, Hawaii, Meteor Crater and New Mexico all in one 2-hour trip around there; that's about what it amounted to."

"You did a great job in it, too," Lind praised. "Hey, listen. When you looked into the craters did you notice any boulder tracks to indicate that many rocks had rolled down, or were there accumulations of boulders at the bottom of these steep slopes?"

"No, not any particular distribution," Conrad replied. "When there were rocks in the bottom, it was in the blocky craters where it looked like the material had always been there." After a pause, he said, "I wasn't really standing in a position to observe any tracks made by the rocks that I rolled down. The rock that I threw in was so small that it didn't go very far anyhow. Now, dust flew and both rocks bounced and rolled depending on how far along it was going down the side of the crater. But it wasn't obvious to me that they were making any tracks."

Lind explained the interest, "The seismologists are trying to get some feeling for whether or not you thought there was a lot of rock rolling that might be causing the signals that they see."

"Most of the rocks that we saw on the sides of craters all had dust around the bottom of them," Conrad pointed out, "and it didn't look as if they had moved for a very long period of time. The majority looked like they were partially buried."

"That's right," Bean agreed. "We didn't see any that looked as if they were going to roll down in the near future."

Lind changed tack, "When you pulled out the core tubes did the holes collapse or did they stay there?"

Bean took it, "The minute we drew out the core tube, the top inch or so crumbled and parts fell in but the sides were still relatively vertical. It was the same with the trenches. As Pete dug a trench, the sides would be almost 90 degrees. Whenever he accidentally tapped the side with his shovel, that part would get knocked off and the other part would remain in place. As long as you didn't touch it, the wall seemed to be happy right there at 90 degrees." Something occurred to Conrad, "Although there wasn't any difference in colour, it seemed to me that there was some layering in the trenches." This discussion of the subsurface reinforced the point that Bean had made earlier in the day that future crews should be provided with better trenching tools.

After concluding this 30-minute technical discussion, Lind read up sports results and updated the crew on their families.

"Pete, Jane said Christopher got a bike for his birthday and within an hour he had a small accident. He didn't have any problems but the bike is a little worse for wear. He's a little disturbed over that. Also, Peter went to Elkin's Lake, deer hunting with the Allens. They're supposed to be back tomorrow afternoon, so, of course we have no report yet on his prowess as hunter. Jane is delighted with the way the flight is going."

"Thank-you, Don," Conrad said.

"Al, Sue reports that they had a real nice luncheon. The kids, of course, are home from school and today is a real nice day here in Houston so they've been having fun around the house. They are really getting anxious for Monday and getting you guys back down here on the Earth."

"Thanks," Bean said.

"We feel the same way, believe me," Conrad noted.

"Dick, Barbara reports Uncle Herb has repaired all the bicycles and the kids are really delighted. The luncheon went off very nicely. Also, Mom and Dad have gone with Aunt Mary to restock your house's larder. Barbara says the children have been particularly good today and she's very happy over that. They're all looking forward to splashdown."

"That sounds like a normal day around my household," Gordon replied.

Lind reported that the preliminary analysis of the steerable high-gain antenna test indicated that, as suspected, the problem was thermal in nature. It had been decided to repeat the test early the following morning.

Conrad complained that they were running out of batteries for the cassette player. As he would say in the post-mission debriefing, "We should have come home in two days instead of three days." One possibility for future crews was to adapt the player to use the spacecraft's 28 V_{DC} supply. A simpler option, as demonstrated by Frank Borman and Jim Lovell during their 14-day Gemini 7 flight in December 1965, was for each astronaut to take a novel and then pass them around.

As the day drew to a close, Lind asked if anyone intended to provide biomedical telemetry overnight. When Conrad said they wouldn't, Lind acceded and noted that it would allow the Flight Surgeon to get a good night's sleep. Conrad said, "He may be interested to hear that we're taking decongestants, but this spacecraft is so loaded with dust I can't believe it. What we brought back from the LM must have been the world's record for dust. I thought it would be cleaned within 12 hours, but we've been cleaning the inlet screens every couple of hours and are still getting junk off of them." Although the dirty items were bagged, the release of lunar dust was evidently ongoing, and as fast as dust accumulated on the inlets to the environmental control system and was cleaned off, more appeared and floated around. The presence of dust did not impair the operation of the vehicle, but it was a nuisance to the crew and was a potential hazard to eyes and lungs.

"The surgeons down here are reviewing their treatment for silicosis so we will be prepared for you," Lind joked.

"What in heaven's name is that?" Conrad asked.

"It's a miner's disease from breathing coal dust," Lind explained.

"Nighty-night," announced Conrad when the crew concluded their brief 10-hour day.

Cliff Charlesworth took over as Flight Director, with Paul Weitz again manning the communications console for the graveyard shift.

PRESS CONFERENCE

The astronauts awakened early on flight day 9, and Conrad's call of "Good morning, guys" took Weitz by surprise.

After a minute, Weitz replied, "Hello, Apollo 12."

"Don't tell me we've got Paul on the horn," Conrad laughed.

"Hey, what are you guys doing?" Weitz joked. "You were supposed to let *us* sleep as long as we wanted this morning."

After Weitz suggested that the crew proceed with breakfast, Conrad called down, "We've got a message we'd like you to send for us."

"Go ahead," replied Weitz.

Conrad said, "This is to Rear Admiral Davis, Recovery Forces, USS *Hornet*: Dear Red Dog, Apollo 12 with three tail-hookers expect recovery ship to make its PIM as we have energy for only one pass. Signed Pete, Dick, and Al."

"Copy," Weitz replied. This was more naval slang, with PIM standing for Point of Intended Movement.

Several minutes later, having taken down the window shades and looked at Earth, Conrad pointed out, "Houston, we just got our first glimpse of you this morning, and there's not very much of you out there!" The spacecraft was 140,000 nautical miles from Earth, which was almost occulting the Sun with only a sliver of the illuminated limb visible.

"Understand," Weitz replied.

The flight plan scheduled 'corridor control' corrections 22 hours and 3 hours prior to the entry interface. When Gordon asked about the situation, Weitz assured him that the spacecraft was "on the glide slope".

After breakfast Weitz read up a flight plan update that would permit engineers to gather more data on the stability of the steerable S-band antenna. Switching to news, he announced, "Among the hundreds of suggestions received by NASA on how to repair the Apollo 12 lunar camera was one calling for the use of a woman's hairpin. I don't know why you guys didn't think of that."

"We didn't have the woman!" Conrad laughed.

A few minutes later Weitz reported, "Cliff Charlesworth and his Green Team are going off-shift for the last time and they said to say they'll see you in Houston."

Conrad asked Weitz to pass on the crew's thanks to Charlesworth's team for their support during the lunar descent. Thinking about the forthcoming quarantine period in the Lunar Receiving Laboratory, he added, "Why not save that flight controllers' blast until after the eleventh of December? I'd sure like to be there."

"That's a promise," Weitz replied. He handed over to Ed Gibson as Pete Frank's team began their shift.

Getting down to business, Gordon terminated the overnight passive thermal control roll and took the sextant sightings for the third navigation exercise of the transearth coast. With this done, he oriented the spacecraft "with its back to the fire" for further tests of the high-gain antenna. During his time, communication was sporadic. After a full hour out of contact, Conrad called down to announce that it was "R&R time" onboard, by which he meant rest and recreation. As previously, the antenna test was terminated after 2.75 hours. The final fuel cell purge and waste water dump of the mission were then performed. As Gibson pointed out, this made the Flight Dynamics Officer happy because it would allow an entire day's worth of tracking to refine the projected parameters of the entry interface knowing that there would be no further perturbations.

"We have a preliminary result of your high-gain test," Gibson announced. In both tests the narrow-beam and reacquisition modes were operated until fluctuations in the uplink and downlink signal strengths were observed. At dropout during the first test, the mode was switched to wide-beam and the signal strength became normal. The second test included acquisition in the wide-beam mode after fluctuations were observed in narrow-beam, and normal signal levels were restored after acquisition. It occurred only in automatic and auto-reacquisition narrow-beam modes. The signal strength was reduced by about the same magnitude in both the uplink and downlink, usually by about 12 decibels. Switching between primary and secondary electronics had no effect. A normal signal could be restored simply by switching to manual and aligning the antenna to Earth. "The malfunction," Gibson said, "has been isolated to the high-gain antenna radio-frequency area, thus eliminating the high-gain antenna, the electronics box, and the S-band transponder. The problem is associated with the dynamic thermal operation of the antenna." It was inferred that the actual fault was a temperature sensitive circuit connection, either a solder crack or a wire break.

Having been told that Bean was catching up on sleep, Gibson asked someone to check his bioharness because the electrocardiogram was not giving a valid reading. "I'm sure he's alive," Conrad laughed. "He only comes around every once in a while for a meal." When Bean reported that the electrical connectors looked okay, Gibson relayed the recommendation of Flight Surgeon John Zieglschmid that he remove and reapply the sensors. Meanwhile, Gordon performed the day's second P23 navigation exercise.

Gerry Griffin took over as Flight Director, with Gerry Carr as Capcom.

"Where've you been all day," Bean asked.

"I just watched the Oilers getting started with Miami," Carr replied, referring to a football game between Houston and Miami.

Mission Control had decided to forgo the first corridor correction, so after making yet another navigation exercise Gordon restored the passive thermal control roll. As he took sextant sightings to update the inertial platform while the spacecraft was in motion, Conrad observed, "That Dick Gordon is getting pretty fancy in PTC, isn't he?"

"Pretty slick," Carr agreed.

"A few more days and I'll understand it," Gordon joked.

As they had lunch Carr provided a sports roundup, including the fact that the Oilers had beaten Miami 32 to 7.

A scheduled television broadcast had been postponed by an hour to improve the line of sight to the Goldstone antenna. When the slow roll reached an angle in which it would be convenient to maintain the high-gain antenna pointing at Earth, Gordon halted the rotation. Carr suggested a test, "On that high-gain, you can go ahead and just go to ReAcq, Narrow Beam, and let's see what happens. If it goes sour on us, we'll have to go Manual." The antenna locked on without difficulty. As the crew set up for the broadcast, Carr read up some more sports results.

"You look great but you're upside down," Carr said when the image appeared on his screen. To enable all three astronauts to appear in-shot they had lined up in the lower equipment bay beyond the end of their couches, with the television camera in position near the side hatch, unfortunately inverted. "You look like a bunch of bats hanging from the ceiling."

After the camera had been adjusted, Carr got down to business. "First of all, I will read you a little statement. The questions you will be asked in this press conference have been submitted by newsmen here at the Manned Spacecraft Center who have

On the way home Dick Gordon (left), Pete Conrad and Al Bean gave a televised press conference in which they answered questions relayed by the Capcom.

been covering the flight. Some of the questions they raise will have been answered in your communications with Mission Control but the public at large has not heard them. The questions are being read to you exactly as submitted by the newsmen, and in an order of priority specified by them. So, here comes question number 1: If you had this mission to fly over again, or were planning another one with your present knowledge, what would you do differently and what equipment would you add or modify specifically in connection with the EVA?"

Conrad went first, "I think we'd work over all the tools, and the tool carrier and the bags. I think it was very good equipment and it operated very well, seeing as we had never attempted to do that kind of work. But now that we've done it, I think we can make some improvements."

Bean added, "I think as far as the PLSS and the OPS and the suits are concerned, you couldn't ask for anything better. But the tools are going to need a little work."

As Carr began the second question Gordon interjected, "I would have wagered a little more with all those people who said I would never be able to find the LM on the lunar surface. In fact, if I had been that smart I would have bet them I'd find the Surveyor also. And I'd retire."

"Roger, Dick," Carr laughed. "Question number 2. Yesterday you said something about the launch into the thunder clouds. Would you, or would you not, consent to launching under those conditions again?"

"I'd go again," Conrad replied.

"We made it this time," Gordon pointed out. "Why couldn't we do it again?"

"Concur," Bean said.

"Question number 3," Carr continued. "Aside from the lightning, what gave you your most apprehensive moment, if any, either before the lunar landing, during your time on the Moon, or afterwards? And if you never had an apprehensive moment, was there ever a time when you may have been a little bit concerned over what was going on?"

Bean noted the current elapsed time, and replied, "Well, how about from liftoff all the way through to a GET of 224 hours 14 minutes 30 seconds?"

"That's a pretty good answer," Conrad agreed. "I think all three of us were a lot calmer through most of the flight than I thought we would be, or any of us thought we would be, except that I think both Al and I were a little bit nervous about ascent from the Moon. I think we were a little tweaked right towards T = 0, but once we got going it really didn't concern either one of us after that."

"The only time I can think of," Bean said, "is one time when we were walking on the lunar surface on the second EVA, I felt my suit pressure kind of pulse. And that PLSS is so good that you never feel any change in pressure as you walk around, or jump up and down, or suchlike. This one time it did, and I took a quick glance at my suit pressure gauge because I thought maybe it was building up or decreasing, some sort of problem where I would have had to use the OPS, but it wasn't on the gauge. So that was the end of it."

Gordon added his own thoughts, "Switching to the command module side, Gerry, everything has gone according to plan. I think that the best thing about my end of the operation is that there have not been any surprises – and I'd like to keep it that way."

"That's kind of nice, no surprises," Carr agreed. "Pete, question number 4 is for you. Out there on the Moon you sounded happy, even euphoric. Some people think that maybe you were on an oxygen high. Were you? And for both you and Al, how did it feel subjectively to be out there?"

"I was very happy but I wasn't on any oxygen high," Conrad replied. "I was very happy because all the work that we had put into that EVA was beginning to pay off. And once we got over the little problem of getting the fuel cask going, I was happy because we were on the timeline – everything was going the way we thought it was going to go. I was having a ball, because it was much easier than the one-g practice we had done in learning how to do that."

Bean expanded, "You're asking about how it feels. I think for about the first 10 minutes that you're out, at least in my case, you find that it is not as hard to move around as you thought it'd be and you're pretty happy about that. But you know that during those first 10 minutes you still must be careful and not overextend yourself, so you're sort of excited trying to get up to speed, get your balance in good shape, and get your movements in good shape so you can start doing the work. And once the first 10 minutes is over and you sort of realise that you now know how to hold your balance and you aren't going to fall down, and everything is working real well, right then you get down to the operational part of it. After that you just press on and get the job done, like Pete said."

"I was in a good humour to start with, seeing we'd landed next to the Surveyor," Conrad added. "That started the thing off right."

"Question number 5. On Apollo 11 Armstrong and Aldrin had to curl up in the corners of the LM to sleep, and complained that they were cold and uncomfortable. You had hammocks and blankets. How did you sleep? And on the subject of sleep, a lot of people are wondering whether you dreamed there on the Moon."

"Well, let's take those in order," Conrad said. "In the first place, we didn't have any blankets. We had the hammocks. And as you may or may not remember, about a week before the flight we found a problem with the boot of my backup suit and all of our suits were returned to the factory so that the boots could be replaced. In the process of doing that, the suits had to be re-rigged when they came back to the Cape. I had to fit my suit without the liquid-cooled garment, because both the flight ones were already packed and you can't put a non-flight one into a flight suit. And as a result, the legs got a little bit too tight. In my hammock that night, I didn't want to take my suit off; it was too dirty in there. I was very uncomfortable. The suit was pressing on the bottom of my feet and my shoulders, and it sounds funny but even by bending your knees you can't get rid of that. If the suit's too short, it's too short. It was about a half an inch too short. Mostly on account of that I only slept maybe four and a half hours. Then Al, very kindly, the next morning, let my suit out for me. That took him about an hour. So that's how I spent my night. As for dreams, I don't dream normally anyhow that I can remember, and I didn't dream there."

Bean added his experience, "I didn't sleep too good on the Moon. Not because we were cold or hot, because we weren't; we had the liquid-cooled garments on and had air running through our suits. And so if we got a little warm we could either turn on the water pump and get a little cool water running to rapidly cool down that way, or get a little air running to cool down. Using those two controls, Pete and I stayed just about the temperature we wanted. And the hammocks were very comfortable. They are pretty long, and it is interesting that when you rig one on Earth you think it will really sag once you get in, but on the Moon you only weigh about 30 or 35 pounds and when I looked at Pete up on his hammock it hardly sagged at all. You just kind of lay there almost horizontal. A real comfortable place to sleep."

"And you didn't dream either, huh?" Carr prompted.

"No, I didn't dream a bit. I woke up and went back to sleep a number of times. Another interesting thing: people have worried about the sound in the LM bothering you. It's fairly noisy in there and there's a couple of pumps that change frequency every once in a while but, all in all, I don't think that was any hindrance to sleep. Do you, Pete?"

"No."

Bean continued, "The one-sixth g is nice. It pushes you down enough so that you feel pressure on your back, or your side, or wherever, but it's not enough to really give you any pressure points in the suit. I think one-sixth is nicer than either zero or one g to sleep in. It's a good happy medium. It's pleasant."

"Question number 6 is for Dick Gordon," Carr announced. "Dick, how does it feel to be alone for a day and a half in orbit around the Moon? And what were you able to observe of Pete and Al's activities on the surface? Also, when the LM was crash landed after rendezvous, were any of you able to see the impact."

"It's a little hard to really express how one would feel, being very close through all that training, being together for 4 days on the way out, and then being left alone to tend to the command module in lunar orbit while they're down there for some 32 hours of the lunar stay. I thought beforehand what it would be like to be completely alone on the back side of the Moon and no contact with any other human being; but surprisingly enough the activities were such that I was awfully busy. I didn't really have time to dwell upon that. And, to be perfectly frank, I was so blasted tired at the end of the day that I could hardly get to bed fast enough to get enough sleep to carry me through the next day's activities, which were busy in themselves because of the photographic requirements that were levied on me in lunar orbit while Pete and Al were working on the surface. I never did observe them personally on the surface, although I did see the LM through the optics (which are right behind us) and I also saw the Surveyor in the crater. I saw both of these objects twice on two different passes. On one pass I put the camera on the sextant. Hopefully I'll have pictures of that for my doubting friends. All in all, I think that describes the activities I went through while Pete and Al were down on the surface. Would you repeat the last part of your question, Gerry?"

"That's about the crash landing, Dick. Did you see it go in?"

"No. None of us saw it go in," Gordon replied. "After we separated, I tracked the LM for a considerable length of time in the optics and thought I had a pretty good state vector for it so that the Auto optics would track automatically. I put the camera on the sextant, hoping it would automatically track the LM into the lunar surface. I don't know whether we were successful with that or not. I have my doubts about it. Certainly, none of us saw it with the naked eye."

"Question number 7," Carr said, pushing on. "You mentioned during the EVA finding three kinds of soil. Will you give a brief description of each – its colour, its texture and so forth – and discuss any problems you had in handling all the different kinds of lunar material."

Conrad took it, "When we say three different kinds of soil, that was a subjective thing in that the colours were all the same. Some soil was firmer than other soil, in the manner in which we sank into it. And the softer soil that we sank deeper into was of a finer grain. This was at Sharp crater, which was about as far away as we could get from the LM. We took samples of these types of soil. The medium-textured one was where we landed on one side of the Surveyor crater. On the other side, when we went down to get to the Surveyor, we found the ground was, I'd say, considerably firmer. We didn't sink in quite as much as over by the LM. Over at Sharp crater was the softest ground, and we sank in the deepest there. Do you have anything to add to that, Al?"

"No, you covered it," Bean replied. However, he spoke about the colours. "One of the real difficult things about the whole EVA, in the geology part of it, was the fact there didn't appear to be any difference in colour among either the rocks or the soils. They all looked about the same. The first day, to me, they all looked sort of a dull grey. I think I described most of the rocks that way, as a dull grey. And the soil is a dull grey. If you look real close, maybe you can find a white rock now and then, or you could maybe disturb something and get a little darker grey. But generally it was all grey. The second day we went out, at least to me, the things that looked grey the first day started to look dark brown, or a tannish brown. One of the most interesting things of the lunar surface operations was how much that colour could change with a change in Sun angle of 7 degrees or so. In fact, when we came upon the Surveyor, you'll recall it was brown and we asked you if it had been painted that way and you said no; it had really been white. When we got up next to it, sure enough it looked brown and the coating on it was the same brown as the soil. Now, I wouldn't be a bit surprised if when we get all those parts back to Houston and under the light of the laboratory they turn out to be kind of a dark grey again. It's going to make geology quite a bit more difficult than on Earth because the colour cues just aren't going to be there; you're going to have to look for texture, fracture, lustre and all those other things to aid you in determining differences in rocks and minerals."

"Let's move on to the next question; it's number 8," Carr said. "Were the Moon's colour, texture and general appearance, as seen from above, as you expected? And is there any place on Earth you know of that looks like the Ocean of Storms?"

Having the most experience of observing the Moon from orbit Gordon took this one, "No. I can't think that there is. It reminds me of desert areas, [...] particularly the back side of the Moon, which is a lot more beat up than the front side. As for the Ocean of Storms, I guess there really isn't any one place on Earth that we can recall."

"The next question is for Pete," Carr announced. "Pete, everybody is wondering about the fall you took on the Moon. Was that accidental or on purpose? How did it feel to fall in the weak lunar gravity, and could you have recovered your footing if Al Bean hadn't been there to help you?"

"I didn't fall on purpose!" Conrad laughed. "I was standing next to Al, trying to pick up a rock that was too big for the tongs. We sort of had a little game we played there of leaning on the tongs and sort of doing a one-armed jabber-do all stretched out. I just sort of rolled over on my side down there on the ground and Al, before I got all the way down, just gave me a shove back up again. I don't think it'll be any problem – the business of falling against a rock and cutting your suit or something. You don't fall that fast. You just wouldn't hit a rock hard enough."

Bean added, "You can lose your balance sort of quickly, particularly if you try to back up and because the ground is uneven you either step into a hole or trip over a rock. But you fall so slowly that it gives you plenty of time almost to turn around, or to catch your footing before you actually get so low that it's too late. I can recall a number of times when I lost my balance. If I'd lost my balance that much on Earth, I would have probably fallen down. But on the Moon, because you start moving so slowly, you are usually able to spin around and bend your knees and recover."

"I think that's another thing," Conrad said. "I saw Al do this two or three times. In trying to bend over to get something, you'd start to fall over. You fall so slowly that you just take a step and keep moving until your feet come back under you again. So it's not that easy to fall over up there, for that matter. And I really don't think there's any problem."

Carr moved on, "I think you've pretty well answered question number 10, but I'll read it anyway and you can add any more thoughts you might have. While you were inspecting the Surveyor spacecraft down there in the crater, you commented about changes in its appearance: the white part seemed to have turned tan, and so on. Will you discuss this further and give us any impressions or conclusions you may have about what caused these changes?"

"Well," began Conrad, "this brown colour is definitely lunar dust that's on it.

And it was evenly distributed all the way around it so I don't think it's dust that we blew as we landed. I think it's accumulated there. It wasn't that easy to wipe off. And the other thing that I think was most apparent to Al and I in cutting the tubes was that in practice (and I'll have to check this) we had the same aluminium tubing as the struts were on the Surveyor, but the tubing appeared much more brittle and much easier to cut up there, so I suspect that some crystallisation or something has happened to the metal in the 31 months it has been sitting there. And the wire bundles that we had to cut, we noticed the insulation had gotten very dry, very hard and very brittle."

"Okay, here's number 11. Do you think that future EVAs can be extended beyond the 4-hour limit? Or do you believe the number of 4-hour EVAs should be increased in order to get more exploration done on each mission?"

Conrad said, "I think you ought to go a longer time on each EVA. We felt badly, sort of, that we got shut off the other day." Early in the planning of Apollo 11 it had been intended that Neil Armstrong and Buzz Aldrin should make two moonwalks. On the first one they would familiarise themselves with the surface environment and collect some samples, and on the second they would deploy a full ALSEP. However, when development delays meant the ALSEP would not be ready in time, the second excursion was deleted. The full program of work was given to Apollo 12 and the moonwalks extended to 4 hours. As Conrad now noted, they had made an early start on the second moonwalk and had 6 hours worth of consumables on their backs. "As far as being tired or anything, we weren't tired and we could have kept on going. We hustled to get back just to make our 4-hour deadline! I think that the big problem is getting suited up, and getting unsuited when you get back in. Doing the work outside is easy. Once you step down the ladder, you're on your way." As Conrad was aware, NASA was already planning longer missions. Referring to these plans, he continued, "I think what you should do is get a long-term PLSS. And if you have a 3-day LM, a PLSS that can stay out for 8 or 9 hours, some way to allow the guy a drink of water and maybe a shot of food, and allow him to sit down and take a little siesta out there for a half an hour in the middle of it, he can do an 8-hour day's work out there. The big thing will be getting it all on, and getting it all off and putting it away when you get back inside."

"I think there's another significant problem Pete didn't mention," Gordon added. "That's the amount of dirt that you bring back in the spacecraft with you. Both Pete and Al, although they'd been in the LM for a considerable length of time before they got back into the command module, still brought back a tremendous amount of dirt and dust in their clothing and on their persons, and I think if you're going to work in that environment for any length of time you are really going to have to tackle this problem of keeping clean."

"Thank-you. That was a good one," Carr said. "Question number 12. For future lunar explorations is a geologist a desirable member of the crew?" This may have been prompted by Conrad's earlier remark that even a professional geologist would find field work on the Moon problematic owing to the ubiquitous dust. The issue of whether NASA should be sending fighter pilots to attempt scientific research was a long-running debate. Amongst the small group of scientists that NASA recruited in

1965 were Jack Schmitt and Ed Gibson. Being a professional geologist, Schmitt was likely to be the first member of this group to reach the Moon. But how long must he wait?

Although Conrad began his reply by saying that "certainly" the program should enable a geologist to reach the lunar surface, he warned, "I'll tell you one thing: it took every bit of knowledge that I had to get that baby down there in the right place. That was no easy task. As a matter of fact, we were discussing this earlier today. I'm a big advocate of the LLTV. It was a tremendous help to me. This is my profession, and it took everything I had to get that LM down in one piece. We have some things to work out to make those tasks easier, but once the transportation system is worked out we'll take the necessary people. There's no doubt that a geologist can do a better job than I can; I'm not a geologist."

"Roger, Pete," said Carr. "This is the last question now. Millions of people who stayed up late one night last week are wondering what happened to that TV camera."

Conrad replied, "All I know is you told me you were getting a picture and then I didn't pay any more attention to it until I heard you talking with Al. We don't know what happened to the camera, but we've brought it back with us so you can find out what is wrong with it."

Having finished the formal questions, Carr asked if the crew had anything of their own to say, and Bean seized the opportunity. "Pete, Dick and I spent a couple of years getting ready for this mission, both backing up Apollo 9 and working on this one. We spent a lot of time sitting around thinking about what our chances were of actually getting to the Moon, and landing there and coming back home. Every one of these missions boils down to about three big things. One, you've got to have trained people that operate the spacecraft and that operate on Earth as flight controllers, and we felt pretty good about that. We have been training hard. We've worked with the flight controllers. And we know they have worked hard. You've got to have a good set of procedures to work by, and people like Bill Tindall and all the men who work with him have spent many long hours and a heck of a lot of effort developing them. And we were pretty happy about that. Now that leaves the hardware: the machinery that's got to work – the Saturn V, and the command module and the LM. That was our sort of big unknown. We knew these are made of millions and millions of parts, and it doesn't take very many parts to go bad before you abort a lunar mission. It's a long chain of events, and any one of them can cause you to come back home early. We, of course, couldn't walk around and check all the parts on any of these things, because we don't know that much about it. We did know the people that worked on it, though. People like Jim Harrington at the Cape, and Buzz Hello there, and Chuck Tringali, our team leader, and a lot of others I can't mention right now. We kind of felt pretty good about the fact that they were handling the gear. We're on the way home now – we'll be back tomorrow – and every bit of this machinery has worked beautifully. We've had a couple of small failures, but none of the equipment that we worried about has shown anything except perfect performance. The fuel cells, for example, are perking along as beautifully as they can, putting out 20 amps apiece and holding their own just perfectly. I think this is a fantastic tribute to the people who designed this equipment, and the people who built it, and the people at the Cape who checked it out. I'm pretty proud of the Apollo 12 mission. We did all the things we were supposed to do. I hope that all of those people there that had anything to do with this hardware, that built it, that designed it, or that checked it out, feel as proud about this mission as I do."

"Roger, Al," Carr replied. "I think I speak for everybody down here when I say that we're all darn proud of it ourselves."

"We have got one last thing to show you and then we'll close," Conrad said. "We wrote a little inscription over the FDAI and signed it." They zoomed the camera in to show the script '*Yankee Clipper sailed with Intrepid to the Sea of Storms, Moon, November 14, 1969*'. With that, they concluded the 37-minute transmission.

Homeward bound, Conrad realised that Apollo 12 had not been as momentous an event in his life as he had expected – he had trained for so long that actually doing it was something of an anticlimax!

As Gordon reinstated the passive thermal control roll, his colleagues took care of housekeeping chores in readiness for the final sleep period, then Conrad signed off, "Nighty night."

"We'll see you in the morning," Carr replied. Shortly thereafter, Glynn Lunney took over as Flight Director.

THE FINAL LEG

Although there were signs that the astronauts had awakened early, Don Lind waited for them to make contact. "Good morning," Gordon announced. "We're just getting cleaned up and eating breakfast."

"Are you ready for the big day?" Lind asked.

"What's happening?" Gordon asked innocently.

"We've got a nice little section of the South Pacific reserved for you," Lind noted. "And we have most of the Navy standing by to pick you up."

"You've cleared certain area at all altitudes, huh?" Gordon mused.

"That's affirmative."

Apollo 12 was 62,500 nautical miles from the Earth, travelling at 7,330 feet per second and accelerating rapidly.

After breakfast Gordon halted the passive thermal control roll, then performed his final P23 navigation and updated the state vector. As he prepared to reinstate the roll, Lind said that Mission Control would soon uplink a much better state vector to supersede that developed onboard. This caused Gordon to take umbrage, "I didn't particularly appreciate your saying 'much better'. You could have said 'a little better', couldn't you?"

"Sorry about that," Lind replied. "It's early in the morning down here."

Several minutes later Lind announced, "I am about to turn you over to the tender care of Paul Weitz, so we'll see you when you get back to the LRL. And have fun in the South Pacific."

"Okay, Don," Gordon replied. "Thanks for all your help."

Pete Frank took over as Flight Director to finish the mission, with Weitz on the communications console in reward for serving so many graveyard shifts.

"Good morning, Paul," Gordon welcomed, and promptly posed a question. Just as the positions of the Earth, Sun and Moon had enabled the crew to view much of the Moon's illuminated hemisphere on the way out, the geometry meant that Earth was very close to the Sun in the sky on the way home, denying them a clear view of their destination. However, the compensation was an unprecedented opportunity to view a solar eclipse as the vehicle passed through the Earth's shadow. It was not a formal objective, but the crew thought they ought to document the sight, so Gordon asked, "Is anybody down there thinking about our photographing this eclipse?" He warned that the only film remaining for the Hasselblad was black-and-white.

"We'll check on it," Weitz promised.

As the crew ran self-tests on some of the equipment for atmospheric entry, Weitz informed Gordon that the final corridor correction manoeuvre would be of the order of 2.5 feet per second. An hour and a half after the query, Weitz relayed the science team's recommendations for photographing the solar corona. "They just want you to take photos coming out of the shadow. Use the Hasselblad with the 80-mm lens, an f-stop of 2.8 and the focus at infinity. They want you to take as many photos as you can, starting about 2 minutes before your sunrise. Initially, you should use a shutter speed of 1 second. As soon as you see any sign of the solar disk you should change to 1/125th of a second and take photos at that setting for 10 seconds. Finally, as the Sun comes up, stop down to f/16 and take a bunch of photos at 1/500th of a second."

With the corridor correction imminent Gordon halted the passive thermal control roll. In reading up the details of the manoeuvre, Weitz concluded with, "We're going around the room now to see whether or not to give you a Go for entry." In fact, of course, Apollo 12 had no alternative than to enter the atmosphere. Shortly thereafter, he supplied the details for that too. Mission Control uplinked the Earth-based REFSMMAT and Gordon took sextant sightings on stars to refine the inertial platform.

An hour later Gordon announced, "We're getting a spectacular view at eclipse." They had removed the strong filter from the sextant and were holding it against the window. "It's unbelievable." From their perspective, the Earth was about 15 times larger than the Sun and the limb was occulting part of the solar disk. "It's not quite a straight line," he noted. "It looks quite a bit different than when you see the Moon eclipse the Sun."

The eclipse had yet to reach totality, and the sight was fascinating. The exposure details Mission Control had supplied were for the faint coronal glow as the Sun was leaving totality, so Bean asked, "Does anyone down there know what we can set the camera at in order to use the Sun filter to take a couple of shots of this eclipse?"

"Stand by and we'll check," Weitz replied.

"They'd better hustle," Bean warned, seeing how quickly things were changing. "It's funny, you cannot see the Earth at all when you just shield your hand from the Sun and look where the Earth should be. It's not there at all."

"Roger, Al," Weitz acknowledged.

Apollo 12 was some 30,000 nautical miles from Earth and travelling at just over 11,000 feet per second.

"Fantastic sight," Bean continued his commentary as the Earth's disk encroached on the last sliver of the Sun. "What we see now is that the Sun is almost completely eclipsed and it is illuminating the entire atmosphere all the way around the Earth."

"We're working on getting a procedure for taking some photographs of it," Weitz assured.

Reasoning that the illumination at the onset of totality ought to be the same as just prior to the end of totality, Bean noted, "Man, it's too light." The scientists, thinking in terms of solar eclipses viewed from the Earth's surface, had evidently thought that once the Sun was fully occulted, the overall light level would diminish to reveal the faint solar corona, but the Earth's atmosphere was refracting sunlight to produce the most spectacular sunset ever witnessed by human eyes. The scientists ought to have known better, because on 24 April 1967 Surveyor 3 had shown what a solar eclipse looks like from the lunar surface, and in particular that the solar corona was washed out by the glow of the atmosphere. "It really looks pretty. You can't see the Earth. It's black, just like space. You can't see any features. All you can see is this sort of purple-blue, orange, and some shades of violet, completely around the Earth."

Conrad pointed out, "It's very interesting, looking at the atmosphere. It has blues and pinks in it but instead of being banded it is segmented, which is very peculiar; I really don't understand why. It may be the difference between over the landmasses and water or something."

"Is it kind of like you would see in the desert in the evening sometimes when you get that blue and pink streaking in the sky?" Weitz asked.

"Yes. Except, like I say, it's segmented," Conrad replied. "About a quarter of the Earth is pure blue, then it becomes pink for about 20 degrees of arc, and then it turns to blue again all the way around the bottom to where it turns pink and finally blue again."

"It's a heck of a time to be without any 70-mm colour film, I'll tell you," mused Bean. "But we have some colour on a 16-mm camera. Have you got a suggestion for the f-stop?"

"We're working on it," Weitz said, although Mission Control was far behind the curve in providing real-time support for the eclipse photography. In the absence of a recommendation, Gordon reported that they had set the movie camera to an aperture of f/1.4, an exposure of 1/60th of a second, and were shooting 1 frame per second.

"It looks like the atmosphere is going to be illuminated for the entire time of the eclipse," Bean warned, implying that they stood little chance of observing the solar corona.

"According to our figures," Weitz said, "you should still be seeing a little piece of the Sun."

"You're absolutely correct," Gordon confirmed. "We still have a little bit of Sun through the horizon on the western limb. But right now the atmosphere of the Earth is completely illuminated all the way around, 360 degrees, and right in the centre it's as black as space itself. This is really spectacular. Have you got any more adjectives for spectacular? I'd like to use some if you have."

Apollo 12's return trajectory produced a solar eclipse, the sheer spectacle of which awed the crew.

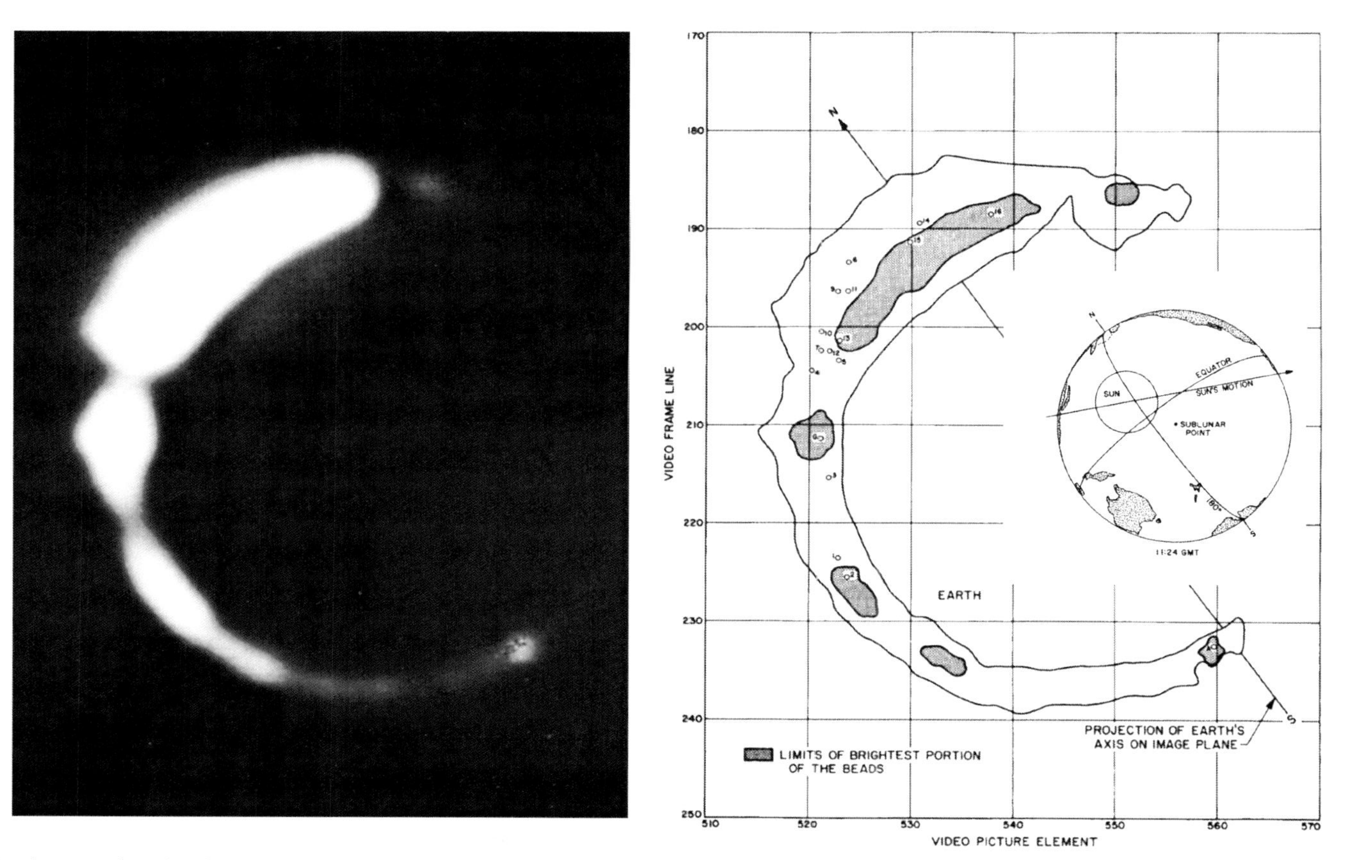

A picture taken by Surveyor 3 on 24 April 1967 showing Earth eclipsing the Sun, with associated exposition of how the Earth's atmosphere refracted sunlight.

"We'll put somebody to work on that, too," Weitz joked.

When Gordon noted that the track of the Sun behind the Earth's disk seemed to be offset from the centre, Weitz promised to have this checked out.

Whereas the scientists were primarily interested in the solar corona, the astronauts were more excited by the occulting planet. As their eyes adapted to the illumination they began to resolve details on the night hemisphere, which was lit by moonlight.

"This has got to be the most spectacular sight of the whole flight," Bean enthused. "Now that the Sun is behind the Earth, we can see clouds on the dark hemisphere. And, of course, the Earth is still defined by this thin blue-and-red segmented band. It's a little bit thicker down where the Sun just set than it is at the other side, but it is really a fantastic sight. The clouds appear sort of a pinkish grey and are scattered all the way around the Earth."

Conrad added, "We haven't been able to distinguish landmasses yet, but we might be able to in a minute when we get better adapted."

"Say, Houston," Gordon called. "We can see lightning in thunderstorms."

"They look sort of just like fireflies down there blinking off and on," Bean added.

When Gordon asked about their range, Weitz said they were some 25,750 nautical miles out, just above the altitude of the geostationary satellites.

"I'd sure hate to run into one up here," Conrad joked.

"It could ruin your day," Weitz agreed.

In fact, because geostationary satellites are in the equatorial plane and Apollo 12's approach trajectory was not there was no chance of a collision.

Just before the crew broke off to make the corridor control manoeuvre, Weitz was able to supply the answer to their query about the Sun's track. The line did not pass directly behind the Earth's disk. "It went behind the western limb of the Earth in the northern hemisphere, and should reappear still in the northern hemisphere."

"Roger," replied Conrad absently, as he prepared to make the burn. A 5.7-second thruster firing produced the delta-V of 2.4 feet per second required to steepen the entry angle from 6.22 degrees below local horizontal to a more desirable 6.47 degrees. Resuming sight-seeing, Conrad said, "We're better night-adapted now, and by golly we can see India, we can see the Red Sea, and we can see the Indian Ocean. It is amazing how well we can see, for that matter. We can see Burma and the clouds going around the coastline of Burma, and we can see Africa and the Gulf of Aqaba. We can just barely distinguish the lights of large towns by the naked eye, and by using the monocular we can confirm that's what we're seeing."

"You may now hold the class record for seeing lights," Weitz noted.

Conrad continued excitedly, "There's a couple of ripdoozer thunderstorms down there that are really, really letting go."

"Can you tell where the thunderstorms are, Pete?" Weitz asked.

"I'd say about 2,300 miles to the southwest of the tip of India. There seems to be a weather system out there, and it has got thunderstorms all the way along it. Venus is just below the Earth, and we can see it quite clearly. Well, you can see all kinds of stars, but Venus is just below the Earth. This is really a sight to behold – to see it at night-time like this."

"The satellite picture does show quite extensive cloud coverage of the area where you are reporting the lightning," Weitz confirmed.

"It looks like just north of India, and I'd say all up through China and Russia, it's completely covered with clouds," Conrad added.

"Understand, Pete," Weitz said.

"Also, right in the centre of the Earth now we have some real bright light shining steadily. Dick's looking at that with the monocular. It's really bright." After Conrad estimated it to be "south of Burma and east of India", Weitz noted that it was at their nadir point. "I can't imagine what that is," Conrad admitted. Weitz said that Mission Control would try to figure it out. It proved to be the glint of the 'full' Moon behind the spacecraft reflecting off the dark ocean.

"Looking at the airglow with the monocular is another sight that isn't like being in Earth orbit," Conrad continued. "It is bright red next to the Earth, and then it's got a green band in it, and then it's got a blue band."

"Would you say these colour bands encircle the Earth now, Pete?" Weitz asked.

"Yes," Conrad said. "But it's not the same all the way around. What I'm seeing is sunrise, really. This is about 40 degrees from the Sun, and there's a bright red band, a sort of a light green band that is very thin, and then a blue one which must be all of the atmosphere."

As totality drew to a close, the increasing glow precluded further study of detail on the night-time hemisphere, leaving only the wait to catch the first glimpse of the solar disk. By now Apollo 12 was 20,202 nautical miles from Earth and travelling at 13,400 feet per second.

In the post-mission debriefing Conrad said of this episode, "We were all caught with our pants down. We should have had good camera settings and film available, because it was certainly a spectacular sight." Gordon concurred, "I feel very strongly about this. I think that someone, the crew as much as anyone, really dropped the ball on this. We knew before launch this was going to occur, and we mentioned it. The people who are interested in this type of thing (if there was any interest in it) were very remiss in not planning further for this particular event. To us, it was one of the most spectacular things we saw. I'm sure there is some scientific value in this type of thing. But the reaction in this regard was virtually nil. The response of the people in Mission Control at the time was extremely poor. The crew was left on their own to come up with guesses on camera settings. Repeated enquiries to the ground took a considerable length of time before any information was forthcoming."

With the sightseeing over, the crew knuckled down to preparing the spacecraft for the final phase of the mission. After half an hour of virtual radio silence, Weitz said, "I have your landing area information, if you're interested."

"Go ahead," replied Conrad.

Weitz read up the forecast. "We're calling for good weather; 1,800 feet, scattered, variable, broken, 10 miles; the wind is out of the east at 15; we've got 3-foot waves on top of 5-foot swells, and they're running about 40 degrees apart. The altimeter is 2,988, which gives a delta-H of plus 38 feet. Your landing time now looks like 20:58 Zulu. Sunrise was at 16:12 Zulu and sunset will be at 04:24. There are some widely scattered showers in the area, less than 10 per cent." Zulu meant Greenwich Mean Time. The target point was 15.817°S, 165.167°W, just to the east of a zig-zag in the International Date Line. The prime recovery ship, the aircraft carrier USS *Hornet*, which in July had retrieved the Apollo 11 crew, would be 5 nautical miles north and 2 nautical miles east of the target point, its position established by celestial fixes and satellite tracking methods, placing it 5.25 nautical miles straight-line distance away. The job was to be directed from the Recovery Operations Control Room in Houston, with the support of the Pacific Recovery Control Center in Kunia on the island of Oahu in Hawaii.

Shortly thereafter, Gordon took sightings for a final P52 to update the inertial platform. With just over an hour remaining to the entry interface, Mission Control uplinked a new state vector into the computer.

A water-glycol coolant system had controlled the command module's temperature throughout the flight by using radiators on the service module for heat rejection, but the service module had to be jettisoned prior to entry. A water boiler linked to an evaporator was used to chill the command module systems immediately prior to entry and to serve as a backup to the primary evaporator during entry.

For entry Gordon flew the left-hand couch, with Conrad in the centre and Bean on the right. With 19 minutes remaining to the entry interface, he pitched the vehicle to check the alignment of the horizon against a scale on his rendezvous window.

The command module had its own reaction control system, which comprised two independent sets of six motors, each delivering 93 pounds of thrust. Like the service module, the propellants were monomethyl hydrazine and nitrogen tetroxide in tanks pressurised by helium. The system was solely to provide attitude control prior to and during atmospheric entry. After arming and checking this system, Gordon yawed the spacecraft 45 degrees out of plane and then, 15 minutes prior to the entry interface, jettisoned the service module. A guillotine severed the connecting wires and tubing in the linking umbilical, and pyrotechnics cut the structural supports. He then yawed the command module back in plane and took up a 'blunt-end forward' and 'heads down' attitude so that once the conical vehicle had entered the atmosphere its offset centre of mass would generate the desired lift vector.

"Clouds and thunderstorms," Bean observed. "Beautiful view out there."

"That horizon isn't much, is it, Al?" Gordon said.

"No," Bean agreed. "It must've been there all along, but it was just grey and you didn't know it."

"I wonder where the service module is?" Conrad said. His view was limited to the circular window in the hatch, so he could not see directly back along their velocity vector.

Gordon was dismissive, "It's out there in the boonies."

There was no radar or aircraft coverage available for the jettison and separation sequences. The discarded module had automatically fired its reaction control system to depletion in a manner designed to achieve a high-apogee orbit. If this had gone to plan, the hulk would probably have been seen by tracking stations. In the absence of confirmation, it was concluded that, as on previous missions, the service module had lost attitude control during the manoeuvre, failed to achieve the planned delta-V, and burned up in the atmosphere.

"This thing's got a perpetual yaw to the left," Gordon complained. "I have got to keep backing off to the right."

"We haven't got anything stuck or anything?" Conrad asked, thinking of a faulty thruster.

"No, it's just doing that for some reason," Gordon replied. "There's the Moon."

"Where?" Bean asked.

"In the lower left-hand corner of my rendezvous window," Gordon said. "Like it's supposed to be." The time that the Moon passed below the horizon was to be used to check their trajectory.

The vehicle accelerated rapidly as it fell to Earth. "We're picking up speed, Al," Conrad noted. "You're up to over 34,000 feet per second."

"That's the fastest I've been this week," Bean observed. They had been travelling faster at translunar injection but that was more than a week ago.

"Oh, man. My ears hurt," Conrad complained, still suffering from a head cold.

"Mine are in bad shape too, the left one," Bean said.

"My left one, too," Conrad said. "Suppose we have leftitis? Left moonitis?"

"Left moonitis," Bean laughed. "We left it, all right."

At this point Apollo 12 passed out of range of the tracking station on the island of Guam, ending the telemetry feed to Mission Control. Weitz promptly tested the first aircraft relay, "Apollo 12, Houston through ARIA."

"Roger, Houston. Read you loud and clear," Conrad replied. "We're watching the moonset."

"Al, turn your camera on," Gordon said onboard. "Maybe you can get a picture of it for a couple of seconds."

Bean had the 16-mm movie camera on a bracket in his rendezvous window ready to document the fireball of atmospheric entry and, later, parachute deployment, but the Moon was not in its field of view.

The fact that the Moon slipped below the horizon at the predicted time provided a measure of confidence that the spacecraft was flying the intended trajectory.

"It won't be long now, babe," Bean mused, with just over 2 minutes remaining to the entry interface. "We're whistling in."

"About 35,481 feet per second now," Conrad pointed out.

Impressed by the rate at which they were passing over the Earth from an altitude of just over 400,000 feet, Bean observed, "Look at those clouds. Cripes, it looks like we're right on them."

At the entry interface the spacecraft was 65.8 nautical miles above local coordinates 13.8°S, 173.5°E, travelling at a space-fixed velocity of 36,116.6 feet per second in a direction 98.2 degrees east of north on a trajectory which dipped 6.48 degrees below horizontal. The transearth coast had lasted 71 hours 52 minutes 52.0 seconds. Entry was flown by the computer, utilising the command module's aerodynamic lift to aim for the target point. The nominal range from the entry interface to splashdown was 1,250 nautical miles. With adequate notice this could be extended to 2,000 nautical miles in order to avoid a storm, but in this case the primary target was clear.

The rapidly increasing deceleration caused the condensate which had accumulated in the apex to 'rain' down on Conrad. "Here comes the water," he announced. As the load reached five g, he laughed, "I'm soaking wet!"

As the vehicle compressed the air in front of it, this produced a shock wave whose ionisation imposed a radio blackout. Bean started the movie camera at 12 frames per second to document the fireball phenomenon for 4 minutes.

"Stand by for V-circular," Gordon called. Once the vehicle had slowed below this velocity, it would be guaranteed to return to Earth. Nevertheless, it was to perform a 'skip' manoeuvre in which the lift vector would be rotated to cause it to gain altitude in order to define the second and final dip into the atmosphere that would lead to the ocean. This 'double dip' was to expose the heat shielding to low heat rates with high heat loads (i.e. lower temperatures applied for a longer time) than would have been produced by a 'straight in' lunar return.

At about this time the *Hornet* established radar contact with the vehicle at a range of 103 nautical miles on an approach bearing of 261 degrees.

At the predicted end of the radio blackout, Weitz tried to establish contact via an ARIA relay, without response. Shortly thereafter, Conrad made his own call, "Hello, Houston. Do you read Apollo 12 out of blackout?"

"Roger, 12," Weitz acknowledged. "Reading you loud and clear."

"It's right on the money," Conrad advised.

As the deceleration load built back up, the computer rolled the vehicle to vary the lift vector and refine its aim for the target. Just in case it came in significantly short or long, HC-130 Hercules designated 'Samoa Rescue 1' and 'Samoa Rescue 2' were in position to drop swimmers who would parachute into the ocean to look after the capsule until the naval recovery force could arrive.

The Earth landing system included the drogue and main parachute systems and a variety of post-landing recovery aids. The apex heat shield would nominally be jettisoned at an altitude of 24,000 feet in order to permit mortars to deploy a pair of reefed 16.5-foot-diameter drogue parachutes. These would be released after they had oriented and decelerated the capsule. Three mortars would deploy pilot chutes, each of which would draw out one of the 83.3-foot-diameter main parachutes. The mains had two-stage reefing for inflation in three steps. The great concern in this case was that the pyrotechnics could have been damaged by the lightning strikes at launch. As Gerry Griffin said in deciding to proceed with translunar injection, if the parachutes had been disabled, that in itself was not a reason to deny the crew the opportunity to fly the Moon. Now it was 'pucker time', because the only way to determine whether the chutes would work was to attempt to deploy them. With the capsule descending essentially vertically through 30,000 feet, the landing system was armed.

"Stand by for drogues," Gordon called. As the appointed time passed, he began to count up the seconds until, some 16 seconds late, the system finally operated.

"We got drogues, Houston," Bean reported. There was no acknowledgement from Weitz, however.

As they descended through 15,000 feet, Conrad called, "Stand by for the mains."

The drogues were jettisoned, and the capsule fell freely for a few seconds until the pilot chutes dragged out the mains.

"Three gorgeous beautiful chutes," Bean called in the blind to Houston.

The prime recovery ship was the antisubmarine aircraft carrier USS *Hornet*. It had launched three fixed-wing E-1B Tracers, twin-engine aircraft with radomes mounted on their backs. One, piloted by Commander Van E. Spradley, had the radio call sign 'Air Boss'. Another was a communications aircraft called 'Relay', and the third was in backup to undertake either function. There were five SH-3D Sea King helicopters airborne. 'Swim 1' piloted by Lieutenant Bill Sherrod was circling 10 nautical miles uprange of the *Hornet* and carried three swimmers whose task was to jump into the ocean and attach the flotation collar and sea anchor to the capsule. 'Swim 2', piloted by Lieutenant Grey Linker, was positioned 15 nautical miles downrange with three similarly equipped swimmers. The third helicopter, called 'Recovery', was piloted by Commander William Aut and carried the decontamination swimmer who would open the hatch of the capsule and assist the crew out. It was to hoist the astronauts using a Billy Pugh net and ferry them to the ship. A Flight Surgeon was onboard this helicopter to make a preliminary health assessment. The fourth helicopter, known as 'Photo', was a photographic platform. A fifth helicopter was available in backup.

As the astronauts worked through the checklist they activated their VHF receiver and immediately heard the Air Boss, stationed 25 nautical miles to the south.

"Air Boss, we read you loud and clear," Bean replied. "We're okay."

"Tallyho. I have a visual," the Recovery helicopter reported to the *Hornet* half a minute later. "He is bearing 135 from me."

With this visual sighting of the capsule, the aircraft left their holding positions in order to begin the recovery activities.

Meanwhile, descending through 5,000 feet with communications intermittent, the astronauts continued working through their checklist.

At 1,500 feet, the *Hornet* announced, "We have you visual. So do the helicopters inbound to your position."

"Roger," came the reply.

At 800 feet a valve that had been opened at 4,000 feet to match the cabin pressure to ambient was now closed in readiness for splashdown.

"Splashdown!" announced Recovery.[7]

The mission lasted 10 days 4 hours 36 minutes 25 seconds from launch to splash. Whereas Apollo 11 had lasted 195.3 hours, Apollo 12 was 244.6 hours. The times spent on the lunar surface were 21.6 hours and 31.5 hours respectively, and those in lunar orbit were 59.6 hours and 89.0 hours respectively. In addition, in the case of Apollo 12 the cislunar coasts lasted longer.

The capsule came down a mere 1.5 nautical miles from the target and 3.5 nautical miles from the *Hornet*. The weather was good: visibility 10 nautical miles, wind 15 knots, wave height 3 feet and swells of about 15 feet. Unfortunately, the capsule hit a

[7] Splashdown occurred at 20:58:25 GMT on 24 November, 4 hours 46 minutes after local sunrise.

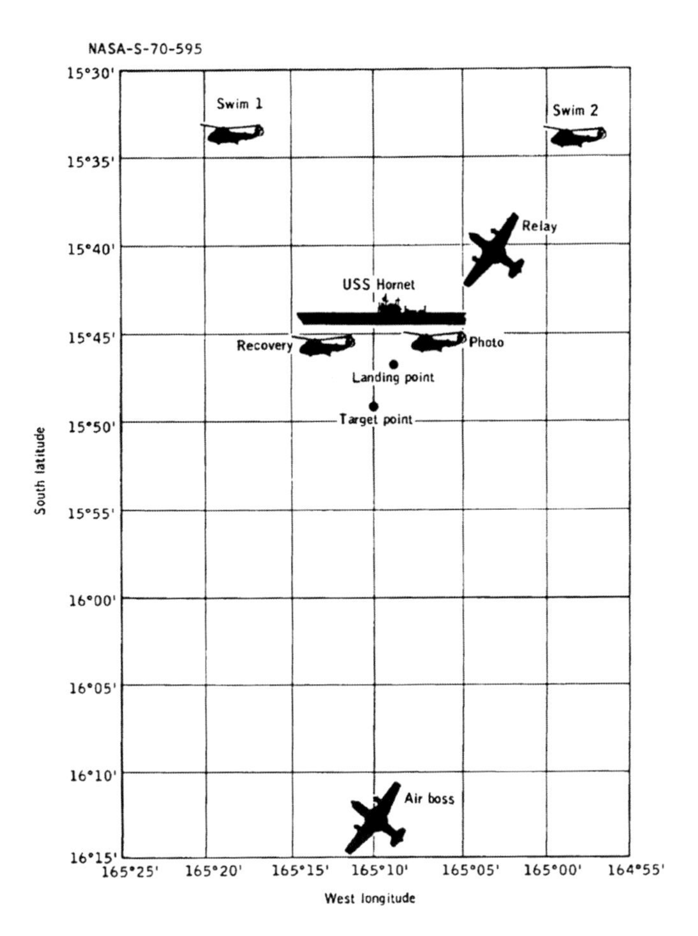

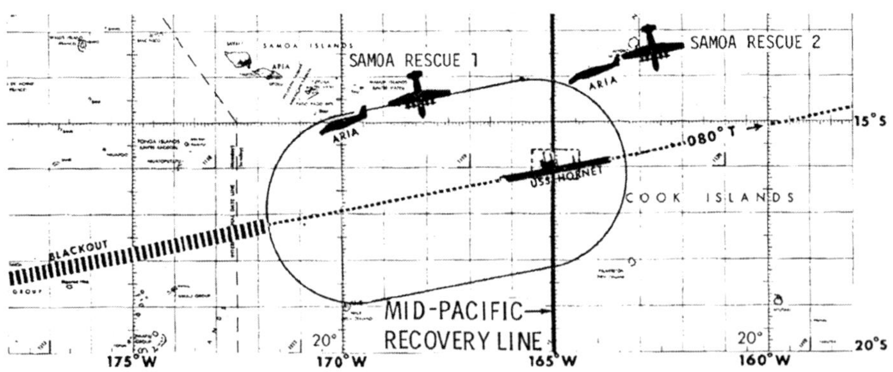

The disposition of the naval recovery force.

The Apollo 12 command module descends under its three main parachutes.

rising wave as it swung on its parachutes, yielding an extremely hard landing. The force of the impact of about 15 g not only displaced portions of the heat shield but also caused the movie camera to separate from its bracket. Bean had his hand poised to push in two circuit breakers to jettison the parachutes at splashdown before they could catch the wind and tip the capsule over, but the camera struck his head with such force that he was momentarily dazed.

Gordon urged, "Hey, Al, hit the breakers." But the capsule flipped over into the Stable II attitude with its apex down. "Al, what happened?"

"Nothing happened," Bean replied. "What are you talking about?"

Conrad looked over at Bean and saw a cut above his right eye, "You're bleeding!"

Bean, surprised, mused, "I must have been knocked out for a few seconds without even knowing it."

As Gordon said in the post-mission debriefing, "We'd looked at this in the flight plan, questioned it, and without really thinking too deeply, decided that if the camera mount could stand the g-load of re-entry then it'd probably stand the impact. It was our mistake; we should have removed it."

AFTER SPLASHDOWN

The recovery aids included a VHF beacon, a VHF transceiver, a flashing beacon, sea dye, an uprighting system, and a connection for the interphone the decontamination swimmer would use to communicate with the crew. The uprighting system consisted of a trio of compressor-inflated gas bags on the apex structure, one on each quadrant except that above the hatch. It took about 5 minutes to flip the capsule onto its base, into the Stable I attitude required to proceed with the recovery operation.

In the case of Apollo 11 each man took a pill prior to entry in order to ward off sea-sickness while awaiting recovery, but the all-Navy crew of Apollo 12 did not. As Conrad related in the post-mission debriefing, "We activated the post-landing environmental control systems as per the checklist. [But] we really didn't need it, as the cabin was plenty cool and even although it was very rough out there nobody had the slightest tendency to be seasick. I would like to comment that I think that had I done this in the Gulf of Mexico in such a rough sea I would have gotten sick. I think that 10 days of zero-g sort of numbs you to motion sickness and none of us even felt queasy. I could hardly get down in the lower equipment bay without feeling queasy in the Gulf egress exercise, and I was hopping all around that cabin out there in a sea three times as rough as it was in the Gulf. It didn't bother me at all."

The rough sea caused other problems, as Gordon debriefed, "When we opened the post-landing vent ducts we took water in through the intakes and that fan just blew it into the spacecraft. After a while, we got tired of getting wet and we just turned the duct off. And when we got too warm I turned it back on to let some air in." Conrad added, "Until they got the flotation collar on, and that sort of stabilised things a little bit."

Five minutes after the capsule righted itself, Swim 1 hovered nearby and dropped Photographers Mate Third Class William R. Pozzi into the water. He was followed by Lieutenant Junior Grade William C. Robertson and Sonar Technician First Class Arles L. Nash. They used bungee lines to affix the flotation collar to the periphery of the capsule, then inflated it using compressed air. Within 10 minutes of entering the water they had the collar on, a sea anchor deployed and a pair of rafts tied alongside.

Meanwhile in the capsule, as Conrad recalled in the post-mission debriefing, "We got the medical kit out and put a Band-Aid on Al so that the whole world wouldn't get upset."

The Interagency Committee on Back Contamination had been established in 1966 to assist NASA in preventing the Earth from being contaminated by lunar materials. It drew its membership from Public Health Services, the Department of Agriculture, the Department of the Interior, the National Academy of Sciences and NASA itself. The first lunar landing crew had been required to don rubber Biological Isolation Garments with integral respirators. When the committee met on 30 October 1969 in Atlanta, Georgia, to review the Apollo 12 plans it decided to allow the astronauts to wear lightweight blue flight suits and face masks to filter exhaled air.[8] In addition, whereas previously the decontamination swimmer had worn the same garment as the astronauts, this time he was permitted to wear a scuba outfit. Major disadvantages of the full isolation garment were that the occupant soon overheated, the mask fogged over, and it was impossible to communicate by voice, leading to everyone stumbling around in rafts essentially blind.

With everything set, Recovery dropped the decontamination swimmer, Lieutenant Junior Grade Ernest Lee Jahncke. He plugged in his intercom to confirm that the crew were ready, and then, just over half an hour after splashdown, opened the hatch, passed in the kits and then closed the hatch. About 10 minutes later, having powered down the command module, the crew swung open the hatch. The rule required the astronauts to don life jackets and for Jahncke to activate the inflators as each man scrambled out.

Conrad was not impressed, as he stated in the post-mission debriefing, "I'm really against inflating these life jackets. Gerry Hammack had the swimmers instructed to pull them on the way out the door, and I object to that strenuously. I don't want some swimmer pulling my life jacket when I'm halfway out the hatch, which he did, and it made me mad." It was not simply the intervention, "it's bad news getting into that Billy Pugh net."

The Public Affairs commentary reported, "Conrad, Gordon and Bean was the exit order from the spacecraft."

With the crew safely out, Jahncke closed the hatch, then used a swab containing a liquid agent to decontaminate each man's garments, the hatch, the adjacent flotation collar, the area around the post-landing vent valve and finally himself.

The Recovery helicopter moved back in and, hovering overhead, lowered a Billy Pugh net to hoist aboard the astronauts, Bean first, then Gordon and finally Conrad. It was a brief flight, as by now the *Hornet* was standing by only 900 yards upwind.

[8] The biological isolation garments were available on the *Hornet* as a precaution in case the crew had reported an inexplicable illness on the way home.

After landing, the helicopter rode the deck elevator down to the hangar, whereupon the astronauts and the Flight Surgeon, Clarence A. Jernigan, walked straight to the Mobile Quarantine Facility. The MQF was a 9-foot-wide 35-foot-long gleaming aluminium commercial travel trailer, minus its wheels, that had been modified for NASA by Airstream and American Standard to provide biological containment. One end was hinged to provide the main door. A technician, Brock 'Randy' Stone, was already in the trailer. It was sealed precisely 1 hour after splashdown, and would be home to these five men until it was delivered to the Lunar Receiving Laboratory in Houston.

As soon as the astronauts were presentable, they received a telephone call from President Richard Nixon, "Hello?"

"Hello, sir," replied Conrad on behalf of his crew.

"I'm just delighted to have this opportunity to welcome you back, and I only wish I could be out there for the splashdown. I can assure you that millions in the United States and around the world were watching. I'm just tremendously proud, personally and speaking also representing the American people, of what you've done. As you know, the night before you took off, we talked on the phone and I invited you and your wives to come to the White House for dinner. I just want to be sure you can make that date."

"Yes, sir, we'll be there," Conrad assured.

"We will expect to see you after you get out of quarantine. Now, there's one other thing I think I should tell you that I notice: you've been responsible for several firsts. You weren't the first on the Moon, but you, I think, Commander Conrad, you were the first to sing from the Moon. Right?"

"I guess, so, sir," Conrad agreed.

"That's right. Also, as a result of your flight we've had the first moonquake, and the first press conference from outer space. Now, after all of those firsts, I think that the nation wants some recognition, and I've been trying to think what would be the best way to recognise you. And over these past 10 days, I noticed Walter Cronkite and the other commentators always refer to you as Commander Conrad, Commander Gordon and Commander Bean, and I, exercising my prerogative as Commander in Chief of the Armed Forces, have decided that you should all be promoted, and that from this day forth you shall be Captain Conrad, Captain Gordon and Captain Bean. Congratulations."

A call was then put through from NASA Administrator Thomas Paine, "How are you doing?"

"Just fine, sir," Conrad replied. "Everybody's in good shape. Al got whacked with the camera when we hit, but otherwise everybody is just fine."

"We're supposed to be in good shape. We have just come off a 10-day vacation!" Gordon quipped.

"You know that our new motto is 'Fly with NASA, we get you there and back on time'," Paine countered.

"That's correct. We go in any kind of weather or anywhere," Conrad pointed out.

After congratulating the astronauts on their promotions, Paine said, "I won't hang on the line now because I know you'll want to talk to your families, but I

The command module initially adopted the apex-down orientation.

Now that the airbags on the command module have flipped it into the apex-up orientation, the Swim 1 helicopter drops swimmers to start the recovery effort.

Attaching the flotation collar to the command module in a rough sea.

The astronauts emerged into the raft assisted by the decontamination swimmer, Ernest Lee Jahncke: first Pete Conrad, then Dick Gordon and finally Al Bean.

An attendant swimmer snaps pictures of the party in the raft.

The Recovery helicopter hoists Al Bean aboard using a Billy Pugh net.

The Apollo 12 astronauts walk the red carpet to the Mobile Quarantine Facility on the hangar deck of the *Hornet*. The plastic tent is for the command module.

Pete Conrad (left), Dick Gordon and Al Bean at the window of the Mobile Quarantine Facility immediately after recovery. Note the Band-Aid above Bean's right eye.

With the Apollo 12 crew safely aboard the *Hornet*, swimmers remain with the command module until it can be hoisted onto the ship.

certainly want to send the appreciation of all of us here at NASA for the tremendous job that you did. The scientists are extremely excited about the results that are coming back [from the ALSEP], and are looking forward to getting the samples you have brought. And everyone of us is just tickled to death with that magnificent job you did. So thanks a lot fellows."

"Well, we enjoyed it sir. It was our pleasure," Conrad replied.

"We sure did," Gordon added.

Paine signed off, "We look forward to seeing you out in Houston in a few days. Have a good Thanksgiving."

After talking to their families, each astronaut was given a comprehensive medical examination. They were in good health, with normal temperatures and weights in the expected range. Bean had a small amount of clear fluid containing air bubbles in the middle ear cavity, but this was eliminated by 24 hours of decongestant therapy. The 1-inch cut above his right eyebrow was sutured, and healed normally. All three men suffered skin irritation at the sites of the biomedical sensors. Gordon was worst, with multiple pustules at the edges and at the centre of each site. Red areas and pustules were also present at all Bean's sensor sites. But because Conrad had removed his sensors 4 days earlier and cleansed the skin and applied cream to the affected areas daily, his lesions were healing. Because the skin reaction to the sensors was the most severe of any mission so far, a study was ordered. It found that Conrad was allergic to a substance in the electrode paste, but a chemical analysis failed to determine the cause of the irritation. No bacteria were cultured from the paste, which contained a substance to inhibit bacteria. However, *staphylococcus aureus* was cultured from the skin of all three men, and this could have accounted for the reported inflammation of the irritated skin.

Within an hour of the Recovery helicopter delivering the astronauts to the *Hornet*, a crane hoisted the command module aboard and placed it onto a dolly, which was wheeled alongside the MQF. Once a hermetically sealed plastic tunnel had been run from the trailer to the capsule, the technician executed the post-retrieval procedures, during which he offloaded the lunar sample containers, Surveyor parts, film, mission logs, etc. The sample containers and film were then passed out through the trailer's decontamination airlock along with crew samples to be sent to Houston. Meanwhile, the command module was sealed and decontaminated for a second time and the transfer tunnel removed, the interior of the helicopter which had retrieved the astronauts was decontaminated, and naval forces attempted to collect the main parachutes, apex cover and drogue chutes in that order of priority.

At 06:40 GMT on 25 November a thermally stable shipping canister with sample return container number 1, some of the film and crew samples left the *Hornet* on a fixed-wing C-2 Greyhound for a flight to Pago-Pago on the island of Samoa, some 330 nautical miles west. On arrival, the cargo was transferred to an Air Force C-141 Starlifter, which promptly took off for Ellington Air Force Base, near Houston. A second Greyhound left at 11:30 GMT with sample return container number 2 and the other film and crew samples, which were transhipped at Pago-Pago to an ARIA for onward flight. On arrival at Ellington, the first shipment was driven to the Lunar Receiving Laboratory, where it arrived at 20:45

GMT on 25 November.[9] The second shipment arrived at 04:48 GMT on 26 November.[10] The contingency sample, the Surveyor parts, and the large rocks in the sample bag, none of which were preserved in vacuum conditions, remained in the MQF.

After a 4-day voyage, the *Hornet* docked at Pearl Harbor on the island of Oahu in Hawaii. The MQF was offloaded to a transporter at 02:18 GMT on 29 November.[11] After a brief welcoming ceremony it was driven to nearby Hickam Air Force Base and loaded onto a C-141 for the flight to Ellington, landing at 11:50 GMT.[12] It was delivered by road to the Lunar Receiving Laboratory 2 hours later. The astronauts nominally had 11 days remaining of the quarantine period. The lunar samples would be quarantined for 50 to 80 days, depending on the results of biological tests. A total of 28 persons, including the crew and members of the medical support teams, were exposed, directly or indirectly, to the lunar material and retained in quarantine. Daily medical observations and periodic laboratory examinations showed no indications of infectious disease related to lunar exposure. No significant trends were noted in any biochemical, immunological, or hematological parameters in either the astronauts or the medical support staff. All personnel were approved for release from quarantine on 10 December, at which time the astronauts returned to their homes.

It fell to the backup crew of Dave Scott, Jim Irwin and Al Worden to organise the welcoming party. They had prepared a 30-minute movie which spoofed the serious documentaries that NASA favoured. It featured film taken by the photographers who routinely attended training activities, and anything funny was considered fair game. Additional scenes were mocked up for the purpose and spliced in. The audio was a facetious voice over that cued clips from the mission audio.

Meanwhile, the command module was offloaded from the *Hornet* and driven to Hickam for deactivation. This primarily involved safing the remaining pyrotechnic systems and draining the propellants from its reaction control system, all performed without ever opening the hatch to the cabin. The deactivated command module was flown to Ellington on a C-133B Cargomaster and delivered to the Lunar Receiving Laboratory at 19:30 GMT on 2 December.[13] Today, it is at the Air and Space Center in Hampton, Virginia.

On 17 December 1969, prior to setting off on their international goodwill tour, the Apollo 12 crew revisited the Cape. With a Navy band playing 'Anchors Aweigh', Kurt H. Debus, Director of the Kennedy Space Center, led them into the transfer isle of the cavernous Vehicle Assembly Building where 8,000 workers had gathered for a ceremony in which the astronauts were presented with colour photographs of their liftoff, plus pebbles from the crawlerway. In addressing the gathering, Conrad said, "We forgive the weatherman for his job, but had we to do it again, I would launch

[9] 2:45 p.m. local time on 25 November.
[10] 10:48 p.m. local time on 25 November.
[11] 4:18 p.m. local time on 28 November.
[12] 5:50 a.m. local time on 29 November.
[13] 1:30 p.m. local time on 2 December.

As the Mobile Quarantine Facility arrived at Ellington Air Force Base in the early hours of 29 November, the Apollo 12 crew were delighted to be greeted by their families (left to right): Barbara Gordon, Jane Conrad and Sue Bean.

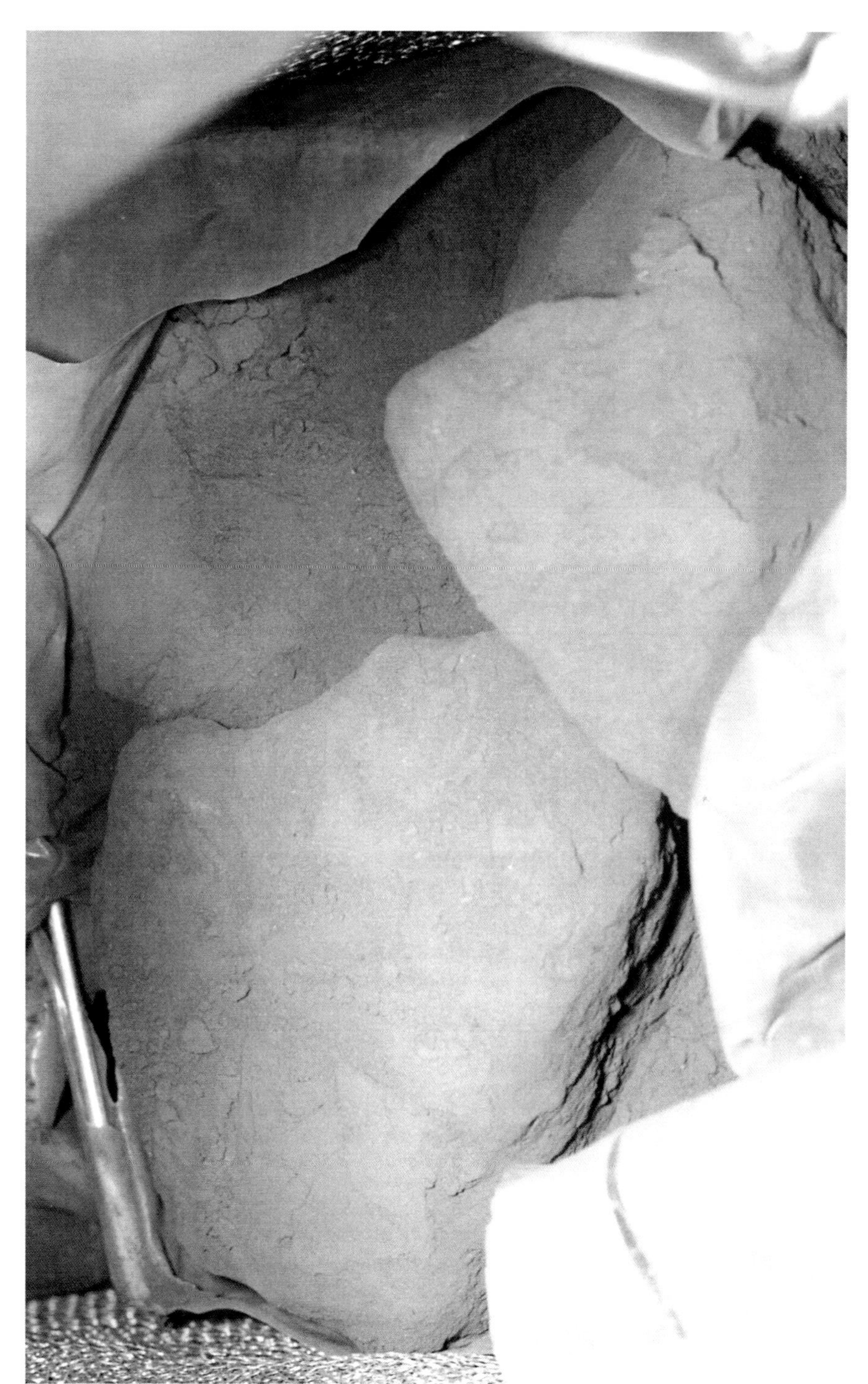

Rocks in one of the Apollo 12 sample return containers on 26 November, soon after it was opened.

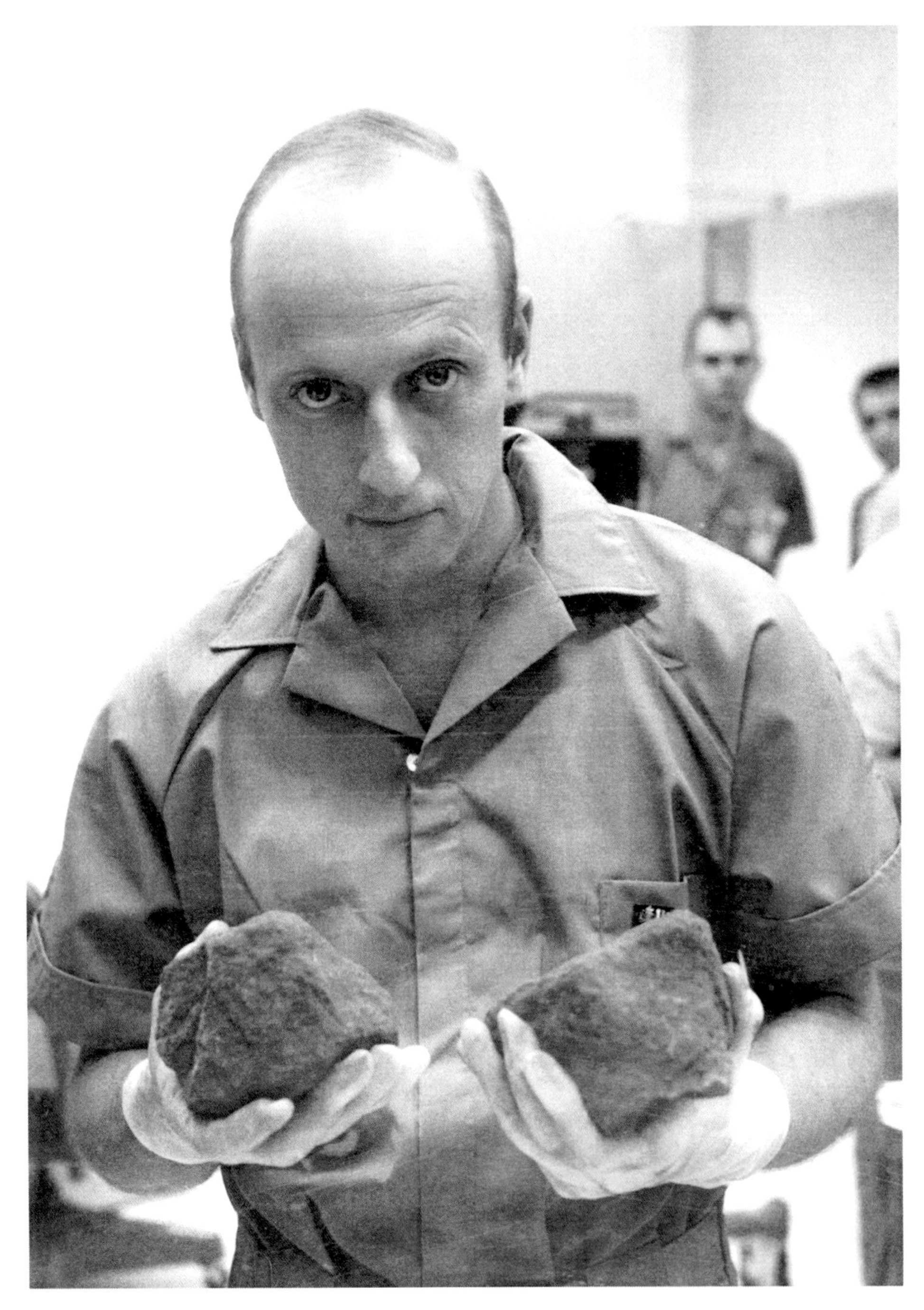

In the Lunar Receiving Laboratory on 29 November Pete Conrad holds two of the lunar rocks that were returned to Earth in a sample bag.

In the Lunar Receiving Laboratory on 29 November Pete Conrad inspects the television camera that he and Al Bean retrieved from Surveyor 3.

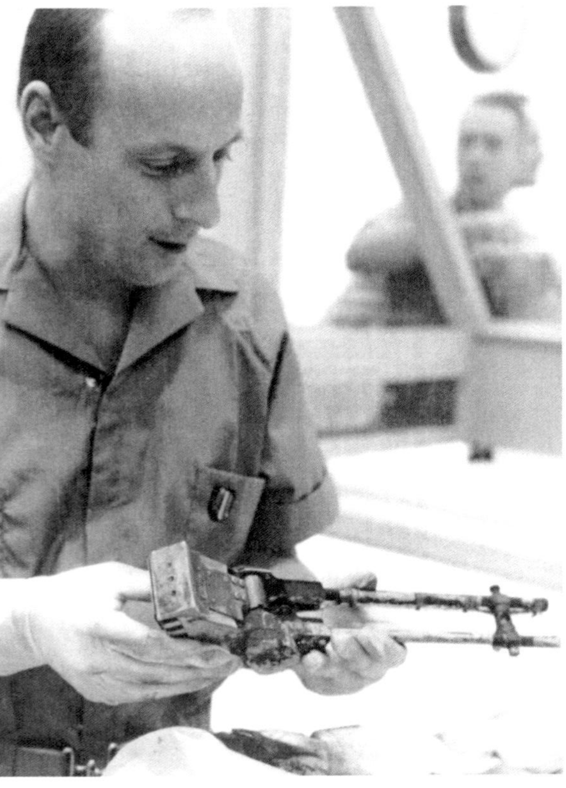

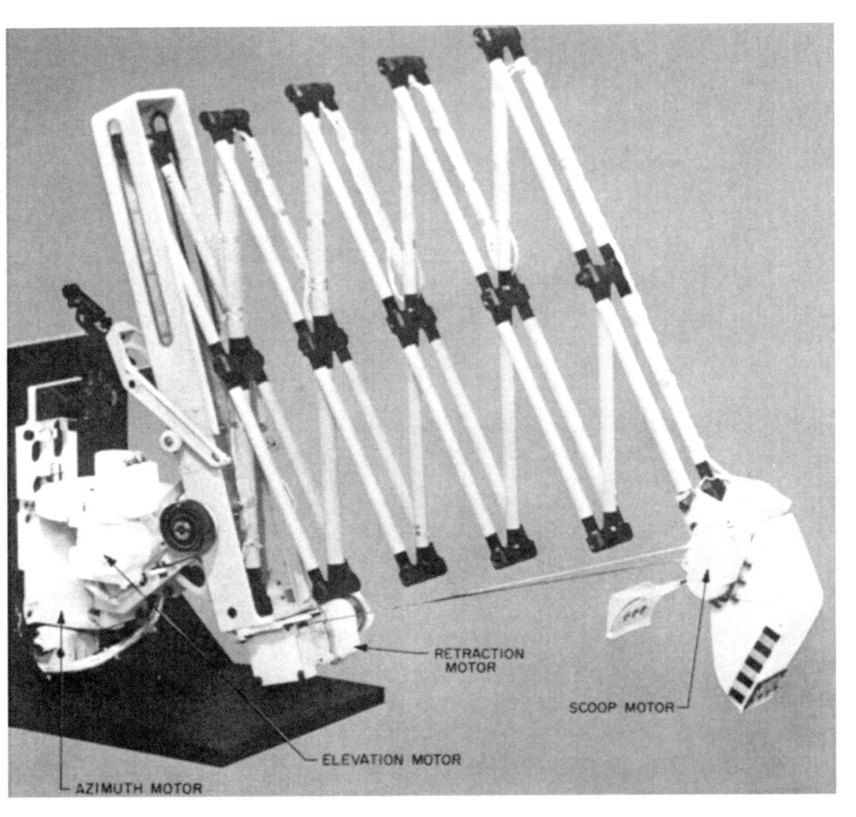

Details of the pantograph robotic arm developed for the Surveyor program, and Pete Conrad with the scoop that he and Al Bean retrieved from Surveyor 3.

The Apollo 12 crew at a presentation in Vehicle Assembly Building of the Kennedy Space Center on 17 December 1969.

Pete Conrad speaks in Vehicle Assembly Building on 17 December 1969.

exactly under the same conditions." Guenter Wendt's pad team had collected a piece of grounding rod from the umbilical tower and cut it into three short sections. These had been mounted with the inscription 'In fond memory of the electrifying launch of Apollo XII' and were presented to the astronauts.

THE FIRST 'ROCK FEST'

After the experience of processing the Apollo 11 material, some of the procedures and apparatus in the Lunar Receiving Laboratory were revised. When the sample return containers arrived, the one containing the selected samples from the first EVA was opened in the vacuum system, but the one containing the documented samples from the geological traverse was taken to the biological laboratory and opened under sterile nitrogen. On 27 November a lunar sample was sent to the Radiation Counting Laboratory. On 30 November the Lunar Sample Preliminary Evaluation Team began the task of describing, photographing and cataloguing the samples. With the benefit of prior experience they were able to work more rapidly, and within one week had determined the basic characteristics of the samples. The sampling objective had been to obtain 60 pounds of representative lunar material in the form of individual rocks, loose fine-grain material and samples in core tubes. The mission actually returned a total of 74.7 pounds of lunar material. The 'first look' revealed that the rocks were larger than those of Apollo 11. On 8 December, the daily report included an outline comparison of the samples from the two lunar landings.

Type of sample	Apollo 11	Apollo 12
fines and chips	24.2 pounds	12.8 pounds
individual rocks	24.3 pounds	61.0 pounds
core sample materials	0.3 pounds	0.9 pounds
Total	48.8 pounds	74.7 pounds

Thus, Apollo 12 returned more rocks and core sample material but less fines. To a first approximation, the fines from both sites were a charcoal-grey granular material that comprised fragments of rock and dust.

Most samples were documented, and in the post-mission debriefing Conrad said, "Our documented sampling went exactly the way we practiced it on Earth." But the procedure they used did not include taking after-sampling shots. As a result, it was not always possible to reliably determine the point of origin for the rocks that were stashed in the large bag of the hand-tool carrier.

The preliminary examination had to be completed by 20 December and the report written by 24 December, so that on 27 December the Analysis Planning Team could start to prepare samples for dispatch to the individual teams who would perform the detailed analyses. The first samples were issued on 7 January 1970, coincident with the First Lunar Science Conference that was held 5–8 January at the Lunar Science

Scientists Harold Urey (left), Gene Shoemaker and Gerard Kuiper.

Institute adjacent to the Manned Spacecraft Center.[14] In attendance were hundreds of scientists, including all 142 of the principal investigators for the Apollo 11 samples. After years of speculation about the nature of the Moon, the opportunity to analyse physical samples had prompted a frenzy of activity in laboratories around the world addressing the chemistry, mineralogy, age, and petrology of the samples, the search for evidence of life, and how the lunar material affected terrestrial life. Although the focus of the conference was on the results of analyses of the Apollo 11 samples, the Preliminary Evaluation Team was able to present its inspection of the material from Apollo 12. The papers by the various teams were submitted to the journal *Science* in early January, refereed and revised over the ensuing fortnight, then published in the 30 January issue, which was given over entirely to these investigations.

To Harold C. Urey, who favoured the 'cold Moon' theory in which the interior was 'pristine' material, the dark maria were the result of impact melting on a vast scale. While Neil Armstrong and Buzz Aldrin were making their moonwalk, Urey was initially concerned when Armstrong reported a vesicular rock, encouraged when Armstrong changed his mind, and then dismissive of Armstrong's report of a rock that he was confident was vesicular. Most of all, Urey was relieved that they did not report the 'frothy vacuum lava' predicted by his main rival, Gerard P. Kuiper, who favoured the 'hot Moon' theory in which the lunar interior was differentiated and the maria represented upwellings of lava through fractures in the floors of major impact scars.

To the dismay of Urey and the delight of Kuiper, the Apollo 11 rocks proved to be a version of basalt rich in magnesium and iron (and therefore described as being 'mafic') which isotopic dating revealed to have crystallised some 3.84 to 3.57 billion years ago. In terms of texture, it was strikingly similar to terrestrial basalt. It was not impact melt. This meant that the Moon had indeed undergone a process of thermal differentiation in which lightweight aluminous minerals migrated up to the surface

[14] These gatherings are now entitled the Lunar and Planetary Sciences Conferences.

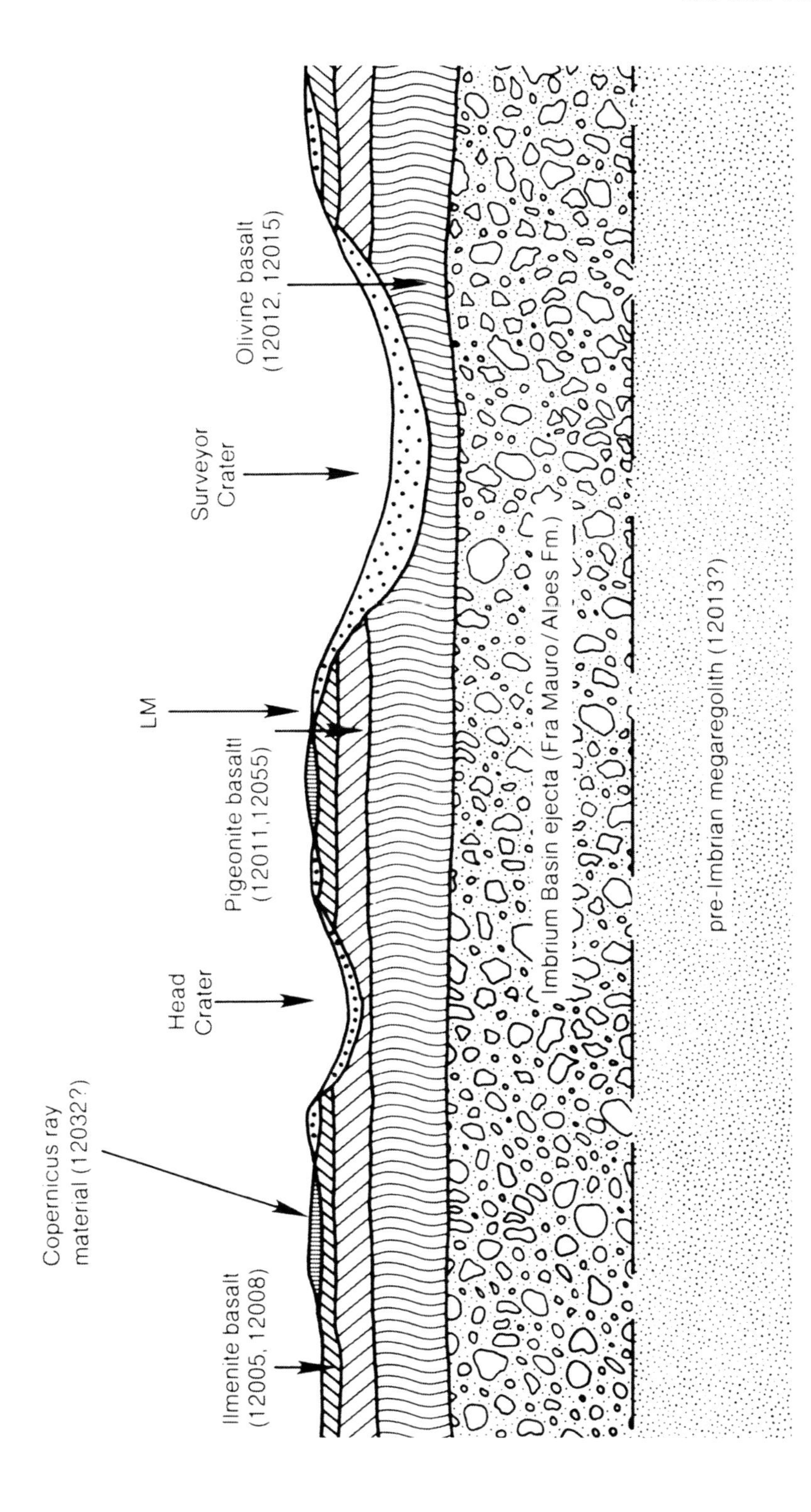

A schematic east-west geological cross-section through the Apollo 12 landing site showing several basalt lava units overlying older ejecta from major impact basins (modified after Rhodes *et al.*, 1977; Wilhelms, 1984). The presence of the Imbrium ejecta layer is inferred. Numbers refer to specific collected samples representative of the various units inferred to be present. Courtesy the Lunar and Planetary Institute and Cambridge University Press.

and the heavier minerals sank. The fact that some of this denser material had later been erupted indicated that the interior had remained 'hot' for a significant period. In comparison to terrestrial basalt, the lunar variety was enriched in titanium.[15] The titanium-bearing mineral, which was new to mineralogists, was named 'armalcolite' in honour of the three astronauts.[16] The most striking fact was the lack of hydrous minerals. The lunar basalt was also deficient in volatile metals such as sodium. The low-alkali (i.e. sodium-depleted) lava would have had an extremely low viscosity, which is why it flowed so readily, and why it left so few 'positive-relief' features. The Sea of Tranquillity was evidently built up by episodic volcanism over a period of several hundred million years. The presence of *two* types of basalt implied either distinct magma reservoirs or a single sustained source whose chemistry had evolved over time.

The loose fragmental 'soil' produced a wealth of data. The majority of the lithic fragments were pulverised basalt. There was little material of meteoritic origin. This was consistent with the 'gardening' process in which a large impact dug up bedrock which was progressively worn down by smaller impacts to yield the seriate regolith. Many of the discrete samples turned out to be consolidated soil, made by the shock of an impact compressing it sufficiently to produce 'glass'. A small residue in the regolith was plagioclase feldspar. Although this is one of the commonest minerals in the Earth's crust, it is rich in sodium. The lunar variety was depleted in sodium but enriched with calcium, making it calcic-plagioclase. A few of the fragments turned out to be sufficiently pure to be anorthosite, this being the name for a rock that is at least 90 per cent plagioclase. But most were diluted by a variety of mafic minerals, and so were more properly called anorthositic gabbro. These were representative of the lunar highlands. This implied that the Moon's primitive crust was composed of anorthositic rock. Further study of the anorthositic fragments yielded an insight into the absence of europium (one of the rare earth elements) in the mare basalt. It turned out that the plagioclase had an affinity for europium, and was enriched in it. The original anorthositic crust had soaked it up so efficiently that the lava that extruded from fractures was from reservoirs deficient in this element. Analyses of isotopes produced by the bombardment of cosmic rays indicated that some lunar rocks had been on (or just below) the surface for at least 10 million years; this was evidence of the slow rate of 'turn over'. One mystery was that radioisotopic dating appeared to suggest the regolith contained fragments that formed 4.6 billion years ago. If the regolith was bedrock eroded by impact activity, how could the overlying material be older than the material from which it was derived? This anomaly was cleared up by some of the samples collected by Conrad and Bean.

After a review of the Apollo 12 transcript, the documentary photography, and the

[15] The Apollo 11 basalt was enriched with as much as 12 per cent titanium oxide. In comparison, the richest terrestrial equivalent is 4.5 per cent titanium oxide.

[16] The name 'armalcolite' was derived from the first letters of the astronauts' surnames. Some years later this mineral was found on Earth, too.

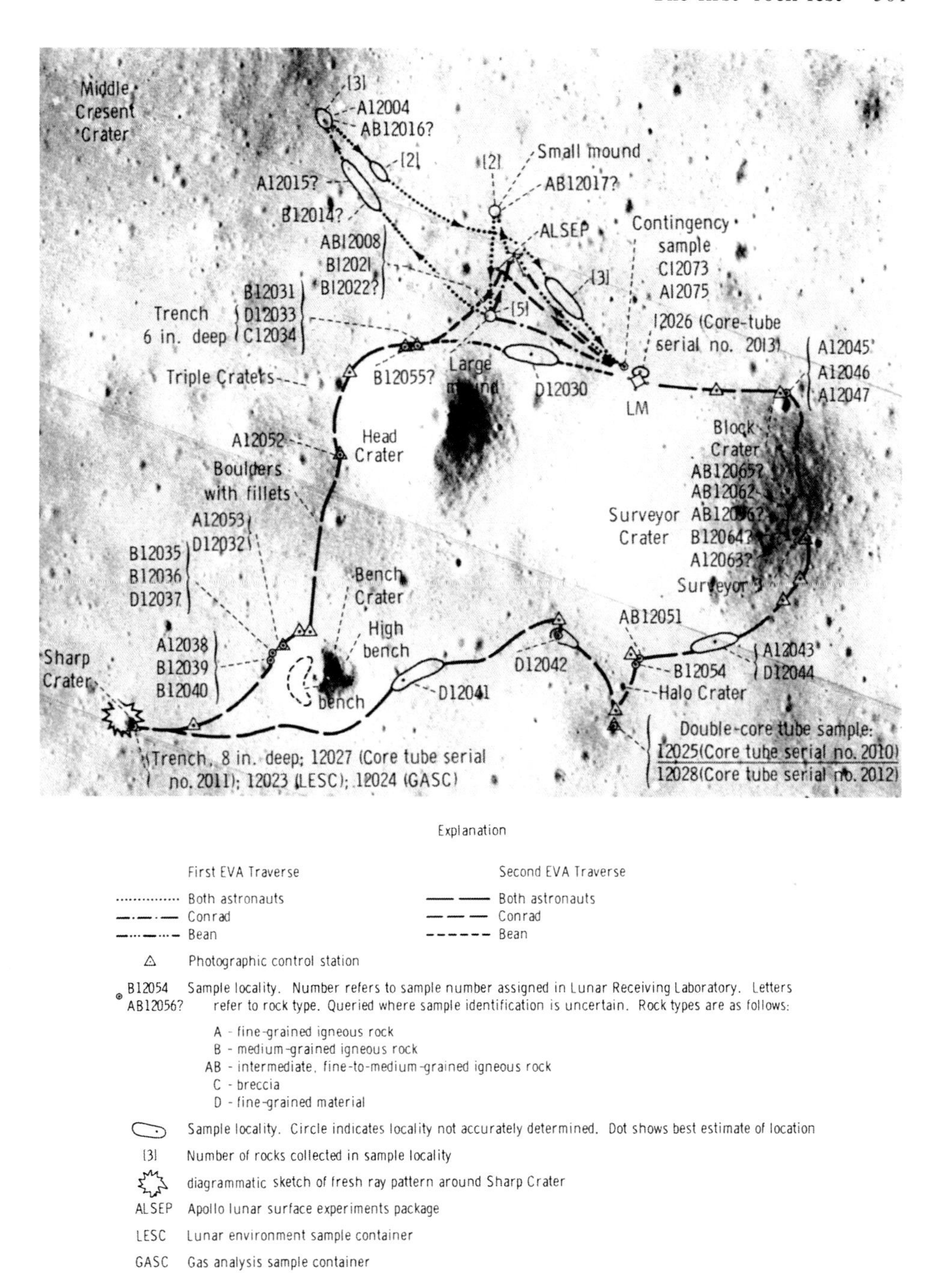

The Apollo 12 Preliminary Science Report published in June 1970 included this map estimating where the individual samples were collected.

individual samples, it was possible to determine the order in which the samples were collected.[17]

The contingency sample was collected early in the first moonwalk about 50 feet northwest of Intrepid. The bag was returned in the cabin environment. It was divided and its various portions assigned sample designations in the range 12070 to 12077:

Sample	Description
12070	1.102-kg of fines.
12071	0.009-kg of chips.
12072	A 0.104-kg porphyritic olivine basalt with phenocrysts of olivine, chromite and pyroxene in a matrix possessing many vesicles and vugs.
12073	A 0.408-kg regolith breccia with a typical mix of lithic fragments and glass. It is the largest rock in the contingency sample. (It was originally 12073 and 12074 as samples of 0.361 and 0.047 kg respectively, but they were combined as 12073, thus eliminating the number 12074.)
12075	A 0.233-kg porphyritic olivine basalt with phenocrysts of olivine and pyroxene in a medium-grained matrix possessing many vugs.
12076	A 0.055-kg porphyritic olivine basalt with phenocrysts of olivine and pyroxene in a fine-grained matrix possessing many vugs.
12077	A 0.023-kg fine-grained crystalline igneous rock.

The selected samples superseded the 'bulk' sample of Apollo 11, and were taken northwest of Intrepid (the furthest 1,100 feet away) during the final part of the first moonwalk. A total of 17 rocks were collected using the tongs and put in one large teflon bag, and 3 larger rocks (of around 2 kg each) were collected using the tongs and put in a second large teflon bag. The bags were then packed with fines using the scoop. One single-length core sample was taken. The two bags and the tube were sealed in a sample return container in lunar vacuum. This container was found to have a slight air leak when probed upon arrival at the Lunar Receiving Laboratory, with the ratio of oxygen to nitrogen indicating that the leakage developed whilst in the spacecraft. It was opened in a vacuum facility, but its contents were exposed to nitrogen at ambient pressure for several hours. As only a few of these samples were supported by cursory photography it was difficult to identify precisely where each rock sample originated. Nevertheless, in most cases their general context was clear.

The sample collected near the small mound north of the ALSEP was:

[17] Sources: 'Apollo 12 Lunar Sample Information', Jeffrey Warner, NASA, TR-R-353, 1970; 'Apollo 12 Preliminary Science Report', NASA, SP-235, 1970; 'Apollo 12 Voice Transcript Pertaining to the Geology of the Landing Site', N.G. Bailey and G.E. Ulrich, USGS, 1975; 'Lunar Sample Compendium', Charles Meyer, NASA (http://curator.jsc.-nasa.gov/lunar/compendium.cfm); 'Lunar Sample Atlas', University of Arizona (http://www.lpi.usra.edu/lunar/samples/atlas/).

Sample	Description
12017	A 0.053-kg medium-grained pigeonite basalt, with a partial coating of glass derived from soil, rock and meteoritic material.

The samples collected near the large mound south of the ALSEP, near the north rim of Head crater, were:

Sample	Description
12008	A 0.058-kg olivine vitrophyre, glassy owing to the rapid cooling of the melt.
12021	A 1.877-kg porphyritic pigeonite basalt with phenocrysts of pyroxene in a coarse-grained matrix.
12022	A 1.864-kg porphyritic ilmenite basalt with phenocrysts of olivine and pyroxene in a medium-grained matrix.
12007	A 0.065-kg porphyritic pigeonite basalt with phenocrysts of pyroxene in a coarse-grained matrix.

The samples obtained on or near the southeastern rim of Middle Crescent crater were:

Sample	Description
12010	A 0.360-kg regolith breccia from the outer rim of Middle Crescent.
12014	A 0.159-kg fine-grained olivine basalt from the "fresh looking" crater about 200 feet from the rim of Middle Crescent.
12015	A 0.191-kg olivine vitrophyre from near the rim of Middle Crescent.
12006	A 0.206-kg medium-grained olivine basalt from near the rim of Middle Crescent.
12012	A 0.176-kg olivine basalt with partially reabsorbed phenocrysts of olivine from near the rim of Middle Crescent.
12004	A 0.585-kg porphyritic olivine basalt with phenocrysts of olivine and pyroxene in a fine-grained matrix from near the rim of Middle Crescent.
12016	A 2.028-kg medium-grained ilmenite basalt from near the rim of Middle Crescent, about to head for home. This is the largest rock in the selected sample.

The samples obtained on the run back to Intrepid were assigned designations, but owing to the lack of documentation it was impossible to match them to locations, so they are listed here simply in numerical order:

Sample	Description
12002	A 1.529-kg porphyritic olivine basalt with phenocrysts of olivine and pyroxene in a medium-grained matrix.
12005	A 0.482-kg ilmenite basalt that contains a high percentage of olivine, a low percentage of titanium oxide and a very high ratio of magnesium to iron. It has a cumulate texture.

12009	A 0.468-kg olivine vitrophyre.
12011	A 0.193-kg porphyritic pigeonite basalt with phenocrysts of olivine and pyroxene in a fine-grained matrix.
12013	A 0.082-kg fine-grained crystalline igneous rock.
12018	A 0.787-kg medium-grained olivine basalt.
12019	A 0.462-kg porphyritic pigeonite basalt with phenocrysts of pyroxene in a fine-grained matrix.
12020	A 0.312-kg medium-grained olivine basalt from approximately 300 feet northwest of Intrepid, heading home.

Back at Intrepid, the following samples were obtained:

Sample	Description
12026	0.101-kg of fines in core tube serial no. 2013.
12001	2,216-kg of fines and chips. It was the < 1 cm fraction of bulk regolith that was collected alongside Intrepid in a bag as filler for the sample return container.
12003	0.300-kg of fines and chips. It was the > 1cm split from 12001, including "a pure bead of glass" and loose material from the bottom of the sample return container. These broken up fines from larger rocks did not make a statistical sample.

The documented samples were obtained on the geological traverse of the second moonwalk. The tongs and scoop were used to fill 13 individual sample bags with 11 rocks and 7 fines. These samples were supported by photography. Ten further rocks were lifted using the tongs and stowed in the large bag of the hand-tool carrier. One single-length and one double-length core sample were taken. Also, the gas analysis sample and the special environment sample were sealed in individual containers. All of these samples except the four largest were sealed in a sample return container in lunar vacuum, but when probed upon arrival at the Lunar Receiving Laboratory this had about 0.5 atmospheres of pressure with a terrestrial ratio of oxygen to nitrogen. All the documented samples were processed in nitrogen cabinets. The four rocks not returned in the container were placed in a large teflon bag which was exposed to the cabin environment. Several samples of wiring and tubing from Surveyor 3 were in a sealed canister, but the other retrieved parts were in a bag exposed to the cabin.

The samples from near the rim of Head crater were:

Bag	Sample	Description
1D	12030	0.070-kg of fines. Glass-covered fragments from a 3-foot crater collected by Al Bean on his way between Intrepid and Head.
3D	12031	A 0.185-kg very coarse-grained pigeonite basalt from the site of the trench excavated near the north rim of Head.
5D	12033	0.450-kg of light-grey fines from the 6-inch-deep trench near the north rim of Head.
6D	12034	A 0.155-kg regolith breccia from the floor of the 6-inch-deep trench near the north rim of Head. Its non-mare composition implied that it was not of local origin.

Tote	12055	A 0.912-kg porphyritic pigeonite basalt with phenocrysts of pyroxene in a medium-grained matrix, taken a short distance west of the trench site, on the way to the triple crater.
Tote	12052	A 1.886-kg porphyritic pigeonite basalt with phenocrysts of pyroxene and olivine in a fine-grained matrix, obtained from the panorama site on the west rim of Head.

Note that Bag 2D was not used and 4D was used later, out of sequence.

The samples from near Bench crater, heading for Sharp, were:

Bag	Sample	Description
Tote	12053	A 0.879-kg porphyritic pigeonite basalt with phenocrysts of pyroxene and olivine in a fine-grained matrix from northwest of Bench.
4D	12032	0.310-kg of fines and chips from northwest of Bench.
7D	12035	A 0.071-kg coarse-grained olivine basalt with a gabbroic texture from northwest of Bench.
7D	12048	0.002-kg of fines accompanying 12035.
8D	12036	A 0.075-kg coarse-grained olivine basalt from northwest of Bench.
8D	12037	0.145-kg of fines accompanying 12036.
9D	12038	A 0.746-kg feldspathic basalt from the west rim of Bench. It was different from the other basalts from Apollo 12, in that it had more feldspar, sodium oxide, aluminium oxide and rare-earth elements, indicating that it was of non-local origin.
10D	12039	A 0.255-kg porphyritic pigeonite basalt with phenocrysts of pyroxene in a coarse-grained matrix from the west rim of Bench.
10D	12040	A 0.319-kg coarse-grained olivine basalt collected together with 12039.

The samples from near Sharp crater were:

Sample	Description
12023	0.269-kg of fines in the special environment sample container from the floor of the 8-inch-deep trench on the east rim of Sharp.
12027	0.080-kg of fines in core tube serial no. 2011 from the floor of the trench.
12024	0.056-kg of "shiny" fragments in the gas analysis sample container from near the east rim of Sharp.

The samples from east of Bench, heading for Halo crater were:

Bag	Sample	Description
11D	12041	0.025-kg of fines including a "glass ball".
12D	12042	0.255-kg of fines from the ground with a "wrinkled texture".

The samples from the rim of a 30-foot crater near Halo were:

Sample	Description
12025	0.056-kg of fines in core tube serial no. 2010, which was the upper section of double-length sample.
12028	0.190-kg of fines in core tube serial no. 2012, which was the lower section of double-length sample.

Note that a rock and related fines were collected from near Halo and placed in 13D, but for some reason this bag was not returned to Earth.

The samples from the southern rim of Surveyor crater were:

Bag	Sample	Description
Tote	12054	A 0.687-kg medium-grained ilmenite basalt with a splash of impact melt of basaltic rather than regolith composition from the southern rim of Surveyor crater.
Tote	12051	A 1.660-kg porphyritic ilmenite basalt with phenocrysts of pyroxene in a medium-grained matrix. It was taken from the ejecta of a fresh 14-foot crater south of Surveyor crater's rim.
14D	12043	A 0.060-kg porphyritic pigeonite basalt with phenocrysts of pyroxene and olivine in a medium-grained matrix from the southern rim of Surveyor crater.
14D	12044	0.092-kg of fines accompanying 12043.

The samples from alongside Surveyor 3 and Block crater on the northeastern rim of Surveyor crater were:

Bag	Sample	Description
Parts	12060	0.021-kg of fines originating from the Surveyor scoop.
Parts	12061	0.010-kg of chips (ten in all) originating from the Surveyor scoop.
Parts	12029	0.007 of fines retrieved directly from the Surveyor scoop.
Tote	12056	A 0.121-kg coarse-grained ilmenite basalt with a gabbroic texture from alongside Surveyor 3.
Tote	12062	A 0.739-kg medium-grained ilmenite basalt from alongside Surveyor 3.
Tote	12064	A 1.214-kg coarse-grained ilmenite basalt from alongside Surveyor 3.

Tote	12063	A 2.426-kg porphyritic ilmenite basalt with phenocrysts of olivine and pyroxene in a medium-grained matrix. It was the heaviest sample from the mission, but was undocumented and hence its origin is uncertain. It may have been collected from alongside Surveyor 3. Alternatively, it could have been lifted from the western rim of Bench.
15D	12045	A 0.063-kg ilmenite vitrophyre from the north rim of Block.
15D	12046	A 0.166-kg medium-grained ilmenite basalt that was collected together with 12045.
15D	12047	A 0.193-kg medium-grained ilmenite basalt that was collected together with 12045 and 12046.
Tote	12065	A 2.109-kg porphyritic pigeonite basalt with phenocrysts of pyroxene and olivine in a very fine-grained matrix. Warner speculates that this is the "grapefruit rock" that Conrad lifted north of Head, but there is no documentation and he seems to have thrown that rock into the crater in order to stimulate the seismometer. It is more likely to have been collected on the visit to Surveyor crater, possibly from Block.

In addition:

Sample	Description
12057	0.650-kg of fines and chips from the bottom of the sample return container that held the documented sample.

In cataloguing the Apollo 12 samples, the Preliminary Evaluation Team noticed that only two of the 34 rock-sized samples (i.e. those whose longest dimension was greater than 4 cm) were breccias, which are mechanically assembled rocks made of fragments of shattered rock bound together by a fine-grained matrix, as compared to about half of those in the case of Apollo 11. The remainder of the Apollo 12 rocks were crystalline.

The crystalline rock was coarser and more texturally diverse than that from the Apollo 11 site, and the titanium content was similar to equivalent terrestrial rocks.[18] Significantly, four kinds of basalt were present in the Apollo 12 samples: olivine basalt, pyroxene basalt, ilmenite basalt and feldspathic basalt. This variety indicated a succession of distinct flows. Nevertheless, the crystallisation dates clustered within a fairly narrow window. The combination of the visually distinct patchwork of flows in this region and the chemical variation in the basalts from different flows within a short interval, suggested that the extrusions resulted from partial melting of pockets of rock by heat from the interior rather than a series of flows from a single evolving reservoir located deep in the mantle.

[18] The titanium dioxide was about 4 per cent in the igneous rocks, but the fine-grained material which contained a proportion of material from elsewhere on the Moon contained up to 8 per cent.

Although the geologists had been keen to visit Surveyor 3 to date Copernicus via its ray, they knew that the context of this site would complicate the exploitation of the basalt ages. Nevertheless, the dates from the Apollo 12 samples had profound implications for lunar history. The first isotopic ratio measurement yielded an age of 2.7 ($\pm$0.2) billion years, which implied that fully *a billion years* had elapsed between the eruption of the basalts at the Apollo 11 site and those at the Apollo 12 site. Even the strongest critics of the proposal that the maria had formed simultaneously were taken aback by such an extended period of volcanic activity. The next result pushed the age up to 3.4 billion years. As the data accumulated, the age converged on 3.2 billion years. But because the stratigraphy of the site was complex, it was difficult to assign this date to a particular point in the sequence which geologists had developed to measure the relative ages of the various depositional units on the lunar surface.[19] Nevertheless, the key point was the 500 million year span in the ages of an eastern and a western mare. Not only did this prove conclusively that the maria were not formed simultaneously, it also meant the process was well established. The Moon was definitely not a 'cold' unevolved body whose low-lying terrain had been filled by a splash of melted rock from a single massive impact early in its history. It was a small planet it its own right. In fact, the degree of chemical differentiation that it had undergone was remarkable.

Geochemist Paul Gast made a surprising discovery in the Apollo 12 basalts: an abundance of potassium, phosphorus and a variety of rare earth elements. Linking their chemical symbols, he coined the label 'KREEP'. After attempting to isolate this material, he realised that it was not present as a distinct mineral. Consequently, the term did not indicate a new type of rock; it was a chemical additive. It is better to use the term as an adjective and say that the Apollo 12 basalts are KREEPy. By way of an 'instant science' explanation, Gast suggested that the additive might have been picked up from the material underlying all the basin ejecta, through which the basalt would have passed on its way to the surface, and therefore might represent the 'non-mare' volcanism that some people thought was common in the highlands. But when the KREEPy basalts proved to be rich in radioactive elements, in particular thorium and uranium, it was realised that this material could not be common in the crust, as radiogenic heating would have prevented the crust from cooling sufficiently to halt volcanism. The mystery of the KREEPy additive was not resolved until more such rocks were returned by later missions. However, the evident enrichment of thorium and uranium resolved the mystery of how the regolith could appear to be older than the bedrock on which it resided. Although most of the regolith at any given location is locally derived, a proportion of it represents ejecta that could have originated from anywhere on the Moon, and the thorium and uranium added to the regolith in this manner had corrupted the process of radioisotopic dating.

After being discarded, Intrepid's ascent stage was sent crashing onto the Moon; it hit at a point 75 km east-southeast of the ALSEP's seismometer, which recorded the

[19] If Apollo 12 had visited Surveyor 1, it would have been a much simpler task to interpret the date of its basalts.

Rock 12065 is a finely grained pigeonite basalt.

lunar crust "ringing" for almost an hour with a signature unlike a terrestrial signal. At the conference in January 1970, Gary Latham, the principal investigator for the seismometer, presented the Apollo 11 results and the early data from Apollo 12. It had initially been difficult to distinguish between a moonquake and an impact, but Intrepid's impact had calibrated the system and it turned out that surprisingly few of the 150 events on record were quakes. It seemed that the lunar crust was brecciated to a depth of about 30 km. In effect, it was a thick 'megaregolith'. As seismic data accumulated, patterns became evident. The *average* rate of events was about one per day, but there was a spate when the Moon was at perigee, which implied that these particular events were triggered by the tidal forces of the Earth's gravity distorting the Moon. However, until later missions had deployed additional seismometers it would not be possible to triangulate and localise the epicentres. Another conclusion was that there were fewer impacts than expected. The seismometer was sufficiently sensitive to detect a grapefruit-sized meteoroid striking within a radius of 350 km. It detected only one such impact per month, on average. These could also be expected to arrive in batches. In orbiting the Sun, the Earth-Moon system periodically passes through swarms of meteoroids left behind by expired comets. But it was evident that there had not been an impact anywhere on the Moon comparable to the crash of the ascent stage. To probe the deep structure of the crust, it was decided that on future missions the spent S-IVB stage of the Saturn V should be sent crashing down onto the Moon.

As to the *origin* of the Moon, for which there were several competing theories, the sample analyses provided evidence to cast doubt on them all! At the press briefing on the final day of the Lunar Science Conference, Robert Jastrow, who had in 1959 played a key role in stimulating NASA's interest in the scientific investigation of the Moon, warned that the flood of hard data from the retrieval of samples rendered "old descriptions in terms of a 'hot' or 'cold' Moon an over-simplification". The theorists would require to become more sophisticated, to reflect the subtlety of the data. Gene Simmons, who led one of the teams that analysed Apollo 11 material, and had just been appointed chief scientist at the Manned Spacecraft Center, summed up: "There is a large amount of undigested data and very little interpretation." He predicted that in the coming months there would be "many revisions of statements as to what it all means".

There was no evidence of any carbon compounds in the samples to indicate a 'chemical signature' of life at (or near) the lunar surface, and there were no micro-fossils. Imagine, therefore, the consternation of the scientists as they disassembled the Surveyor television camera and found *streptococcus* bacteria in the insulation. It was unlikely to be post-flight contamination, but may well have be been introduced during the assembly process, in which case the bacteria had survived the harsh lunar environment.

FUTURE PROSPECTS

On 25 May 1961 President John F. Kennedy challenged his nation to achieve the goal, before the decade was out, of landing a man on the Moon and returning him safely to the Earth. This was achieved by Apollo 11 in July 1969. However, because that mission was essentially an engineering demonstration, the ambitious scientific plans were deferred. Apollo 12 therefore performed what had once been expected of the first landing: two moonwalks involving geological sampling and the deployment of a scientific station on the lunar surface. Furthermore, the procedures designed to achieve a pinpoint landing, which enabled Intrepid to set down within easy walking distance of its objective, cleared the way for more adventurous missions.

Some time after the Apollo 12 mission, its astronauts got together to discuss their respective futures. Conrad briefly speculated that he might become the first man to make a second lunar landing by flying the final mission of the program, which was sure to be very ambitious in terms of its site and surface activities. Bean wondered if *he* would command that mission if Conrad moved on. Concluding that the prospects of a return to the Moon were poor, both decided to try for the Skylab space station that was the only part of George Mueller's Apollo Applications Program to receive funding. In early 1971 Deke Slayton picked Conrad to command the first Skylab mission and Bean to command the second, although their assignments were not made public until January 1972. With no prospect of a lunar mission, Gerry Carr, Paul Weitz and Ed Gibson joined Skylab. Gordon opted to remain in the rotation by serving as backup commander for Apollo 15 in the hope of walking on the Moon as commander of Apollo 18.

As Apollo 11 was flying home in July 1969, NASA decided to launch Skylab on a Saturn V. At that time, it was hoped to continue production of this launch vehicle. By the end of the year it was clear that budget considerations made this impractical, and a launcher was taken from the lunar program. The cancellation of Apollo 20 was announced on 4 January 1970 by George Low, who had recently been promoted to Deputy Administrator. He also said that because Skylab was scheduled for launch in late 1972, Apollo 13 and Apollo 14 would fly in 1970, Apollo 15 and Apollo 16 in 1971, and Apollo 17 in early 1972. It was intended to use Saturn IB launch vehicles to dispatch three manned missions to Skylab in 1972 and 1973, then send Apollo 18 and Apollo 19 to the Moon in 1974. But on 2 September 1970 budgetary pressure obliged NASA to cancel the two lunar missions which had been scheduled to follow Skylab. Gordon therefore served as backup commander for Apollo 15 knowing there was only a slim chance he would reach the Moon. Although Gene Cernan's situation after serving as backup commander for Apollo 14 made him the obvious candidate for Apollo 17, Gordon rested his hopes on the fact that his lunar module pilot was none other than Jack Schmitt, the only professional geologist in the astronaut office. It would be up to Deke Slayton to decide who would fly the final lunar mission.

Glossary

acid A substance which liberates hydrogen ions in solution. An acid reacts with a base to make a salt and water (only). A substance with acid properties (opposite to an alkali; a base) is said to be acidic.

albedo The reflectivity (expressed as a percentage) of a material; a dark material has a low value. On average, the Moon reflects 7 per cent of incident light. The darkest asteroids reflect only 1–3 per cent, so they are as dark as coal.

alkali A soluble hydroxide of a metal. A substance with alkali properties (opposite to those of an acid) is said to be alkaline. The light univalent metals (lithium, sodium and potassium) are called alkali metals; the bivalent metals (beryllium, magnesium and calcium) are alkali earth metals.

anorthosite A rock comprising more than 90 per cent plagioclase feldspar. It is a slow-cooled plutonic rock with a conspicuous crystalline structure. It is the initial lunar crust that formed as the magma ocean cooled, and (being alkali poor) is calcic plagioclase feldspar. Anorthosites are usually referred to as ferroan after the iron-rich compositions of their olivine and pyroxene minority constituents, as opposed to the more magnesium-rich varieties of these mafic minerals.

basalt A lava possessing a homogeneous ground-mass with a texture so fine that its crystals are not resolvable to the naked eye; the extruded form of gabbro. The albedo of lunar basalt is related to its chemical composition. The lighter variety is plagioclase feldspar enriched by mafic minerals (typically either pyroxene or olivine, which are mutually incompatible) in roughly equal proportions, but the darker variety is only about 30 per cent plagioclase feldspar. The very darkest lava is also enriched with titanium.

base A substance that reacts with acid to produce a salt and water (only). A substance with this property is said to be basic.

bedrock The intact layer of rock immediately below the regolith. It is a regional geological unit in the overall stratigraphic structure.

bench An inflection on a slope (such as a mountain flank or a crater wall) which creates a relatively flat surface.

D.M. Harland, *Apollo 12 – On the Ocean of Storms*, Springer Praxis Books,
DOI 10.1007/978-1-4419-7607-9, © Springer Science+Business Media, LLC 2011

boulder In terms of lunar field geology, a 'boulder' was defined as a rock that had at least one dimension in excess of a metre.

breccia A mechanically assembled rock. A regolith breccia is a clod of soil made by the shock of an impact compressing the regolith. A fragmental breccia is made up of angular fragments of many other shattered rocks, but with little or no melt.

core The central part of a planetary body that has undergone differentiation, with the iron-rich material sinking to form a metallic nucleus.

cosmic ray The rain of charged particles in space. The solar wind contributes ions of the lighter elements, but there is also a flux of relativistic nuclei of heavier elements from beyond the Solar System.

crust The outer part of a planetary body that has undergone differentiation, with a low-density silicate-rich material forming a crystalline envelope.

crystal As a silicate-rich lava cools, its minerals precipitate from solution to create a homogeneous rock. The atoms link up to create a regular lattice, the shape of which depends on the elements involved. Certain elements can fit into a specific shape, and are called 'compatibles'; all others are called 'incompatibles'. There can be elements in common, but as soon as a crystal starts to take on a particular form the incompatibles are excluded. Because different crystals form at different temperatures, the magma chemistry evolves and becomes progressively enriched by a succession of different incompatibles. If gas is released into the solution and becomes trapped, the resulting rock will have cavities. If the rock cools rapidly, the cavities will be spherical and have smooth glassy walls, but if it cools slowly some of the metals in the gas may form crystals on the wall of the cavity, partially filling it and making its shape irregular. A spherical cavity is referred to as a vesicle and an irregular one as a vug. If the molten magma solidifies too rapidly for a lattice to form, the result is a homogenous mass of glass (obsidian is the glassy form of rhyolite, for example).

feldspar See: plagioclase feldspar

gabbro A general term for plutonic rocks made of plagioclase and clino-pyroxene that crystallised sufficiently slowly to form granular crystals; its extruded form is basalt.

gardening When a small meteorite strikes the lunar surface at high speed, it either vaporises or melts both itself and the target material. The material immediately beneath the point of contact is compressed sufficient to create regolith breccia. Below this, the shock merely serves to shatter coherent structure. Large impacts excavate bedrock which is then broken down into ever-smaller pieces by the relentless rain of meteoroids.

glass If a high-viscosity melt solidifies into a state with no internal order, the result is glass rather than a crystal. Glass can also be made by an extreme pressure-pulse, such as occurs at the point of a meteoritic impact.

grain The individual crystals in a rock, the texture of which varies from coarsely to finely grained.

granite A silicic gabbro with a granular texture; its extruded form is rhyolite.

igneous Of, or relating to, material formed from a liquid state.

ilmenite An iron-titanium oxide mineral; $FeTiO_3$. It is a major constituent of the mare basalts, and its presence explained why the maria are so dark. The fact that the proportions of this mineral were different at each mare site visited explained why the various maria had different albedos.

lava The extruded form of magma.

mafic A mineral enriched in iron and/or magnesium at the expense of silicon and aluminium. Being denser, these minerals (such as olivine and pyroxene) formed deep in the lunar magma ocean and subsequently rose through the deep faults in the basin floors and were extruded as the mare basalts.

magma ocean As the Moon accreted, its surface was molten to a depth of several hundred kilometres. Over a period of several hundred million years, the buoyant plagioclase feldspar differentiated from the silicate-rich mantle and crystallised as the primitive crust.

magnesian suite A family of plutonic rocks that is distinguished from the ferroan anorthosites by a high fraction of magnesium-rich silicates. Unlike the homogeneous anorthositic crust, the magnesian suite is chemically diverse, indicating that the mantle reservoirs had distinctive characteristics.

mantle That section of a planetary body that is sandwiched between the crust and the core. It is the main source of magma for volcanic activity.

mascon Analysis of the Doppler variation of the signals from the Lunar Orbiters led to the discovery that the Moon's gravitational field is uneven. The variations in gravity were interpreted as mass concentrations (mascons) associated with the circular maria which contain dark maria. Because mascons perturb low orbits, a spacecraft must manoeuvre to maintain a specific orbit.

olivine A mafic silicate; $(Mg,Fe)_2SiO_4$. It is not as common as pyroxene in lunar basalt, and can be absent.

phenocryst A conspicuous mineral crystal in a porphyritic rock.

pigeonite A mineral in the clino-pyroxene group, and occurs as phenocrysts in volcanic rocks.

plagioclase feldspar This group of minerals form 60 per cent of the outer 15 km of Earth's crust. In German, 'feldspar' means 'field crystal'. Plagioclase feldspar is a subset of the feldspar group; $(Ca,Na)(Al,Si)_4O_8$. Much terrestrial plagioclase is sodium-based, but the Moon is deficient in sodium and its plagioclase is calcic. Feldspathic laths make a light-coloured rock. Anorthosite is extremely pure plagioclase and is white; norite is plagioclase enriched by pyroxene; troctolite is plagioclase enriched by olivine.

plutonic rock A coarsely grained igneous rock that solidifies underground; strictly, at *great* depth, because a rock that solidifies just beneath the surface is referred to as hypabyssal.

porphyry An extremely finely grained basaltic matrix with conspicuous mineral crystals known as phenocrysts.

pyroxene A mafic silicate; $(Ca,Fe,Mg)_2Si_2O_6$. It is the most common mafic mineral in lunar basalt.

rare earth elements A group of elements that are rare in Earth's crust: lanthanum, cerium, neodymium, praseodymium, promethium, samarium, europium, gladolinium, terbium, erbium, thulium, dysprosium, holmium, ytterbium and lutetium.

ray The light-toned material that is deposited in a radial pattern around a newly formed crater. It is bright because it is finely pulverised rock and the fractured crystals are very reflective.

regolith The proper name for a 'soil' composed solely of rock fragments, forming an unconsolidated mass of debris on top of bedrock. As a result of the gardening process, the lunar regolith has a seriate distribution of fragment sizes.

rhyolite A light-coloured silica-rich viscous basalt; the extruded form of granite.

silica Crystalline silicon dioxide; SiO_2. It has a high melting point and forms a variety of silicates in rock.

silicate Silica can join with calcium, aluminium, iron, magnesium, sodium and potassium to make a variety of minerals featuring the silicon tetrahedron (SiO_4).

solar wind The plasma that flows out from the Sun. It is mostly ionised hydrogen (free protons and electrons) but also contains heavier nuclei.

terminator The line of longitude corresponding to sunrise or sunset on the lunar surface. As the Moon orbits Earth, it turns on its axis and the terminators track at a corresponding rate.

vesicle As magma approaches the surface, its pressure rapidly diminishes and the volatiles in the melt boil out of solution. If the gas is unable to escape, it makes spherical cavities in the rock. If it cools rapidly, the minerals line the cavity with glass. If the magma is under directional stress as this solidifies (i.e. it is flowing) then the vesicles may be distorted.

vitrophyre An igneous rock whose matrix is glassy.

volatile elements Those elements that boil out of solution at low temperatures. The Moon is strikingly deficient in volatiles such as sodium, sulphur and chlorine.

volcanic rock Igneous rock that solidifies after being extruded onto the surface.

vug If the gas in a vesicle cools sufficiently slowly for the volatile minerals to recrystallise, this will partially fill the cavity, making it irregular.

Index